ELECTRIC AND HYBRID VEHICLES

ELECTRIC AND HYBRID VEHICLES

TECHNOLOGIES, MODELING AND CONTROL: A MECHATRONIC APPROACH

Amir Khajepour
University of Waterloo, Canada

Saber Fallah
University of Surrey, UK

Avesta Goodarzi
University of Waterloo, Canada
Iran University of Science and Technology, Iran

WILEY

This edition first published 2014

Registered office
John Wiley & Sons Ltd, The Atrium, Southern Gate, Chichester, West Sussex, PO198SQ, United Kingdom

For details of our global editorial offices, for customer services and for information about how to apply for permission to reuse the copyright material in this book please see our website at www.wiley.com.

Library of Congress Cataloging-in-Publication Data applied for.

ISBN 9781118341513

Set in 10/12pt TimesLTStd-Roman by Thomson Digital, Noida, India.

1 2014

To our students, whose enthusiasm and hard work are a constant source of inspiration, and to our families, without whose endless patience and support, this book might never have been written.

Contents

Preface

Concerns over the environment, public health and the availability of fossil fuels have forced the establishment of aggressive emissions regulations, such as the U.S. 2020 CAFE Standards, and have triggered momentous changes in global automotive strategies. New technologies and products are now required to enhance fuel efficiency and reduce harmful emissions, without sacrificing performance, cost-efficiency and safety.

Vehicle electrification and hybridization have been increasingly recognized as the most promising road transportation solutions to both the global energy crisis and the increasingly stringent requirements related to environmental protection and vehicle safety. However, the electrification of automotive systems presents significant design challenges, specifically related to drivetrain systems, chassis design and layout, multidisciplinary power management and optimization, system integration, and vehicle dynamics and control.

Electric and hybrid electric vehicles (EVs and HEVs) are complex mechatronic systems; their design requires holistic consideration of vehicle and tire dynamics, powertrain, electric motors and batteries, and control and estimation modules that are integrated through millions of lines of computer code. Several books have already been published that outline very well the electrical aspects of EV and HEV platforms. In this book, we have expanded upon these early works to present a more comprehensive perspective that combines electrical, control, and dynamics in systems-level design. It places new emphasis on how dramatically vehicle dynamics and, subsequently, our understanding of conventional vehicle design is changed by electrification.

This book is structured to address both senior undergraduate and graduate level courses, and can serve as an excellent reference for anyone with a background in dynamics, electrical, and control engineering. The content is sufficiently broad to allow course instructors the opportunity to tailor the material according to students' backgrounds. Several introductory chapters provide important background information on vehicle technologies in propulsion, powertrain, body and chassis, and the evolution of automotive technology design from conventional vehicles to the HEV and EV models we see on the road today. Students with electrical and control engineering backgrounds, but limited experience with automotive and mechanical engineering applications will benefit from these initial chapters, while students with stronger automotive backgrounds will benefit from later chapters that focus on HEV and EV power management optimization and vehicle control.

In this book we have tried our best to ensure students are presented with a balance between building a solid conceptual understanding and developing procedural skills related to automotive design. Examples are presented throughout that encourage students to apply their theoretical knowledge to real EV and HEV design challenges and considerations specific to these vehicles. End-of-chapter problems are provided for further practice and to facilitate a better understanding of the materials.

Acknowledgments

This book would not have been possible without the help of many people. We are particularly grateful to Joan Ang, Gaurav Pokharel, Amir Ostadi, and Ivanna Ramnath for their careful review and meticulous editing of each book chapter; John Chen, Kenan Habib, and Azadeh Zandieh, who drew most of the figures; and Soheil Fard, Reza Zarringhalam, and Ayyoub Rezaeian, who prepared the examples and problems. We also thank John Wiley & Sons, Ltd. for providing the opportunity, encouragement, and support throughout this book project.

1

Introduction to Vehicle Propulsion and Powertrain Technologies

The advent of the internal combustion engine has significantly influenced human life. As the main propulsion technology used in vehicles, the internal combustion engine has become an integral part of modern life. However, as internal combustion engine vehicles increase in number, they constitute one of the largest sources of air pollution and greenhouse gas emissions. This chapter introduces currently available propulsion technologies, as well as their advantages and disadvantages. The chapter begins by providing a brief history of internal combustion engine vehicles, then reviews the environmental challenges associated with combustion engine emissions. The rest of the chapter discusses the benefits of emission control technology, alternatively-fueled propulsion, and advanced powertrain technologies.

1.1 History of Vehicle Development

Vehicles have a long and varied history. In this section, we highlight few key events [1–5]. In 1769, Nicolas-Josef Cugnot and M. Brezin designed and built the first self-propelled vehicle, a steam-powered motor carriage capable of a maximum speed of 6 km/hr. However, even when modified for faster speeds, its heavy mass hindered the vehicle's performance. In 1807, the invention of the internal combustion engine (ICE) by François Isaac de Rivaz created new possibilities. This engine generated propulsion energy by using a mixture of hydrogen and oxygen. Several other engineers developed designs for the ICE, all of which were commercially unsuccessful because they lacked the fuel necessary to safely facilitate internal combustion.

Jean Joseph Etienne Lenoir invented the first successful gas engine 53 years later. After numerous modifications and improvements on Lenoir's design, the brothers Charles and Frank Duryear built the first gasoline-powered car in 1893, a design that was ready for road trials. In 1901, the German engineer Ferdinand Porsche manufactured a car that was powered by an internal combustion engine and hub-mounted electric motors (Figure 1.1). This was one of the first hybrid vehicles on record.

In 1904, Henry Ford developed the first assembly line manufacturing plant for gas-powered vehicles. As the twentieth century progressed, the automotive industry began to develop

Electric and Hybrid Vehicles: Technologies, Modeling and Control: A Mechatronic Approach, First Edition.
Amir Khajepour, Saber Fallah and Avesta Goodarzi.

Figure 1.1 The early 1900's Lohner-Porsche, the first hybrid vehicle on record. *Source:* Reproduced by permission of Porsche Cars North America, Inc.

rapidly, and motor vehicles were soon available in steam, electric, and gasoline versions. While gas-powered vehicles are prevalent in our world today, electric vehicles (EVs) were more popular than other vehicle alternatives in the early 1900s. The primary reason for its popularity at this time was functionality. Unlike gasoline vehicles, electric vehicles were without the engine-related vibration, smell, and noise. However, gasoline vehicles require manual gear shifting, which was regarded as a difficult component of driving at the time. Likewise, EVs were also preferable to steam-powered vehicles because they were capable of longer ranges on a single charge, and were more convenient during colder weather. Under similar weather conditions, steam-powered vehicles suffered from start-up times of up to 45 minutes.

The electric vehicles continued to be attractive until the 1920s, with peak production occurring in 1912. However, improvements in intercity road quality propelled the need for ICE vehicles, which were capable of operating at longer distances. At the same time, the discovery of oil reduced the price of gasoline, making internal combustion vehicles more affordable to consumers. Moreover, the invention of the electric starter made the use of the internal combustion vehicles more convenient, whereas the long recharge time of electric vehicles and the expensive large battery packs made the electric vehicles less attractive to consumers. As a result of these developments, the market for EVs gradually disappeared by the 1930s.

Moving forward in time, cars became less of a luxury and more of a necessity for everyday life. While the increased use and development of internal combustion vehicles have changed city landscapes and lifestyles, scientists began to worry about the long-term environmental effects of ICE exhaust emissions. Most concerns related to the influence of vehicle emissions on the onset of global warming and excessive greenhouse gas production. Furthermore, oil prices began to rise around this time, along with increased public awareness of the limited supply of oil resources. This led to a pressing need for alternatively-powered vehicles, which began to surface in the 1970s.

The afore-mentioned issues also led governments to take several legislative and regulatory actions to moderate oil production dependency and reduce the causes of air pollution. Likewise, several organizations from around the world began to work with vehicle manufacturers to create conditions more favorable to the development of electric vehicles. As a result of these changes, the automotive industry began to renew their attempts to develop electric vehicles.

In 1974, the American company, Sebring-Vanguard, designed the CitiCar, the first mass-produced electric car, and continued its production until 1977. From 1977–1979, General Motors Company spent over $20 million in electric car development and research, with a stated objective to produce electric vehicles by the mid-1980s. Similarly, Peugeot and Renault designed an electric vehicle that could drive at 100 km/h, with a travel range of 140 km. In 1989, Audi unveiled the first generation of the experimental Audi Duo, a petrol engine/electric hybrid concept vehicle. This car had rear wheels driven by a 12.6 hp electric engine, and front-wheel drive powered by a 2.3 litre 5-cylinder engine with a 136 hp output. In 1996, General Motors designed and developed an electric motor vehicle called the EV1, which had a top speed of 130 km/h and a range of 130 km.

Improvements in electric vehicle performance motivated the automotive industry to develop and market mass-produced electric vehicles. However, the electric vehicle market could not achieve success with consumers, and was largely failing by the end of the 1990s. The primary reason for the failure of EVs was their limited performance capability as compared to gasoline-powered vehicles. Moreover, the low price of oil at the time, as well as the EVs' high initial costs, high maintenance costs, and infrastructure shortage contributed to their commercial lack of success.

However, a new type of motorization appeared in the late 1990s: hybrid technology, which combined the internal combustion engine with an electric motor. Toyota made the Prius, the first mass-produced hybrid electric vehicle (HEV), which was launched successfully in Japan in 1997. Shortly after, in 1999, the Honda Insight was launched in both the United States and Japan. These two vehicles pioneered the hybrid vehicle concept, and led to a shift in the market perception of alternative fuel vehicles.

Since then, other car manufacturers have designed and produced a variety of fuel-efficient vehicles using vehicle electrification technology. Ford introduced its first hybrid vehicle in 2004 with the Escape SUV hybrid model. Likewise, General Motors introduced the Silverado and the Sierra in 2004 as their first hybrid vehicle models. Currently, countries such as Brazil and China are competing to increase their shares in the worldwide market of EVs and HEVs.

Nevertheless, the share of HEVs worldwide is still quite low when compared to IC engine vehicles. In 2004, the share of HEVs was 0.25% whereas in 2007, the total production of 541 000 HEVs was only 0.8% of the production of light vehicles worldwide. Anticipated production numbers of HEVs predict an increase to 1.7 million by 2014. The increasing share of HEVs overall reflects the accelerating pace towards greater electrification of vehicles, and eventually, the production of zero emission vehicles with better efficiency [6].

1.2 Internal Combustion Engine Vehicles (ICEVs)

Most modern vehicles create propulsion power through an internal combustion engine. An internal combustion engine generates propulsion power from the combustion of fuel and an oxidizer in a confined cylindrical space known as a combustion chamber. The engine then takes the heat energy generated by the combustion process and converts it into mechanical work based on the principle of energy conservation.

The oxidizer of an IC engine is typically oxygen, which is sufficiently available in the Earth's atmosphere. The most common fuels used in an IC engine are gasoline and diesel;

however, other fuels such as hydrogen, methane, and propane are also used. The exothermic reaction of the fuel with the oxidizer results in the production of high-temperature and high-pressure gases. The expansion of these gases applies force to a piston inside the combustion chamber. Subsequently, the linear motion of the piston is transferred to the wheels through the crankshaft and the vehicle transmission system. The advantageous features of IC engines are a high power-to-weight ratio and excellent fuel energy density.

Dutch physicist Christian Huygens first proposed the design concept of a working internal combustion engine in 1680. Almost 130 years later, Swiss engineer François Isaac de Rivaz realized Huygens' vision by inventing an unsuccessful version of the IC engine. His internal combustion engine attempted to propel automobiles by burning a fuel mixture of hydrogen and oxygen for power. While de Rivaz's attempts failed to achieve success, the efforts of English engineer Samuel Brown were successful. In 1826, Brown developed a hydrogen-fueled combustion engine that was industry-compatible. Brown's engine had separate combustion and working cylinders with four hp.

The first functioning and successfully designed gas-powered internal combustion engine was invented in 1860 by Jean Joseph Etienne Lenoir. Fueled by coal gas, this engine was a double-acting, electric spark-ignition capable of running continuously. In 1863, Lenoir further improved this engine model so that it was able to run on petroleum using a primitive carburetor. Following these developments, engineers continued to invent and modify variations of the IC engine. However, the most significant contribution was the invention of the four-stroke engine by Nicolaus August Otto in 1876. The patented "Otto Cycle Engine," a universally implemented, practical four-stroke IC engine was soon in all liquid-fueled automobiles. A four-stroke engine still powers most modern cars and trucks.

In 1885, German engineers and design partners, Gottlieb Daimler and Wilhelm Maybach modified the Otto Cycle Engine by reducing its size while increasing its speed and efficiency. Their engine was small, lightweight, and quick; it used a gasoline-injected carburetor and had a vertical cylinder. In 1886, as a separate work and without any knowledge of Daimler's and Maybach's work, German engineer Karl Benz patented the first gas-fueled automobile powered by a four-stroke engine. Benz's engine designs were the pioneers for modern vehicles. Rudolph Diesel made another key contribution towards improving the efficiency of IC engines by inventing and patenting the diesel engine in 1892.

Currently, gasoline and diesel engines are the main power sources for the majority of road vehicles. Most cars and light-duty vehicles use gasoline engines while heavy-duty vehicles, buses, and some passenger cars use diesel. Diesel engines are more fuel-efficient and more powerful at lower speeds than gasoline engines. However, they are also noisier, heavier, and more difficult to start in cold weather conditions. While gasoline engines have an easier time starting in the cold, they are still prone to problems under extreme conditions.

Both gasoline and diesel engines convert the chemical energy of fuel into motion through a four-stroke process, and are available as either reciprocating or piston engines. A reciprocating engine includes a number of cylinders (combustion chambers), each containing a piston that moves up and down. Each piston connects to the crankshaft (Figure 1.2) through a connecting rod, which rotates the crankshaft by using the reciprocating (up-and-down) motions of the piston. The crankshaft features offset axis sections called crankpins. While the upper end of the connecting rod attaches to the piston with a joint, the bearing offset sections attach to the end of the connecting rod. The design of the crankshaft causes the reciprocating motion of the pistons to translate into a rotary motion. In manual transmissions, the rotary motion of the crankshaft

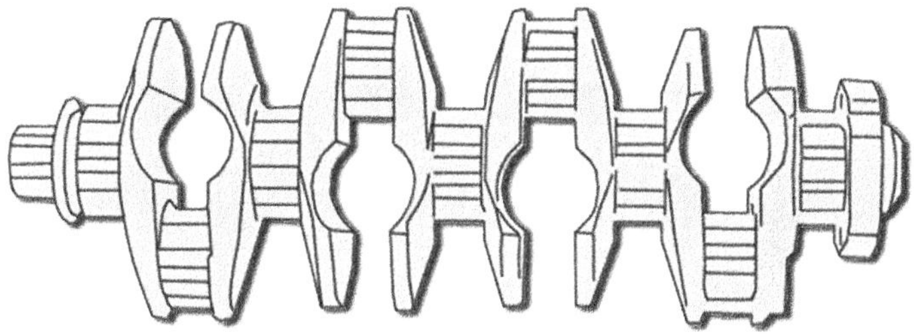

Figure 1.2 A vehicle engine's crankshaft

transfers to the drivetrain through the flywheel; in automatic transmissions, the motion transfers through a torque convertor.

The cycles of engines with more than one cylinder are arranged evenly for smooth operation. The number of cylinders used in an engine varies, based on engine performance and specification. Generally, engines with 4–8 cylinders power vehicles; however, higher performance vehicles may have up to 16 cylinders, while some small cars or motorcycles use engines with only 1 or 2 cylinders. Engines with more cylinders provide a higher engine capacity for the vehicle. Engine capacity or engine displacement is the total cylinder volume swept by all of the pistons in a single movement. Increasing the diameter of the piston and lengthening the stroke increases engine capacity. Engines with greater capacities can achieve greater power and torque at the cost of increased fuel consumption. Higher numbers of cylinders make it possible to use smaller and lighter cylinders for a given fuel mass, resulting in a smoother engine operation. However, increased cylinder numbers result in a heavier mass and more internal friction between pistons and cylinders. This may adversely affect the overall efficiency and performance of the vehicle engine.

Turning on an engine requires an electric starter motor to rotate the crankshaft. The rotary motion of the crankshaft causes some of the connecting rods to push the pistons upward, thereby compressing the mixture of fuel and air. The combustion propels the engine to start working, and as long as the vehicle moves, the inertia of the crankshaft causes the pistons to move up and down inside the cylinders [7].

1.2.1 The Four-Stroke Gasoline Engine

A gasoline internal combustion engine converts the chemical energy of fuel to mechanical energy through a process called the four-stroke cycle. This process is also known as the Otto cycle in honor of its inventor, Nikolaus Otto. Each four-stroke cycle makes two engine revolutions. Cylinders, pistons, valves, and spark plugs are the primary components of the engine during the combustion process. As illustrated in Figure 1.3, the four-stroke process includes the intake stroke, compression stroke, power stroke, and exhaust stroke as described below:

1. *Intake stroke*: This is the first stroke of the cycle. Initially, the intake valve opens and the piston is at the top of the cylinder. The piston subsequently slides down the cylinder to create a low-pressure environment. While the outlet (exhaust) valve remains closed, the low-pressure environment draws a measured volume of fuel and air mixture inside the cylinder. When the piston reaches the bottom of the stroke (to the maximum volume position), the intake valve closes and this stroke ends.

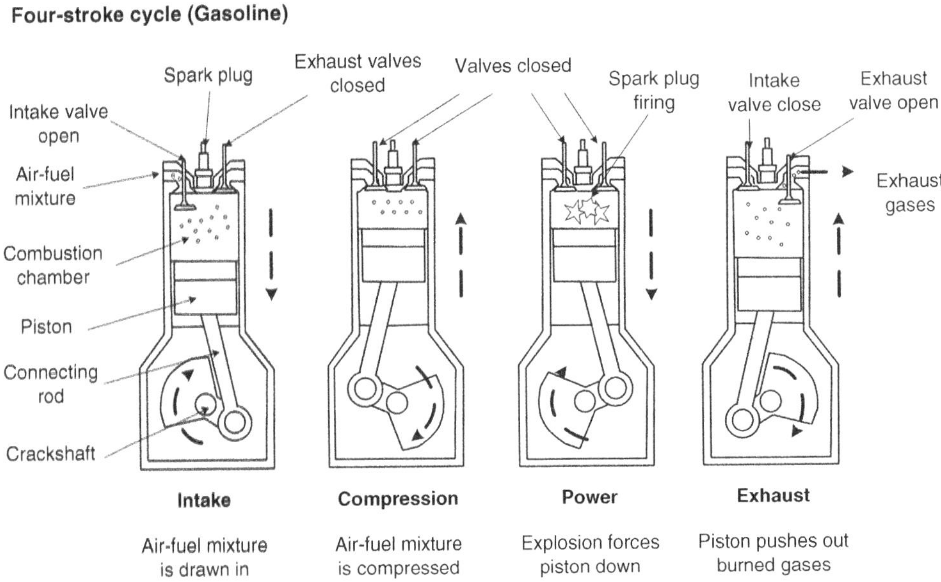

Figure 1.3 The gasoline (Otto) engine cycle

2. *Compression stroke*: While both the intake and outlet valves are closed, the piston starts moving upwards and squeezes the air–fuel mixture from the top of the cylinder. The stroke finishes when the piston reaches the minimum volume position. The air–fuel compression results in an increase in pressure, temperature, and fuel mixture density.
3. *Power stroke*: Once the piston reaches its maximum compression ratio (the top of the cylinder), the compressed air–fuel mixture ignites with a spark plug. This stage generates lots of heat, and the resulting high temperature and pressure gases push the piston downward. The intake and outlet valves remain closed during this stroke. Gasoline engines, occasionally referred to as SI engines, use the spark ignition (SI) system. The SI system is an electrical system that includes a lead-acid battery and an induction coil. The system provides a high-voltage electric spark to ignite the air–fuel mixture inside the cylinder. The battery is recharged on-board using an alternator driven by the engine.
4. *Exhaust stroke*: This is the fourth and final stroke of the cycle. The leftover combustion by-products and gases in the power stroke, called the exhaust, are vacated during the exhaust stroke. Once the piston reaches the bottom of the cylinder at the end of the power stroke, the outlet (exhaust) valve opens and the exhaust stroke begins. The crankshaft pushes the piston upwards to expel the exhaust from the cylinder through the outlet valve. At the end of this stroke, the exhaust valve closes. As the exhaust valve closes, the intake valve opens and the sequence cycle repeats.

1.2.2 The Four-Stroke Diesel Engine

Like the gasoline engine, the diesel engine is also a four-stroke engine, though it operates slightly differently. Rather than a spark ignition system, diesel engines use a compression heating ignition system in which the ignition of the air–fuel mixture occurs because of compression instead of a spark plug. In other words, ignition occurs when the diesel fuel is sprayed into the cylinder filled with compressed air.

When compared to gasoline engines, additional advantages of the diesel engine include its superior fuel economy, higher compression ratio, and enhanced durability. The desirable fuel economy is a result of its properties, high compression ratio, and lower fuel price. The higher compression ratio produces better thermal efficiency, consequently resulting in highly effective mechanical work. Diesel fuel is also cheaper than gasoline when measured in fuel volume. Moreover, despite both fuel types having similar amounts of energy in specified weights, the amount of energy in a specific volume of diesel fuel is still 10% greater than that of gasoline. However, the high compression ratio of a diesel engine results in very high pressure, which consequently produces a high level of stress on engine materials. Furthermore, diesel engine components such as cylinders, pistons, rods, and valves are much heavier and thicker, which causes the diesel engine to be sluggish. Another issue with the diesel engine is its complicated fuel injector system. Since diesel fuel does not evaporate easily, high-pressure injection nozzles are required to produce the appropriate air–fuel mixture. As illustrated in Figure 1.4, the four-strokes of the diesel cycle are intake stroke, compression stroke, power stroke, and exhaust stroke as described below:

1. *Intake stroke*: The intake valve opens and draws air in the cylinder while the piston slides downward. The intake valve closes when the piston reaches the maximum volume.
2. *Compression stroke*: The piston pushes upwards and compresses the air to about 1/16th of its initial volume. Air temperature can reach up to 1000 °C in this stage.
3. *Power stroke*: Diesel fuel enters the cylinder by the fuel injector just before the peak of compression. The air–fuel mixture starts burning due to the high pressure and temperature within the combustion chamber, and the resulting thermodynamic energy pushes the piston down. Throughout the course of this stage, both intake and exhaust valves remain closed.

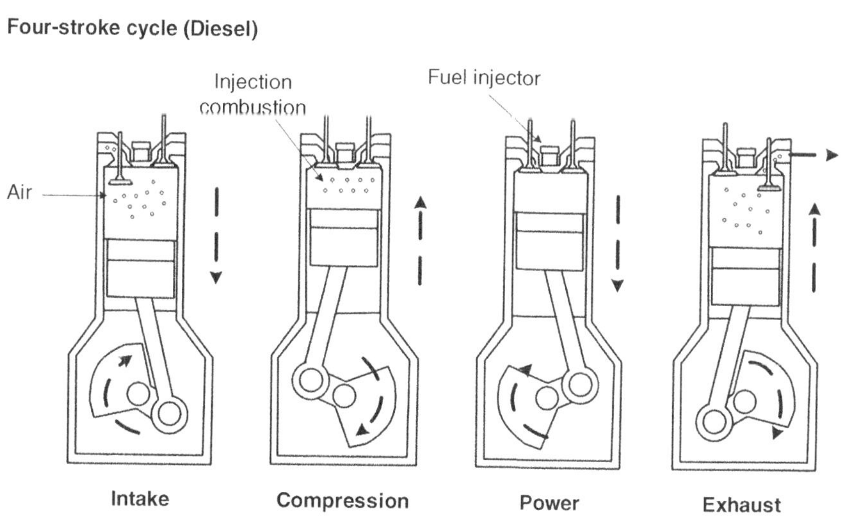

Figure 1.4 The diesel engine cycle

4. *Exhaust stroke*: The exhaust valve opens and the piston slides upwards while the combustion gases emits through the exhaust valve. Once the piston reaches the head of the cylinder, the exhaust valve closes and the intake valve opens, restarting the cycle.

1.2.3 ICE Performance Characteristics

An established performance baseline evaluates and compares several parameters and variables of IC engines (ICEs). Assessing and comparing efficiencies is a method of evaluating the performance of an engine. This includes aspects such as fuel efficiency, thermal efficiency, work efficiency, and engine emissions, among others. The following subsection discusses the most significant parameters, including power-to-weight ratio, air-to-fuel ratio, power-to-volume ratio, and volumetric efficiency.

1.2.3.1 Power-to-Weight Ratio

Power-to-weight ratio is the engine's power output to the weight of the vehicle; it evaluates engine performance, regardless of vehicle size. Generally, a power-to-weight ratio value is a compromise between comfort, fuel economy, and emissions. Since engine power is a function of its speed, the power-to-weight ratio varies based on the speed range of the vehicle. When the vehicle is at a standstill, the ratio is at zero; it rises when the vehicle accelerates. The ratio reaches its peak value as speed increases, and begins declining as the speed continues to rise.

Power-to-weight ratio has a direct relation to the maximum acceleration of a vehicle. Essentially, a greater power-to-weight ratio directs a vehicle to accelerate faster than a vehicle with a lower ratio. For example, consider two vehicles that have the same engines but different power-to-weight ratios. If additional conditions such as energy loss amounts are identical, the vehicle with the greater ratio is always faster. As such, decreasing the weight of vehicles with higher levels of power can result in enhanced fuel economy and reduced emissions. Diesel vehicles generally have a smaller power-to-weight ratio in comparison to gasoline vehicles. This is because diesel vehicles require a heavier engine (and additional components) to resist the operating pressure caused by the high compression ratio.

1.2.3.2 Air–Fuel Ratio

Both gasoline and diesel fuels are a composition of hydrocarbons (made from hydrogen, oxygen and carbon). They react with the oxygen available in the air to kindle burning. In an internal combustion engine, the air (oxygen) available in the cylinder can burn a portion of fuel. If the amount of fuel is greater than the air available (rich fuel mixture), some unburned fuel will remain after combustion occurs. The vehicle expels the unburned fuel into the environment through the exhaust valves and tailpipe, consequently polluting the air and environment. On the other hand, if the amount of air is more than the amount of fuel (lean fuel mixture), there will be more oxygen available in the exhaust gases produced. Note, in a gasoline engine, lean fuel mixtures may cause engine knocking, an abnormal combustion in which the remaining air–fuel mixture in the chamber detonates due to high-pressure and heat. Engine knocking produces aggravating noises, reduces engine efficiency, and may damage engine materials. Additionally, a lean air–fuel mixture may also generate more nitrogen–oxide pollutants. Ultimately, the engine performance is a function of the direct relationship between an engine's air flow and its fuel requirements.

In engine design, the definition of air–fuel ratio is a measure of the air–fuel mixture quality. This ratio measures the weight ratio between air and fuel in the air–fuel mixture within a combustion chamber at any given moment. A desirable air–fuel ratio is one that allows the engine to achieve maximum power and the best fuel economy while producing the least emissions. However, these objectives are conflicting since the fuel requirements of an engine change, based on variations in temperature, load, and speed conditions.

The stoichiometric air–fuel ratio is an air–fuel mixture with a certain weight of air that oxidizes available fuel without leaving behind excess oxygen or unburned fuel. This ratio is 14.7 for gasoline engines and 14.5 for diesel engines. A mixture that is less than a stoichiometric ratio is a rich mixture, while a mixture greater than a stoichiometric ratio is a lean mixture. Gasoline engines need lean mixtures to attain optimal fuel economy through minimum fuel consumption while they need rich mixtures to suppress combustion knock and to attain maximum power (Figure 1.5). Idle heavy loads and high-speed conditions require a rich mixture and normal cruising and light load conditions require a lean mixture. The air–fuel ratio needed to start an engine is approximately 9, and the best possible fuel economy is attainable with an air–fuel ratio of 16. Acceleration, idling, and full power requirements need an air–fuel ratio of approximately 12.

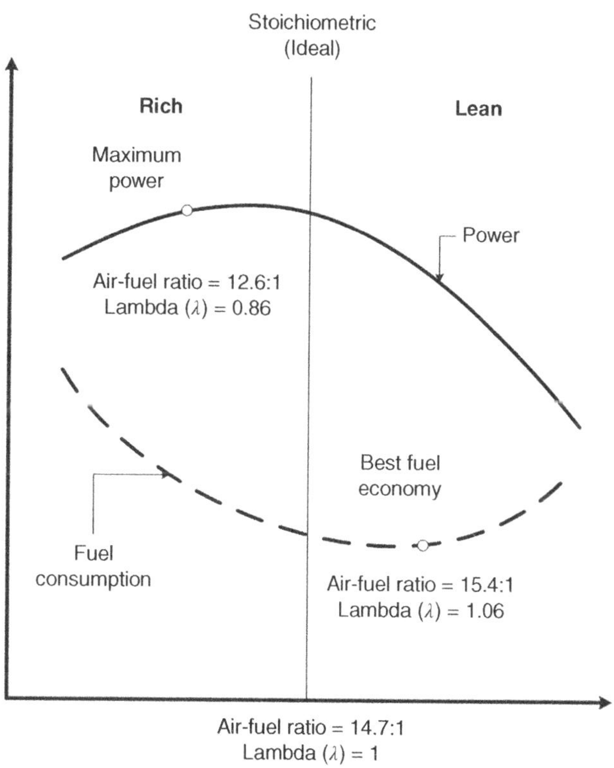

Figure 1.5 Effects of air–fuel ratio variation on the fuel economy and power generation of a gasoline engine

The throttle and fuel injector regulate the air–fuel ratio in a gasoline engine vehicle. The throttle is a valve that varies the engine airflow, while the electronic fuel injector sprays a measured amount of fuel into the cylinder. The function of the throttle is to adjust the amount of air entering the intake manifold. The vehicle driver controls this mechanism through the throttle (accelerator) pedal.

Older vehicles (pre-1990) used a carburetor to mix air and fuel. A carburetor is a mechanical device that regulates the amount of fuel drawn into the airstream based on the speed and pressure of airflow entering the engine. Fuel injection systems replaced carburetors and they are no longer used in production vehicles due to their lower fuel efficiency and higher emission rates.

In modern vehicles, the Engine Control Unit (ECU) determines the amount of fuel that the injector sprays into the intake manifold. The ECU calculates this amount based on information obtained from sensors installed within the vehicle. Oxygen, mass airflow, throttle position, and coolant temperature sensors send the most significant data. The exhaust pipe houses the oxygen sensor, which detects rich and lean mixtures by monitoring the amount of available oxygen. The throttle position sensor monitors the throttle pedal position. The mass airflow sensor measures the air mass entering the intake manifold. The coolant temperature sensor measures the engine temperature to determine if it has reached the appropriate operating conditions. When the driver steps on the throttle pedal to demand more power from the engine, the mass airflow sensor sends the ECU information about the amount of air mass entering the engine. The ECU subsequently injects more fuel into the intake manifold based on this information. Likewise, the ECU also acquires temperature data from the coolant temperature sensor. If the engine has not reached proper working temperature, the ECU accordingly injects more fuel to help heat up the engine. In cruise and light load conditions, the ECU keeps the air–fuel ratio close to stoichiometric proportions based on the amount of oxygen leaving the exhaust pipe, as measured by the oxygen sensor. On the other hand, high speed/high load conditions require a proper rich air–fuel mixture. In these circumstances, the injector feeds a measured amount of fuel regardless of the information given by the sensors. This variation is necessary to ensure a higher margin against detonation.

With some exceptions, diesel engines generally draw an uncontrolled amount of air into the cylinders because they do not have a throttle valve or throttled system. Regulating the amount of fuel injected at the end of a compression stroke controls the speed and power. Unlike gasoline engines, which can run on rich, lean, and stoichiometric fuel mixtures, diesel engines mostly run on a lean mixture. The combustion temperature of lean fuel mixtures is lower than that of stoichiometric mixtures because it burns less fuel. Lower temperature environments lessen the amount of heat lost in the engine, allowing more heat energy to convert into mechanical work by pistons during combustion.

1.2.3.3 Power-to-Volume Ratio

The power-to-volume (compression) ratio is the ratio between the generated power and volume of air–fuel mixture at the beginning and end of the compression stroke. Essentially, the compression ratio represents the compressibility of the air–fuel mixture within the cylinder of a vehicle engine. This fundamental specification is important when evaluating the performance of an engine. Generally, a higher compression ratio is more desirable because it provides the vehicle with greater power and increases engine efficiency. As such, an engine with a higher compression ratio generates more mechanical energy for a specific mass of air–fuel mixture. Moreover, a high compression ratio may also indicate that the fuel can burn completely, thus reducing the exhaustion of by-products. A disadvantage of increasing the compression ratio is that the engine becomes more prone to engine knocking. However, a lean air–fuel mixture, gasoline with lower octane ratings (poor quality), and knock sensor malfunctions can also trigger this phenomenon.

1.2.3.4 Volumetric Efficiency

In most reciprocating engines, the air–fuel mixture is drawn into the combustion chamber during the intake stroke due to low-pressure environment that is created by the downward motion of the piston. The amount of air inhaled by the engine is usually less than the theoretical amount of air that an engine can receive under atmospheric pressure due to factors such as intake system components, cycle time limitations, friction losses, and leaks. The amount of air that can be drawn to an engine is important since for a specific air–fuel ratio, higher amounts of inhaled air means more fuel can be combusted and consequently more energy can be converted to the output power.

Volumetric efficiency measures the effectiveness of an engine's intake process and provides a ratio of the amount of air–fuel drawn in the combustion chamber and the theoretical maximum. The theoretical maximum is the actual capacity of the combustion chamber under static conditions. Higher volumetric efficiency of an engine results in an increase in engine speed and overall power. The volumetric efficiency of an engine is affected by parameters such as the compression ratio, fuel type, air–fuel ratio, engine speed, air–fuel mixture temperature, and the pressure ratio of exhaust to intake manifold. Moreover, volumetric efficiency can be enhanced through the use of larger valves, multiple valves, variable valve timing, and force induction systems such as supercharging or turbocharging. Larger valves pull in a greater amount of airflow at the expense of heavier weight while multi-valve engines take advantage of two or more smaller valves at the expense of higher complexity.

1.2.4 ICE Vehicle Emissions

Vehicle emissions are one of the key contributors to environmental air pollution and global climate change. Tank-to-wheel and well-to-wheel emissions are the most common ways to discuss motor vehicle emissions. Tank-to-wheel refers to emissions produced during vehicle operation and during fuel combustion. On the other hand, well-to-wheel emission refers to discharges during the production and distribution of fuel, as well as throughout vehicle operation. The subsequent section discusses tank-to-wheel emissions and their effects on the global environment.

As mentioned earlier in this chapter, IC engines generate propulsion power by combusting air and a fuel, usually gasoline or diesel, within a combustion chamber. Gasoline and diesel fuels are a mixture of hydrocarbons (made of hydrogen, oxygen and carbon atoms), while air is mainly composed of nitrogen (N_2) and oxygen (O_2). In a perfectly operating engine with ideal combustion conditions, hydrocarbons would react with oxygen to produce water vapor (H_2O) and carbon dioxide (CO_2), while nitrogen would pass through the engine unaffected. However, depending on the operating conditions and the fuel-air ratio, an incomplete combustion procedure may occur, causing the vehicle to emit pollutants as well.

The amount of pollution that a vehicle emits depends on many factors. The most important factors include fuel rate consumption, driving conditions (e.g., the speed, acceleration, and load on the vehicle), the type of fuel used (e.g., gasoline or diesel), and the technology used to control emissions (e.g., catalysts), among others.

The main emissions from vehicles are carbon dioxide (CO_2), carbon monoxide (CO), hydrocarbons (HC), particulate matters (PM), nitrogen oxides (NO_x), nitrous oxide (N_2O), and methane (CH_4). The reaction of oxygen with existing sulfur and carbon fuel impurities can also result in sulfur monoxides (SO) and sulfur dioxide (SO_2) emissions, which contribute to acid rain formation. Vehicle emissions belong to two categories: greenhouse gas emissions and air pollution emissions. Greenhouse gas emissions contribute to climate change, particularly global warming, whereas the term air pollution emission refers to harmful smog-forming pollutants released by vehicles. Carbon dioxide, nitrous oxide, and methane are greenhouse gas emissions, whereas hydrocarbons, particulate matters, nitrogen oxides, sulfur monoxides, and sulfur dioxide are air pollution emissions [8–15].

1.2.4.1 Greenhouse Gas Emissions

Greenhouse gases trap heat in the atmosphere by absorbing and emitting radiation within the thermal infrared range. This process is the fundamental cause of the greenhouse effect. However, the high concentration level of these gases contributes to climate change – particularly global warming.

Global warming is a term used to describe increases in the average atmospheric temperature near the Earth's surface, as well as in the troposphere, both of which contribute to changes in global climate patterns. The most significant consequences of global warming are melting glaciers, rising sea levels, flooding, gully erosion, desertification, and extreme weather conditions. Although there are a variety of natural sources that emit greenhouse gases, scientists have observed elevated levels of these gases in recent decades, especially in regards to carbon dioxide and methane. This increase is widely attributed to vehicle emissions and other human activities that involve burning fossil fuels. The primary greenhouse gases in the Earth's atmosphere include carbon dioxide (CO_2), methane (CH_4), nitrous oxide (NO_x), and water vapor (H_2O).

- *Carbon dioxide*: During the combustion process, hydrocarbons in fuel react with oxygen in the air to create water vapor and carbon dioxide. Even if perfect combustion were to occur, the sheer volume of vehicles worldwide would still release astonishing amounts of CO_2. Carbon dioxide is the principal greenhouse gas warming the Earth. Vehicles are the second largest source of carbon dioxide emissions (coal-burning power plants are first), generating nearly 1.5 billion tons of carbon dioxide annually.
- *Nitrous oxides*: Nitrous oxides are a product of the reaction that occurs between nitrogen and oxygen during fossil fuel combustion. Nitrous oxides are approximately 130 times more effective in trapping atmospheric heat than carbon dioxide over a hundred-year period. Though fuel nitrogen content produces nitrous oxides, the primary source of this emission is the pollution control device (catalytic converter) used in modern vehicles. The amount of emissions released by a vehicle largely depends on its fuel type, technology, and maintenance and operating points. For example, a catalytic converter – the device used to remove pollutants from the vehicle exhaust – can potentially promote the formation of N_2O. The highest volume of emissions occurs when the catalyst is not fully functional and when exhaust gases are at low temperatures. Drive cycle, vehicle speed, and catalyst age are also contributory factors to the production of nitrous oxides emissions.

- *Methane*: Methane emissions occur because of incomplete fuel combustion. The primary factors that affect the volume of methane emissions are the emission control system, fuel type, engine design and tuning, and vehicle age. Additionally, methane emission increases at lower ambient temperatures because liquid fuel does not completely vaporize (and thus, burn) at lower temperatures. Methane emission is at its lowest when the catalyst is fully functional, and when exhaust gases are at their hottest.

1.2.4.2 Air Pollution Emissions

Air pollution has become a major environmental concern because of the continual worldwide increase in automobile production. Currently, vehicle-induced air pollution has reached startling levels in both developed and developing countries. Health authorities have confirmed that harmful air pollutants can cause a variety of lung-related illnesses such as asthma, emphysema, and bronchitis, while also increasing the risk of cancer. The following emissions are involved in air pollution:

- *Carbon monoxide*: Carbon monoxide (CO) forms during rich fuel mixture combustion due to a lack of sufficient oxygen. Specifically, the combination of carbon atoms in a hydrocarbon fuel with one oxygen atom, rather than two, forms this emission. Chemical kinetic effects also result in the production of small amounts of carbon monoxide under lean conditions. CO is colorless, odorless, highly poisonous, and extremely detrimental to human health. Vehicles emit high volumes of CO when the air–fuel ratios within the engine are too rich. This tends to occur as soon as the vehicle starts because the engine is not yet at optimal operating conditions. Additionally, carbon monoxide emissions increase at higher altitudes because the amount of oxygen in the air is not sufficient for perfect combustion. Other factors that increase these emissions include leaky injectors, high fuel pressure, and improperly closed-loop catalyst controls. Transportation sources such as cars and trucks are the primary contributors of carbon monoxide, exceeding 90% of total CO emissions in certain locations (e.g., urban areas).
- *Hydrocarbons*: Hydrocarbons are another major vehicle emission. During incomplete combustion, some fuel molecules remain unburned or partially burned and are exhausted into the atmosphere. The reaction of exhausted hydrocarbons with other compounds in the atmosphere produces ground-level ozone, a key component of smog. Hydrocarbons are toxic and, with prolonged exposure, can cause a variety of human illnesses such as liver disease, lung disease, and cancer. The main factors that contribute to high amounts of hydrocarbon emissions are improper ignition timing, improper air–fuel ratio, and low air temperature levels. Additional reasons for excessive hydrocarbon emissions include malfunctions with ignition components, air injection components, and catalytic converters. Fuel evaporation during refueling and crevice volumes (such as the space between the piston and cylinder wall) can also result in hydrocarbons being emitted into the air.
- *Particulate matter*: Vehicle exhaust contains a mixture of small solid particles and liquid droplets. This mixture is particulate matter and it forms during combustion, as well as after carbon-containing molecules condense into their solid form. Pollutants found in particulate matter vary in size and include substances such as carbon, sulfur, and nitrogen compound metals. Vehicles emit particulate matter referred to as fine particulate matter or $PM_{2.5}$ because of their 2.5-μm diameter. These pollutants are small enough to travel deep into the

lungs of humans and can adversely affect the heart. Diesel engines produce much higher rates of particulate matter pollutants than gasoline engines.

- *Nitrogen oxides*: During combustion, high temperatures and pressures inside the combustion chamber may result in a reaction between nitrogen and oxygen in the air. This reaction results in the production of nitrogen oxides, including nitric oxide (NO) and nitrogen dioxide (NO_2). Moderate and heavy load conditions usually result in peak production of these emissions due to the high level of combustion pressures and temperatures that occur under these circumstances. However, light operating conditions such as cruise or light throttle operations may also produce small amounts of nitrogen oxides. Nitrogen dioxide causes a variety of health and environmental concerns, specifically in regards to ozone and smog. Nitrogen oxide emissions become part of particulate matter formation through chemical reactions in the atmosphere.

1.2.4.3 Idling Emissions

Vehicle idling is perhaps the most significant factor in generating air pollutants and wasting natural resources. In addition to these concerns, idling can also cause damage to the engine. When idling, the fuel is below peak temperatures, resulting in incomplete combustion and fuel residue build-up on cylinder walls and spark plugs. Consequently, this build-up can significantly diminish the efficiency of an engine. Furthermore, idling and its associated consequences pose a significant health risk to humans. Drivers and passengers inside idling vehicles are subject to concentrated exposure of pollution because there is no airflow available to dissipate it. Children are even more vulnerable under these circumstances because they breathe faster and inhale a higher ratio of air in comparison to their body weight.

In addition to traffic jams, the main causes for idling in both diesel and gasoline vehicles are individual habits and a general misunderstanding of engine functionality. Some of the reasons for individuals to intentionally idle the engine include: using it as a means to keep the engine warm during cold weather, attempting to maintain adequate battery voltage for the use of electrical devices, and using the air-conditioner or heater for personal comfort. A common misconception is that engine idling should be used to warm up the vehicle prior to driving in cold weather. In reality, the most effective way to heat up the engine is simply by driving the vehicle. The catalytic converter cannot operate below certain temperatures and will function better when the vehicle is moving. Idling also causes the vehicle to release higher levels of emissions. Driving the vehicle quickly after starting the engine while avoiding significant acceleration or high speeds during the first few kilometers is more effective in heating up an engine. This will ensure that the vehicle reaches optimal operational temperatures with minimum amount of fuel.

1.2.4.4 Gasoline Engine Emissions vs. Diesel Engine Emissions

The majority of emissions from diesel-fueled vehicles are particulate matter, nitrogen oxides, and hydrocarbons, whereas gasoline-fueled vehicles release mostly carbon monoxide, hydrocarbons, and nitrogen oxides. The amount of emissions released by a vehicle depends on a number of factors, including operation conditions, environment, and engine temperatures, fuel quality, and most importantly, air–fuel ratio. Figures 1.6 and 1.7 illustrate the effect of air–fuel ratio variations on gasoline and diesel vehicle emissions. In these figures, λ represents a

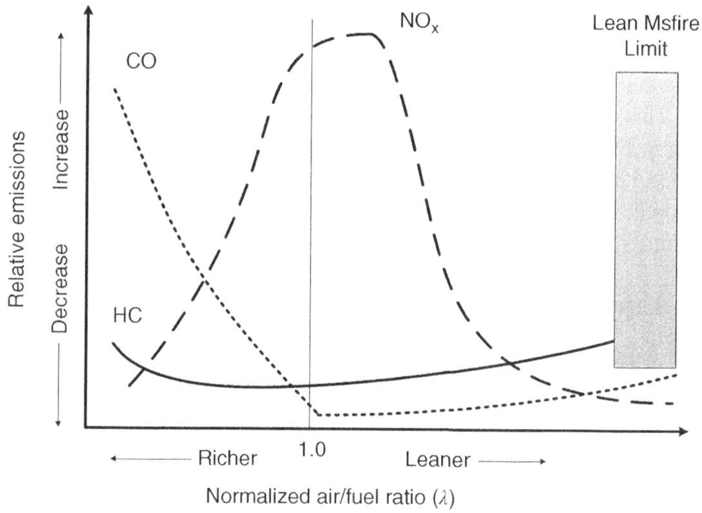

Figure 1.6 Effect of air–fuel ratio on gasoline vehicle emissions

normalized ratio based on the proportion of air–fuel ratio and stoichiometric air–fuel ratio. As illustrated in Figure 1.6, the amount of emissions produced under lean mixtures is less than the amount produced by rich mixtures in a gasoline engine. However, lean mixtures can cause engine knocking and reduce engine power. Furthermore, lean mixtures generate high amounts of nitrogen oxide emissions when they have an air–fuel ratio near stoichiometric proportions. This is because a stoichiometric ratio produces the high temperature gases necessary to form

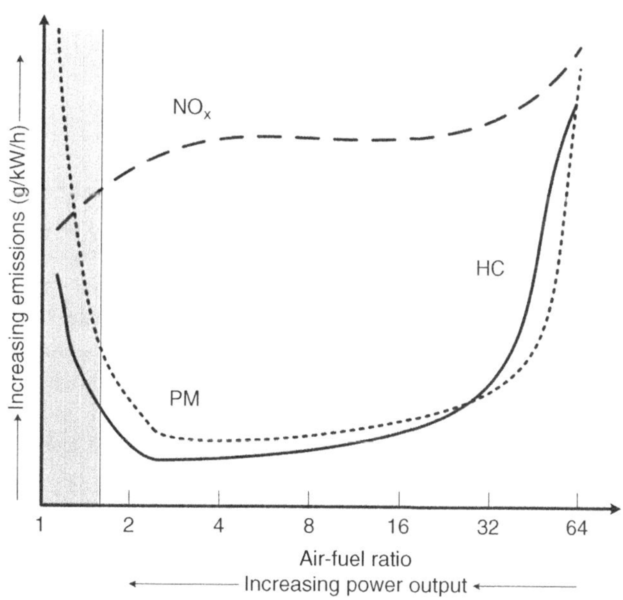

Figure 1.7 Effect of air–fuel ratio on gasoline vehicle emissions

nitrogen oxide. As such, the designs of gasoline engines need to meet the conflicting requirements necessary for efficiency, power, and emissions. As Figure 1.7 illustrates, diesel engines have greater amounts of hydrocarbons and particulate matter at high air–fuel ratios (lean mixtures). This occurs because after combustion, the heat generated by lean mixtures is not high enough to burn out residual hydrocarbons. The burning of low air–fuel ratios (rich mixtures) in a diesel engine increases particulate matter emissions because it lacks the oxygen necessary for oxidation. A common problem with gasoline and diesel engines is that reducing nitrogen oxide emissions results in the increase of hydrocarbon emissions.

1.3 Vehicle Emission Control Technologies

The advances in engine design, air–fuel mixture preparation, and proper ignition timing can considerably reduce emissions in modern ICE vehicles. However, these advances often do not meet increasingly stringent emission policies and regulations. An obstacle towards further improving the release of vehicle emissions is the conflicting specifications of engine complexity, fuel efficiency, power, and emission requirements. As such, emission control technologies are important in normalizing engine emissions to standard levels without jeopardizing vehicle and engine performance.

Emission control systems are designed to reduce the amount of air pollution emitted by a vehicle. Types of tank-to-wheel emissions of ICE vehicles are tailpipe exhaust emissions, evaporative emissions, and crankcase emissions. Tailpipe exhaust emissions refer to emissions that are expelled into the air through the exhaust pipe. These emissions usually contain hydrocarbons, carbon monoxide, nitrogen oxide, and particulate matter. Evaporative emissions refer to fuel vapors that escape into the air through the fuel tank or during refueling. Likewise, crankcase emissions are unburned or partially burned fuels vented to the engine compartment.

Gasoline and diesel engines use several technologies to control emissions, with the most important ones being catalytic converters, exhaust gas recirculation, crankcase emission control, and evaporate emission control. While the structures may differ, the main functions of these control technologies are similar for both engine types [16–21].

1.3.1 Advanced Engine Design

Emissions from an ICE can be reduced by improving the engine design and controlling the combustion process. Engine design should be optimized for efficiency and performance under a variety of driving conditions, while minimizing its emissions. Properly controlling the engine and combustion process variables such as ignition timing, air–fuel ratio, the volumetric efficiency, and the compression ratio can also significantly reduce the level of engine emissions. Advanced control technologies such as fuel injection systems, electronic engine units, controlled air induction systems, variable valve timing, and turbocharging systems have been shown to be effective in reducing emissions of both diesel and gasoline vehicles. Generally, the most important of these are variable valve timing and turbocharging systems. The subsequent section discusses these two technologies in detail.

1.3.1.1 Variable Valve Timing

"Valve timing" is the time interval within which a valve is open, while the time interval in which the intake and exhaust valves simultaneously open is called a "valve overlap time" or "timing of breathing." Proper controls of valve timing and valve overlaps play an important role in reducing engine emissions, improving engine fuel economy, and enhancing the output power. For example, a slight delay in the closing of the intake valve pushes some air–fuel mixture back into the intake manifold by the piston during the compression stroke. This action subsequently increases the intake manifold pressure and results in a richer air–fuel mixture during the next cycle. What this sequence of events means is that delayed intake valve closing results in better fuel economy and lower nitrogen oxide emissions during partial load conditions. However, these benefits come at the expense of a slight loss in peak engine torque. In contrast, closing the intake valve earlier during normal combustion circumstances produces lower pressure within the cylinder during the compression stroke. Consequently, this reduces the amount of work required from the piston. The early intake valve closing also produces better fuel economy and lower nitrogen–oxide emissions. However, closing the intake valve earlier also increases hydrocarbon emissions due to low temperatures caused by low-pressure conditions. Another disadvantage of prematurely closing the intake valve is the reduction in engine performance at high-speed conditions. Engine performance suffers under these circumstances because the intake valve closes before the maximum amount of air–fuel mixture enters the cylinder.

In general, reductions in overlap result in a smoother idle and more slow-speed torque, while increases in overlap produce more power and better engine breathing. However, the disadvantages of more overlap include rough idling and high exhaust emissions. Most conventional vehicles use an engine with fixed valve events, in which valves open and close at fixed times during engine strokes, independent of engine load and speed conditions. The flexibility of variation in valve events that are subject to speed and load conditions allows the engine to operate more efficiently over its operating range and conditions.

The design of variable valve timing aims to improve engine performance by controlling the timing of valves through different operating modes. This technology makes it possible to control the stream flow of intake and exhaust gases coming into and out of the combustion chamber with variable valve events. It also allows the engine to achieve optimal power and torque across a wider range of engine speeds with lower emissions. In diesel engines, the combination of variable valve timing and exhaust gas recirculation can significantly reduce the hydrocarbon and nitrogen–oxide emissions. However, this reduction comes at the expense of an increased amount of particulate matter in the exhaust.

1.3.1.2 Turbocharging Systems

The primary function of a turbocharger is to increase the amount of oxygen inhaled into the combustion chamber by compressing the air intake. This compression results in enhanced volumetric efficiency, reduced particulate matter and hydrocarbon emissions, and improvements in fuel efficiency. However, due to the increase in combustion pressure, turbochargers can also increase the rate of nitrogen–oxide emissions. A turbocharger is composed of a turbine, a compressor, and a center housing/hub rotating assembly (see Figure 1.8). Hot exhaust gases drive the turbine and they leave the engine through the exhaust ports. The captured kinetic energy of the exhaust gases drives the compressor through a shaft in the turbo housing. The blades of the compressor draw ambient air inside and accelerate it back into the engine. Before the intake air enters the intake manifold, the compressor increases its mass, compresses it, and increases its pressure.

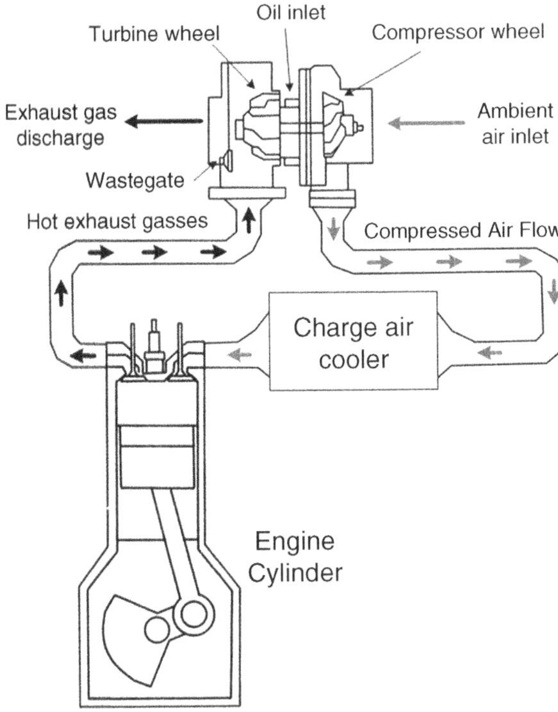

Figure 1.8 A schematic of a turbocharging system

Turbochargers are capable of providing volumetric efficiency greater than 100% because in these engine types, the intake manifold pressure exceeds atmospheric pressure. However, the engine efficiency comes at the expense of reduced power from the cylinders. This reduction occurs because the turbine in the exhaust causes the exhaust stroke of the engine to push harder against higher backpressure. The two major problems of turbochargers are turbo lag and high intake air temperature. Turbo lag is a sluggish (delayed) response from the engine during the initial push on the throttle pedal. It occurs due to the initial inertia of the turbocharger. Essentially, at low engine speeds, the exhaust gas flow is not strong enough to push the turbine quickly. The compression of the intake air results in increased air pressure, which in turn causes the temperature to rise. Excessive intake air temperature and pressure may cause detonation (engine knocking) or pre-ignition phenomena, which reduces the output power. Pre-ignition is an abnormal combustion in which high pressure and temperature cause the air–fuel mixture to ignite earlier than it would by spark plug fire. To cool down the intake air, an intercooler located in the middle of the piping between the turbocharger's compressor and the engine's air intake valve is integrated. The waste-gate usually regulates and limits air intake pressure. The waste-gate controls air pressure by bypassing some of the exhaust gas flow before entering the turbine as it reaches the intake pressure threshold.

An advanced type of turbocharger technology is the variable turbine geometry turbocharger, also known as a variable geometry turbocharger or a variable nozzle turbine. This type of turbocharger contains a set of adjustable vanes in the turbine housing. These vanes are modified in angle throughout the engine speed to maximize and control boost pressure over a wide range of engine operations. The vanes guide the exhaust flow towards the turbine while an actuator adjusts their angles. The purpose of a variable geometry turbocharger is to partially close the vanes at low-speed conditions. Partially closing these vanes directs exhaust gas towards the turbine and pushes the turbine blades to the right angle. This will cause the turbine to spin faster before the engine reaches proper speed. At higher speeds where the exhaust flow is sufficiently strong, the vanes open completely to capture the high kinetic energy of the exhaust gas. This reduces the exhaust pressure in the turbocharger by removing the waste-gate from the circuit.

A variable geometry turbocharger reduces turbo lag at low engine speeds, provides cleaner exhaust gas, and improves fuel economy while maintaining power and performance. Furthermore, these turbochargers also effectively reduce particulate matter emissions from diesel engines by providing lean combustion in the engine. Diesel engine vehicles mainly use variable geometry turbochargers; they have achieved some success with gasoline engine vehicles as well. The high temperature of exhaust gases is the main obstruction preventing complete integration of variable geometry turbochargers within gasoline engines.

1.3.2 Catalytic Converters

Positioned in the exhaust pipe, catalytic converters are emission control devices that convert unsafe exhaust emissions into harmless compounds through a combination of catalysts. A catalyst is a chemical material that increases the rate of a chemical reaction without changing itself in the process. Platinum, rhodium, and palladium are the most commonly used catalysts.

The catalytic converter plays a significant role in vehicle emission reduction, which is why it is the main emission control technology used in vehicles. However, leaded fuels adversely affect the efficiency of catalytic convertors. These fuels form deposits that coat the catalyst and prevent proper contact between exhaust gasses and catalysts. The catalytic converters used in gasoline vehicles include oxidation (two-way) catalysts and oxidation–reduction (three-way) catalysts, while diesel vehicles use diesel oxidation catalysts, selective catalytic reduction, and nitrogen–oxide adsorber catalysts. Important factors that affect the performance of a catalytic converter are catalyst temperatures, mixture air–fuel ratio, and hydrocarbon mix.

1.3.2.1 The Two-Way Catalyst

Two-way catalysts use an oxidation process to convert carbon monoxide and hydrocarbons to carbon dioxide and water. Vehicles used the oxidation catalyst from the mid-1970s until the 1980s, but the invention of the three-way catalyst has rendered it obsolete in modern vehicles.

1.3.2.2 The Three-Way Catalyst

Almost all modern vehicles have three-way catalysts. They convert nitrogen oxide back into nitrogen and oxygen, and convert carbon monoxide and hydrocarbons into water and carbon dioxide. When the air–fuel ratio is close to stoichiometric, it is possible to achieve the maximum conversion efficiency for all three pollutants. In these circumstances, three-way catalysts are capable of oxidizing hydrocarbons and carbon monoxide while reducing nitrogen–oxide emissions. A fuel-ratio range between 0.98 and 1.003 can achieve a high conversion efficiency. However, this range is quite narrow and limiting. For leaner mixtures, the efficiency of the catalyst in conversion of hydrocarbons and monoxide increases, however, the catalyst is inefficient at reducing nitrogen–oxide emission. In contrast, for richer mixtures, the conversion efficiency of nitrogen–oxide emissions increases while the efficiency of carbon monoxide and hydrocarbons decreases.

An ECU helps keep a three-way catalyst at its optimal operation conditions, which greatly depends on the catalytic temperature. In other words, high conversion efficiency can be achieved if the catalytic temperature reaches a certain value known as "light off temperature," which is usually in the range of 200 °C–300 °C. By contrast, the efficiency of catalysts can decrease because of overheated conditions, which can occur due to improper engine tuning, inadequate rich air–fuel mixtures, or cylinder misfiring.

1.3.2.3 Diesel Oxidation Catalyst (DOC)

The lean combustion of diesel engines prevents the use of three-way catalysts. As such, most diesel engines use a diesel oxidation catalyst (DOC), which converts carbon monoxides and hydrocarbons to carbon dioxide and water through an oxidation process, while also reducing soot mass. However, DOCs are inefficient in reducing nitrogen oxides and particulate matter. Additionally, they are capable of oxidizing sulfur dioxide that exists in diesel exhaust (specifically in heavy-duty vehicles). The oxidation of sulfur dioxide forms sulfate particles, which in turn increase the amount of total particle emissions. As such, DOCs need to be designed in a way that balances a sufficient reduction of hydrocarbons and carbon monoxides with reasonable amounts of sulfur-dioxide. The efficiency of catalysts can be improved if they are used in conjunction with other emission control technologies. The effect of DOCs on fuel consumption is not substantial. Light-duty diesel vehicles frequently use DOCs, although they have been shown to be effective in heavy-duty vehicles as well.

1.3.2.4 Selective Catalytic Reduction (SCR)

Selective catalytic reduction (SCR) is one of the most efficient emission control technologies. SCR helps to reduce nitrogen oxide emissions to near zero levels through a catalytic reaction while also providing good fuel economy and durability. It is possible to apply SCR to all types of diesel-powered vehicles (light, medium, heavy-duty) without compromising engine power and performance. SCR functions by injecting diesel exhaust fluid (DEF), a liquid containing urea and water, into the exhaust stream, which is just ahead of the SCR converter. Urea is a nitrogen compound that hydrolyses into ammonia (NH_3) when heated. In the presence of a catalyst, the ammonia reacts with the exhaust stream to produce nitrogen and water vapor. However, some un-reacted ammonia may be released into the atmosphere as well. This is known as an "ammonia slip" and can be caused by over-injected DEF in the gas stream, low temperature conditions, or a degraded catalyst.

A highly controlled system regulates the amount of injected DEF, it adequately distributes the ammonia in the gas stream and provides a consistent gas velocity for the catalyst. The control system works either as an open-loop or closed-loop system. The open-loop system uses a nitrogen oxide estimation algorithm to estimate the amount of nitrogen oxide available in the exhaust stream. The algorithm uses operation conditions such as engine speed, exhaust temperature, and load to determine the amount of DEF to inject. The closed-loop system obtains data through a sensor that measures nitrogen oxide concentration; from this it calculates the amount of DEF to inject. A concern during the conversion process of this catalyst is the possible formation of sulfate. As such, it is important to use low sulfur fuel to avoid further particulate emissions and to enhance the catalyst efficiency. A disadvantage of the SCR system is its dependency on a tank to store DEF liquid, if the tank runs dry, the SCR system stops functioning.

1.3.2.5 Nitrogen–Oxide (NO_x) Adsorber Catalyst

As opposed to gasoline engines, diesel engines usually run at lean fuel mixture conditions. However, recently, the design of lean-burn gasoline engines suits passenger vehicles as well. A lean-burn gasoline engine provides a higher compression ratio, better performance, superior fuel efficiency, and lower carbon monoxide emissions than that of conventional gasoline engines. These benefits, however, come at the expense of higher production of nitrogen–oxide emissions. These emissions occur as a result of the increased levels of oxygen in the fuel mixture and exhaust gases. Since a conventional three-way catalyst does not effectively reduce nitrogen oxides under lean conditions, the mass production of such vehicles has been a subject of great debate.

A NO_x adsorber (known as a lean NOx trap or LNT) is a catalyst designed to reduce nitrogen–oxide emissions; it is most effective in lean-burn engines. An LNT system stores nitrogen–oxide emissions in a catalyst under lean conditions. It then catalytically reduces the trapped nitrogen oxide to nitrogen when the engine is running under rich conditions. The cycle in which a catalytic reaction between excess unburned hydrocarbons and nitrogen oxide converts the trapped nitrogen oxide to nitrogen is called the nitrogen–oxide regeneration cycle. The operating temperature, system responsiveness, and diesel sulfur content are the main factors affecting the catalyst efficiency. Like other catalysts, LNT offers superior performance at higher operating temperatures and lower sulfur fuels. The main advantages of LNT technology include high efficiency in nitrogen oxide emission reduction, low light-off temperature, and a cost-effective system. However, LNT adversely affects vehicle fuel economy because of the fuel requirements and burn conditions of the nitrogen oxide regeneration cycle. While the effectiveness of LNT under stoichiometric conditions is less than that of three-way catalysts, it still provides significant benefits under lean conditions.

1.3.3 The Diesel Particulate Filter (DPF)

A diesel particulate filter (DPF) collects and removes diesel particulate matters or soot from the exhaust of the engine by passing exhaust gases through the walls between numerous channels (see Figure 1.9). A DPF accumulates soot and particulate matters over time; however, since

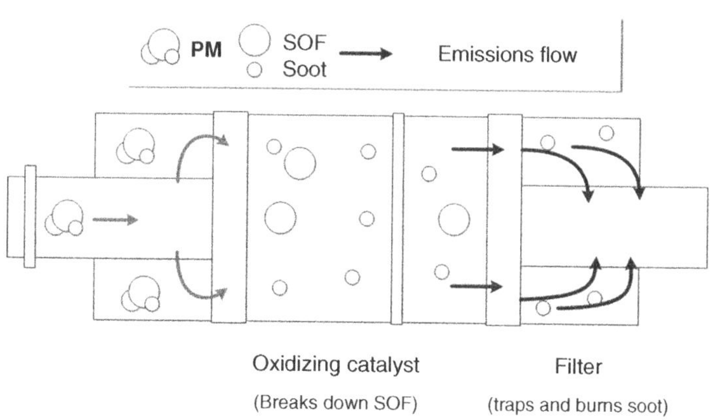

Figure 1.9 A schematic of a passive diesel particulate filter

filters have limited capacity, they need to be removed or cleaned in order to avoid backpressure blockage on the engine. If left dirty, this blockage may result in engine damage or destruction.

Some filters are designed for single use and can easily be discarded or replaced. For other filters, it is necessary to remove the accumulated particulate matters. A practical approach to dispose of the trapped materials is to burn or oxidize the particulate matters within the filter once the exhaust gases reach a certain temperature. The process of cleaning by burning the trapped materials is called filter regeneration.

A passively regenerated filter is a filter regenerated by available exhaust heat or by using a catalyst. On the other hand, an actively regenerated filter is one that uses active means during the filter regeneration process. For example, actively regenerated filters can either use the injection of diesel fuel into an upstream of DOC, or utilize a fuel burner that heats the filter to particulate matters combustion temperatures. Passive regeneration cannot be used if the exhaust temperature is lower than the threshold, whereas in active regeneration techniques, it is possible to integrate various engine controls to provide filter regeneration conditions on demand. Integrating a combination of passive and active strategies can ensure the completion of filter regeneration under all possible vehicle operating conditions. Additionally, reduction in particulate matters and nitrogen oxide emissions can be simultaneously achieved if a diesel particulate filter is used in conjunction with exhaust gas recirculation, nitrogen–oxide adsorber catalysts, or selective catalytic reduction.

1.3.4 Exhaust Gas Recirculation (EGR)

Figure 1.10 depicts a schematic of an exhaust gas recirculation (EGR) for a gasoline engine. This technology reduces nitrogen oxide emissions in both gasoline and diesel engines during operating periods in which high combustion temperatures occur. The principal function of an EGR is to recirculate a controlled portion of exhaust gases into the engine combustion chamber

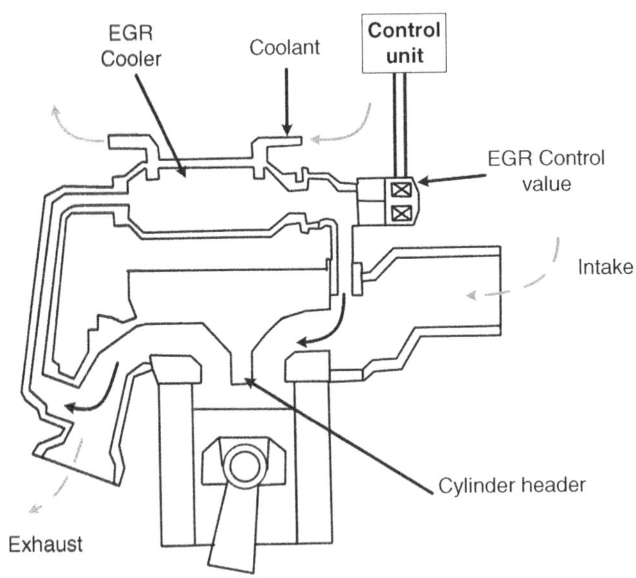

Figure 1.10 Schematic of the EGR loop for a gasoline engine

through a valve. This technology recirculates some exhaust gases to the intake manifold where the temperature of recirculated exhaust gases usually decreases by passing gases through an intercooler. The recirculated exhaust gases subsequently mix with the incoming intake manifold air, resulting in a diluted air–fuel mixture. This dilution causes a lower heat release and a lower peak cylinder temperature, which subsequently reduces nitrogen oxide formation. The recirculated exhaust gases must mix homogeneously with incoming air to have a consistent EGR distribution per cylinder. Improper EGR distribution may cause one cylinder to receive higher amounts of EGR, resulting in a higher rate of particulate matters emissions. Conversely, cylinders receiving low amounts of EGR produce higher rates of nitrogen oxide emissions. The EGR system prolongs the engine life because it reduces the peak of combustion temperature.

In a gasoline engine, an EGR system provides different types of exhaust gas flow. For example, EGR systems use a high flow of exhaust gases during operating conditions with high combustion temperatures, such as cruising and mid-range acceleration. Likewise, it provides a low flow of gases during low speed or light load conditions. The EGR system stops recirculating gases at full power demands as the recirculation process reduces engine operating efficiency, and also during idling to avoid higher level of emissions.

It is important to precisely regulate the amount of recirculated exhaust gases. This amount is a trade-off between a reduction in nitrogen oxide emissions and engine efficiency. Excessive recirculation of exhaust gases degrades the engine performance and efficiency, while minimal recirculation results in engine knocking and a decrease in EGR efficiency. Modern vehicles utilize ECU, sensors, and servo-driven EGR valves to balance engine efficiency and vehicle drivability by regulating the recirculation flow rate.

In diesel engines, reducing nitrogen–oxide emission occurs at the cost of an increase in particulate matters' emissions. The two main types of EGR systems used by diesel engines include high pressure loop EGR and low-pressure loop EGR (Figure 1.11). A high-pressure loop EGR diverts the exhaust gases before they reach the turbocharger turbine section. Conversely, the low-pressure loop EGR captures the exhaust gases after passing through

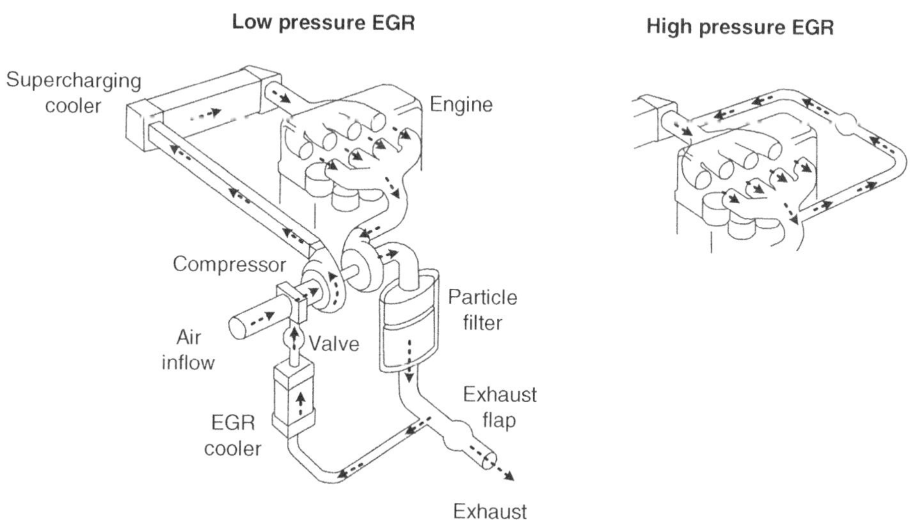

Figure 1.11 The EGR loop for a diesel engine

the turbocharger and the diesel particulate filter. Incorporating a diesel particulate filter or an oxidation catalyst in an EGR loop helps reduce the amount of particulate matters re-routed to the combustion process.

Integrating a combination of high- and low-pressure loop systems with a variable geometry turbocharger can enhance the EGR system efficiency over a wide range of operating conditions. The low-pressure EGR recirculates exhaust gases at low engine speeds and loads, while high-pressure EGR functions under higher engine speeds and loads. An optimized combination of the technologies will provide a more effective and efficient reduction of nitrogen oxide.

1.3.5 Crankcase Emission Control System

Blow-by gases are gases that leak into the crankcase during power or compression strokes while an engine is running. They are composed of mostly unburnt or partially burnt hydrocarbons and combustion by-products, which leak through the clearances between the piston rings and the cylinder walls.

The ventilation of blow-by gases helps prevent pressure build-up and increases engine longevity. In uncontrolled crankcase emission, a road draft tube releases blow-by gases and other vapors directly into the atmosphere. Hydrocarbon and particulate matters emissions are the classifications of these gases, and, if they enter the cabin, can be a significant source of emission exposure for drivers and passengers.

The crankcase emission control system, shown in Figure 1.12, is designed to re-burn the blow-by gases by diverting them into the intake system of a gasoline or diesel engine. When a gasoline engine vehicle is moving, fresh air flows to the crankcase through an air inlet path called the breather. The incoming air transfers and recirculates the blow-by gases from the crankcase into the intake system, allowing it to mix with an incoming stream of air.

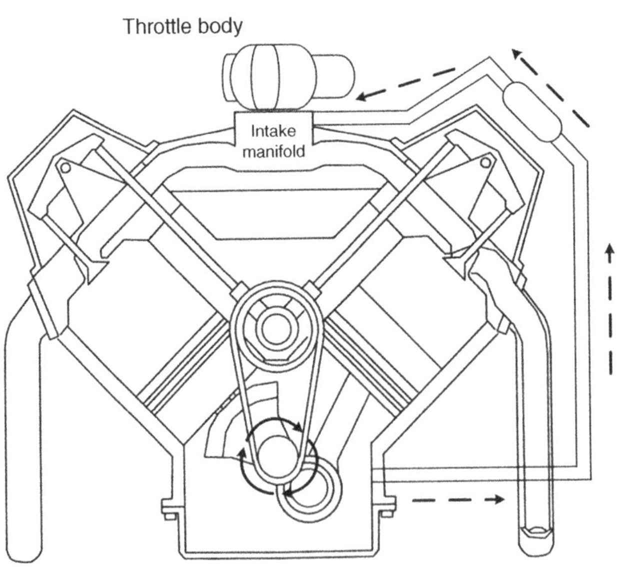

Figure 1.12 Crankcase emission control system of a gasoline engine

A check valve adjusts the gas flow between the crankcase and the intake manifold; the positive crankcase ventilation (PCV) valve controls the intake system pressure of the check valve. Then the diverted mix of air and blow-gases is injected into the combustion chamber where it is re-burned. Valve malfunctions may result in the blockage of blow-by gases and since the gases are under pressure, they may find other ways to escape from the crankcase to other parts of the engine compartment. These malfunctions can cause damage to engine parts, and result in oil leaks and sludge formation within the engine.

In a diesel engine, a multi-stage filter collects and returns the emitted lube oil to the engine's sump. Then, filtered gases reroute with a balanced differential pressure to the intake system. The crankcase emission control system of diesel vehicles often includes filter housing, a pressure regulator, a pressure relief valve, and an oil check valve.

1.4 Vehicles with Alternative Fuels

Although control emission technologies cost-effectively reduce internal combustion engine emissions, a narrow range of operating conditions limit their maximum efficiency. Furthermore, malfunctioning or improper tuning can dramatically degrade their performance or even result in increased vehicle emissions.

The development of vehicles operating with cleaner alternative fuels has received increasingly greater attention from industries and governments in the twenty-first century. Alternative fuels have the potential to reduce fossil fuel dependency and reduce pollutant emissions. Unfortunately, the cost of such vehicles or their operating costs are significantly higher than that of conventional gasoline/diesel vehicles in many countries. Nevertheless, vehicles powered with alternative fuels are more cost-effective and economical in some territories. The economic benefits of these fuel types largely depend on fuel consumption and government policies such as fuel tax and fuel price. The most significant types of alternative fuels used in vehicles include natural gas, liquid petroleum gas (propane and butane), biodiesel, and hydrogen.

1.4.1 Natural Gas Vehicles (NGVs)

As a clean-burning alternative to diesel/gasoline fuel, natural gas is a combustible mixture of hydrocarbon gases. It primarily consists of methane (approximately 80%), but also includes lesser amounts of propane, ethane, and butane. It is possible to use it in vehicles in the form of Compressed Natural Gas (CNG) or, less commonly, Liquefied Natural Gas (LNG). Light and medium-duty vehicles generally use CNG, whereas heavy-duty vehicles use LNG.

Dedicated, retrofitted, bi-fuel, and dual-fuel engines are all possible designs for a natural gas vehicle (NGV). Both dedicated and retrofitted engines operate merely on natural gas fuel. A bi-fuel engine runs on either natural gas or gasoline while a dual-fuel engine functions on a combination of natural gas and diesel fuel. A retrofitted engine is an engine modified from an engine designed for gasoline vehicles, while dedicated engines are those engines specifically designed and optimized to run on natural gas. Bi-fuel vehicles use two tanks to store fuels separately, allowing the engine to run on one fuel at a time. Conversely, dual-fuel vehicles burn a blend of two fuels stored in a single tank. In comparison with dedicated and dual-fuel vehicles, bi-fuel vehicles offer fuel flexibility at the expense of passenger/cargo space. Dedicated and bi-fuel engines are spark-ignited while dual-fuel engines need a diesel igniter to operate. The air-gas mixture in the cylinder of dual-fuel engines ignites with the injection of a small amount of diesel fuel, which self-ignites. Natural gas vehicles produce significantly lower nitrogen oxides, particulate matters, and carbon dioxide emissions than conventional gasoline/diesel vehicles because of their low carbon content and high compression ratio.

Since the air mixture in the combustion chamber of a NGV is completely gaseous, the emissions associated with cold start ignitions do not exist. However, because the main constituent of natural gas is methane, these vehicles emit considerably more methane than gasoline and diesel vehicles. The stoichiometric ratio of CNG is 17.2, higher than that of diesel and gasoline fuels. Gasoline or diesel engines retrofitted to operate on CNG may not provide the amount of emission benefits as a dedicated engine. Burning LPG in an engine optimized for gasoline/diesel fuel stoichiometric conditions will result in a rich-fuel condition.

Economically, CNG is usually cheaper than gasoline due to the more stable market and abundance of resources. The Energy Protection Agency (EPA) in the United States selected the Honda Civic GX – designed as a dedicated CNG vehicle – for eight years till 2012 as the cleanest-burning combustion engine vehicle. Although most major automotive industries offered vehicle models powered by CNG, the Honda Civic GX is the most commercialized CNG vehicle currently available in the global market.

Despite its advantages, the market for NGVs is very narrow because of its limitations in design technology, fuel infrastructure, and driving range. The primary issue is onboard fuel storage. Currently, CNG is stored in a high-pressure cylinder, and LNG is stored in an insulated tank. Both storage systems are considerably heavier, more expensive, and bulkier than storage systems for equivalent amounts of gasoline or diesel, thereby reducing the vehicle's power-to-weight ratio. Furthermore, NGVs require frequent refueling in comparison to their gasoline and diesel counterparts for similar travel distances due to lower volume energy density of natural gas in comparison with gasoline and diesel. The performance of these vehicles is comparable to gasoline and diesel vehicles when the CNG tank is under full pressure. However, when the pressure falls below a certain threshold, NGVs display limited performance at high speeds and uphill settings. Another important obstacle preventing the widespread commercialization of NGVs is the lack of refueling infrastructure for natural gas. Moreover, the use of gas storage cylinders that have limited shape flexibility is inconvenient for vehicle designers. Despite clean performance and abundant resources, NGVs still run on unsustainable fuel resources.

Due to economic factors, most NGVs use retrofitted engines. Unfortunately, the performance of retrofitted engines is not optimal compared to dedicated engines that are designed with the express aim of operating with natural gas fuels. In addition, as a result of extremely limited access to CNG refueling stations, most retrofitted NGVs are bi-fuel. Currently, fleet vehicles such as taxi cabs, transit vehicles, and school buses are the main applications of natural gas-fueled technologies.

1.4.2 Liquefied Petroleum Gas Vehicles (LPGVs)

As a by-product of natural gas and refined crude oil, liquefied petroleum gas (LPG) is composed of propane and butane. This fuel is also known as propane and autogas, and is currently the most prevalent alternative fuel. LPG is a fuel for conventional vehicles while producing lower levels of toxic and smog-forming air pollutants compared to fossil fuels.

The engine technologies behind LPG-fueled vehicles are similar to those of NGVs. The design of an LPG engine can be as a dedicated or bi-fuel engine, though it is possible to convert conventional combustion engines to use LPG as well. However, the use of LPG in dual-fuel diesel engines is uncommon due to its poor knock resistance.

Converting a gasoline or diesel engine to operate with LPG is cheaper than converting it to operate with natural gas because of lower fuel tank costs. Another advantage to CNG is the convenience of carrying LPG onboard the vehicle. The principal function of a LPG engine is similar to that of a spark-ignition engine. Like natural gas, LPG can function with a higher compression ratio, which results in increased power output and reduced exhaust emissions. LPG engines generate lower amounts of emissions under cold start conditions and perform better in low-speed and light-throttle conditions as well. This is because LPG injects into the combustion chamber as a vapor and the air mixture within the chamber is completely gaseous. The power output and torque of LPG engines are similar to gasoline engines. As such, LPG-driven vehicles can perform comparably to gasoline vehicles in terms of climbing slopes and traveling in mountainous areas.

Unlike gasoline engine vehicles, LPGVs produce near-zero particulate emissions, very little carbon monoxide, and only moderate amounts of hydrocarbons. Additionally, the carbon monoxide emissions of LPG vehicles are typically lower as well, while nitrogen oxide emissions are similar to the amount released from gasoline vehicles. LPGVs produce much higher levels of carbon monoxide and hydrocarbon emissions compared to NGVs. Like CNG, LPG has a stoichiometric ratio ($A/F = 15.7$) higher than that of gasoline/diesel fuel. As such, a retrofitted LPG engine may not be as efficient as a dedicated engine in terms of emission benefits.

Toyota made the first LPG vehicles in the 1970s. Despite the advantages of LPG-vehicles, they have a limited share of vehicles on the road. Like NGVs, the main obstacles preventing widespread commercialization of LPG-equipped vehicles are driving range, vehicle fuel availability, and infrastructure. The current LPG vehicles suffer from limiting vehicle space due to the additional tank required for the LPG fuel. With a similar volume of fuel, LPG vehicles travel shorter distances compared to diesel and gasoline vehicles due to the lower volumetric energy density of LPG, making it necessary for the vehicle to refuel frequently. Moreover, LPG is also a by-product of substances with limited resources. Additionally, despite producing lower level of emissions compared to gasoline/diesel-fueled vehicles, LPG vehicles still produce pollutants that affect air quality. Using a three-way catalyst or an oxidation catalyst in conjunction with an LPG engine further reduces the emission level.

1.4.3 Biodiesel

Biodiesel is a natural alternative fuel for compression-ignition (diesel) engines that can be derived from a variety of renewable biological energy sources, including vegetable oil, animal fat, and cooking oil. Biodiesel is the result of a chemical process known as trans-esterification. This process involves extraction of the basic compounds of biodiesel such as esters from natural oil resources through a catalytic chemical reaction with methanol. Currently, the primary feed stocks used in the production of biodiesel fuels are soybeans, canola oil, and yellow grease. Biodiesel fuel or their blends in most diesel engines work with little to no modification to the engine. Usually, the concentration of biodiesel in fuel blends is denoted by a system known as the "B" factor. For example, the denotation of 100% (pure) biodiesel is B100 while a diesel fuel with 20% biodiesel and 80% petro-diesel is known as B20.

The expected use of biodiesel in pure or blended form reduces carbon dioxide, particulate matters, and hydrocarbon emissions of diesel engines, in addition to slightly increasing nitrogen–oxide emissions. These actions occur because of the oxygen available in biodiesel, which allows the fuel to burn more completely. Moreover, nitrogen–oxide emission control technologies can eliminate or significantly reduce the nitrogen oxide emissions that the combustion of these fuels emits. In addition to reducing emissions, biodiesel fuel offers other important advantages compared to fossil fuels. Some of these benefits include: renewability (significantly beneficial for the environment), limited sulfur–dioxide production (can directly reduce the formation of acid rain), safer disposal of extraneous fuel (biodiesel fuel is non-toxic and degrades faster than petroleum diesel fuel), and better lubricant properties (biodiesel fuel is a very effective lubricant and acts very much like a solvent, capable of cleaning the engine).

Despite these benefits, biodiesel fuel still has a number of disadvantages. The main problem is a conflict with the engine warranty policies of major auto manufacturers, who have not yet accepted the use of blends more concentrated than B5. This lack of legal policy restricts the widespread use of biodiesel fuel. Additionally, biodiesel is more expensive to purchase on a recurring basis. They have lower energy content and fuel economy than fossil fuels – approximately 10% lower for B100 and 2% lower for B20 blends. The B100 blend raises further concerns due to its potentially adverse impact on engine durability over prolonged periods of usage, as well as its suitability for lower temperatures.

1.4.4 Hydrogen

Hydrogen is an alternative fuel for vehicle propulsion when burnt in an internal combustion engine using the combustion of a mixture of hydrogen and air. Hydrogen fuel is a decarbonized energy carrier that does not naturally occur in pure form. Hydrogen can be obtained mainly either by converting combustible molecules into hydrogen or by consuming electric energy. The common methods used to produce hydrogen include water electrolysis, natural gas reformation, or partial oxidation and steam reformation of other fossil fuels. Hydrogen production is available domestically from renewable sources such as wind, solar, and nuclear. Currently, natural gas most often functions to generate hydrogen. While there are environmental consequences of producing hydrogen from reformed fossil fuels, studies comparing the well-to-wheel emissions of hydrogen and petroleum productions have shown overall reductions in ozone and GHG while using hydrogen. However, an even more distinct advantage would be to use renewable energy to generate hydrogen. Hydrogen is appealing because it can reduce or eliminate carbon dioxide emissions from vehicular propulsion systems by deprecating the use of fossil fuels. Regardless of the amount of emissions released from the formation of hydrogen, hydrogen-fueled ICE vehicles can potentially have near-zero tailpipe emissions. The only harmful emission released from these vehicles would be nitrogen–oxide emissions, which occur because of the high temperature from hydrogen oxidation during the combustion process. Integrating nitrogen–oxide emission control technologies is the most convenient way to reduce this type of emission in hydrogen vehicles.

Since a hydrogen-fueled engine burns fuel similarly to conventional gasoline engines, a slight modification to a gasoline-based internal combustion engine can create a hydrogen-fueled engine. François Isaac de Rivaz designed the first hydrogen engine in 1807. However, it was not until 1970 that Paul Dieges added a patented modification to Rivaz's design. This modification allowed the engine to run using hydrogen, and eliminated much of the negative environmental effects. Modified internal combustion engines (e.g., Mazda's Wankel engine) are a low-cost and viable alternative technology. A number of automakers, including BMW, Mazda, and Ford, have separately launched production-bound hydrogen ICE vehicles. Some critics assert that hydrogen is not an ideal fuel solution for ICE engines because it increases the propensity for pre-ignition. However, certain experiments by auto-makers have completely dismissed these concerns. Currently, there are even suggestions for a gradual withdrawal from conventional fossil fuels to make way for complete reliance on hydrogen fuels.

A major concern regarding the use of hydrogen as a fuel is its storage. The most common ways to carry hydrogen on a vehicle includes storing it as a compressed gas, liquid, or in the form of metal hybrids by using chemical storage. Reforming liquid fuel sources such as methanol, ethanol, natural gas, and liquid petroleum gas can produce hydrogen on-board a vehicle. However, doing so requires a process to convert the existing liquid fuel sources into hydrogen. These requirements increase the overall cost while also adding maintenance costs to the system. Moreover, reforming fuel sources also releases carbon dioxide. Nevertheless, the amount of emissions released by engines operating on reformed fuel is much less than that of gasoline-powered engines. However, the volumetric energy density of compressed hydrogen gas is also much smaller than that of gasoline. As a result, hydrogen-powered vehicles have shorter traveling distances and less torque output.

Compressed hydrogen requires a heavier chassis and greater amount of space to hold equivalent amounts of gasoline. This increases the weight of the vehicle, resulting in a lower power-to-weight ratio. However, with the same fuel mass, this engine can run at higher efficiency levels than a gasoline combustion engine. This increased efficiency is a result of the higher compression and leaner air–fuel ratios of hydrogen fuel. In general, the issues associated with hydrogen engines are production, distribution, storage, and refueling. While sustained research and development efforts are bringing hydrogen-powered vehicles closer to widespread commercialization, the development of these vehicles is still in its early stages. The increased usage of hydrogen vehicles can substantially reduce fossil fuel dependency.

1.5 Powertrain Technologies

The powertrain system of a vehicle converts one or multiple forms of energy to mechanical energy that is used in vehicle wheels for propulsion. The efficiency, function, and design of vehicle powertrains are crucial in the vehicle's fuel consumption, emission release, comfort, and weight. The powertrain of a conventional vehicle includes the engine, transmission, and drivetrain. The engine generates mechanical power by burning fuel that is transmitted to the drivetrain through the transmission. The transmission system adjusts the speed and torque output of the engine through a variety of gears and shafts, while the drivetrain transfers the adjusted torque and speed to the wheels through the differential, axle, and driveshaft. The powertrain components and subsystems need to be designed and optimized to provide superior vehicle performance, driving comfort, fuel economy, emission reduction, and minimal noises. There are various powertrain configurations, for example, it is possible to mount the engine at the front or rear of the vehicle. With the exception of some light vehicles and buses, most vehicle types employ a front-mounted engine. The powertrain also determines which wheels are driven on a vehicle, with options such as: rear-wheel drive, front-wheel drive, or multi-wheel drive [22–28].

1.5.1 Rear-Wheel Drive Powertrains

Figure 1.13 illustrates a rear-wheel drive powertrain in which the engine power transfers to the rear wheels through the transmission, drive (propeller) shaft, differential, and axle. The front of

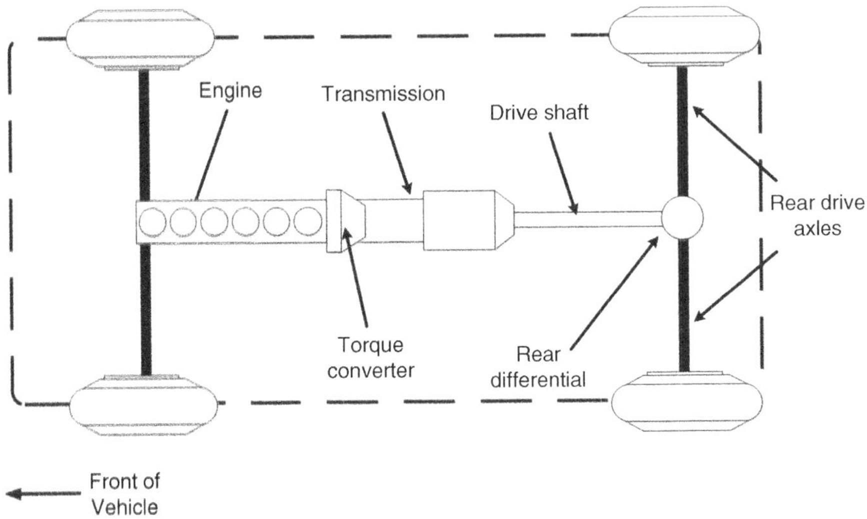

Figure 1.13 Rear-wheel drive powertrain configuration

the vehicle contains the engine, while the transmission is behind the engine. Usually, the engine and transmission are in alignment with longitudinal installation. This type of system provides a good catalytic converter configuration because of its long exhaust system. It also provides more space for the engine, making it more suitable for powerful vehicles. The proper weight distribution that occurs in a rear-wheel drive system results in greater vehicle handling and performance ability. Additionally, the load transfer to the back of the vehicle allows this system to provide better traction on the driven wheels during heavy acceleration. The driveshaft and other components needed to transfer the power to rear wheels result in reduced fuel economy and engine power. Furthermore, the additional weight of these components also increases the weight-to-power ratio of the vehicle. Luxury and racing vehicles usually have rear-wheel drive integration.

1.5.2 Front-Wheel Drive (FWD) Powertrains

As Figure 1.14 shows, in a front-wheel drive (FWD) powertrain, the engine and transmission system are at the front of the vehicle. As such, the engine power transfers directly to the front wheels through the transaxle. The transaxle consolidates the transmission, differential, and the associated components of the driven axle in one unit.

The front-wheel drive arrangement provides better fuel economy and less engine power loss than rear-wheel drive because it contains less moving components. During heavy acceleration, traction loss may occur due to load transfer towards the back of the vehicle. However, FWD vehicles as they deliver the load on steered and driven wheels, have improved road-holding and traction. Moreover, this arrangement provides more space for the cargo and passenger cabin at the expense of reduced space for the engine and other front-located components (e.g., steering mechanism, engine compartments).

There are two types of FWD powertrain systems, transverse or longitudinal, depending on how the vehicle's engine and transmission are mounted. In transverse powertrains, the engine

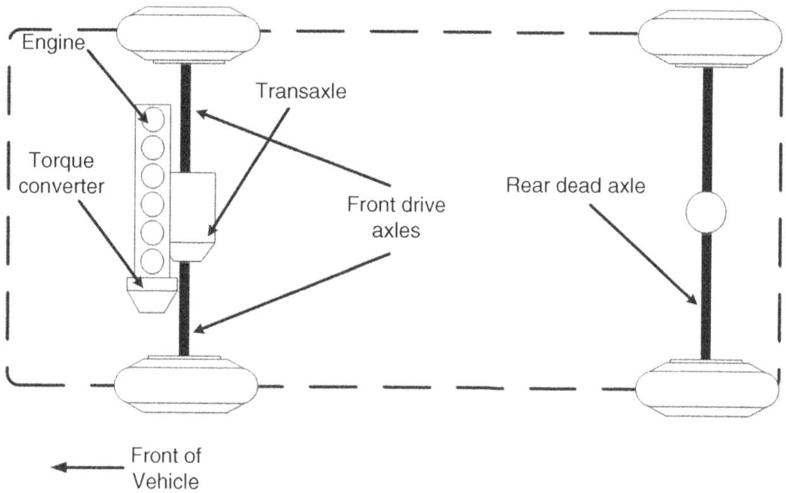

Figure 1.14 Front-wheel drive transverse powertrain configuration

and transmission are mounted parallel to the front axle. Conversely, in the longitudinal powertrain, the engine and transmission are parallel to the longitudinal vehicle axis. The integration of each system also depends on the engine and transmission size, as well as the available space. The transverse front-wheel drive powertrain is found in the majority of modern small and mid-sized vehicles.

1.5.3 Multi-Wheel Drive Powertrains

A multi-wheel drive powertrain delivers the engine torque to all four wheels of the vehicle. In some configurations, the torque transmits to both axles when all the wheels are continuously driven. In other configurations, the engine power transmits to one of the two axles, while another axle can engage manually or automatically when needed. In a multi wheel-drive, the engine torque transfers to the axles through a power-split differential system such as a center differential, transfer case, or another type of gear set. In this power-split system, the torque of each axle routes to the wheels on each side through each axle differential.

Multi-wheel drive systems are further available as four-wheel drive and all-wheel drive (AWD) systems. In four-wheel drive systems, the power-split differential may contain one or more sets of low-range gears with the ability to select two-wheel drive or four-wheel drive operation. The low-range gears reduce the vehicle speed in order to increase the torque routed to the axles. This helps when driving in extreme off-road conditions or on rough terrain. Conversely, the power-split differential of all-wheel drive systems does not include the low-range gears, and it does not offer a manual switch from four-wheel drive to two-wheel drive. Generally, multi-wheel drive systems provide better traction on all road conditions, better acceleration in low gear, and better handling performance. However, these advantages come at the expense of a very complex powertrain system and reduced fuel economy. Figure 1.15 illustrates a schematic of a multi-wheel drive powertrain configuration.

There are also other less common powertrain configurations that vehicles can use, such as a rear-mounted engine rear-wheel drive, or a rear-mounted engine four-wheel drive. These

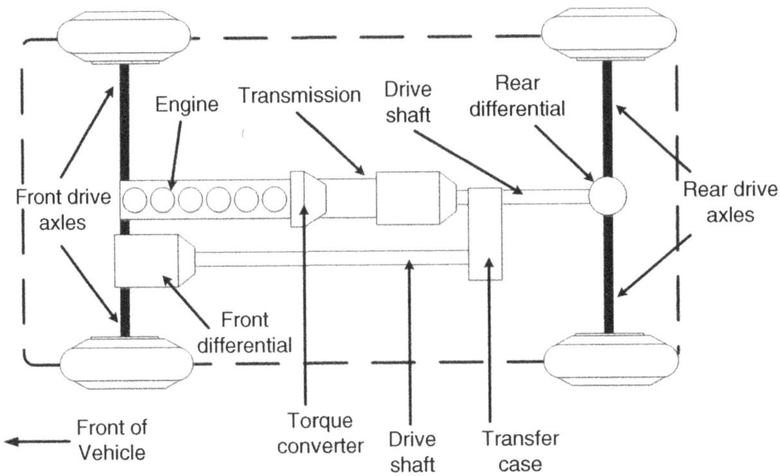

Figure 1.15 Multi-wheel drive powertrain configuration

configurations are rare in modern vehicles because they make the vehicle prone to instability, particularly during cornering.

1.6 Transmission Systems

A transmission system manages the output torque and power of the engine and sends it to the drive wheels through a number of gears or drive-ratios. This system connects to the engine through a flywheel, a heavy disk used to make the rotational output speed of the engine uniform. The engine only provides a limited range of torques and speeds, which is insufficient for some working conditions such as starting, stopping, acceleration, or deceleration. Essentially, the engine provides speed and torque up to a certain value, while the wheel road speeds and moving torque amount vary, based on driving and load conditions. For example, a high torque and low speed are required to move a vehicle from a stop knowing that the engine cannot provide the required torques at low speed conditions. In this case, the transmission system increases the level of torque delivered to the drive wheels by reducing the engine speed to match the lower wheel speed torque requirements. In contrast, high-speed cruise conditions require much less torque to move the vehicle. In this setting, the engine power and torque are directly transmitted to the drive wheels through the transmission system while holding the high speed of the engine's. Additionally, a transmission system also provides positional options for the vehicle to be in neutral, to reverse, or to park. The neutral position allows the driver to disengage the engine from the drive wheels, while the reverse position allows the vehicle to move backwards. The park position locks the driveshaft to prevent the vehicle from rolling. Some of the transmission types used in modern vehicles include: manual, automatic, semi-automatic, and continuously variable transmissions.

1.6.1 Manual Transmission/Transaxle Systems

The manual transmission system is the first type of transmission that was integrated into a vehicle while the manual transaxle system developed many years later with the introduction of front-wheel drive vehicles. Manual transmission systems and manual transaxle systems adjust the engine power and speed through a set of gears that can be shifted manually by a driver through a clutch pedal and shift knob. Typically, manual transmissions are used in powertrains in which the engine and drive wheels are located on two ends of the vehicle (e.g., front-mounted engine rear-wheel drive or four-wheel drive systems). Likewise, manual transaxle systems are used in powertrains in which the engine and drive-wheels are located on the same vehicle end (e.g., front-mounted engine front-wheel drive). These systems are similar in many ways – they both function to efficiently transfer engine torque to the drive wheels. The primary difference between the two is that a transaxle system transfers engine torque while also providing the differential action to the drive wheels, whereas the differential for a manual transmission system is separate and is connected to the transmission through the driveshaft. The mistakenly interchangeable use of the terms transmission and transaxle is common. The main components of a manual transmission/transaxle include the clutch, the gears, the shafts, and the synchronized collars:

- *The clutch*: The clutch is a coupling device with the primary function of engaging or disengaging the engine from the transmission system, as necessary. It regulates torque transfer from the engine to the gears, and makes gear shifting easier and smoother. The driver typically operates the clutch by using a foot pedal.
- *The gears*: Gears are a major component of the manual transmission/transaxle system. They are mounted on two separate shafts, and mesh with one another by sliding in and out of engagement. Gears with different sizes and different numbers of teeth mesh together in order to change the engine torque and speed by transmitting the rotating motion from one shaft to another. The gears provide functionalities such as magnifying the engine torque while reducing engine speed or vice versa, transferring the torque while keeping the engine speed, or changing the torque direction. Driving gears and driven gears are the two types of gears on a manual transmission/transaxle system. The driving gears receive their rotation from input shaft. In manual transmissions, the driven gears transfer the rotation to the output shaft, whereas in manual transaxles, this rotation transfers to the differential. The gear ratio reflects the ratio of the input and output torques of a transmission system. The gear ratio is a function of the number of teeth meshing on the driving and driven gears, as well as their rotational velocities. Most modern transmission/transaxle systems have at least four forward gear ratios. Higher gear ratios belonging to the first and second gears provide higher torque and lower speed to the drive wheel, while lower gear ratios belonging to the third and fourth gears transfer the engine torque with lower or no magnifications. Usually, the fourth gear provides a direct drive at which the input and output shafts rotate at the same speed when the gear ratio is one. If the rotational speed of the driving gear is less than that of the driven gear, the gear ratio is less than one. This is the overdrive gear ratio, supplied by the fifth and upper gears. In addition to less engine wear, overdrive ratios lead to enhanced fuel economy. This fuel efficiency occurs because the overdrive ratio causes the drive wheels to turn faster than the engine at high speeds in which the torque requirement to keep vehicle moving is at its minimum.
- *The shafts*: The shafts that construct the structure of a manual transmission system are the input shaft, output shaft, and counter shaft. The splines of the input shaft engage with those of the clutch; the engine power is transferred to the transmission system through this shaft. The output shaft transfers the regulated torque to the differential. The countershaft – also

known as the layshaft – is an intermediate shaft that transmits the motion from the input shaft to the output shaft. The driving gears affix and attach to the countershaft; the driven gears mount onto the output shaft and can rotate freely, independent of the output shaft rotation. The input and output shafts are two different shafts that are linked together through the countershaft. In most designs, the input and output shafts lie along the same shaft and have the option of combining in a single shaft called the main shaft. However, these shafts rotate separately at different speeds while still linked together through the countershaft.

The shaft design of a manual transaxle system is different. Typically, this system includes two shafts known as the primary and secondary shafts. The primary shaft receives the torque from the engine while the secondary shaft transmits the regulated torque directly to the differential inside the transaxle by using an additional gear. Unlike the fixed gears of the secondary shaft, the gears on the primary shaft can rotate independent of the shaft rotation.

- *Collars*: The function of a collar varies based on its intended system type. In a manual transmission, it connects or disconnects the gears to the output shaft; in a transaxle system, it connects or disconnects the gears to the primary shaft. A collar slides between the space of two gears and can be engaged in one gear at one time. As such, if there are more than two gears in the transmission/transaxle system, the required number of collars increases. The process of connecting or disconnecting the collar and gears may cause noises and grinding because of the different shaft speeds and rotating gears. In modern transmission/transaxle systems, a synchronizer is used to synchronize the rotational speed of the gear and shaft while shifting in order to prevent gears from grinding or clashing while they are engaged.

  Usually, cars and SUVs are equipped with a transmission/transaxle system with four to six forward gear ratios, as well as a reverse gear ratio. However, the transmission/transaxle systems of some cars can provide up to eight gear ratios. For heavy-duty vehicles such as trucks, the transmission/transaxle design provides at least nine gear ratios since these vehicles need to have a wider range of power band from the engine. Most manual transmission/transaxle systems have the flexibility of choosing any forward gear ratio while driving; however, certain vehicles (e.g., racing cars) also allow sequential gearshifts. This is a sequential transmission system. The main advantages of manual/transaxle systems include its simple structure and high power transmission efficiency. However, their function also requires effort and skill from the driver.

1.6.2 Automatic Transmission/Transaxle Systems

General Motors introduced the first automatic transmission system in the early 1930s. Currently, this system is prevalent in most modern vehicles, especially passenger cars. Unlike a manual transmission/transaxle system in which gear selection and shifting are manual functions by the driver, an automatic transmission/transaxle system provides automatic gear shifting through a control system. This control system is electrical or mechanical and operates through a fluid coupling device called the torque converter. As the vehicle moves, gear selection and shift decisions are a function of: engine load prediction, vehicle speed, throttle position, and other specific factors. The fundamental difference between an automatic and manual transmission/transaxle system is the gear sets and engine connection. Automatic systems utilize a torque converter and planetary gear sets rather than a clutch, and gear sets operating by sliding in and out of engagement. The automatic transmission/transaxle system offers more comfortable driving than a manual transmission/transaxle at the cost of a more complicated and expensive structure. The early designs of automatic systems were less fuel-efficient than their manual counterparts. However, as a result of continuous development, modern automatic transmission/transaxle systems can now provide fuel economy similar to that of manual systems. The primary components of an automatic transmission/transaxle system are the torque converter, planetary gear sets, clutches, and the hydraulic control system unit.

1.6.2.1 The Torque Converter

A torque converter is a hydraulic mechanical device that replaces the function of a traditional manual transmission clutch. The main functions of the converter include transferring engine output power to the gear unit, making the gear shifts smoother, providing torque multiplication at low speed conditions, and avoiding engine stalls in the process of stopping or while the vehicle is idling.

The torque converter contains special oil and has a donut-shaped case that contains three sets of vanes (see Figure 1.16). These vanes are the impeller, the turbine, and the stator. Bolts secure the impeller to the flywheel of the engine, while the turbine is attached to the main shaft of the gear unit. A hydraulic pump forces the oil fluid into the rotating impeller vanes. The centrifugal force created by the impeller pumps the fluid towards the turbine vanes. The fluid striking on the turbine vanes causes the turbine to rotate. This rotation directly affects the gearset because it connects to the turbine. At low speed conditions, the fluid returns towards the impeller after striking the turbine vanes. However, the rotational direction of the returned fluid is opposite to that of the impeller, which results in engine power loss. This loss occurs because the impeller needs to redirect the returned fluid. A stator is located between the impeller and the turbine, and increases the efficiency of the torque converter under such conditions. It is mounted on a fixed shaft through a one-way clutch, allowing it to rotate with the direction of the impeller. When the stator is locked by a one-way clutch, it redirects fluid flow returning from the turbine to the rotational direction of the impeller before hitting the impeller vanes again. The angles and direction of the stator vanes cause the pressure of the returned fluid to increase before reaching the impeller. The impeller once again pumps out fluid towards the turbine, but with even higher pressure. This pressure results in torque multiplication and makes the automatic transmission/transaxle powerful enough to provide a high torque at low speeds. The fluid speed and pressure escalate as the impeller speed rises, resulting in an increase in turbine speed. The high turbine speed consequently prevents fluid redirection towards the impeller and allows more engine torque to be directed toward gear unit. When the turbine speed increases, the one-way clutch starts unlocking the stator and allows it to rotate. The operating point at which the impeller, stator, and turbine are rotating at the same speed is the coupling phase, where torque multiplication does not occur.

Fluid slippage in a torque converter makes it impossible to achieve a perfect coupling phase. Specifically, it causes engine power loss, which negatively affects the high-speed efficiency of automatic transmission/transaxle systems. Modern torque converters are equipped with an internal locking clutch, known as a converter lock-up clutch. These locking mechanisms eliminate fluid slippage during vehicle cruising by locking the torque converter components. They also improve the fuel efficiency of automatic transmission/transaxle systems and reduce the amount of heat that the system generates. The lock-up clutch disengages at low speeds to

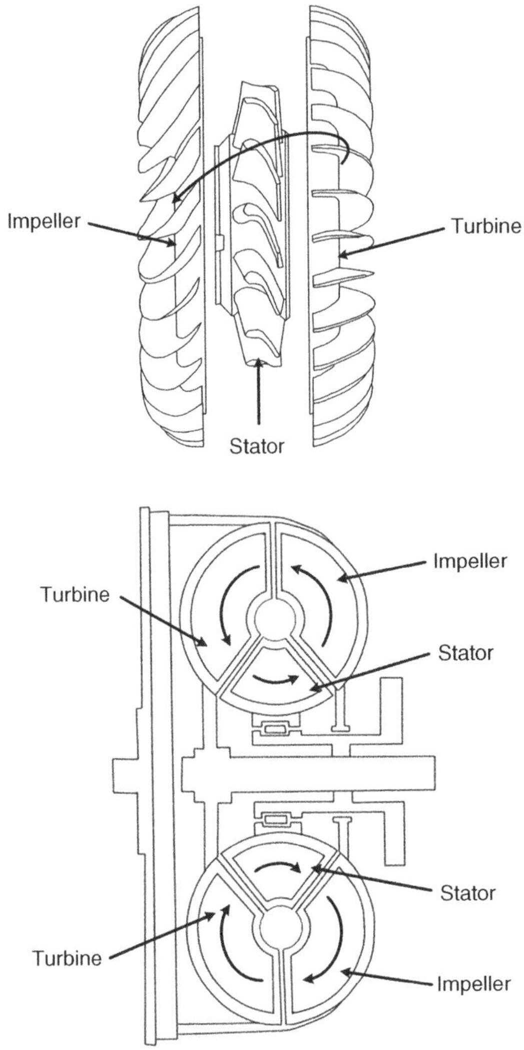

Figure 1.16 A schematic of the torque converter

prevent engine stalls. An electrical solenoid controls the function of the lock-up clutch. The fluid from the impeller still strikes the turbine vanes while the vehicle is idling; however, it does not carry enough kinetic energy to rotate the turbine. This causes the gear set to disconnect from the engine in an attempt to prevent the engine stalling.

1.6.2.2 Planetary Gear Sets

The function of regulating the output engine torque in an automatic transmission/transaxle system occurs through one or more set of gear sets with a special structure, called planetary gear sets. A basic planetary gear set is composed of three different-sized gears: the sun gear, the planet gears, and the ring gear. The sun gear is mounted on a shaft placed at the center of the gear unit while the planet gears (connected to the planet carrier) rotate around it. A ring gear with internal teeth encircles the sun and planet gears.

The planetary gears are always in mesh with sun and ring gears and rotate between these two gears. Each part of the planetary gear set rotates by engine power or is locked by a series of clutches and bands. The motion combinations and power manipulation of these gears provide different gear ratios. For example, a low ratio occurs if the sun gear locks while the planet gears and ring gear rotate. Likewise, when the planetary carrier is the driven component, a gear reduction occurs. In contrast, if the planetary carrier is the driving component, this results in overdrive. A planetary gear set acts as a solid unit when two of its gears lock, resulting in a direct drive.

It is impossible to achieve all the required gear ratios with a single planetary gear set. This is because gear manipulation is limited with one set, as it only provides two forward gear ratios and one reverse. Higher gear ratios can be achieved by using compound planetary gear set configurations in automatic transmission/transaxle systems. These configurations have a structure similar to a basic planetary gear set, they can have more than one of each gear type (sun, ring, etc.). For example, a Ravigneaux compound planetary gear set has two sun gears and two sets of planet gears with a common ring gear, while a Simpson compound planetary gear set has two sets of planet gears and two ring gears with a common sun gear. The Wilson planetary gear set and the Lepelletier planetary gear set are other possible configurations with more complex structures.

1.6.2.3 Clutches and Bands

Clutches and bands are used in the gear set unit as holding devices. What distinguishes the automatic transmission/transaxle system from the manual transmission/transaxle system is the controlled hydraulic in the gear set unit. Gear shifting is typically manipulated through these components, which are controlled by a mechanical hydraulic control system or electrical hydraulic systems. During gear shifting, clutches and bands work by making specific gears of a planetary gear set fixed and motionless. They do this by locking it to the transmission/transaxle case or to the shaft it is mounted upon, while allowing other gears to rotate. Hydraulic control valves activate clutches and bands which operate based on information received from sensors about the road speed, throttle position, and engine load.

1.6.2.4 Hydraulic Control Unit

The hydraulic control unit directs the gear shifts to correspond with the vehicle operating conditions by sending the pressurized hydraulic fluid to a series of clutches and bands. A pump driven by the torque converter pressurizes and circulates transmission fluid throughout the transmission/transaxle to activate various elements. The pressure built by the pump is known as the line pressure and is the activation source of the various transmission/transaxle components. The vacuum modulator, the throttle valve, and the governor valve provide the required information to the transmission control system by adjusting the line pressure. The governor provides the engine speed to the transmission control system while the throttle valve determines the throttle position. The vacuum modulator sends the engine load information to the transmission control system.

In modern automatic transmission/transaxle systems, the hydraulic pressure is electronically controlled in accordance with the information obtained from input sensors, used in place of governor and throttle valves as well as the vacuum modulator. Using electronic control systems improves the efficiency of the transmission/transaxle systems by providing more precise gear shifting in accordance with the engine demands, thereby resulting in better fuel economy and lower emissions.

1.6.3 Automated Manual Transmissions (AMTs)

As previously mentioned, a simpler structure, better fuel economy, and superior efficiency are the most desirable aspects of a manual transmission while being more comfortable to drive and easier to use are the best features of an automatic transmission system. An intermediate solution bridging these two systems is the automated manual transmissions, also known as semi-automatic transmission systems. The design of the system takes advantage of the best features of both transmissions.

In this transmission, the driver has the option of using a fully automated mode or shift manually by using switches or levers. Two main variations of the automated transmission system include sequential manual transmission and dual-clutch transmission. The automated transmission system provides better fuel efficiency, lower emissions, lower energy loss, and a more comfortable ride than either manual or automatic transmission systems.

1.6.4 Continuous Variable Transmissions (CVTs)

Unlike the limited number of gear ratios that automatic and manual transmission/transaxle systems offer, a continuous variable transmission (CVT) system provides an unlimited number of gear ratios within the upper and lower limits of their operating range. CVTs do this by properly adjusting the geometrical configuration of the system without using a set of gears. A specific type of CVT is the infinitely variable transmission (IVT) system, in which the gear ratio ranges from zero to the highest ratio. The zero ratio decouples the engine load from the driveshaft, thus making idling possible.

Generally, CVT transmission systems allow the engine to function at an optimal and efficient range of operation under varying driving conditions. This efficiency results in enhanced fuel economy, reduced emissions, and improved vehicle performance. Since there are no gears in CVTs, the gear ratio refers to the speed ratio of the input and output shafts of the transmission system. Additionally, the continuous and smooth gear change of CVTs offers more comfort, superior acceleration performance, and a lightweight transmission system with a compact structure. Conversely, the negative aspects of CVTs include its noisier performance and lower torque capacity. Due to technological advances and the aforementioned advantages, these transmission systems have found their way into the automotive industry. Currently, several car manufacturers have integrated CVT technology into their products.

There are various types of CVTs currently available, including the spherical, hydrostatic, magnetic, ratcheting, cone, radial roller, and toroidal CVTs. A general way to classify CVTs is by determining whether they are mechanical or non-mechanical systems. Mechanical systems are further available as friction and ratcheting systems, while non-mechanical systems can be sub-divided between hydraulic and electrical. Currently, the most commonly used CVT systems are variable-diameter pulley or pulley-based CVTs, toroidal or roller-based CVTs, and hydrostatic CVTs.

1.6.4.1 Pulley-Based CVTs

Pulley-based CVTs – classified as friction-based mechanical CVTs – are the most commonly integrated CVT designs in vehicle powertrain systems. The main components of this design are two variable-diameter pulleys and a high-density rubber or metal belt. One pulley, known as the driving pulley, receives power from the engine through the crankshaft, while the second pulley, called the driven pulley, transfers the adjusted power to the differentials through the driveshaft. A hydraulic cylinder actuates pulleys of the CVT. Each pulley is composed of a pair of cones facing each other where one of the cones is fixed while the other one can move along its center axes. One of these cones is stationary, while the other one can move along its center axes. A V-belt rubber/metal belt with a fixed length connects the pulleys to each other and transfers the power from the driving pulley to the driven one. For vehicle applications, it is more common to use metal belts, which do not have the limitations of rubber belts. Hydraulic cylinders, sensors and microprocessors are other components of a pulley-based CVT. They adjust the pulley diameter based on the engine load and driving conditions.

CVTs change the gear ratio by adjusting the diameter of the driving and driven pulleys on which the belt rides. For example, if the moveable cone is forced to move closer to the fixed cone, the belt operating between them rides higher to increase the diameter of the pulley. In contrast, when the moveable cone is forced to move far away from the fixed one, the belt rides lower in the groove, resulting in a decrease in pulley diameter. Electronically controlled movements of the cones in each pulley make the diameter adjustments. The fixed belt length means that increasing the diameter of one pulley requires a reduction in the diameter of the other pulley. The diameter changes between the pulleys allow the vehicle to operate on an unlimited number of gear ratios. For instance, the smaller diameter of the driving pulley causes a lower speed of the driven pulley, thus producing a lower gear ratio. Similarly, the larger diameter of the driving pulley with respect to the driven one results in a higher gear ratio.

The main disadvantages of pulley-based CVTs include excessive build-up of heat and sensitivity to wear because of the friction generated at the contact points. Lubricating oil can help resolve these issues at the cost of reducing the friction and torque capacity of the system. In addition, slippage in the drive belt may occur in the high torque demand condition such as when a vehicle accelerates from a stop at peak torque, thereby resulting in power loss.

1.6.4.2 Toroidal CVTs

Like pulley-based CVTs, toroidal CVTs are also friction-based. The primary advantage of toroidal CVTs is their parallel power flow, which allows the system to have a high torque capacity and compact design. The main operation of a toroidal CVT is similar to that of a pulley-based CVT. However, unlike the belts and pulleys used by pulley-based CVTs, toroidal CVTs use discs and power rollers or wheels (Figure 1.17). In toroidal CVTs, one disc connects to the engine crankshaft while the other one connects to the driveshaft. Power transfers from the driving disc to the driven disc through the rollers or wheels located between them, which spin around the horizontal axis. The rollers located between the discs adjust the gear ratio based on how they are angled relative to the discs. As such, even a slight tilt in the roller/wheel angle around the vertical axis can incrementally adjust the gear ratio. For example, lower gear ratios can be achieved when the rollers/wheels are in contact with the driving disk near the center while they are in contact with the driven disk near the rim.

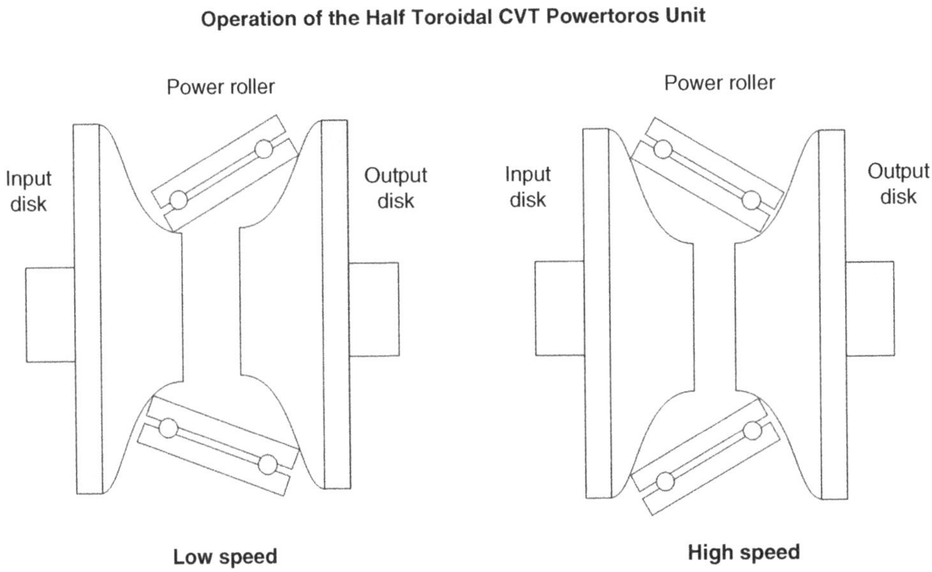

Figure 1.17 Operation of the half-toroidal CVT

Two variations of toroidal CVTs are the full toroidal and the half-toroidal CVT. The difference between these two subtypes is the position of the tilting rotation center. In a half-toroidal CVT (Figure 1.18a), the power roller tilting rotation center is outside the center of the cavity between the discs. As shown in Figure 1.18b, in a full toroidal CVT, however, the titling rotation center of the power roller coincides with the center of cavity between discs. Ultimately, half-toroidal CVT designs are more efficient than full-toroidal drives at the

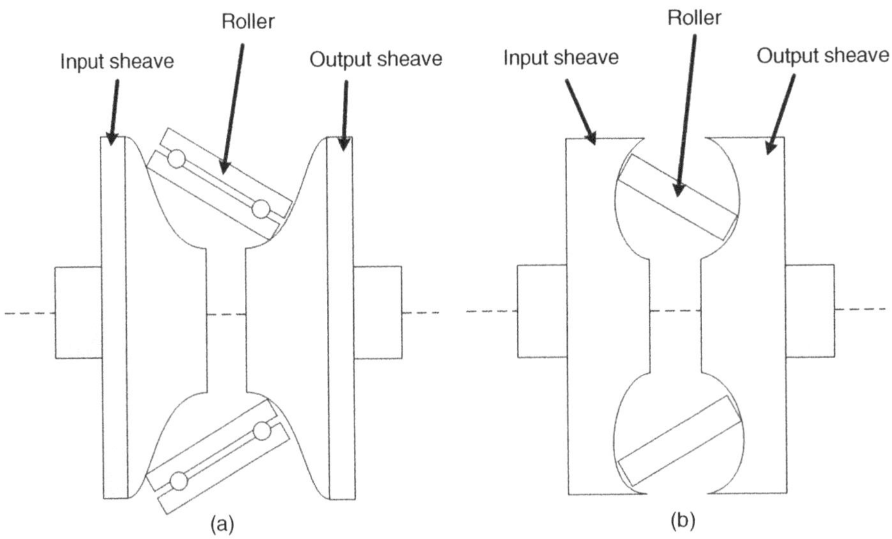

Figure 1.18 Half-toroidal (a) and full-toroidal (b) CVT types

expense of a lower torque capacity. The disadvantages of toroidal CVTs are similar to those of pulley-based CVTs.

1.6.4.3 Hydrostatic CVTs

A hydrostatic CVT is a mechanical CVT with at least one variable-displacement pump and one motor. The hydrostatic pump connects to the engine and converts the output mechanical power of the engine into fluid power. Through the hydraulic motor, this power is converted into rotational power at the output shaft.

Proper control of the hydraulic flow can change the gear ratio. Usually, hydraulic CVTs combine the use of a planetary gear set and clutches to improve the overall efficiency of the transmission system. Typically, the hydraulic system transmits power at low speeds while the planetary gear set transfers power in high-speed conditions. Hydrostatic CVTs are capable of transmitting more torque at the expense of sensitivity to contamination and higher costs. Heavy-duty agricultural and off-road vehicles often have this type of transmission system.

1.7 Drivetrain and Differentials

The drivetrain is the last unit in a powertrain system that transfers the engine power to the road. The main components of the drivetrain unit include the driveshaft, the differentials, axle shafts, and wheels. Of these, the differential is the most important component. The driveshaft is the input to the differential, while the axle shafts are its output. A differential is a gearbox that directs and adequately distributes the engine power to the driving wheels. Additionally, it allows the driving wheels on either side of the vehicle to independently rotate at different speeds. This is important in improving the handling performance of the vehicle during cornering. When cornering, the outside wheels need to spin faster than the inside ones because the inside wheels travel across a shorter distance. Otherwise, the wheels would lock into the same rotational speed, resulting either in the inside or outside wheel skipping and losing traction. This lack of traction would cause unexpected handling difficulties, tire damage, and strains on the drivetrain.

Note that the sum of the split power between the left and right wheels is equal to the total power received from the engine. Similarly, the average sum of the rotational speed of the wheels is equal to the driveshaft's rotational speed. In other words, if the rotational speed of one wheel is rising, the speed of the opposite wheel must compensate by an equal reduction.

Usually, a vehicle has one differential unit that is located in the middle of either the front or the rear-driving axle, depending on the vehicle type. All-wheel drive and four-wheel drive vehicles typically have three differentials, with two of them located on the front and rear axles in order to direct engine power to the right and left wheels. In this arrangement, the third differential splits the engine power between the front and rear axles. There are various designs of differentials integrated into vehicles based on vehicle performance and functionality. The most important differentials include open differentials, limited slip differentials, locking differentials, and transfer case differentials.

1.7.1 Open Differentials

Figure 1.19 is a schematic of an open differential system. Open differentials, also known as standard differentials, are the most common differentials in modern vehicles. They include two

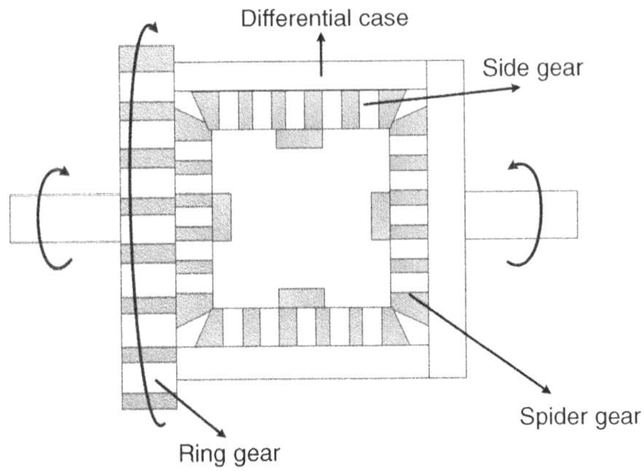

Figure 1.19 Schematic of an open differential system

side gears, a set of spider gears (known as planetary gears), the ring gear, and the differential case. The driveshaft rotates the ring gear bolted to the differential case. The spider gears then rotate with the differential case and can also rotate along their own shaft axes. Each side gear is connected to an axle shaft and is driven by the spider gears. These spider gears allow the outside and inside wheels to move at different velocities during cornering.

On paved roads, the open differential system works well and provides adequate handling performance. However, if the wheels encounter unequal traction conditions, power routes to the wheel with the least amount of traction. This causes unexpected handling performance for the driver and can hinder the vehicle's motion. For example, if one wheel spins on a slippery surface while another is on a dry surface, all the power is directed towards the wheel on the slippery road, leaving the other wheel to receive minimal or no power. The design of limited slip differentials and locking differentials addresses the limitations of open differentials on uneven traction conditions.

1.7.2 Limited Slip Differentials

The design of limited slip differentials resolves the limitations of open differentials in conditions where traction is lost. However, resolving this issue also comes at the expense of a more complex system. "Limited slip" refers to the function of the system in limiting wheel spin. In this design, the differential operation is limited when the traction loss of any wheel exceeds a certain threshold. It then redirects power to the wheel with more traction by up to 80%, thereby maintaining rotational power in both wheels. Limited slip differentials are used mainly in off-road vehicles since it is possible for one of the wheels to lose traction in off-road conditions. In addition, high-speed performance vehicles utilize these differentials to improve vehicle handling during high-speed cornering, when the different velocities of the left and right wheels may lead to higher amounts of differential slip.

Limited slip differentials have various designs; the most important ones include mechanical, viscous coupling, and electronic limited slip differentials. Mechanical limited slip differentials are based on the interaction between two (or more) mechanical parts. Clutch, cone, and helical gear limited slip differentials belong in this category. In the viscous coupling variation, the differential slip is controlled by the hydrodynamic friction from fluids with high viscosity.

These differentials are less efficient than the mechanical ones because power loss occurs during activation. However, in modern vehicles, electronic limited slip differentials are used to activate the mechanical components of the system. An electronic limited slip differential utilizes different sensors and microcomputers in order to monitor the wheel slip and vehicle motion.

1.7.3 Locking Differentials

The design of locking differentials, also known as lockers, enables the wheels on an axle to lock relative to each other, avoiding the difference in speed between two wheels on the axle, regardless of the amount of traction on each wheel. Automatic and selectable lockers are two variations of locking differential systems. The automatic differentials lock both wheels on the axle under normal conditions, while they unlock the differential for certain conditions such as cornering. Automatic lockers automatically lock and unlock the differential system, thereby causing fast-changing and strange handling characteristics on the street. The function of the differential is noisy and increases tire wear. Off-road four-wheel drive vehicles are the main applications of the automatic lockers and typically, they are integrated on the rear axles only, since front axle installations can cause difficulties in steering. In contrast, a selectable locker functions as an open differential in normal driving conditions; however, the driver has the option of locking both wheels on the axle. When locked, both axles will then turn at the same speed irrespective of the road surface. The locking process can occur through compressed air, mechanical cable, electric actuator or hydraulic fluid. It is possible to integrate selectable lockers in both front and rear axles without sacrificing vehicle handling performance and tire wear. Off-road vehicles are the main applications of such differentials. The integration of the selectable differentials on on-road vehicles is limited due to the high cost and complexity of installation.

1.7.4 Transfer Case Differentials

Typically, four-wheel and all-wheel drive vehicles have three differentials, two of which are mounted on the front and rear axles, whereas the third one, called the transfer case, is located between the front and rear output shafts to the final drive units. Attached to the rear of the transmission and also to the front and rear axles through two separate driveshafts, the transfer case and center differentials are mechanical devices which split the power between the front and rear axles while varying the distribution of power between the front and rear axles smoothly and over a certain range. In addition, they adjust the speed differences in front and rear axles. The transfer case receives the power directly from the transmission and directs it to either the rear wheels or both the front and rear wheels. The transfer action may occur through a set of gears or through a chain-driven device. Through the transfer case, the driver has the option of selecting two-wheel drive or four-wheel drive mode by a shifter. However, some vehicles accomplish this electronically through a switch. Some transfer cases are not selectable such as all-drive vehicles.

Problems

1. What is the principle of energy efficiency in internal combustion engines and how does it differ between diesel and gasoline engines (explain in detail)?
2. What is the difference between engine knocking and pre-ignition phenomena?

3. What is the stoichiometric ratio of CNG, LPG, hydrogen, ethanol, and methanol? How can it affect the engine performance?
4. What is the difference between a supercharger system and a turbocharger system?
5. Why do some modern diesel engines use a throttle valve in their fuel system? Explain how this valve reduces vehicle emissions.
6. What is a carburetor? Provide reasons explaining why it has been replaced by controlled fuel injection systems.
7. Investigate the effects of ignition timing on the emissions of a gasoline engine.
8. What is the contribution of vehicular air pollution on your city and country? Conduct research on your government's policies on air pollution reduction.
9. How does the waste-gate in a turbocharger regulate the intake air pressure? Explain.
10. Describe the operating characteristics of a three-way catalyst.
11. Describe the operating characteristics of diesel catalytic converters.
12. Conduct research on new advances in gasoline and diesel engine emission control technologies.
13. Research and find other alternative fuels available on the market. Discuss the advantages and disadvantages of these options.
14. What is the market share of vehicles with alternative fuel systems? What are the main obstacles preventing these vehicles from mass production in your local area?
15. What is the difference between a transfer case and a center differential?
16. How does a manual transmission system work? Explain.
17. Describe the structure and function of a clutch in a manual transmission/transaxle system.

References

1. http://www1.eere.energy.gov/vehiclesandfuels/avta/light_duty/fsev/fsev_history.html.
2. http://www.hybridcars.com/history/history-of-hybrid-vehicles.html.
3. http://www.sustainable-mobility.org/month-issue/decoding/histoiry-of-the-electric-vehicle-along-development-process.html.
4. http://inventors.about.com/library/weekly/aacarsgasa.htm.
5. http://www.speedace.info/internal_combustion_engine.htm.
6. Jong, R. Ahman, M., Jacobs, R., and Dumitrescu, E. (2009) *Hybrid Electric Vehicles: An Overview of Current Technology and Its Application in Developing and Transitional Countries*, UNEP, Washington, DC.
7. Taylor, C.F. (1985) *The Internal-Combustion Engine in Theory and Practice: Combustion, Fuels, Materials, Design*, 2nd edn, MIT Press, Cambridge, MA.
8. Wallington, T.J., Kaiser, E.W., and Farrell, J.T. (2006) Automotive fuels and internal combustion engines: a chemical perspective. *Chemical Society Reviews*, **35**, 335–347.
9. http://www.epa.gov/.
10. http://www.energy.gov/.
11. http://www.eia.gov/.
12. http://www.iea.org/.

13. Lipman, T.E., and Delucchi, M.A. (2002) Emissions of nitrous oxide and methane from conventional and alternative motor vehicles. *Climate Change*, **53**, 477–516.
14. https://www.princeton.edu/ssp/64-tiger-cub-1/64-data/combustion-chemistry.pdf.
15. Wallington, T.J., Sullivan, J.L., and Hurley, M.D. (2008) Emissions of CO_2, CO, NO_x, HC, PM, HFC-134a, N_2O and CH4 from the global light duty vehicle fleet. *Meteorologische Zeitschrift*, **12**(2), 109–116.
16. http://www.meca.org/galleries/default-file/MECA%20Diesel%20White%20Paper%2012-07-07%20final.pdf.
17. http://ec.europa.eu/enterprise/sectors/automotive/files/pagesbackground/emission_control/vol_1-emission_control_technology_en.pdf.
18. http://www.aecc.be/en/Technology/Catalysts.html.
19. http://www.volvoscr.com/default.aspx?pageid=3362.
20. http://www.autoshop101.com/forms/h61.pdf.
21. http://www.ika.rwth-aachen.de/r2h/index.php/Hydrogen_Internal_Combustion_Engine.
22. Reimpell, J., Stoll, H., and Betzler, J.W. (2001) *The Automotive Chassis: Engineering Principles*, 2nd edn, Butterworth-Heinemann, Oxford.
23. Wagner, G. (2001) Application of transmission systems for different driveline configurations in passenger cars, SAE Technical Paper 2001-01-0882, 2001, DOI: 10.4271/2001-01-0882.
24. Kluger, M., and Long, D. (1999) An overview of current automatic, manual and continuously variable transmission efficiencies and their projected future improvements, SAE Technical Paper 1999-01-1259, 1999, DOI: 10.4271/1999-01-1259.
25. Johanson, C., and Duffy, J.E. (2010) *Automatic Transmissions and Transaxles, A2*, 3rd edn, the Goodheart-Willcox Co., Inc, ISBN: 978-1-60525-203-2.
26. Iannelli, L. (2011) Automated transmission systems: the impact of control technology, in *IEEE Control Systems Society* (eds T. Samad and A.M. Annaswamy), available at: www.ieeecss.org.
27. Renius, K.T., and Resch, R. (2005) Continuously variable tractor transmissions. Paper presented at ASAE – the Society for Engineering in Agricultural, Food, and Biological Systems, the 2005 Agricultural Equipment Technology Conference, Louisville, Kentucky.
28. Imanishi, T., and Machida, H. (2001) Development of POWERTOROS unit half-toroidal CVT(2): comparison between half-toroidal and full-toroidal CVTs. *NSK Technical Journal Motion and Control*, **10**, April.

2

Electric and Hybrid Powertrain Technologies

2.1 Introduction

For the past few decades, internal combustion engine vehicles and conventional powertrain systems have dominated the transportation industry. Growing concerns about the environmental repercussions of vehicle emissions began a global effort to develop next generation transportation solutions with increased efficiency, and reduced greenhouse gas and air pollution emissions.

Electric vehicles (EVs), depending on their source of electric energy, have the potential to be zero emission vehicles and have potential as a future alternative to internal combustion vehicles. However, despite considerable improvements, electric vehicles still have a limited driving range due to the limitations of the current electrical storage technologies.

To address this problem, hybrid vehicles combine the advantages of two different energy sources. The hybridization of energy sources can increase energy security, improve fuel economy, lower fuel costs, and reduce emissions. Moreover, there are many different ways of implementing hybridization such as: hybrid electric vehicles, plug-in hybrid electric vehicles, hydraulic hybrid vehicles, and pneumatic hybrid vehicles. Implementing all of these can boost internal combustion engine efficiency and recover energy lost during braking.

Electric and hybrid electric vehicles present many new and complex design challenges that are absent in the well-established conventional automotive technologies and transportation systems. In addition to striving to meet increasingly strict requirements for fuel efficiency and economy, vehicle safety and performance, and environmental protection, the automotive industry is expending considerable effort and resources to address the challenges associated with developing hybrid electric and electric vehicles. This chapter provides a better understanding of electric and hybrid vehicles and the benefits and design challenges that accompany their development.

Electric and Hybrid Vehicles: Technologies, Modeling and Control: A Mechatronic Approach, First Edition.
Amir Khajepour, Saber Fallah and Avesta Goodarzi.

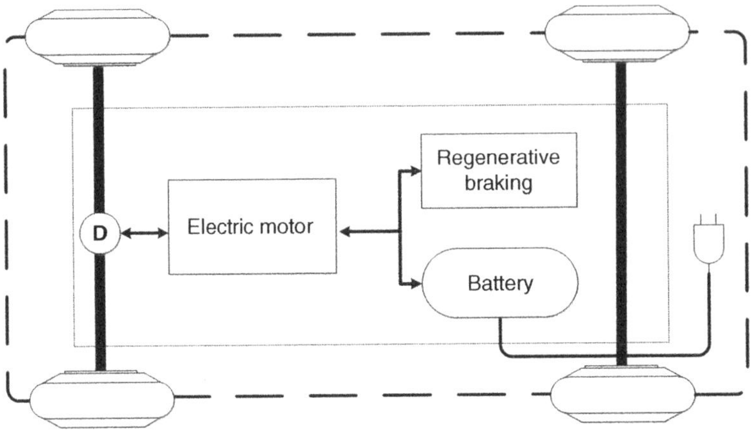

Figure 2.1 Battery electric vehicle. *Note:* D = differential

2.2 Battery Electric Vehicles (BEVs)

Figure 2.1 depicts a schematic of a battery-powered electric vehicle (BEV). BEVs use one or more electric motors for propulsion and batteries to store electricity. The batteries store energy to power all of the electrical systems in the car. The batteries can recharge from grid electricity at recharging stations, house outlets, non-grid sources such as solar panels, or by using onboard recuperative energy systems. BEVs can potentially emit zero greenhouse gases and air pollutants; however, the type of electricity generation (solar, wind, coal, etc.) determines the well-to-wheel emission. Nevertheless, even if the electricity that charges the batteries comes from a CO_2 emitting source such as a coal-powered plant, the amount of CO_2 emitted from a BEV is about one-half to one-third less than what a gasoline-powered vehicle emits. Additionally, electric vehicles have a "tank-to-wheels" efficiency that is three times greater than that of a gasoline vehicle.

In addition to the environmental benefits, there are other advantages of using electric vehicles compared to conventional ICE vehicles. Electric vehicles can deliver at least 75% energy efficiency, while internal combustion engines can be as low as 15% [1]. Also, it is more cost-effective to maintain an electric vehicle because there are less mechanical or emission control components. For example, BEVs do not have a muffler, catalytic converter, tailpipe, and gas tank. Likewise, the clutch assembly and the transmission system are usually replaced with the electric motor drive system to control the torque of the motor. The main components of the drive system of an electric vehicle are an accelerator pedal, an electric motor drive/controller, batteries, and traction electric motors.

Despite these benefits, there are a few disadvantages associated with BEVs to consider. Currently, a key concern regarding BEVs is the low energy and power density of batteries compared to liquid fuels. Another concern is the recharging time. Although it is significantly cheaper to recharge a battery pack than to refuel a tank, the speed of recharging (usually 4–8 hours at home and a minimum of 30 minutes at charging station) can be a long wait.

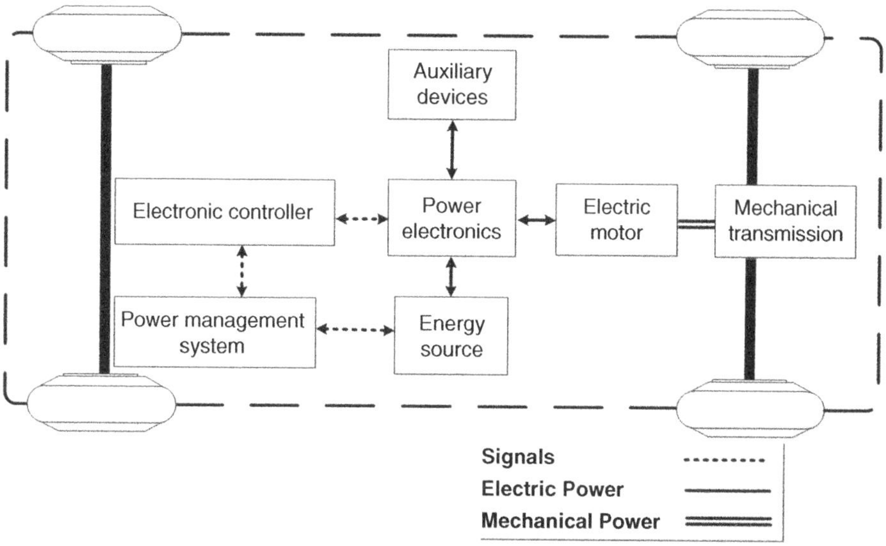

Figure 2.2 Schematic of a BEV powertrain

Additionally, due to the limited capacity of EV battery packs, the driving range currently possible on a single battery charge is shorter than that of a conventional vehicle running on a full tank of gasoline. Even with improvements in the battery storage capacity and lifespan, battery packs are still very expensive and heavy.

2.2.1 The BEV Powertrain Configuration

The electric powertrain system functions to provide the required propulsion power in accordance with the commands of the driver by interfacing the batteries with the vehicle wheels at a high efficiency. Generally, an electric powertrain contains electrical and mechanical modules (see Figure 2.2). The electrical module consists of subsystems such as traction electric motors, power electronics (e.g., charge controller, inverters, and converters), and power sources/storage (e.g., batteries, capacitors, flywheels). The mechanical module consists of the transmission system (optional), the differential system (optional), and driving wheels. The electric control module, consisting of three functional modules (sensors, interface circuitry, and microprocessors), optimizes and monitors the powertrain performance. The sensors acquire measurable quantities such as temperature, speed, torque, current, and voltage, and convert these quantities to electronic signals. The interface circuitry conditions these signals to the appropriate level before sending them to the microprocessors. A control unit – referred to as the power management control system – controls the application of the traction electric motor power and optimizes the power flow between the powertrain components in order to achieve the maximum driving range. Another important function of the controller is to maximize the energy recuperation during vehicle braking [2]. Chapter 8 discusses the concept of a power management system and its design procedure.

The powertrain of BEVs is simpler than that of ICE vehicles. Power is transmitted to the driving wheels through a few select components. Another key advantage of electric

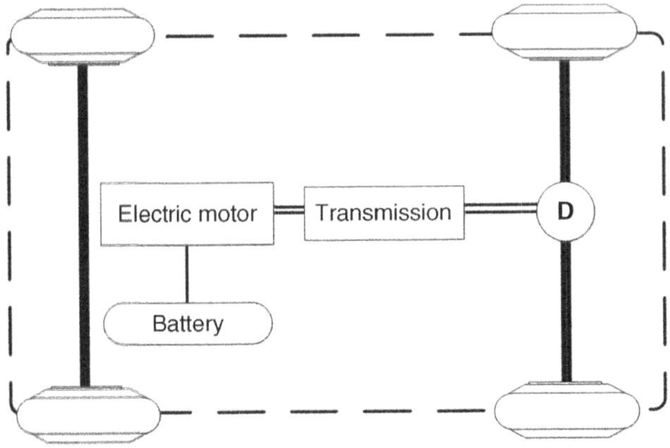

Figure 2.3 Schematic of a converted rear-wheel drive electric powertrain system. *Note:* D = differential

powertrains over their ICE counterparts is the capability to recuperate the kinetic energy wasted during braking, as well as the potential to generate and store energy while traveling downhill. In these situations, the motor operates as a generator to produce electric power. In addition, the rapid dynamics of electric motors in BEVs enable accurate control of wheel torque, thereby achieving better handling performance. Moreover, this advantage also allows stability and safety control systems such as active cruise control, collision avoidance, and vehicle stability control to perform faster, resulting in better vehicle safety.

The two types of electric powertrain systems are converted and dedicated. In converted systems, an electric motor and batteries replace the ICE engine and fuel tank, while the rest of the powertrain components, including the transmission and differential systems, remain the same. Figures 2.3 and 2.4 respectively depict converted rear-wheel drive and front-wheel drive electric powertrain systems. In such systems, the electric motor and batteries are mounted on the vehicle chassis, and the power transfers to the wheels through the transmission system. A clutch or torque converter, depending on the use of a manual or automatic transmission, can connect or disconnect the power flow from the electric motor to the driving wheels. Much like conventional powertrains, the transmission system adjusts the electric motor torque outputs.

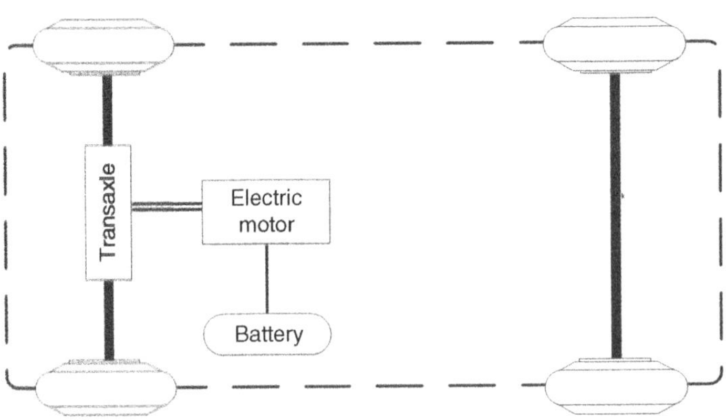

Figure 2.4 Schematic of a converted front-wheel drive electric powertrain system

Vehicles integrated with converted powertrain systems suffer from heavy weight, vehicle performance degradation, and complexity. Additionally, they have less flexibility to be converted to all-wheel drive systems than their dedicated counterparts.

On the other hand, dedicated electric powertrains integrate into vehicles where the body and chassis design meet the structural requirements associated with BEVs, and take advantage of the greater flexibility of electric propulsion systems. The different variations of dedicated BEV powertrains are a function of the number of battery packs, motors, and component arrangements. However, depending on the number and location of the motors, the categorization of the electric powertrains can be as in-wheel motors or out-wheel motors. In an out-wheel configuration, one or more electric motors are mounted on the vehicle chassis and provide the traction forces to the driving wheels either directly or through gear units and differentials, while an in-wheel configuration provides the traction forces through motors located inside the driving wheels, without requiring any mechanical component. Both out-wheel and in-wheel configurations can be converted to an all-wheel drive powertrain system. The proper arrangement and configuration of an electric powertrain system depend on different factors such as: performance expectations (e.g., the acceleration, maximum speed, climbing capacity, etc.), the vehicle constraints (weight, payload, etc.), and energy sources (batteries, capacitors, flywheels, etc.). Some of the possible electric powertrain configurations for the propulsion system of a BEV are as follows:

- *Out-wheel motor rear-wheel drive:* This configuration consists of an electric motor, fixed gearing, and a differential. Typically, the fixed gearing is a planetary gear set with a fixed forward gear ratio, along with one reverse. Unlike conventional ICE vehicles, modern EVs can easily achieve the desired torque-speed characteristics for vehicle motion through a combination of an electric motor and fixed gearing. Generally, the electric motor has constant power at a greater speed range. In comparison to converted electric powertrains, this configuration has a simpler structure, lighter weight, and a smaller size due to the removal of the complex transmission system and its clutch or torque convertor. Figure 2.5 shows this powertrain configuration.

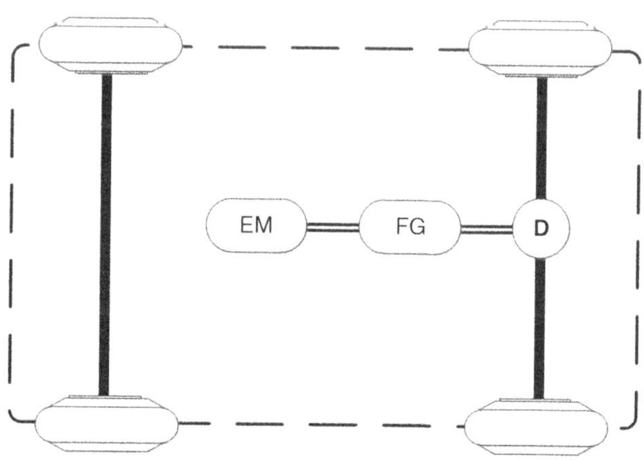

Figure 2.5 Out-wheel motor rear-wheel drive on an electric powertrain configuration. *Note:* EM = Electric Motor, FG = Fixed Gearing, D = Differential

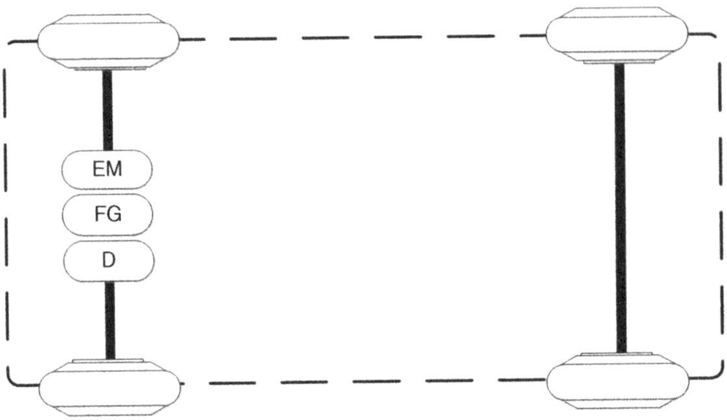

Figure 2.6 Out-wheel motor front-wheel drive on an electric powertrain configuration. *Note:* EM = Electric Motor, FG = Fixed Gearing, D = Differential

- *Out-wheel motor front-wheel drive:* The principle of this configuration is analogous to the front-engine front-wheel drive of a conventional vehicle, in which all the powertrain components compact and integrate on the front axle (see Section 1.5.2). This configuration consists of an electric motor, fixed gearing, and a differential, all of which integrate into a single assembly. Figure 2.6 shows this powertrain configuration.
- *Dual out-wheel motors front-wheel drive:* In the afore-mentioned configurations, the motion of the left and right wheels are separated from each other by the mechanical differential. In contrast, this design has the differential removed; instead, separate electric motors and fixed gearings drive the wheels. This configuration provides better vehicle stability since electric motors can accurately control individual wheel torques. However, the use of an additional electric motor, power converter, and fixed gearing increases the initial cost of the system. Figure 2.7 shows this powertrain configuration.

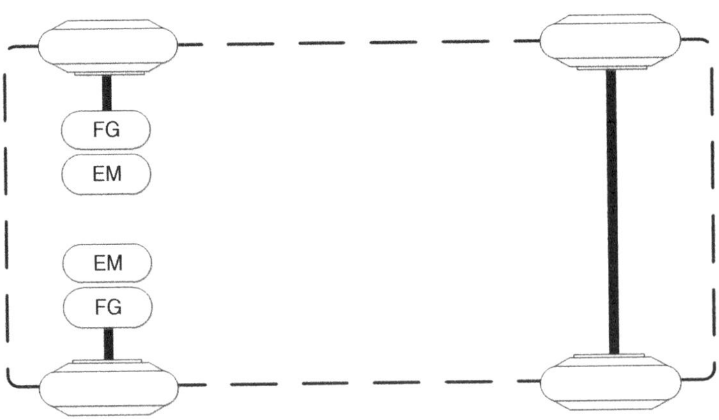

Figure 2.7 Dual out-wheel motors front-wheel drive on an electric powertrain configuration. *Note:* EM = Electric Motor, FG = Fixed Gearing

Figure 2.8 In-wheel motor all-wheel drive on an electric powertrain arrangement. *Note:* EM = electric motor

- *In-wheel motor drive:* In-wheel motor technology can minimize or even eliminate the mechanical components of an electric powertrain system. In this design, electric motors are inside a pair or all of the wheels, with or without a fixed gearing. Fixed gearing, along with a high-speed inner-rotor electric motor, enables the desired wheel speed to be achieved by reducing the motor speed. On the other hand, it is also possible to install a low-speed outer-rotor electric motor inside a wheel without a fixed gearing by mounting the outer-rotor onto the wheel rim. In this arrangement, the electric motor directly controls the wheel torque (and hence, the vehicle speed). The arrangement, shown in Figure 2.8, represents an in-wheel motor all-wheel drive arrangement, in which achieving the benefits of front-wheel drive, rear wheel-drive, and all-wheel drive configurations happens easily by turning the related electric motors on or off. Despite its advantages, implementation of an in-wheel motor likely will induce a heavier combined tire-wheel-suspension system. This characteristic is undesirable in relation to vehicle dynamics and stability control. Furthermore, the electric motors need to have high resistance against excessive lateral and longitudinal loads and water intrusion.

2.2.2 Electric Traction Motors

Electric motors convert the electrical energy from energy sources into mechanical energy in order to provide the required traction force for the vehicle motion. The electric motors of a BEV must satisfy a wide range of driving requirements, such as frequent starting and stopping, high-rate of acceleration/deceleration, low-torque high-speed cruising, high-torque low-speed hill climbing, and moving the vehicle from a standstill. The type, size, weight, and performance of an electric motor in a BEV depend on the overall powertrain specifications. These specifications include single or multiple-motor configuration, fixed or variable transmission, and whether the motor is geared or gearless. However, the primary requirements and specifications associated with the proper selection of electric motors for a BEV are as follows [2,3]:

- to provide sufficient maximum torque, typically four or five times that of the rated torque for temporary acceleration and hill-climbing;

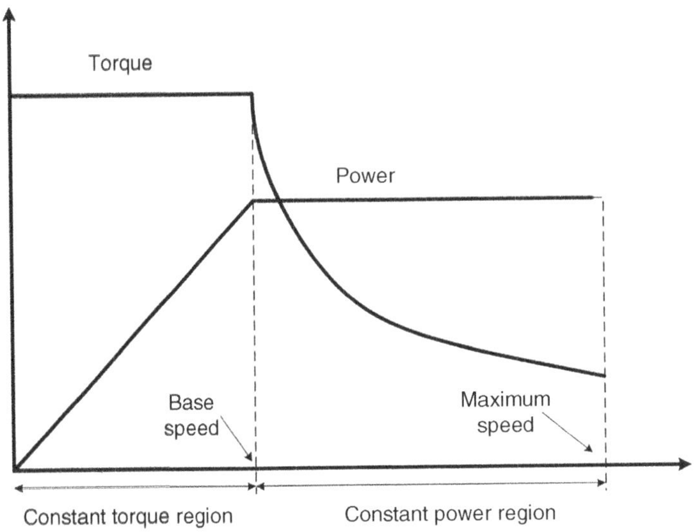

Figure 2.9 Typical traction electric motor characteristics

- to provide high efficiency over wide speed and torque ranges for the reduction of total vehicle weight and the extension of driving range;
- to provide high controllability, high steady-state accuracy, and good dynamic performance;
- to provide sufficient robustness against high temperature, bad weather, and frequent vibration;
- to provide high efficiency for regenerative braking.

The afore-mentioned specifications reflect the important role that the speed-power (torque) characteristics of the traction motor have in the design of an electric powertrain. The common traction electric motors adopted in BEVs usually have the characteristics shown in Figure 2.9. At a low-speed region, the electric motor offers a constant torque (rated torque) over the entire speed range while reaching the base speed. While passing the base speed (high speed region), the motor provides a constant power, and the torque begins decreasing with speed. The ratio between the maximum speed and the base speed represents the motor effectiveness on an electric powertrain performance. An electric motor with a higher speed ratio can provide higher maximum torque, resulting in higher initial acceleration and an improvement in gradeability performance. However, each type of electric traction motor has its limited speed ratio.

There are various ways to classify traction electric motors adopted in BEVs; however, a general classification divides the electric motors into the Direct Current (DC) and Alternating Current (AC) motors. The proceeding sections discuss the performance, limitations, and constraints of both motor types.

2.2.2.1 DC Motors

Typically, DC motors have a set of coils (field), a rotor (armature), a commutator, and an optional brush (see Figure 2.10). In DC motors, the set of coils generates magnetic forces

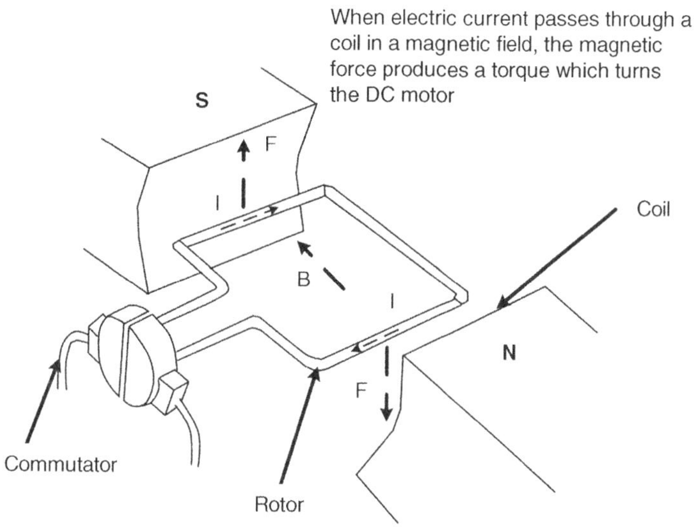

Figure 2.10 A schematic of a brushed DC motor

that provide the torque. The rotor is mounted on bearings and turns inside a magnetic field. The commutator acts like a switch to supply voltage to a revolving armature from the stationary brush assembly and makes the rotor turn, thereby providing the mechanical power. Brushes make contact with the commutator to make the connections. DC motors provide the sufficient traction requirement because of their torque-speed characteristics. Also, they have a simple speed control system which lowers the cost of the DC motor/controller combination. The best use of these motors is for short bursts of acceleration. However, they suffer from being heavy, low efficiency, low reliability, and high maintenance. This is because the brush contact results in wear and tear, which requires periodic brush replacement. The DC motor/controller system is still popular today because it keeps the cost down on some electric vehicles.

The most common DC motors used in BEVs are:

- series wound DC motors
- shunt wound DC motors
- separately excited DC motors.

2.2.2.2 AC Motors

Similar to DC motors, AC motors have a set of coils (field) and a rotor (armature). In these motors, there is no need for a commutator and brushes due to the periodic nature of the alternative current. Compared to DC motors, AC motors have the advantages of higher efficiency, higher power density, lower operating costs, lighter weight, and are no maintenance needs. The main drawback of these motors is the cost of the power electronics that are needed to convert (invert) DC from the battery into AC.

Table 2.1 EV motors' characteristics

Type	Advantages	Disadvantages
Induction motors	Reliable	High loss
	Low maintenance	Low efficiency
	Low cost	Low speed ratio
	Robust in severe operating conditions	Control complexity
Synchronous motors	Small size	Low speed ratio
	High power density	Low output power
	Low cost	Control complexity
	High efficiency	
Switched reluctance motors	Simple construction	Acoustic noise problem
	Outstanding torque-speed characteristics	Control complexity
	Simple control	Torque ripple
		Special converter topology

The AC motor candidates with the greatest potential for electric powertrains include:

- induction motors (IM)
- synchronous motors (PM brushless motor)
- switched reluctance motors (SRM).

Of the AC motors mentioned above, large vehicles such as SUVs and transit buses mostly use induction motors. On the other hand, synchronous motors are suitable for small and moderate vehicles such as passenger cars, as well as for in-wheel motor applications. For example, the Honda Insight and the Toyota Prius use 10 kw and 50 kw synchronous motors, respectively. The Chevrolet Silverado, the Chrysler Durango, and the BMW X5 also use induction motors.

SRMs are widely considered to have significant potential for BEV applications due to their outstanding torque-speed characteristics. However, their disadvantages greatly outweigh the advantages, rendering their BEV application unfeasible at this time. Table 2.1 summarizes the advantages and disadvantages of each of these motors.

2.2.3 Energy Sources and Storages

2.2.3.1 Energy Sources

The energy sources in BEVs supply the electrical energy required for vehicle motion and store the regenerated vehicular kinetic and potential energies. The performance of BEVs is highly dependent on the cost, durability, and performance of the energy sources. Energy sources are the main limitation preventing the mass production of BEVs because currently there is no single energy source technology that can sufficiently meet all the performance requirements of

BEVs. Different parameter definitions quantify the performance of energy sources on a BEV. The most important parameters are shown below [2]:

- *Energy and coulometric capacities*: The theoretical energy capacity of an energy source refers to the total amount of energy that can be stored in the device measured by watt-hours (Wh). Another measurement of electric energy source capacity is coulometric capacity (Ah), which represents the total Amp-hours available when the energy source discharges, and is a measure of the size of the device. When assessing the suitability of a BEV energy source, the energy capacity is a more important factor than the coulometric capacity because the total available amount of energy is important in a BEV driving range. The theoretical and coulometric capacities represent the actual energy content and size of the energy sources. However, it is not applicable to electrochemical batteries, since the complete depletion of batteries may cause permanent damage to the batteries. To prevent this, the battery discharge is stopped at a defined open circuit voltage (where the battery is in a disengaged state) level. This voltage level is the cut-off voltage. Thus, the energy capacity and coulometric capacity of batteries before they reach the cut-off voltage are termed as being the useable energy and useable coulometric energy, respectively.
- *State-of-charge (SoC) and depth-of-discharge (DoD)* are two measures of the energy capacity available from energy sources. As a percentage of the maximum energy capacity, the SoC denotes the amount of remaining energy capacity in a BEV energy source. In other words, the function of the SoC in a BEV is the equivalent to that of a fuel gauge in ICE vehicles. Alternatively, DoD, measured as a percentage of maximum capacity, is the amount of discharged energy capacity from the BEV energy source. DoD is the inverse of SoC; as one increases, the other decreases.
- *Energy densities*: There are two measures of energy density: volumetric energy density and gravimetric energy density. The volumetric energy density refers to the amount of usable energy capacity per unit volume in watt-hours/liter (Wh/L), known as energy density. On the other hand, the gravimetric energy density is the amount of usable energy capacity per unit mass in watt-hours/kilogram (Wh/kg), known as specific density. Despite commonly using these terms interchangeably, the latter is more important than the former in the quantification of a BEV energy source because the performance and the desired driving range of each BEV are limited to its maximum weight. The energy density affects the usable space, which is sometimes less important in a BEV performance and driving range.
- *Power densities*: The energy rate (in watts) available to deliver per unit mass or volume is the power density. The gravimetric power density (W/kg) and the volumetric power density (W/L) are two possible definitions. Similar to energy densities, the terms for gravimetric power and volumetric power densities are specific power and power density, respectively. Again, the former is more important than the latter in BEV applications because specific power plays an important role in the evaluation of a BEV power source to achieve the desired amount of acceleration and hill-climbing capability.
- *Energy efficiency*: Typically, the definition of the energy efficiency of a BEV energy source, providing that it additionally acts as energy storage, is the ratio of the output between the electrical energy during discharging to the input electrical energy during charging. This efficiency is different from the charge efficiency, the definition of which is the ratio of discharge coulometric energy (Ah) to the charge coulometric energy (Ah). The energy

efficiency is a more important factor than charge efficiency when assessing energy source effectiveness in BEV applications.

- *Cycle life*: The definition of cycle life is usually the number of deep-discharge cycles that occur before failing to meet specific performance criteria. The cycle life represents the life of the BEV energy sources based on the specifications of energy storage. Typically, the nominal cycle life of an energy source is estimated for a specific operating condition. The rate and depth of the DoD, as well as the environmental conditions such as temperature and humidity, affect actual cycle life. Usually, the citation of cycle life of an energy source includes the percentage of DoD. For instance, a battery can claim to offer 400 cycles at 100% DoD or 1000 cycles at 50% DoD.
- *C-rate*: This parameter denotes the rate at which an energy source discharges or charges relative to its maximum capacity. Its definition point normalizes the discharge current against energy source capacity, which may be different between types of specific power sources. 1C (nC) is the capacity rating (Ah) of an energy source, meaning that the discharge current will discharge the entire device in one hour (1/n h).
- *Cost*: The cost is another key factor in making energy sources of BEVs capable of competing with their ICE counterparts. Two terms associated with the power/energy sources are the initial manufacturing cost and the maintenance (running) cost, where the former is generally more dominant than the latter.

2.2.3.2 Energy Storage Systems

The *energy storage subsystem* provides electric energy when the primary energy sources in a vehicle cannot meet loads on their own. This is done by storing surplus energy generated by power sources or recovering energy through regenerative braking.

An energy storage device can be categorized by two key factors: the maximum amount of energy the device is able to store per unit volume, and the rate at which energy can be transferred to and from the storage device. It is desirable for a device to have both high energy storage capabilities as well as high-power deliverability. Typically, energy storage systems with high energy density are capable of delivering continuous power with minimal reduction to their lifespan and tend to have greater energy storage capabilities while suffering from low power density. In contrast, energy storage systems with high power density are capable of delivering pulse power while tending to have lesser energy storage capabilities. For example, batteries have a high-energy content, but lack high power capability. In addition, batteries cannot operate at high power levels for a prolonged period of time without reducing the life expectancy of the battery, meaning rapid deep discharge cycles result in frequent battery pack replacement. Ultra-capacitors, on the other hand, are high power devices, but have a much lower energy density in comparison to batteries.

Typically, city driving involves frequent accelerations and decelerations, respectively resulting in short power bursts and high magnitude currents from regenerative braking. Rapid accelerations demand peak power from the energy storage system, while rapid deceleration transfers peak power to the energy storage system. Thus, combining multiple energy storage systems makes it possible to utilize the advantages of each energy source. For example, by combining an energy source with high energy capacity and another energy source with high power delivery capabilities, it is possible to create an energy system that has both high energy capacity and high power delivery capabilities. Moreover, adding a high power capacity storage device such as an ultra-capacitor or a flywheel may help in the reduction of high power stress on the battery. Using hybrid energy storage systems in a well-thought-out configuration makes it possible to use power whenever necessary and to maximize the lifespan of both devices. However, having multiple energy storage systems and energy sources also means that there must be a method to control the power sharing between the systems. For instance, in order to meet the power and energy load of a BEV at a given point in time, it is mandatory to split the load requirement between the vehicle's energy storage devices and its energy sources. For successful vehicle operation, the sum of the power extracted from each storage unit and energy source must meet or exceed the power requirement at a particular moment in time. Managing the power and energy of a hybrid energy storage system poses many challenges because the differences between the technologies in terms of physical, electrical, and chemical characteristics result in different power, energy, voltage characteristics, and charge–discharge methods. Due to these differences, the interactions between these systems require careful study and verification.

Various propositions exist for the propulsion systems of BEVs for different types of energy sources and storages. The energy sources actively used in BEVs include rechargeable electrochemical batteries (called "batteries"), ultra-high-capacitance capacitors (called "ultra-capacitors"), and ultra-high-speed flywheels (called "flywheels"). All devices are energy storage systems capable of storing the energy during charging. Currently, as well as in the near future, batteries will endure as the dominant energy source in BEV applications due to their technological maturity and acceptable cost.

2.2.3.3 Batteries

Despite the design and development of different types of batteries for BEV applications, they still cannot achieve BEV performance requirements. Additionally, trade-offs between energy density, power density, cycle life, and cost limit the use of each battery type to specific BEV applications. Thus, there is no clear choice for the best battery technology that is suitable for all BEVs. Lead-acid batteries, nickel-metal hybrid (NiMH) batteries, lithium-ion batteries, nickel-zinc (Ni-Zn) batteries, and nickel-cadmium (Ni-Ca) batteries are used in BEVs. As discussed below, each of these battery technologies has its own advantages and disadvantages [2,4].

- *Lead-acid battery:* So-called VRLA, the invention of this battery dates back to more than two centuries ago. Currently, lead-acid batteries are available in production volumes due to their mature technology and low cost. This battery technology is an attractive candidate in BEV applications due to its many advantages, which include low cost, fast recharge capability, high specific power, robustness against severe temperature variation, and its availability in a variety of sizes and design. However, they suffer from low specific energy due to the weight of its lead collectors, low energy density, limited cycle life, high cut-off voltage, and lack of long-term storage.
- *Nickel-metal hydride (Ni-MH) battery:* The Ni-MH, available on the market since 1992, continues to develop and gain repute as the near-term choice for BEV applications. The main advantages of the Ni-MH battery include its high specific energy and energy density, which are twice that of lead-acid batteries. Furthermore, these batteries also offer fast recharge capability, long cycle life, wide operation temperature ranges, and environmental friendliness due to their recyclability. The main drawback of this technology is its high initial cost.

Table 2.2 Key parameters of BEV batteries

	Specific energy (Wh/kg)	Energy density (Wh/l)	Specific power (W/kg)	Cycle life (Cycles)
VRLA	30–45	60–90	200–300	400–600
Ni-MH	60–70	130–170	150–300	600–1200
Li-ion	90–130	140–200	250–450	800–1200
Ni-Zn	60–65	120–130	150–300	300
Ni-Cd	40–60	80–110	150–300	600–1200

Source: [2]

Additionally, the cycle life of a Ni-HM battery is sensitive to high current discharges. The best operation performance of the Ni-MH battery is achievable if it is discharged 20–50% of its rated capacity.

- *Nickel-Zinc (Ni-Zn) battery:* This type of battery is suitable for BEV applications due to its high specific energy and power densities, low cost materials, deep cycle capability, and relatively wide operating temperature range (−10 °C–50 °C). However, its short cycle life limits its use for BEV applications.
- *Nickel-Cadmium (Ni-Cd) battery:* Regardless of its high initial cost and low specific energy, this battery has the advantage of a long cycle life, as well as rapid recharge capability, a wide operating temperature range (−40 °C–85 °C), low self-discharge rate, excellent long-term storage, and a variety of sizes and designs. Despite the recyclability, cadmium in the battery can pollute the environment if not properly disposed of.
- *Lithium-Ion (Li-Ion) battery:* Since its introduction in 1991, it has garnered much attention for its ability to be used in BEV applications and is considered to be the most promising type of rechargeable battery, despite being still in a stage of development. Li-ion battery technology offers very high energy density, good high-temperature performance, high specific power, high specific energy, and long cycle life. Additionally, the Li-ion battery is recyclable, and thus an environmental-friendly technology. The high costs and high self-discharge rates are two disadvantages of Li-ion batteries. Currently, Li-ion batteries are superior to other battery technologies in terms of applications to BEVs.

Table 2.2 summarizes the key parameters of the batteries mentioned above. The specific power and cycle life ratings are given at 80% DoD and at C/3-rate. Table 2.3 compares the electric specifications of some BEVs available on the market. Each of the vehicles analyzed is the 2012 model.

2.2.3.4 Ultra-capacitors

The frequent start/stop operation of BEVs adversely affects the life cycle of batteries due to the high variation of the battery discharge profile. Typically, BEV batteries have high specific energy that is suitable for a reasonable driving range and cruising at a reasonable speed. However, they have significant difficulty providing sudden bursts of power for rapid acceleration and hill-climbing, due to their relatively low specific power density. Note that the high peak power demand deteriorates the battery life, and the low specific power density of

Table 2.3 Comparison of electric specification of 2012 BEV models

	Mitsubishi i-MiEV	Ford Focus	Nissan Leaf	Tesla model S
Class size	Subcompact	Compact	Mid-size	Large
Drive	Rear-wheel	Front-wheel	Front-wheel	Rear-wheel
Battery	360 V Li-ion	350 V Li-ion	360 V Li-ion	350 V Li-ion
Motor	66 kw PM, brushless	107 kW AC Induction	80 kw PM, brushless	260 kw AC Induction

Source: [5]

a battery jeopardizes the regenerative braking system because the battery cannot efficiently absorb the regenerated energy. One way to overcome these issues is to increase the capacity of batteries at the cost of deteriorating vehicle performance.

Ultra-capacitors are an energy (storage) source device with a high specific power density, high efficiency, and long cycle life, with the disadvantage of having low specific energy density. They can provide high output power within a short period of time, thereby making them suitable for high power demand conditions. Furthermore, their high specific power density makes them an excellent candidate for braking energy recovery. As mentioned in the previous section, it is possible to take advantage of both the ultra-capacitor's high specific power density and the battery's high energy storage capability by using a combination of the battery and the ultra-capacitor. This combination significantly improves the vehicle performance, and helps reduce the battery cycling duty, extends the battery life, downsizes the battery, and provides higher energy recovery. However, the integration of ultra-capacitors and batteries (often referred to as a hybrid energy system) requires additional power electronics, thus resulting in an increase in initial vehicle costs.

In this hybrid energy system, the ultra-capacitor must deliver additional power that exceeds the average power delivered by the battery. However, if the power demand is greater than the output power feasibly achievable by the ultra-capacitor, it is fair to expect the battery to provide the required amount of power. The ultra-capacitor maintains its charge via the recovered energy or by the battery energy during periods of reduced power demand. When the ultra-capacitor is in a full-charge situation, the battery should be directly recharged using the recovered energy. Therefore, it is necessary to accurately control the power flow between ultra-capacitors, batteries, motors, and power electronics with a power management system to maximize the benefits of a hybrid energy source.

2.2.3.5 Flywheels

Oerlikon Engineering Company in Switzerland created the first passenger bus exclusively powered by a large flywheel over a quarter of a century ago. When introduced to the world, the flywheel weighed around 1500 kg. It operated at 3000 rpm, and was recharged using electricity at each bus stop. Traditionally, the flywheel consisted of a large steel rotor. This rotor weighed hundreds of kilograms and collectively spun at the order of ten hundreds of rpm. This traditional design is quite a contrast to the design and development of modern flywheels, which consist of a lightweight composite rotor weighing only tens of kilograms but rotate at the order of ten thousands of rpm.

Flywheels can achieve the potential energy storage requirements for EV applications. Specifically, these devices can feasibly satisfy the requirements regarding high specific energy, high specific power, long cycle life, high energy efficiency, quick recharging, limited maintenance, and cost effectiveness. If flywheels are used in a hybrid configuration (with the primary source in a BEV) as an auxiliary energy source, they function to store energy in a mechanical form. This storage occurs during periods of cruising or regenerative braking. Additionally, the flywheel also simultaneously generates electrical energy during this period in an effort to satisfy the power demands that occur when the vehicle is starting, accelerating, or climbing a hill. Ultimately, the flywheel is unique in that in addition to providing load leveling for the primary source of energy in a BEV, it can also feasibly function as the sole energy source.

The features of flywheels include high specific energy, high specific power, high efficiency for conversion between electrical and mechanical energies, and the ability to be easily adaptable to long-term energy storage. There are numerous benefits to using flywheels as an auxiliary energy source in hybrid configurations with the battery for BEVs. Many of these benefits are quite similar to that of ultra-capacitors. These benefits include:

- the ability to decouple the requirements regarding specific energy and specific power of the battery, thus optimizing the battery's specific energy density (i.e., the cycle life);
- the ability to increase usable energy, in addition to the endurance and cycle life of the battery; this is possible because the high-rate power demand and high-current discharge are drastically reduced by the load leveling effect of the flywheel;
- rapid short-term recharges with high efficiency during periods of low power demand or regenerative braking;
- extended vehicle range, which occurs due to a combination of load leveling the primary energy source, and improved energy recovery during regenerative braking;
- the ability to know the exact amount of energy stored, with a simple measurement of the rotation speed.

An ultra-high-speed flywheel has the potential to solely power an EV, which eliminates the need for a battery. This application can have numerous long-term benefits, including having a higher specific energy and a higher specific power than any battery. In fact, the specific power of the flywheel may even be greater than that of the ICE. Moreover, the practically unlimited cycle life of the flywheel resolves the issue of cycle life limitations that plague other energy sources. Despite the significant above-mentioned benefits, flywheels suffer from two major problems. The first relates to the possible gyroscopic phenomena during cornering or during traveling on a road with banking angle. The generated gyroscopic forces in these conditions adversely affect the vehicle maneuverability. The second issue concerns failure containment. In the case of an accident or flywheel malfunctioning, the high power stored in the flywheel will be released and can severely damage the other vehicle components. Also, flywheels lose a large amount of energy in a short period of time and their integration into BEV powertrain systems is complicated. In 2013, Volvo announced a flywheel system was fitted to the rear axle of its S60.

2.2.4 *Power Electronic Converters*

In ICEVs, a 12-volt auxiliary battery system supplies all the auxiliary units such as lights, horns, radio, power windows, etc. This auxiliary battery is charged with an alternator coupled to the engine. The conditions are different for BEVs for three reasons. First, BEVs do not have an alternator to maintain the auxiliary battery charge. Second, the intended design of accessories in BEVs may be set to work at different voltage levels, such as 24, 48, and 120, in order to achieve the minimum power consumption. Third, with the development of advanced traction motors (such as AC motors and other electronic systems), both AC and DC electrical powers can work simultaneously, while the main batteries offer DC electric power.

In order to avoid these issues, BEVs use power electronic converters in order to supply all the auxiliary loads on board the vehicle. These components control the power flows between power sources, loads, and power buses. For example, a bidirectional DC–AC converter can control a motor/generator machine based on a control signal. When running the electric machine as a generator, the DC–AC converter will serve as a rectifier to output DC current. When running the electric machine as a motor, the DC–AC converter acts as an inverter to convert the DC bus voltage in three-phase voltage. In other words, the main functions of power electronic converters are to efficiently convert DC voltage of an energy source to the different levels of the DC output voltage. Furthermore, they distribute electric power to auxiliary accessories, and to provide proper power management between different energy sources and storage elements, which is a necessary function. Typically, power electronic converters include DC–DC converters, known as DC choppers, and DC–AC converters, known as inverters. Compact size, lightweight, reliability, and high efficiency are the main specifications of converters for electric vehicles [6,7].

A DC–DC converter supplies the required power loads for BEVs that function at different voltages, and charges the on-board auxiliary battery to a fully charged state (it acts as an alternator in a conventional ICEV). In a BEV, a DC–DC converter may also function to supply the driver of a DC traction motor. BEVs use both unidirectional and bidirectional DC–DC converters. Unidirectional DC–DC converters supply on-board loads such as sensors, driving control systems, power windows, windshield powers, and safety equipment, whereas bidirectional DC–DC converters work as regenerative braking, a battery charger, and backup power.

A DC–AC inverter receives DC power from the batteries and converts it to AC to supply the high power AC electric traction motor. On the other hand, an AC–DC rectifier can function in an electric powertrain to change alternating current into direct current during regenerative braking or charging the battery by an auxiliary engine.

2.2.5 *Power Bus*

The power bus is a DC link between sources and loads. In traditional ICE vehicles, loads directly connect to power sources while in BEVs loads and power supplies connect indirectly using power buses, the power control units (PCUs), and power converters. Such vehicles contain two direct current buses: one designated as the high-voltage power bus, the other as the low-voltage power bus. The high-voltage bus provides electrical energy for propulsion motors and other high-power loads, whereas low-power accessory loads such as lamps, micro-controllers and small motors connect to the low-voltage bus.

The two DC power buses link with one or more DC–DC converters to transfer energy back and forth. It is good to note that there are no set voltages for the "high" and "low" voltage buses; instead they depend on the vehicle configuration. The occasional inclusion of a third power bus functions to provide power for external plug-in appliances or tools. In addition to addressing power requirements, the use of multiple power buses at different voltages also addresses safety requirements.

2.2.6 *Regenerative Braking System*

A vehicle's braking performance is an important consideration in vehicle safety. A proper braking system must be able to quickly reduce the vehicle's speed, while also maintaining the vehicle direction as controlled by the steering wheel. Specifically, reducing the vehicle's speed requires the system to supply adequate braking torque on all wheels. In a conventional ICEV, typically, a friction brake system helps to decelerate or stop a moving vehicle by creating friction between the brake pads and the wheels. This braking process results in considerable energy lost due to creating excessive heat energy. A unique feature of BEVs that distinguishes them from conventional ICEVs is their ability to recover and retain substantial amounts of braking energy through a regenerative braking system. This system recoups the kinetic energy of the vehicle during braking and converts it back into electrical energy through proper control of the electric motors that operate as generators. Receptive energy sources within BEVs, such as batteries, ultra-capacitors, or flywheels, store the converted electrical energy. This process helps extend the driving range of the vehicle to improve overall vehicle efficiency.

The regenerative braking system typically operates when the driver presses the brake pedal while decelerating, reducing speed before stopping, or when releasing the accelerator pedal to generate the same deceleration feeling as in ICEVs. In the afore-mentioned instances, the electric motors operate as generators. The torque created by this reversal resists the forward momentum and eventually stops the car. If the regenerative braking torque is not enough to meet the driver's demand for braking torque, the conventional friction braking system is combined with the regenerative braking system. When the regenerative braking torque is at its maximum, the friction braking system provides the additional braking torque required to decelerate the vehicle (Figure 2.11). Additionally, the friction braking system provides the total

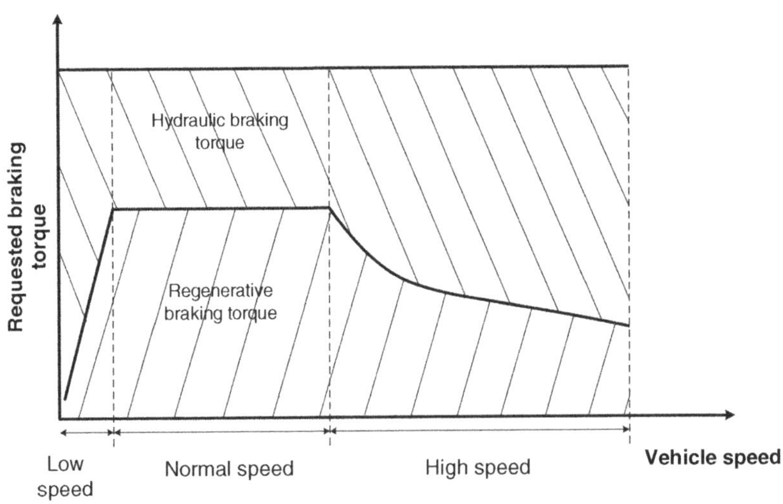

Figure 2.11 Regenerative and hydraulic braking system

desired braking efforts if the receptive sources have a full charge and regenerative braking is no longer applicable.

As seen in Figure 2.11, the braking torque of the regenerative braking system decreases while the vehicle is coasting at high speeds. At high speeds, the electric traction motor acting as a generator operates in constant-power mode, resulting in lower torque capability. At low speeds, the maximum braking torque is small and hence the friction braking is used to stop the car.

These operating conditions and limitations of a regenerative braking system indicate the requirements of proper design and control of the hybrid braking system to ensure adequate braking performance of the vehicle, while also functioning to recover as much braking energy as possible.

2.3 Fuel-Cell Electric Vehicles (FCEVs)

Using batteries as the only energy source in BEVs is limiting in a number of ways. Most importantly, this reliance on batteries results in limited driving ranges and extended recharging times. Recently, the application of fuel-cells as an energy source in electric vehicles has received great attention. Fuel-cell vehicles (see Figure 2.12) use fuel-cells to generate electricity from hydrogen fuel though gasoline-like liquids and methanol have fueled some fuel-cells. Unlike batteries, ultra-capacitors, and flywheels, a fuel-cell is not a storage device. It is an energy source unit where generated electricity either provides direct power to the traction motors or remains stored in the on-board energy storage for future use. In electric vehicles, the use of fuel-cells is superior to the use of batteries for a number of reasons. These reasons include a shorter reactant feeding time (compared to the long recharge times of batteries), a longer lifetime, and less maintenance requirements.

The most prominent advantage of FCEVs is their driving range, which is similar to that of a vehicle operating with an ICE. This is because what determines the FCEV range is the amount of hydrogen fuel available in the fuel tank, independent of the fuel-cell size. In fact, the relevance of fuel-cell sizes is contingent on the required power level of the FCEVs. An

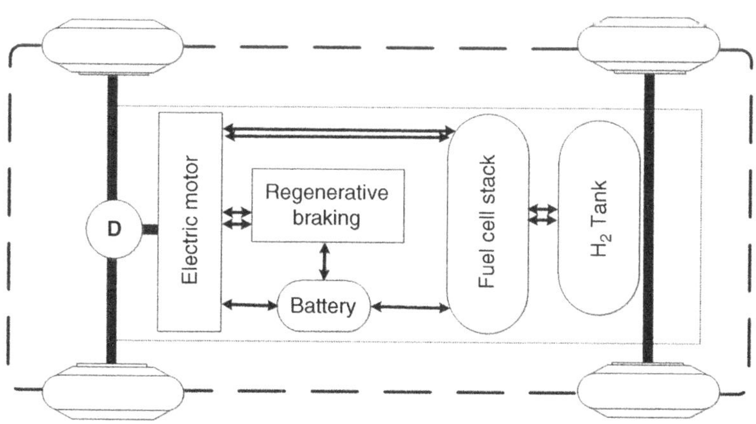

Figure 2.12 A fuel-cell electric vehicle

extremely low or no emissions level is another notable benefit of using hydrogen fuel-cells to propel a vehicle. In fact, a fuel-cell running on pure hydrogen emits no carbon dioxide; the only by-products are heat and water. If renewable energy is used to generate hydrogen, fuel-cells virtually eliminate GHG emissions.

As a result of the electrochemical reaction that converts energy stored in hydrogen to electrical energy, the efficiency of fuel-cell electric vehicles is inherently greater than that of combustion-based propulsion systems. By capturing wasted heat and reintroducing it into the chemical process of the cell, the efficiency of fuel-cells can exceed 85%.

Operating at temperatures below freezing is a concern in the scheme of mobile fuel-cell applications. The solidification of the water-based internal components of the cell would render it inoperable, and most cell designs are not robust enough to endure storage in sub-zero conditions. During operation, however, the fuel-cell generates enough heat as a by-product of the chemical processes to keep itself running at an acceptable temperature.

The introduction of Honda's first fuel-cell vehicle, the FCX, in 1999, and the subsequent unveiling of the second generation FCX Clarity in 2007 marked the beginning of one of the more successful ventures in the hydrogen vehicle industry. This vehicle uses 8.3 pounds of hydrogen gas stored in two compressed aluminum tanks (5000 psi) with a total capacity of 41 gallons. The FCX Clarity saw limited marketing beginning in June 2008 in the United States, and later that year was available in Japan. Currently, the FCX Clarity is available for lease in southern California, where 16 hydrogen re-fueling stations are open to the public. Honda also stated that mass production of the vehicles could begin as early as 2020.

The ongoing development of the FCX fuel-cell vehicle is an important first step towards mass production of fuel-cell vehicles. Mercedes Benz has initiated an analogous lease-based testing program, also in southern California, with the hydrogen-powered F-Cell car. The F-Cell has an estimated range of 190–240 miles. Finally, the introduction of the Chevrolet Equinox fuel-cell car constitutes a third test-drive project evaluating the potential and overall impact of hydrogen propulsion systems. GM plans to prepare this vehicle for more thorough testing in commercial fleets like taxi services.

Similar to hydrogen fuel engine vehicles, the primary concern associated with FCEVs is related to the difficulties of storing and transporting hydrogen fuels. As such, storing hydrogen onboard a fuel-cell vehicle continues to be a challenge for auto-makers. The high pressure of the storage vessel and inevitably large investment in a fueling infrastructure are other factors that are limiting the widespread usage of fuel-cell vehicles for the time being. The durability, reliability, and cost of implementation, along with the storage, production, and delivery of hydrogen itself, are other main limiting factors.

In general, the hydrogen infrastructure necessary to facilitate a shift to hydrogen-powered vehicles consists of pipeline hydrogen transport and filling stations. In the event that a station is not near a hydrogen pipeline, compressed storage tanks could provide the supply. Attempts to avoid the substantial costs associated with installing an entirely new hydrogen distribution infrastructure have resulted in investigating the possibility of simply utilizing the existing pipeline network (put in place for natural gas) after extensive modification. In any case, all levels of government must adopt new regulations and safety standards in order to enable the commercialization of hydrogen in consumer products. The method of re-fueling a fuel-cell car is dependent on the specific type of fuel used. In the case of a hydrogen-powered cell, it would be necessary to create some kind of connection between the car and the station dispenser that would maintain a sealed system to avoid safety hazards.

When considering emissions on a larger scale, it is important to consider the variety of production methods used in the production of hydrogen. For example, the most common method consists of gaseous hydrogen produced in a centralized plant via steam methane reforming from natural gas. The second most common method is similar to the first, except that hydrogen production occurs in its liquid form rather than its gaseous one. Another method involves producing gaseous hydrogen by electrolyzing water with electricity in the immediate vicinity of the re-fueling stations. In all cases, fewer GHG emissions relative to gasoline vehicles will occur. Gaseous methods of production generally result in fewer emissions than liquid, except for the cases where renewable energy sources are used [8].

2.3.1 *Fuel-Cell Technologies*

The powertrain of an FCEV is analogous to that of a BEV shown in Figure 2.2, with the exception of an added fuel-cell stack. All fuel-cells are comprised of an anode, a cathode, and an electrolyte that permits charges to pass between the two charged sides of the cell (Figure 2.13). As electrons attract to the anode from the cathode, they produce direct electrical current. Each individual cell produces very little electricity; thus, they typically function in conjunction with one another (that is, "stacked") in parallel or series circuits to maximize the amount of voltage and output current. The majority of fuel-cells are powered by hydrogen. This procedure operates by directly feeding hydrogen to the fuel-cell system, or alternatively, by reforming hydrogen-rich fuels such as methanol, ethanol, and hydrocarbon fuels.

There are several types of fuel-cells that are available on the market or now under development. There are different ways of classifying current fuel-cells, such as the type of chemical reactions that take place within the cell, the type of catalysts required, the type of fueling, and the operating temperatures. However, a typical classification by type includes the type of electrolyte employed, including polymer electrolyte membrane, direct methanol,

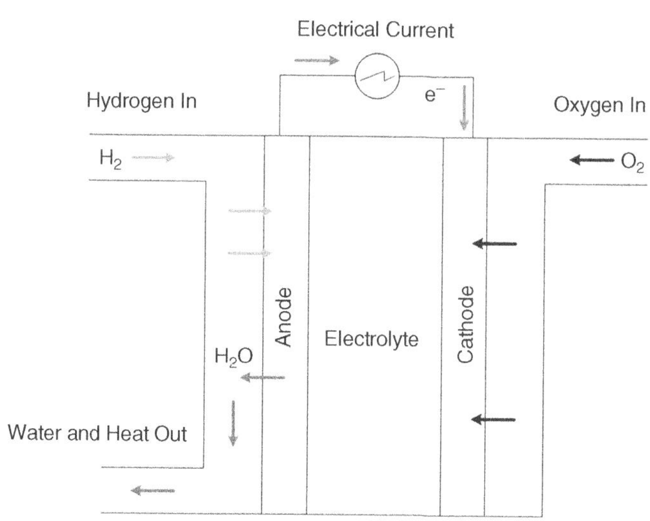

Figure 2.13 Schematic of a fuel-cell

alkaline, phosphoric acid, molten carbonate, and solid oxide. The following sub-sections explain the advantages, limitations, and potential applications of each fuel-cell [2,9,10].

2.3.1.1 Polymer Electrolyte Membrane Fuel-Cells (PEMFCS)

A PEMFC makes use of a thin, permeable polymer electrolyte sheet and porous carbon electrodes containing platinum catalyst. This cell has a higher power density and subsequently lower weight than other fuel-cell types, and operates at relatively low temperatures of around 60 to 80 °C. This results in faster start-up times and better durability. Due to fast start-up time, low sensitivity to orientation, and desired power-to-weight ratio, PEMFCs are widely considered the most viable option for transportation applications, such as passenger cars and buses. PEMFCs can produce an output power of around 50–250 kW. They are also safe for users to operate because it is a solid, flexible electrolyte that cannot leak or crack. Furthermore, the system does not need any corrosive material to function, requiring only a hydrogen source and air for the oxygen supply. There are, however, some disadvantages of using PEMFC. Due to the use of platinum – a very rare metal – as a catalyst, the cell is very expensive to use. Another issue associated with PEMFCs is in regards to water management. In a PEMFC, water polarizes into its constituent components: hydrogen and oxygen. However, a certain water level is necessary to guarantee proper operation. In a PEMFC, if a membrane is too wet or too dry, the functionality of the PEMFC drops due to improper polarization. Moreover, the platinum catalyst is very sensitive to CO poisoning. CO may enter into the fuel-cell with the entering air, and water polarization or reformed fuel use can create it. Thus, it is possible that the system will require additional CO-reducing reactors, further adding to the overall cost. Similar to hydrogen-fueled ICEVs (Section 1.4.4), the hydrogen fuel-cell suffers from on-board hydrogen storage and low energy density of hydrogen.

2.3.1.2 Direct Methanol Fuel-Cells (DMFCs)

Pure methanol powers direct methanol fuel-cells (DMFCs) instead of hydrogen, and these are a subclass of PEMFCs. They contain a thicker membrane and a greater density of catalysts than PEMFCs. Using these fuel-cells enables methanol to be mixed with steam and fed directly to the fuel-cell anode. DMFCs operate at between 50 °C and 120 °C with a low efficiency of up to 40%. Despite low efficiency, they benefit from high specific energy density, which is more important for vehicle applications than efficiency. Due to the high specific energy density of methanol compared to hydrogen, direct methanol fuel-cells do not suffer from the fuel storage issues that plague other hydrogen fuel-cells.

Another benefit of using methanol is that it is easier to transport and supply to the public, based on the existing infrastructure currently available for current liquid fossil fuels such as diesel and gasoline. Slow dynamic behavior, water management, and low specific power density are issues associated with DMFCs. Additionally, instead of plain water, the by-products of DMFCs are carbon dioxide and water – though the emitted carbon dioxide is more than gasoline or diesel fuel. If their power density and energy conversion efficiency increase while their costs reduce, DMFCs can be a potential candidate as an energy source of electric vehicles. Nonetheless, direct methanol fuel-cell technology is relatively new and the research and development associated with DMFCs is around 3–4 years behind other fuel-cell types powered by pure hydrogen.

2.3.1.3 Alkaline Fuel-Cells (AFCs)

With its active application in the US space program, alkaline fuel-cells (AFCs) were among the first fuel-cell technologies ever developed. Due to its high conductivity, AFCs use a solution of potassium hydroxide (alkaline) as the electrolyte. They can also utilize a variety of non-precious metals as catalysts at the anode and cathode, thus significantly reducing overall cost and making them more attractive for electric vehicle applications. The operating temperature of high-temperature AFCs is between 100 °C and 250 °C. With recent developments, newer AFC designs are capable of operating at lower temperatures, in the range of 23 °C–70 °C. Low cost, high performance, reasonable efficiency, and lower working temperatures are the main benefits of AFCs over other fuel-cell technologies. However, there are some major challenges associated with the widespread application of AFCs.

First, the possibility of the AFC being poisoned by carbon dioxide (CO_2) is a cause for concern. That is, even small amounts of CO_2 existing in the inlet air or hydrogen can negatively affect the operation of these cells. Due to this limitation, it is necessary to purify the hydrogen and oxygen used in the cell; a costly procedure. Likewise, this inclination towards poisoning also reduces the lifetime of the cell, further adding to the cost. Second, AFCs are prone to the risk of leakage of potassium hydroxide, which is highly corrosive. Third, AFCs consist of a complicated structure, which includes a circulation pump, a heat exchanger, and an evaporator. Lastly, AFC stacks have demonstrated the capability to maintain stable operation for more than 8,000 operating hours, while AFCs need to reach much higher operating times if they are to be realistically economical for electric vehicle applications. Largely due to issues with material durability, achievement of this task has not occurred, further placing a significant obstruction in the way of the commercialization of AFCs.

2.3.1.4 Phosphoric Acid Fuel-Cells (PAFCs)

Considered the "first generation" of modern fuel-cells, PAFCs use both liquid phosphoric acid as an electrolyte, as well as porous carbon electrodes that contain a platinum catalyst. Like PEMCs, PAFCs rely on an acid electrolyte. Unlike PEMCs, which are prone to "poison" from carbon monoxide, PAFCs are more resilient to the impurities in fossil fuels that must reform into hydrogen.

When used for the co-generation of electricity and heat, PAFCs are 85% efficient. However, they show less efficiency at strictly generating electricity (37%–42%), a rate that is only slightly more efficient than combustion engines (33%–35% efficiency). PAFCs also have less power density than other fuel-cells with the same weight and volume. Due to their lack of power, PAFCs tend to be typically large and heavy. Moreover, like PEMCs, PAFCs require a platinum catalyst, thus resulting in increased costs of production.

Another major drawback related to PAFCs is its freezing point, which is around 42 °C. Keeping the fuel-cell above its freezing temperature requires additional hardware, thus adding to the cost, weight, and complexity of the system. Due to the high freezing temperature, considerable energy must heat up the cell to its operating temperature and keep it warmed up. In addition, frequently turning the vehicle on and off results in significant energy loss, thereby making it unsuitable for city driving conditions. PAFCs could, however, supply power for larger vehicles, such as buses.

2.3.1.5 Molten Carbonate Fuel-Cells (MCFCs)

Molten carbonate fuel-cells (MCFCs) are high-temperature fuel-cells in which non-precious metals can function as catalysts at the anode and cathode. In MCFCs, a molten carbonate salt mixture (such as lithium-potassium carbonate or lithium-sodium carbonate) functions as the electrolyte. The main advantage that MCFCs have over AFCs, PAFCs, and PEMFCs is the use of a hydrocarbon instead of pure hydrogen, thus being able to work without requiring an external reformer to convert more energy-dense fuels to hydrogen. In fact, the high-temperature operation (600 °C–700 °C) of MCFCs allows internal reforming, wherein the conversion of hydrocarbon fuels to hydrogen occurs within the fuel-cell itself, thus ultimately reducing the overall cost of MCFCs.

Additionally, MCFCs are not sensitive to carbon monoxide or carbon dioxide "poisoning." The benefits of internal reforming and resistance against impurities make MCFCs advantageous for vehicle applications due to the availability of hydrocarbon fuels. One of the most prominent drawbacks of current MCFC technology is durability. The high temperature of operation and the corrosive electrolyte used cause the breakdown and corrosion of the component to accelerate, thus resulting in decreased cell life. Moreover, high-temperature operation results in a slow start-up and requires significant thermal shielding to retain heat and for safety. Generally, MCFCs benefit from a low-cost catalyst, reasonable efficiency, and low sensitivity to poisoning, while also suffering from slow start-up, reduced material choice due to high temperature, corrosive electrolytes, and slow dynamic performance.

2.3.1.6 Solid Oxide Fuel-Cells (SOFCs)

Solid oxide fuel-cells (SOFCs) rely on a hard, non-porous ceramic compound such as Yttria-stabilized zirconia as the electrolyte. SOFCs are distinct in that the solidity of the electrolyte allows the cells to take a different configuration from the plate-like configuration common in other fuel-cell types. The temperature at which SOFCs operate is generally quite high, in the range of around 1000 °C (1830°F). This high temperature also provides benefits and drawbacks similar to those of MCFCs. For example, due to the temperature, there is no need for a precious-metal catalyst, thus reducing the overall cost. Furthermore, the heat allows SOFCs to internally reform fuels. This enables the use of a variety of fuels, and does not require a reformer, which reduces costs.

The disadvantages of SOFCs include its slow start-up times, as well as its requirement for thermal shielding. Further adding to the complications is the fact that thermal shielding may be acceptable for utility applications, but not for transportation and small portable applications. Likewise, high operating temperatures result in strict durability requirements on the materials. As a result, one of the primary technical challenges for SOFCs include developing materials that have a low-cost value while still being highly durable under cell-operating temperatures.

Unlike other fuel-cells, SOFCs are quite resistant to sulfur. An additional benefit of SOFCs is their resistance to carbon monoxide (CO) poisoning. On the contrary, they may even use the substance as fuel, thus allowing SOFCs to use gases made from coal. The characteristics of the afore-mentioned fuel-cells are summarized in Table 2.4.

Table 2.4 Typical characteristics of fuel-cells

Fuel-cell Type	Operating Temperature	Efficiency (%)	Advantages	Disadvantages
PEMFC	50 – 100 °C typically 80 °C	60	• Low corrosion and electrolyte management problems • Low temperature • Quick start-up	• Expensive catalysts • Sensitive to fuel impurities • Low temperature waste heat
AFC	90 – 100 °C	60	• Fast dynamic performance • Low cost components	• Sensitive to CO_2 in fuel and air • Electrolyte management
PAFC	150 – 200 °C	40	• Higher temperature benefits • Increased tolerance to fuel impurities	• Pt catalyst • Long start-up time • Low current and power
MCFC	600 – 700 °C	45 – 50	• High efficiency • Fuel flexibility • Flexibility of using variety of catalysts • Suitable for hydrocarbon fuels	• High temperature corrosion and break-down of cell components • Long start-up time • Low power density
SOFC	700 – 1000 °C	60	• High efficiency • Fuel flexibility • Can use a variety of catalysts • Solid electrolyte • Suitable for hydrocarbon fuels	• High temperature corrosion and break-down of cell components • Long start-up time and limits

Source: [10]

2.4 Hybrid Electric Vehicles

In spite of having zero emissions, the economic limitation of production of pure electric vehicles results from the associated battery costs, and battery performance management. Similarly, the use of fuel-cells in electric vehicles is still in its early stages of development. Since these issues present significant technical challenges, hybrid vehicles provide an ideal transitory stage before the zero emission vehicles can be prepared for mass production. A definition of a hybrid vehicle (HV) is one that derives propulsion from multiple distinct types of energy sources, with the possibility of recovering friction energy using recuperation power systems. An example of such a system is a regenerative braking system, which captures and stores energy lost during wheel braking in either a mechanical or an electrical capacitor. One of

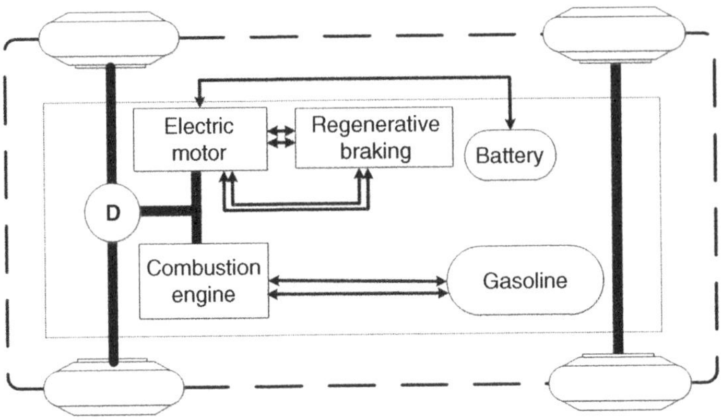

Figure 2.14 Hybrid electric vehicle

the propulsion power sources functions as the main power source and the secondary propulsion power source in these hybrid vehicles assists the primary power source.

Hybrid electric vehicles (HEVs) combine the electric motor and high-voltage battery of a purely electric vehicle with the internal combustion engine of a conventional vehicle. In other words, a hybrid electric vehicle system utilizes the advantages of both an internal combustion engine and an electric energy source. This combination allows for considerable improvement in vehicle efficiency. The electric motor supplies the additional power for the vehicle's traction requirements and regenerates power during regenerative braking. HEVs can be effective in a variety of applications, and therefore appear in all sectors of transportation, from commercial to consumer vehicles. Compared to ICE vehicles, anticipated advantages of hybrid technology include enhanced fuel economy, reduced emission and noise, longer engine and brake system life, and lower operating costs.

The precise amount of tailpipe emissions released by HEVs varies by the individual vehicle specifications, type of hybrid power system used, and fuel. For further emission reduction, it is possible to use an alternative fuel instead of gasoline. For example, in 2009, Hyundai Motor Company started developing a domestic market for the Elantra LPI Hybrid, in which the engine run on liquefied petroleum as a fuel. However, at this moment, the majority of HEVs run using gasoline or diesel.

As shown in Figure 2.14, on a fundamental level, HEVs comprise of a power unit, a propulsion system, and an energy storage system. Compression and spark ignition engines with fossil fuels, alternative fuels, and fuel-cells are possible power unit technologies. The propulsion system transfers the generated power to the wheels by a mechanical mechanism, electric motor, or by a combination of both mechanical and electrical components. Possible energy storage systems are ultra-capacitors, flywheels, and, most commonly used, batteries. HEVs are independent of grid electricity, and batteries are recharged onboard by electricity from engine-driven generators and from brake energy recuperation.

2.4.1 Degree of Hybridization

In terms of vehicle hybridization, there are three main categories of hybrid system: micro-hybrid, mild-hybrid, and full-hybrid. A micro-hybrid is a vehicle in which an electric machine functions

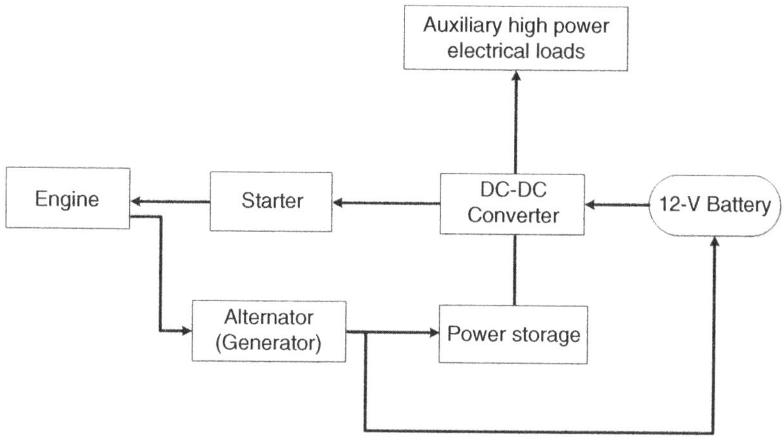

Figure 2.15 Schematic of a micro-hybrid system

in applications such as stop/start and regenerative braking, but not to supply additional torque when the engine is running. To date, most micro-hybrid applications have been focused on small gasoline engines, but the achievement of stop/start in a diesel engine is a considerably greater challenge due to the much higher starting torque requirement. The BMW 1 series and the Smart Fortwo, manufactured by Mercedes, are examples of commercialized micro-hybrid vehicles. Figure 2.15 illustrates a schematic of a micro-hybrid system.

A mild-hybrid has an electric motor-generator integrated to provide up to approximately 10% of the maximum engine power in the form of additional torque. Mild HEVs (see Figure 2.16) improve the drawbacks of fossil-fuel vehicles, where engine efficiency is not

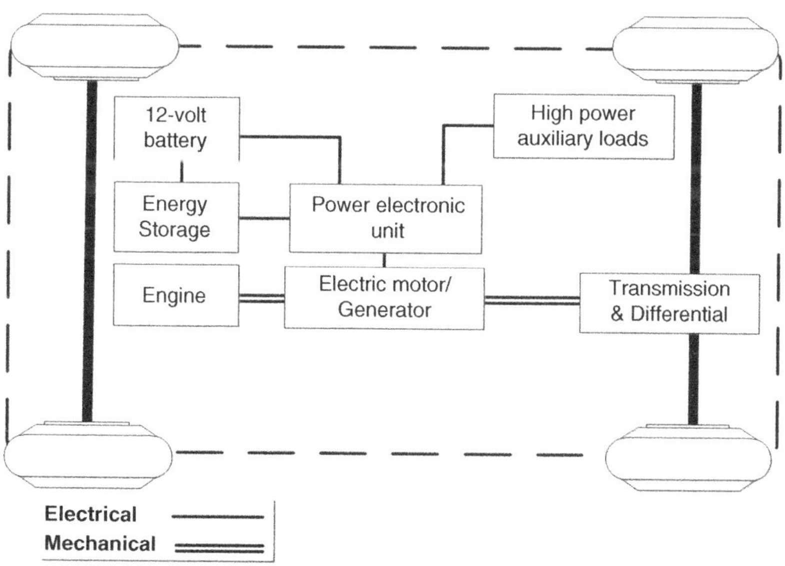

Figure 2.16 Schematic of a mild-hybrid electric powertrain

at its maximum levels. They do so by utilizing a large motor/generator in a parallel configuration with an ICE. The ICE consumes more fuel and emits more emissions when coasting, braking, or idling. Mild HEVs can provide some level of power assistance to the engine whenever the engine is not efficient, and are able to provide engine start–stop functionality. Moreover, during idling, they can turn off the engine and use electric power instead.

An exclusive electric propulsion mode is not possible in such vehicles. To replenish the batteries, it is possible to integrate low level of energy recuperation systems, such as regenerative braking systems. Compared to conventional ICE, mild HEVs improve vehicle emissions, especially in high stop–start traffic, even though their fuel consumption is less efficient than full HEVs.

General Motors fitted a mild-hybrid system in some of its products, such as the Chevrolet Malibu Hybrid and the Saturn Ayra Green Line. This mild-hybrid system is the Belt Alternator Starter (BAS). Using a 36–48 volt system, BAS supplies power for the start-up motor on demand and provides an auxiliary source of power to vehicle electronic accessories. In this system, the ICE starts through a motor-generator unit and a belt-drive. Upon starting, the engine charges the batteries by driving the motor-generator, while the regenerative braking system can recharge the system battery.

A full-hybrid is one in which the electric motor typically provides at least 40% of the maximum engine power as additional torque. Full HEVs utilize a larger electric motor and battery, thus providing the flexibility for better engine sizing. In other words, the electric drive can facilitate the downsizing of the engine by adding enough power to maintain operation at a higher percentage of rated torque (higher efficiency).

Usually, the size of batteries and motors in full HEVs meets acceleration requirements, whereas the engine design satisfies the minimum specified gradeability. Compared to ICEs, full HEVs improve vehicle fuel consumption and reduce vehicle emissions at the expense of adding complexity to the powertrain structure, while also increasing the cost–weight of the vehicle.

In full- and mild-hybrid electric powertrains, it is possible to combine the power from the electric motor and the engine to define proper energy flow routes and control ports. Although there are several powertrain architectures of hybrid electric vehicles, the main ones include series hybrid, parallel hybrid, power-split (parallel-series) hybrid, and complex hybrid. These configurations of HEVs particularly differ in regards to how power is transmitted to the wheels, and the definition of the connection between the powertrain components. Table 2.5 compares the functionalities of the afore-mentioned HEVs.

Table 2.5 Comparison of HEV functionalities in terms of degree of hybridization

	Engine Start/Stop	Regenerative Braking	Motor Assist	Electric Drive
Micro-hybrid	Yes	Slight	Slight	No
Mild-hybrid	Yes	Yes	Yes	No
Full-hybrid	Yes	Yes	Yes	Yes

2.4.2 Parallel Hybrid Configuration

Parallel hybrid configuration is the most popular configuration in HEVs. This configuration mechanically couples the mechanical and electrical powers, allowing their use either simultaneously or independently. Usually, in parallel hybrid powertrains, the mechanical and electrical power outputs are combined by using mechanical devices such as torque-couplers and speed-couplers. A mechanical torque-coupler adds the torques of the combustion engine and the electric motor using a gearbox unit or a pulley/chain assembly. Torque-couplers work in all types of hybridizations, from micro to full hybrid systems.

In contrast, speed-couplers can be in the form of a planetary gear unit, they combine the combustion engine and electric motor powers. An electric motor with a floating stator can work as a speed-coupler as well. The main benefit of hybrid powertrains with speed-couplers over those with torque couplers is the decoupled speeds of the two power plants, granting the opportunity to choose the speed of both the power plants independently. During the ICE power and battery charging mode, the mechanical coupler divides the engine power into two portions: one to drive the vehicle, and the other to charge the battery [9].

In a parallel hybrid arrangement, it is possible to shut off the ICE and only run the electric motor from the battery pack (as in a full electric vehicle). In this parallel hybrid, the ICE has the ability to recharge the battery during less intense power driving cycles, thereby making it more efficient for highway driving than the stop–start conditions of city driving. To be concrete, classifications of the operating modes of a parallel hybrid electric powertrain are:

- *Engine-alone traction mode:* In this mode, the combustion engine generates the total required power for vehicle motion while the motor is off. This mode is used when the engine is running at or near its optimal operating conditions, such as when cruising.
- *Electric-alone traction mode:* In this mode, the engine is off and the electric source supplies the total required power for vehicle motion. This mode is on when the ICE efficiency is low, such as when starting the vehicle, or when the vehicle speed is low (e.g., during reverse gear).
- *Hybrid mode:* In this mode, the wheels receive power from both power sources. This mode is activated in conditions requiring more power such as when accelerating or during high speed driving.
- *Engine traction and battery charging mode:* In this mode the power generated by the ICE is more than the power required for vehicle motion. In this case, the additional power recharges the battery by switching the electric motor to operate as a generator.
- *Regeneration mode:* In this mode, the kinetic energy the vehicle wastes while braking, or during downhill motion recuperates through the regenerative system of the powertrain.

The parallel configuration provides increased efficiency and performance for highway driving, long-distance trips and cruising because the structure allows both the electrical and mechanical power sources to work simultaneously, while also providing the flexibility to switch between power sources.

Although the parallel hybrid system typically includes only one motor that can act as a generator when needed, a parallel hybrid configuration involves a complicated and expensive transmission and drivetrain systems. Some of the parallel hybrid vehicles in the market are the Chevy Malibu Hybrid, the Honda Insight, the Ford Escape Hybrid, and the Honda Civic Hybrid.

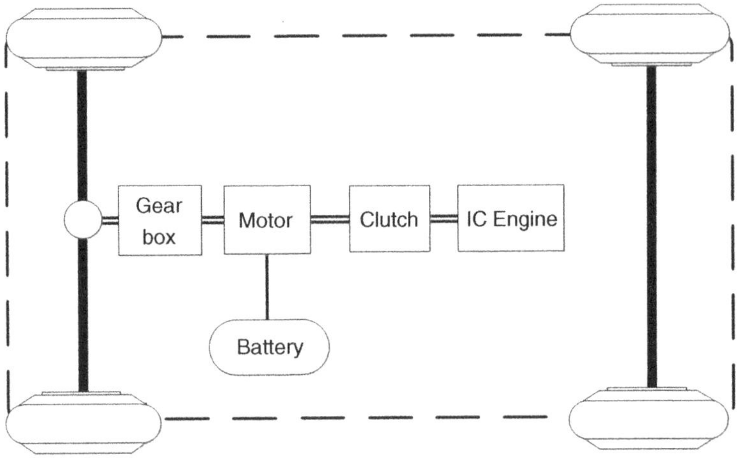

Figure 2.17 Parallel single-shaft powertrain

Depending on the traction requirements, powertrain component limitations, and engine characteristics, there are different architectures of parallel hybrid powertrains. Each arrangement meets the vehicle performance specifications and fuel economy improvement targets according to the power ratings of the motor and engine, as well as the sizing of the energy storage unit. The following subsections describe the most important configurations of parallel hybrids [9].

2.4.2.1 Parallel Single-Shaft Hybrid Powertrain

In this arrangement, the combustion engine, the electric motor, and the transmission system are mounted upon the same shaft where the motor and engine operate at the same speed (Figure 2.17). The rotor of the electric motor functions as the speed-coupler and is mounted upon the engine shaft, between the engine and the final drive. The stator is mounted upon the outer transmission housing or in a separate intervening housing. This arrangement has a very simple and compact structure that is straightforward to control. The transmission can be manual, automatic, automated manual, or a continuously variable transmission. Currently, Honda have integrated this arrangement of parallel hybrid powertrain with a CVT in its Insight and Civic models. The design of the single-shaft parallel system can work with or without a clutch when the engine is turned off. The Honda designs do not use a clutch, but Nissan developed the Infiniti M35 HYBRID in which two clutches are linked in parallel, one to the motor and one directly to the engine and transmission.

Depending on the position of the motor/generator in parallel single-shaft hybrid systems, classification of such systems is as pre-transmission and post-transmission hybrid parallel. Pre-transmission hybrid powertrains are those configurations in which the electric motor/generator connects to the engine output shaft between the engine and transmission system. In such systems, both the engine and motor can deliver power to the driving wheels through the transmission system. A power management unit controls the mechanical and electrical powers in order to meet the instantaneous power requirements of the vehicle. Figure 2.18 shows a schematic of a pre-transmission parallel hybrid system. The pre-transmission arrangement is

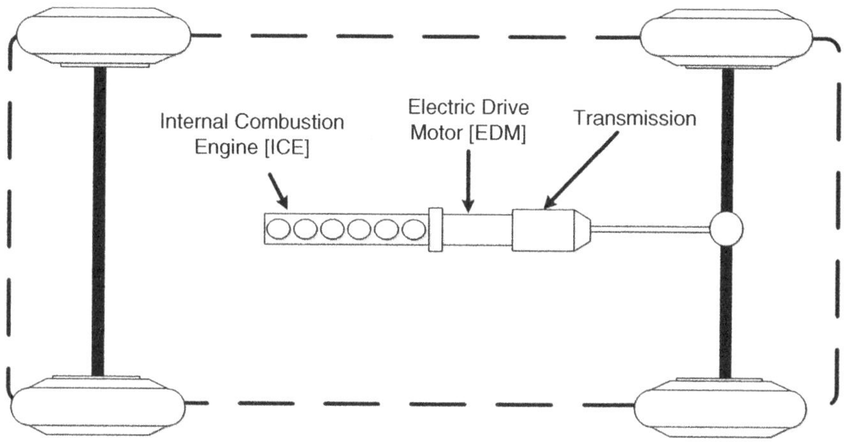

Figure 2.18 A schematic of a pre-transmission parallel hybrid powertrain

commonly used in mild-hybrid electric powertrains, where the motor is not able to provide sufficient traction force to propel the vehicle.

Hybrid designs highly favor the pre-transmission parallel architecture because it allows torque multiplication through the transmission, thereby making electric power more effective when providing assistance to the engine during high-power demands, in addition to utilizing a smaller electric motor/generator. It also allows an electric mode during low and medium power demands, as well as allowing the battery to charge while the vehicle is idling.

In contrast, post-transmission hybrid powertrains (Figure 2.19) couple the electric motor to the conventional powertrain between the transmission and final drive. In order to operate at a wide range of vehicle speeds, a post-transmission parallel architecture requires either a dedicated transmission system to match the motor/generator to the different vehicle loads, or an electric motor/generator that can provide sufficient power in the case of high torque demands. It is possible to use a single planetary gear set to match the motor/generator to the

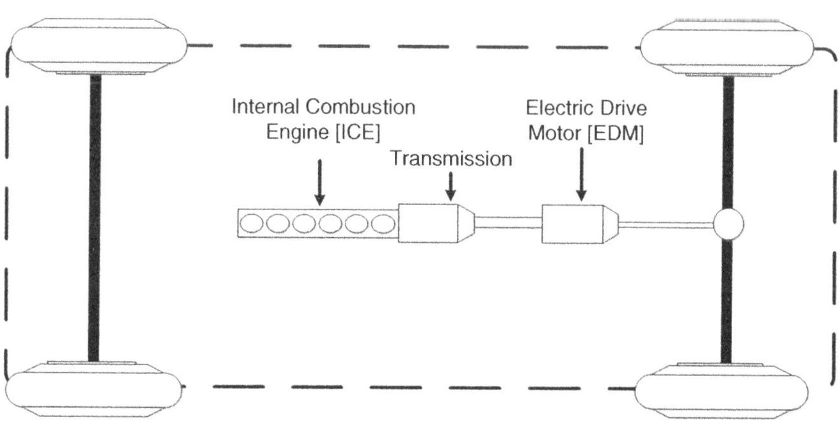

Figure 2.19 A schematic of a post-transmission parallel hybrid powertrain

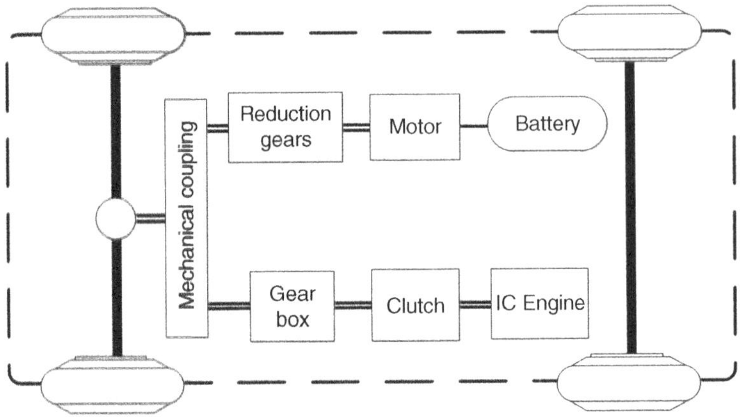

Figure 2.20 A double-shaft parallel hybrid powertrain

different vehicle loads. However, since the system does not have a dedicated transmission, it requires a larger motor/generator and supply electronics to deliver the required traction forces, especially at low speed conditions. The main advantage of post-transmission systems is the efficient recuperation of the braking energy due to the physical location of the traction motor.

2.4.2.2 Parallel Double-Shaft Hybrid Powertrain

In this arrangement, the engine and motor have separate shafts where the combustion engine and electric motor powers can provide torque to the wheels separately. The combustion engine and electric motor outputs couple together with a mechanical torque-coupler or speed-coupler. One variation of such an arrangement (shown in Figure 2.20) uses two transmission systems, one located between the engine and the torque-coupler, and the other located between the electric motor and the torque-coupler. Although it is possible to use either a single-ratio or a multi-ratio transmission system for each power source, a reduction gear typically functions to take advantage of the electric motor's high torque characteristics at low speeds, while a transmission system with different or infinite gear ratios helps the combustion engine operate at its optimal region.

In this design, the transmission allows both the electric motor and the combustion engine to operate at or near their optimal operating conditions, resulting in superior performance and better overall efficiency. However, the integration of two transmission systems makes the powertrain structure complicated and expensive.

In another variation of this arrangement (shown in Figure 2.21), the transmission system location is between the torque-coupler and the drive shaft. The transmission system regulates the power outputs of both the combustion engine and the electric motor with the same ratios. The main application of this arrangement is in vehicles using a small engine and motor. A transmission system with different gear ratios enhances the tractive effort at low speeds.

A parallel hybrid powertrain with separated driven axles is another double-shaft configuration. In this design (shown in Figure 2.22), the electric motor powers a pair of wheels on one axle, and the combustion engine drives the other pair of wheels on the other. The tractive efforts from two power sources combine through the road. Regenerative braking recharges the

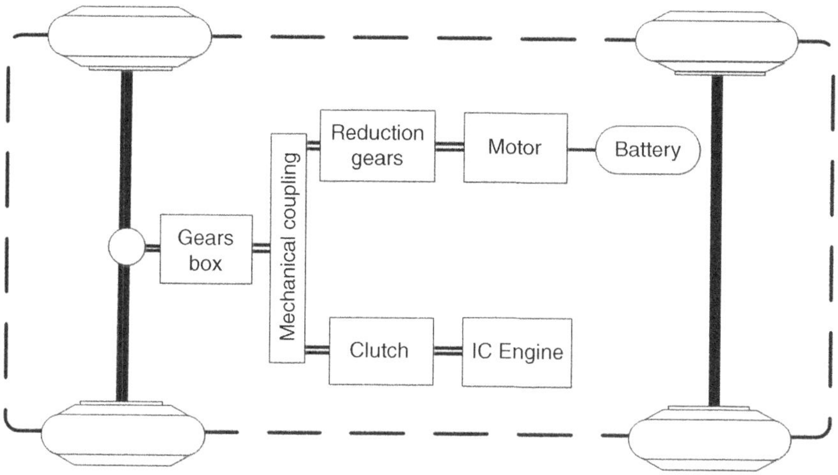

Figure 2.21 Another type of double-shaft parallel hybrid powertrain

batteries, or power transfers from the engine to the batteries through the road surface. The operating characteristics of this design are analogous to those of parallel double-shaft designs, in which both transmissions can be a single-ratio or a multi-ratio. The separation of electric and conventional powertrains are advantageous because it makes the overall structure simple, offers the benefit of a four-wheel drive, and provides the advantages of a conventional vehicle, if needed. Nevertheless, separated powertrains occupy spaces needed for the passenger and luggage, and do not offer the opportunity to charge the batteries from the engine when the vehicle is at a standstill. The space issue can be resolved partially by using a single reduction gear instead of a multi-ratio transmission system, in addition to integrating in-wheel motors instead of the out-wheel electric motor. In-wheel motors, however, will introduce the issues discussed in Section 2.2.1 on the BEV powertrain configuration

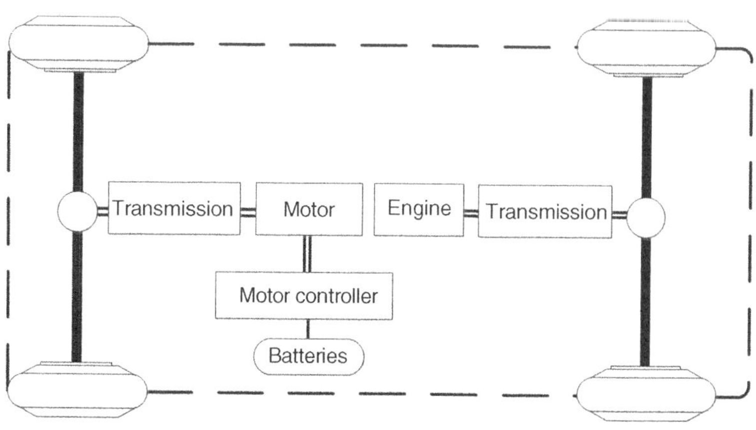

Figure 2.22 A separated axle torque combination parallel hybrid powertrain

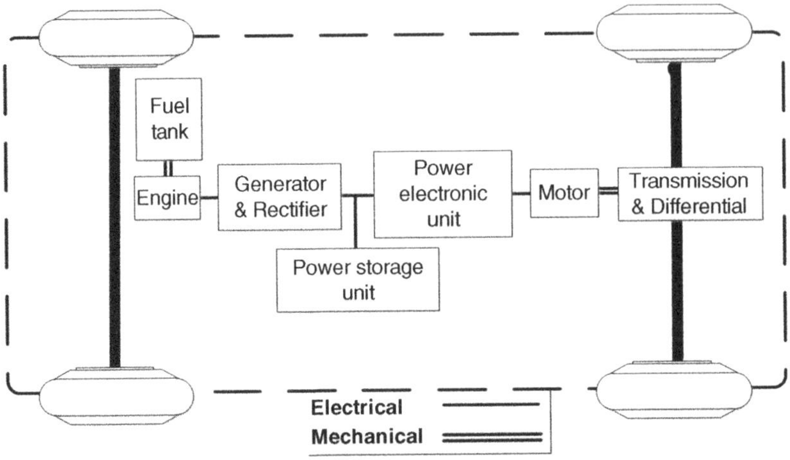

Figure 2.23 A series powertrain structure

2.4.3 *Series Hybrid Configuration*

In this configuration, an electric motor provides electric traction, while a small ICE drives the electric generator, providing power for the driving electric motors, and efficiently charges the battery. In other words, the propulsion system of a series HEV is an electric motor with batteries that can be charged through a generator driven by an ICE. In this architecture, since there is no direct link between the engine and the driving wheels, the combustion engine can operate in its most efficient range, regardless of the vehicle load and speed.

The substantial difference between the series and parallel configurations relates to the electric motor and transmission systems. Although it is possible to integrate a torque amplification device in order to improve the performance of the electric motor and reduce its size, the motor used in series configurations must be significantly more powerful than those used in parallel configurations (Figure 2.23). This is because it must be capable of providing all the necessary tractive power. Parallel HEV motors, on the other hand, must only provide boost power to supplement the engine if and when necessary. The parallel configuration requires a more expensive, multi-gear transmission to drive all the wheels. On the hand, the motor in a series HEV is capable of providing high torque over a wide speed range with no stepped gear ratio changes. A series hybrid configuration converts the combustion engine's mechanical power to electrical power and then back to mechanical power, resulting in more power losses than a conventional powertrain with the same combustion engine. However, this drawback can be compensated for by the recuperated power captured from the vehicle's kinetic or potential energy through the regenerative braking system. This feature makes the series configuration more efficient for stop-and-go city driving conditions.

Similar to parallel hybrid electric powertrains, series hybrid powertrains have different operation modes, as described below:

- *Engine-alone traction mode:* In this mode, the combustion engine only provides the total required power for vehicle motion through the generator while the electric motor acts as an electric transmission between the engine and the driving wheels. The batteries neither provide nor receive any power from the drivetrain.

- *Electric-alone traction mode:* In this mode, the engine is off and the electric source supplies the total required power for vehicle motion.
- *Hybrid mode:* In this mode, the engine–generator and batteries simultaneously deliver traction power to the wheels.
- *Engine traction and battery charging mode:* In this mode of operation, the engine–generator provides the traction power while charging the batteries.
- *Regeneration mode:* The regenerative braking system charges the batteries during braking, decelerating, and downhill travel. In this mode, the traction motor operates as a generator and the engine-generator is not active.
- *Battery-charging mode:* In this mode, the traction motor stops while the batteries are charged by the engine–generator.
- *Hybrid battery-charging mode:* Both the engine–generator and traction motor operate as a generator and charge the batteries simultaneously.

2.4.4 Power-Split Configuration

The design of power-split hybrid or series-parallel hybrid systems takes advantage of the best features of both series and parallel hybrids, and provides a more efficient hybrid powertrain system. In comparison with a parallel hybrid powertrain, a series-parallel uses an additional electric motor that primarily functions as a generator (Figure 2.24). On the other hand, when compared to a series hybrid powertrain, the series-parallel connects the engine to the final drives through a mechanical link, thus allowing the engine to drive the wheels directly. In other words, these systems split the power from the engine between two paths: one transfers the power to the wheels through a mechanical gear system, and the other transfers the power to the

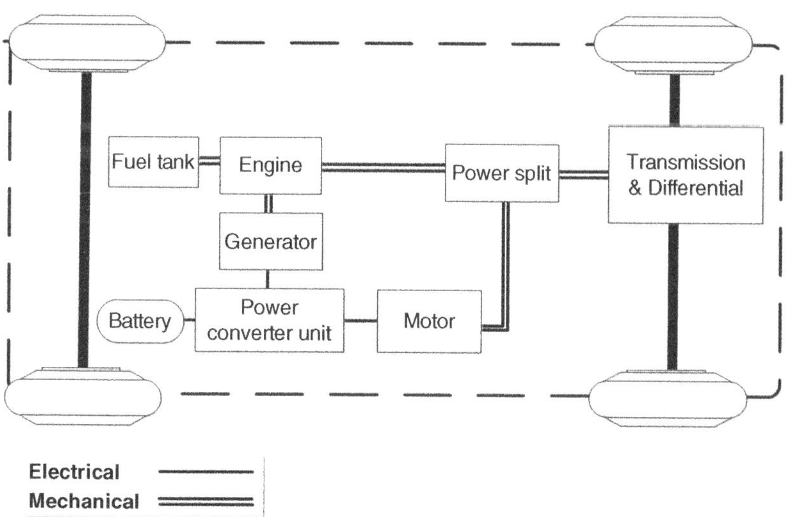

Figure 2.24 A series-parallel powertrain structure

wheels through a generator and an electric motor. This means that the electric motor and the combustion engine can work independently or in synergy to optimize efficiency. Furthermore, the system is capable of generating electricity through the electric motor as a generator while simultaneously driving the wheels. The main benefit of this system over the series system is that it allows the decoupling of engine power from the electric motor, therefore providing more flexibility in power control. Using this structure, it is possible to compensate for the deficiency of conventional combustion engines by providing the required torque at lower speeds by utilizing electric motors. In low-load, low-speed situations, the electric motor can power the wheels, while working with the engine till it reaches its optimum working conditions. Thus, the engine used in a power-split is smaller, less flexible, and more efficient than a conventional ICE. Drawbacks of this hybrid powertrain system include its complexity and a complicated power management system. The Toyota Prius and the Ford C-Max use this powertrain structure.

There are two groups of a series-parallel hybrid powertrain in operating mode of a series-parallel hybrid powertrain: ICE-heavy and electric-heavy. The operation mode of ICE-heavy is similar to that of a parallel hybrid powertrain, in which the engine is more active than the electric motor. The operation mode of electric-heavy is similar to that of a series hybrid powertrain, in which the electric traction motor is more active than the engine.

A series-parallel hybrid powertrain uses a power split device to effectively combine the engine driving power and the motor driving power. A planetary-geared, electrically controlled, continuous variable transmission (E-CVT), and magnetic continuous variable transmission (M-CVT) are the main devices used to combine the powers.

2.4.4.1 Electric-Continuous Variable Transmission (E-CVT)

E-CVTs typically function as a power split device in hybrid powertrains to combine the engine power and the electric motor/generator through a planetary gear set. In fact, an E-CVT benefits from relative motions among three axes of a planetary gear set, in which the shafts are either the input or output shaft. As mentioned in Section 1.6.2.2, the use of a planetary gear set makes it possible to achieve a combination of either two input shafts and one output shaft, or a combination of one input shaft and two output shafts. It is possible to use the E-CVT with a fixed-ratio drive-chain to connect the output power of the motor/generator to the final drive. In an E-CVT (see Figure 2.25), the traction motor, motor–generator, and engine shafts are coaxial where the ring carrier connects to the traction motor, the planet carrier connects to the engine, and the sun carrier connects to the motor–generator. The ring carrier transfers the traction motor power to the final drive only during "electric mode," while it can provide additional traction power to the wheels during "hybrid mode." The sun carrier is used during regenerative mode or for starting the engine.

In addition to the ability to continuously vary the speed of the engine, the E-CVT has a smooth and continuous performance during the regenerative braking process. However, it suffers from low power density, low efficiency, and noise problems. High speed or high load conditions can magnify these issues.

2.4.4.2 Magnetic-Continuous Variable Transmission (M-CVT)

A magnetic continuous variable transmission (M-CVT) achieves torque transmission and variable gear ratios by magnetic materials and without mechanical contact. The composition of

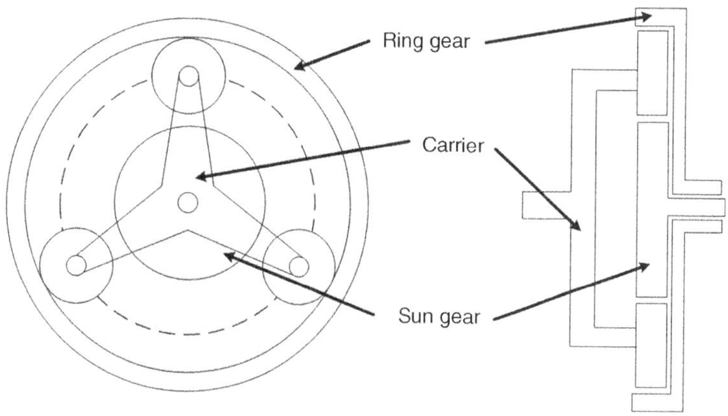

Figure 2.25 Topology of an E-CVT

M-CVTs (Figure 2.26) includes: three coaxial magnetic rotors, namely, the outer rotor, the inner rotor, and the modulating rotor, as well as a stator and electronic control unit [11]. This device splits the power flow from the engine into two paths. One path generates power by sharing the outer rotor of the magnetic gear with the rotor of the traction motor, while another path forms by sharing the inner rotor of the magnetic gear with the rotor of the motor-generator. Two consecutive converters, which are separately coupled to the stators of the motor and the generator, control the generated power flow. The proper locking of the rotors, which activate and deactivate the magnetic field according to the engine load and driving conditions, provide a continuous gear ratio change and a proper power split. The outer rotor transfers the traction motor power to the final drive only during "electric mode," while it can provide additional traction power to the wheels during "hybrid mode." The inner rotor works during regenerative mode, or for starting the engine.

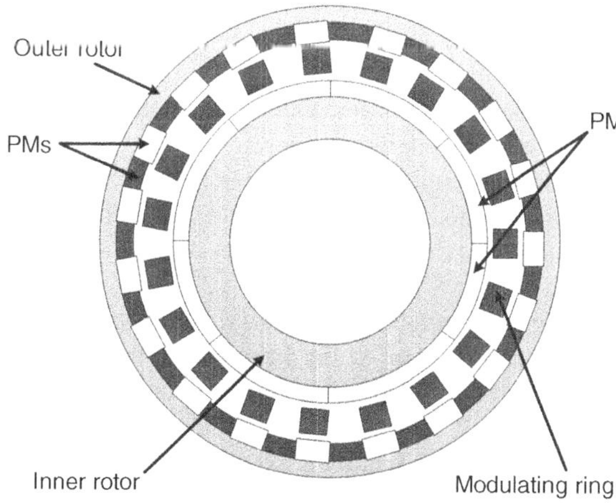

Figure 2.26 Topology of an M-CVT

M-CVTs belong to the category of electrical CVTs. Generally, M-CVTs benefit from low noise level, low maintenance/repair costs, low energy transfer losses, no lubrication needs, and good controllability. The main concern associated with such systems relates to the high voltage requirements and costs. M-CVTs have a strong potential use in hybrid and electric vehicles, with emerging new power electronics for continuously variable voltage, continuously variable frequency, fast induction current control, and digital overall control systems.

2.4.5 *Compound Hybrid Configuration*

The structure of a compound hybrid configuration, also known as a complex hybrid, is similar to that of a series-parallel hybrid. The difference relies on the bidirectional functionality of the electric motor in a compound powertrain and the unidirectional functionality of the generator in a series-hybrid powertrain [2]. In other words, the second electric motor in the compound hybrid powertrain can act as both a generator and a traction motor, thus providing a bidirectional power flow. This bidirectional power flow offers more operation modes to the system than a series-parallel one. This is especially true for the three-propulsion power operating mode, which can be obtained due to the engine and two traction electric motors. A series-parallel hybrid powertrain cannot offer this operating mode.

The main application of a compound hybrid powertrain is vehicles with a dual-axle propulsion system, where the front-wheel axle and rear-wheel axle are separately driven. The powertrain system of such vehicles is lighter, more fuel-efficient, and less noisy compared to a conventional all-wheel drive vehicle, due to the removal of mechanical components, such as the propeller shaft and the differentials.

Two power traction configurations are possible in a compound hybrid vehicle. In the first, a hybrid powertrain drives the front-wheel axle, while an electric motor drives the rear one. In the second, a hybrid powertrain drives the rear-wheel axle, while an electric motor drives the front. The operating modes of a front-wheel hybrid rear-wheel electric compound hybrid powertrain include [2]:

- *Start-up mode:* In this mode, the engine is off, while the battery provides electrical power to both the front and rear electric motors to individually drive the front and rear axles.
- *Electric mode:* In this mode, the battery provides the power to only the front electric motor to propel the front axle, while the engine and rear electric motor are off. This operation mode occurs during light load driving conditions.
- *Hybrid mode:* In this mode, the engine and two electric motors are on. The engine and front electric motor work together to propel the front axle, while the rear electric motor simultaneously drives the rear axle. This mode is on during heavy load conditions, such as full-throttle acceleration.
- *Engine traction and battery charging mode:* In this mode of operation, the engine power is divided to provide the traction power to the front-wheel axle and to charge the battery through the electric motor, which is switched to its generator mode. This mode is suitable for normal driving conditions.
- *Regeneration mode:* The regenerative braking system charges the batteries during braking, decelerating, and downhill travel. In this mode, the two electric motors operate as generators to simultaneously recharge the battery.

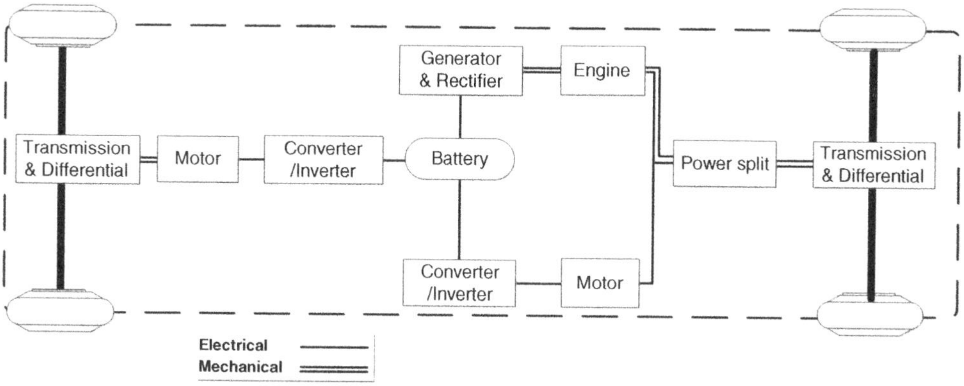

Figure 2.27 Configuration of a rear-wheel hybrid front-wheel electric compound hybrid powertrain

- *Axle-balancing mode:* This mode is unique to dual-axle systems, especially compound hybrid powertrains. In this mode, the traction torques are interchanged between the front and rear axles where the rear electric motor drives the rear axle, thereby achieving axle balancing.

The operation modes of a rear-wheel, hybrid front-wheel, electric compound hybrid powertrain (Figure 2.27) are similar to those of a front-wheel hybrid, rear-wheel electric compound, hybrid powertrain.

2.5 Plug-in Hybrid Electric Vehicles (PHEVs)

Plug-in hybrid electric vehicles (Figure 2.28) combine the advantages of hybridization with the chance to travel using only electricity provided by the grid for a certain time, rather than just through internal recharging systems. Similar to HEVs, the energy used to power a PHEV comes from both the internal combustion engine and a battery pack; however, the primary power source of PHEVs is electricity, while the ICE may be used as the secondary power source. Likewise, the drivetrain of a PHEV can be built in a parallel, series, or power-split

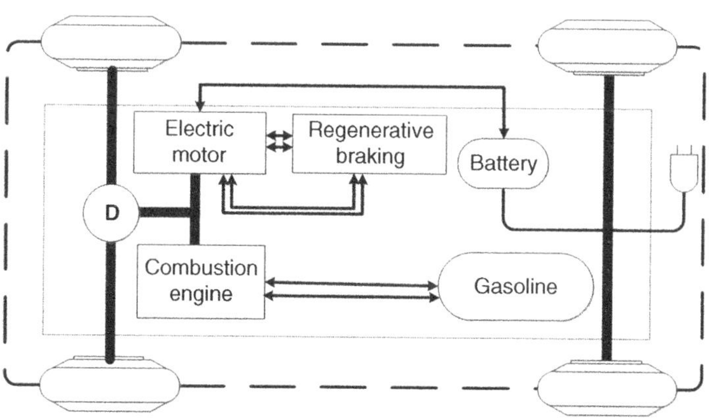

Figure 2.28 A plug-in hybrid electric vehicle

structure, with the possibility of charging the batteries on-board and also from an electrical outlet. Plug-in hybrid electric vehicles provide the opportunity to reduce reliance on fuel while retaining the driving range of conventional vehicles.

Advantages of PHEVs over BEVs include its reduced dependence on recharging infrastructure, and thus, lowered costs. As such, it might be feasible to produce PHEVs at higher volumes in the early years of its development. While PHEVs need smaller batteries than pure electric vehicles, they will require much larger batteries than today's hybrid electric vehicles. Unlike today's HEVs, PHEVs will also have the capability of repeated deep discharges. This is in contrast to HEVs, which usually operate at a near-constant "state-of-charge" mode, and as a result, cannot undergo deep discharge–recharge cycles. Furthermore, since battery capacity is still far below that of electric vehicles, it requires more power-oriented battery configurations in order to bring power up to the levels necessary for operation when the vehicle is idle, or when providing a burst of acceleration.

PHEVs can be emission-free if they run on electricity mode. Like BEVs, emission calculations must also account for electricity production methods if the battery recharges via an external power source. In spite of this, average emission from electricity generated in power plants is less than that produced by diesel or gasoline engines. Additionally, the emission from such power plants relies on the production efficiency and the actual mix of sources used to produce fuel. Depending on the battery capacity, the distance traveled on electricity is usually between 20–80 km. Generally, a PHEV labeled PHEV'X' indicates the distance the vehicle can travel when in electricity mode. For example, "PHEV20" stands for a plug-in hybrid vehicle capable of running on electricity for 20 km, whereas a "PHEV80" can travel up to 80 km. The Chevrolet Volt is the first PHEV manufactured by GM for mass production. It uses the series power configuration [12].

Major drawbacks of PHEVs relate to batteries, which increase the total mass of vehicle, and thus adversely affect the performance of the vehicle. In addition, batteries are expensive, which results in an increase in the final cost of PHEVs. Tables 2.6 and 2.7 compare the characteristics, functionalities, benefits, and limitation of PHEVs to other electric vehicles.

Table 2.6 Comparison of PHEV characteristics with other electrified vehicles

	Propulsion	Energy Storage system	Energy source
BEV	• Electric motor drive	• Battery • Super-capacitor	• Grid electricity • On-board electricity charging
FCEV	• Electric motor drive	• Battery • Super-capacitor • Hydrogen tank	• Hydrogen • On-board electricity charging
HEV	• Internal combustion engine (primarily) • Electric motor drive (secondary)	• Battery • Super-capacitor • Fuel tank	• Gasoline/diesel • On-board electricity charging
PHEV	• Electric motor drive (primarily) • Internal combustion engine (secondary)	• Battery • Super-capacitor • Fuel tank	• Grid electricity • Gasoline/diesel • On-board electricity charging

Source: [13]

Table 2.7 Comparison of advantages and disadvantages of a PHEV with other electrified vehicles

	Advantages	Disadvantages	Major Issues
BEV	• Zero local emission • High energy efficiency • Independent of fossil fuels • Commercially available	• Relatively short range • High initial cost	• Battery size and management • Charging facilities • Cost • Battery life time
FCEV	• Zero/low local emission • High energy efficiency • Independent of fossil fuels (if not using fossil fuel to produce hydrogen)	• Relatively short range • High initial cost • Under development	• Fuel-cell cost • Fuel-cell life cycle and reliability • Hydrogen production • Hydrogen distribution and infrastructure • Cost
HEV	• Low local emissions • High fuel economy • Long driving range • Commercially available	• Dependence on fossil fuels • Higher cost than ICEVs	• Battery sizing and management • Control, optimization, and management of multiple energy sources
PHEV	• Minimum local emissions • Higher fuel economy than ICEVs • Longer driving range than BEVs • Commercially available	• Slightly dependence on fossil fuels • Higher cost than ICEVs	• Battery sizing and management • Control, optimization, and management of multiple energy sources

Source: [13]

2.6 Hybrid Hydraulic Vehicles (HHVs)

With a principal operation similar to that of electric hybrids, hybrid hydraulic vehicles (HHVs) use two propulsion power sources: an IC engine and a hydraulic motor. In such vehicles, hydraulic hybrid systems transfer energy from the engine to the wheels using fluid power. An accumulator, a reservoir, and a hydraulic drive (pump/motor) are the main components that establish the drivetrain of HHVs [12]. Filled with nitrogen gas, the accumulator is a high-pressure hydraulic fluid vessel that stores energy and compresses nitrogen gas using hydraulic fluid. The low-pressure reservoir is a storage tank for the hydraulic fluid after the hydraulic drive uses it. The hydraulic drive pressurizes and transfers hydraulic fluid from the reservoir to an accumulator when it operates as a pump or vice versa when it operates as a hydraulic motor. The dual functionality of the hydraulic drive enables it to act as a motor when it rotates the wheels using pressurized fluid, and to act as a pump when it pressurizes the fluid using the vehicle's momentum. Compared to HEVs, the motor/pump in HHVs replaces the motor/generator, and the hydraulic accumulator replaces the batteries. Similar to BEVs, the drives may be one pump/motor driving a pair of wheels through a differential, or one per each wheel, also known as an in-wheel motor vehicle.

Power in a hydraulic hybrid system completely originates from the recuperation of lost friction energy. During braking, these vehicles capture kinetic energy and convert it to

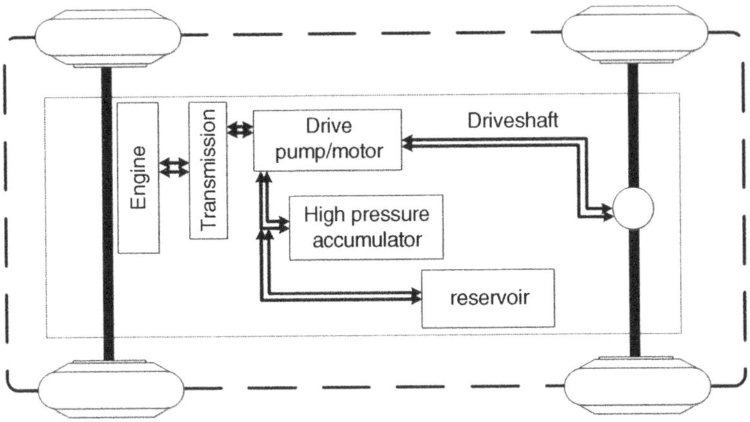

Figure 2.29 A parallel hydraulic hybrid powertrain structure

pressurized hydraulic fluid and store it in the accumulator. When accelerating, this stored energy works to move the vehicle. With current technologies, it is possible to recuperate up to 30% of lost friction energy in HEVs. In comparison, HHVs are capable of regenerating more than 70% of this energy. The high-efficiency regenerative braking system makes HHVs a clean, efficient, and cost-effective alternative to conventional stop-and-go vehicles. In addition to fuel efficiency, HHVs enhance the emission efficiency of the vehicle by significantly reducing carbon dioxide and nitrogen oxide emissions.

In HHVs, the hydraulic components can be configured in a parallel or series with an ICE. In a parallel configuration (Figure 2.29), engine power directly transmits to the wheels by a conventional transmission, while the accumulator and reservoir are connected in parallel to the conventional powertrain through the hydraulic pump/motor. This configuration assists the ICE in accelerating by providing power to the driveshaft, and recuperates the friction energy during braking. During cruising, the combustion engine solely turns the wheels. As a drawback, in this configuration, it is not possible to shut off the engine while idling. Heavy-duty vehicles are currently the only practical application of this configuration.

In a series configuration (Figure 2.30), hydraulic power stored in the accumulator is directly transmitted to the wheels while decoupling the ICE from the wheel loads. In other words, this configuration does not have the conventional powertrain, and the vehicle is propelled by the transfer of high-pressure fluid from the accumulator to the reservoir through the drive acting as a motor. The series configuration involves fewer mechanical components than a parallel configuration, thus resulting in better efficiency. In addition, this configuration switches the ICE off when it is not at optimum working operations, such as during light acceleration or short cruising. However, the engine will turn on when the pressure level drops below a certain threshold, and assist the hydraulic system to complete its circuit. The regenerative braking performance is similar to the parallel system.

In comparison with BEVs, HHVs have simpler structure with less complexity. In addition, HHVs have a higher efficient regenerative braking system and are more cost-effective, and do not use significantly advanced technologies. However, the size and noise of hydraulic systems continue to present technical challenges. The main obstacle in the widespread integration of

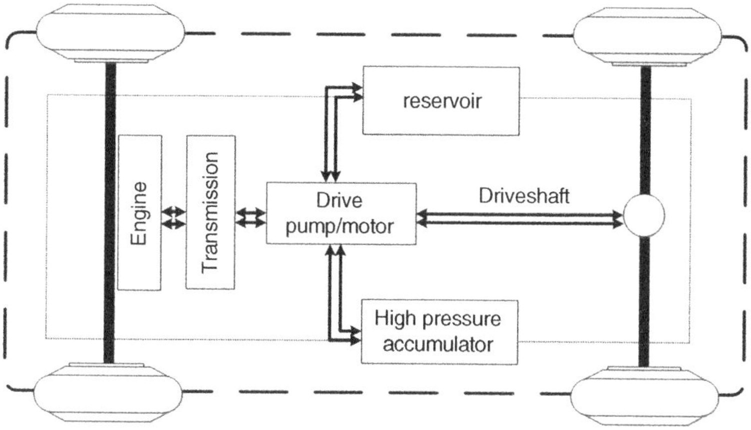

Figure 2.30 A series hydraulic hybrid powertrain structure

hydraulic hybrid systems in vehicles (especially passenger cars) is its inability to support electrical accessories, such as air-conditioning, radio, etc., while the engine is off.

2.7 Pneumatic Hybrid Vehicles (PHVs)

Pneumatic hybrid vehicles are based on the same principles as hybrid electric vehicles. They employ two energy sources, fuel and pressurized air, to propel the vehicle. During braking, the kinetic energy of an air hybrid vehicle is converted to pressurized air by running the same ICE in the compressor mode. Pneumatic hybrid engines can have four modes of operation: Compression Braking (CB), Air Motor (AM), supercharged, and conventional internal combustion. In this design, when the driver applies the brake pedal, fuel is shut off, and the engine operates as a two-stroke air compressor, storing the vehicle's kinetic energy in the form of pressurized air in the reservoir tank [14].

There are various uses for the energy stored in the tank. The first option is to run the internal combustion engine as an air motor. In the air motor mode, the charging valve between the air tank and engine opens, and the pressurized air runs the engine as a two-stroke air motor. Usually, this mode operates during low load engine conditions to avoid high fuel consumption.

The second option of using the stored braking energy is to run the engine in air motor mode to start it up. It is possible to activate start-up mode after a long stop to avoid a cold start, or after a short stop to avoid engine idling, resulting in a lower engine fuel consumption and emission. In addition, stored energy can run the engine accessories. Some attempt has been made to remove all or some of the engine accessories from the engine to avoid excessive power losses, particularly at high engine speeds. In an air hybrid engine, removing some or all engine accessories is possible, as it will run by an auxiliary air motor, which feeds off of air stored in the tank through series or parallel configurations.

In a series configuration (Figure 2.31), the shaft that runs the engine accessories is connected to the engine by a clutch. This shaft also passes through an air motor. If the tank pressure is high enough to run the engine accessories, the clutch disengages and the air motor runs all or some of the engine accessories. If the tank pressure is not high enough, the clutch engages, and the air motor works in idle mode while the engine runs the accessories.

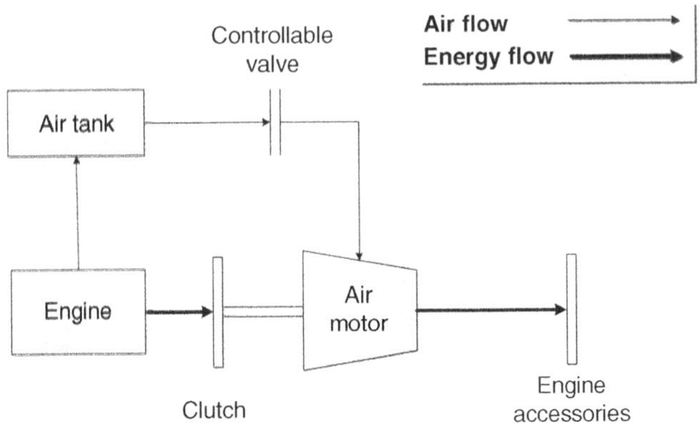

Figure 2.31 A series configuration for running engine accessories in PHVs

In a parallel configuration (Figure 2.32), air motor and engine output shafts are connected to the main shaft of the accessories through a planetary gear. If the tank pressure is high enough, the air motor clutch engages and the engine clutch disengages. This way, the air motor runs all or some of the accessories. If the air tank pressure is not high enough, the air motor clutch disengages, the engine clutch engages, and the engine runs all the accessories.

Figure 2.33 depicts the supercharged mode. This mode is activated when the desired torque is high. In this mode, the engine supercharges by the pressurized air from the tank, and there is an increase in the mass of fuel and air entering the engine cylinders, resulting in high engine torque and power. In contrast to typical supercharged engines, which exhibit a low efficiency at low speeds, air hybrid engines can be supercharged at any operating point, thanks to the air stored in the tank. The conventional mode activates when the desired load is moderate or the tank pressure is relatively low or empty.

In contrast to other alternatives, such as electric hybrid or hydraulic hybrid engines, which require a secondary powertrain system, air hybrid engines use the same engine as the secondary powertrain, thus reducing the extra mass. Although the air hybrid concept seems to be simple,

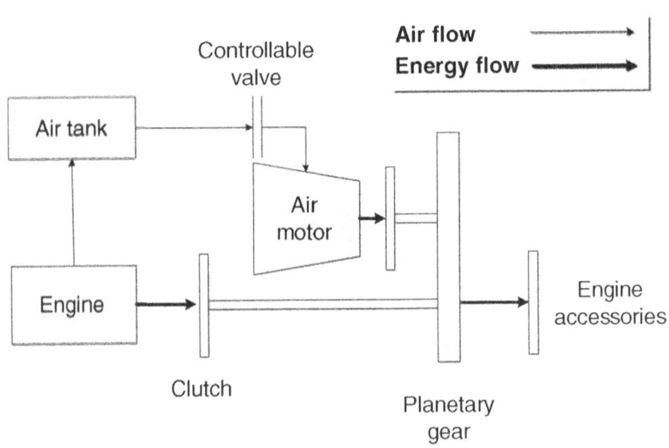

Figure 2.32 A parallel configuration for running engine accessories in PHVs

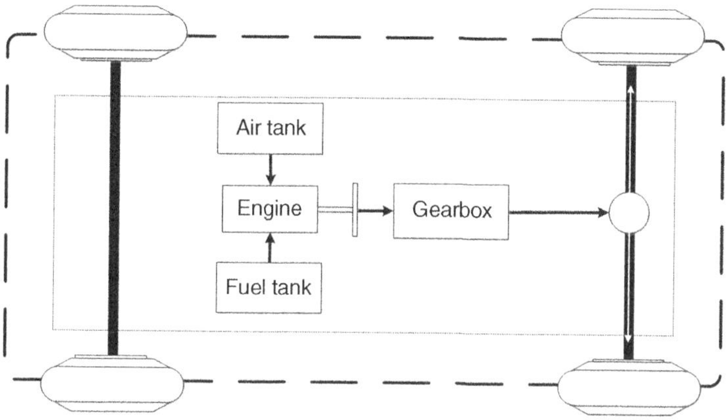

Figure 2.33 Energy flow in supercharged mode

there are some practical issues to resolve before accepting the concept as a hybrid powertrain solution. One of the most important challenges of air hybrid engines is the poor energy-storing capacity of the system due to the low air pressure when the engine is used as a compressor. The maximum storing pressure is a function of the engine compression ratio that limits the energy density of the stored pressurized air. New developments in this area have removed this problem [14].

Yet another challenge in the implementation of air hybrid engines is the inevitability of using flexible valvetrains. Since an air hybrid engine has different operational modes, it requires a flexible valvetrain to implement the concept. This could increase the engine complexity and costs.

2.8 Power/Energy Management Systems

One of the challenging tasks in hybrid or electric vehicles is the proper utilization of on-board stored energies. This requires the adoption of an intelligent power management system (EMS) in order to manage the energy consumption and maximize the energy recuperation of the vehicle. The term "power management" refers to the design of a high-level control algorithm that determines the proper power flow between the vehicle powertrain, the generator, the motor, and the battery. The power source of the electric and hybrid electric vehicles is the batteries that can charge from regenerative braking, the grid or directly from the engine power. Regenerative braking increases the performance of electric and hybrid electric vehicles. Therefore, based on the drive cycles, either the engine or regenerative braking could charge the batteries while driving. In the case of HEVs and PHEVs, since the efficiency of internal combustion engines is dependent on the engine's speed and load, the energy storage system should be charged during the high-efficiency operation of the engine in a drive cycle, if needed. This is the role of the power management controller to maximize the overall efficiency of the system. Generally, the EMS must realize the following functions:

- to optimize the system energy flow;
- to minimize vehicle emissions (in the case of hybrid vehicles);

- to estimate the residual available energy and driving range;
- to maximize the regenerated energy from braking;
- to prioritize real-time power requests from the vehicle auxiliary electrical loads in the case of electrified vehicles;
- to efficiently allocate the power from energy sources.

The combination of these factors required to maximize the overall efficiency of electric and hybrid electric vehicles makes designing a power management control system a challenging task.

There is a slight difference between power management and energy management of a vehicle. Fundamentally, energy is accumulation (integration) of power over a certain period of time. The energy management deals with energy consumption and recuperation over a trip, whereas power management deals with power distribution and power flow control between electric or hybrid powertrain components at instant time to satisfy the power demands. Often, the terms power management and energy management are incorrectly seen as interchangeable.

The power and energy management problem has two parts, the first is how to divide the available power between loads, the second is being able to integrate and arbitrate multiple energy storage systems. There is currently no consensus among researchers and the industry as to the best way of addressing the power and energy management problem.

Managing the power and energy of a hybrid energy storages system poses many challenges because the differences in the physical, electrical and chemical characteristics result in different power, energy, voltage characteristics and charge–discharge methods. Due to these differences, the interactions between these systems require careful study and verification. Chapter 8 describes the concept of power/energy management control and its design procedure in detail.

2.9 Summary

The aim of this chapter was to present an overview on the state of current electric and hybrid vehicles, including BEVs, FCEVs, HEVs, PHEVs, HHVs, and PHVs.

A BEV utilizes electric propulsion technologies, typically composed of a traction electric motor, power electronics, and energy sources. The major factors in a BEV design are system integration, component sizing, energy management, system optimization, and body design. At the moment, the critical factor hindering the mass production and commercialization of BEVs is the lack of an electric energy source that can demonstrate high specific energy and energy density, high specific power and power density, fast charging and deep discharging capabilities, long cycle and service life and high efficiency, and adequate safety, and cost effectiveness.

FCEVs use fuel-cell technology to generate the required traction power for vehicle motion. Fuel-cells are electrochemical devices that generate electric energy through chemical reactions. Unlike batteries, fuel-cells are not capable of storing energy, and can only generate energy through electrochemical reactions. Quiet operation, zero to very low emissions, rapid refueling, fuel flexibility, and durability are the main advantages associated with FCEVs. Although fossil fuels can work as the fuel of a fuel-cell system, hydrogen seems to be the ideal nonpolluting fuel for BEFCs due to its high specific energy and its byproduct – water. BEFC technology is still in its early stages of development, and fuel-cell technologies require further research and development.

By adopting both the ICE and an electric motor, HEVs and PHEVs are excellent transient technologies from conventional ICE vehicles to zero emission vehicles. The classifications of current powertrain technologies of HEVs and PHEVs are: (1) series hybrid; (2) parallel hybrid; (3) series-parallel hybrid; and (4) compound hybrid. The major issues associated with HEVs and PHEVs are: managing multiple energy sources, component sizing, minimum system costs, and good driving performance [13].

Rather than electrification, likely candidates for the hybridization of ICE vehicles are HHVs and PHVs. These technologies are still in their early stages and require significant effort in terms of research and development.

Problems

1. Explain the principal operation of a DC motor.
2. Explain the principal operation of an induction motor and discuss its advantages and disadvantages in HEV and BEV applications.
3. Explain the principal operation of a synchronous motor and discuss its advantages and disadvantages on HEV and BEV applications.
4. Explain the principal operation of a switched reluctance motor and discuss its advantages and disadvantages on HEV and BEV applications.
5. Explain the general structure of a battery electric vehicle.
6. Conduct research on the types of batteries used in electric vehicles and explain the reasons for their use.
7. Conduct research on available battery technologies other than those mentioned in the text, and compare their potential in electric vehicle applications.
8. Conduct research on available vehicles that use a flywheel as either their primary or secondary energy source.
9. Conduct research on the technologies used on flywheels, and their potential for application in electric vehicles.
10. Compare the structure of unidirectional and bidirectional DC–DC converters and explain their limitations.
11. Conduct research on other available fuel-cell technologies and compare the advantages and disadvantages with the fuel-cell technologies mentioned in the text.
12. Explain the function and structure of an E-CVT.
13. Explain the function and structure of an M-CVT.
14. Find the PHEVs available on the market and compare their characteristics.
15. What are the benefits and limitations of a hydraulic hybrid vehicle?
16. What are the benefits and limitations of a pneumatic hybrid vehicle?

References

1. Besselink, I., Van Oorschot, P., Meinders, E., and Nijmeijer, H. (2010) Design of an efficient, low weight battery electric vehicle based on a VW Lupo 3l. Paper presented at EVS-25, Shenzhen, China.
2. Chan, C.C., and Chau, K.T. (2001) *Modern Electric Vehicle Technology*, Oxford University Press, Oxford.
3. Zeroulia, M., Benbouzid, M.H., and Diallo, D. (2006) Electric motor drive selection issues for HEV propulsion systems: a comparative study. *IEEE Transactions on Vehicular Technology*, **55** (6).
4. Khaligh, A., and Li, Z. (2010) Battery, ultracapacitor, fuel-cell, and hybrid energy storage systems for electric, hybrid electric, fuel-cell, and plug-in hybrid electric vehicles: state of the art. *IEEE Transactions on Vehicular Technology*, **59** (6).
5. http://www.fueleconomy.gov/feg/evsbs.shtml.
6. Bellur, D.M., and Kazimierczuk, M.K. (2007) DC-DC Converters for Electric Vehicle Applications. Proceedings of Electrical Insulation Conference and Electrical Manufacturing, 2007.
7. Emadi, A., Williamson, S.S., and Khaligh, A. (2006) Power electronics intensive solutions for advanced electric, hybrid electric, and fuel-cell vehicular power systems. *IEEE Transactions on Power Electronics*, **21** (3).
8. www.energy.gov.
9. Ehsani, M., Gao, Y., and Emadi, A. (2010) *Modern Electric, Hybrid Electric, and Fuel-Cell Vehicles: Fundamentals, Theory, and Design*, 2nd edn. CRC Press, Boca Raton, FL.
10. http://www1.eere.energy.gov/hydrogenandfuelcells/fuelcells/fc_types.html.
11. Jian, L., and Chau, K.T. (2009) A novel electronic-continuously variable transmission propulsion system using coaxial magnetic gearing for hybrid electric vehicles. *Journal of Asian Electric Vehicles*, **7** (2), 1291–1296.
12. www.epa.gov.
13. Chan, C.C., Bouscayrol, A., and Chen, K. (2010) Electric, hybrid, and fuel-cell vehicles: architectures and modeling. *IEEE Transactions on Vehicular Technology*, **59** (2), 589–598.
14. Fazeli, A. (2011) Development of a novel air hybrid engine. PhD thesis. University of Waterloo, Canada.

3

Body and Chassis Technologies and Design

3.1 Introduction

At a first glance, discussing the body, chassis, and its associated systems may not seem relevant to electric vehicles (EVs) and hybrid electric vehicles (HEVs) because the primary differences between these vehicles and conventional internal combustion engine vehicles (ICEV's) relate to their powertrain systems. However, there are numerous challenges associated with the body, chassis, and chassis systems of electric and hybrid electric vehicles that should be noted separately.

One of the main challenges is optimizing energy consumption for EVs and HEVs, which have limited on-board energy. As such, constructing lightweight body, chassis, and chassis systems should be a primary focus of the design process. Likewise, accommodating a heavy and oversized battery pack into the chassis and body design is another challenge. For these reasons, it is important to study the design and technology behind the body, chassis, and chassis systems.

This chapter has two main objectives:

1. to provide general knowledge about the body, chassis and chassis systems of automobiles;
2. to highlight this information in relation to EVs and HEVs.

In order to achieve these objectives, this chapter addresses body chassis and chassis systems in the following sections discussing both general and specific information regarding EVs and HEVs.

3.2 General Configuration of Automobiles

As illustrated in Figure 3.1, a car can be divided into two main parts: the platform and the upper body. The definition of a platform varies based on the car manufacturer, but it generally refers to the chassis structural members, chassis systems, and the powertrain. The design and configuration of the platform includes the most significant technological, functional, and

Electric and Hybrid Vehicles: Technologies, Modeling and Control: A Mechatronic Approach, First Edition.
Amir Khajepour, Saber Fallah and Avesta Goodarzi.

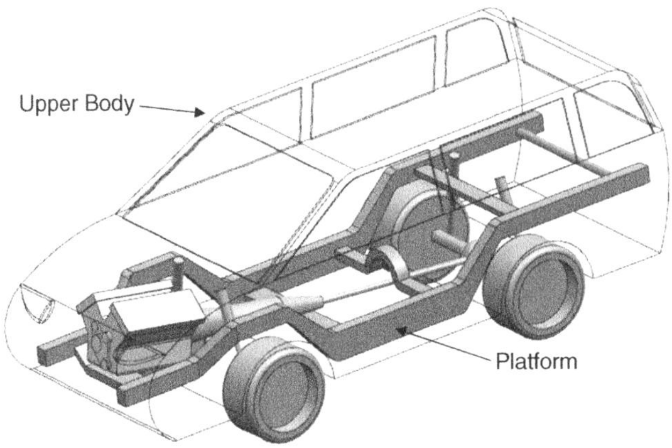

Figure 3.1 General configuration of an automobile

costly components of the vehicle. These components help determine the vehicle's general size, strength, and performance.

In some vehicles (mainly commercial and heavy-duty vehicles), the chassis is an independent ladder frame on which the body, chassis systems, and powertrain are directly mounted. This configuration is known as body-on-frame; and it allows the vehicle platform to be physically separated from the vehicle body as an independent system.

In contrast, the unibody construction of most passenger cars consists of a chassis structure integrated into the body floor. This configuration does not allow separation of the platform and body as two independent systems.

Designing and producing a platform are significantly more challenging, costly, and time-consuming, than developing an upper body. As a result, car manufacturers try to keep platform designs unchanged between different generations of cars, and share these designs between different car models. A modern car is often based on combining upper bodies of different styles and classes with a shared platform design. Platform consolidation significantly reduces the time and cost associated with vehicle development.

Example 3.1

The French car manufacturing group, PSA Peugeot Citroën spent €655 000 000 to develop a specific platform called PF2. The company then spent €442 000 000 more to develop a variety of upper body designs for the Peugeot 307, one of their most successful models. The estimated cost for the entire project was around €1 100 000 000.

Shortly after, PSA Peugeot developed another model known as the Citroën C4 on the PF2 platform at a cost of €442 000 000. As such, the company thus developed one car for €1 100 000 000, but was able to produce two cars for €1 500 000 000, averaging around €750 000 000 per car. If they were to produce a third distinctive model on this platform, the costs associated with each model would average only €500 000 000 [1].

As previously mentioned, the car platform includes the chassis structural parts and systems, as well as the vehicle powertrain. Whether the chassis is an independent ladder frame or an integrated unibody configuration, its structural members play an essential role in the strength, noise and vibration characteristics, and crashworthiness of the vehicle.

Chassis systems are directly connected to the chassis. They include the suspension, steering, and brake systems, and significantly affect performance features such as comfort, handling, and safety. Cars exhibit either comfort or sporty characteristics based on the design and adjustment of the chassis systems.

The powertrain produces and transfers driving power to the vehicle wheels; it also determines important factors such as vehicle acceleration, speed, gradeability, fuel consumption, and emission levels.

Lastly, the upper body forms the exterior and interior styles of the vehicle. It functions as a protective barrier for vehicle occupants, and also houses trim parts and body systems.

3.3 Body and Chassis Fundamentals

3.3.1 General Packaging

There are three different types of general packaging in automobiles: three-box, two-box, and one-box. The three-box configuration is the most popular, and is typically used by sedan cars. As seen in Figure 3.2(a), a three-box configuration vehicle consists of three individual boxes: the front box for powertrain and other mechanical components; the central box for passengers; and the rear box for luggage and cargo.

Hatchbacks, station wagons, sport utility vehicles and vans often use a two-box configuration. It features a separate box for powertrain and a common box for passengers and cargo (Figure 3.2(b)).

Some light commercial vehicles, such as full-size vans, are constructed in a one-box configuration. As Figure 3.2(c) illustrates, the vehicle body in this configuration is based on a single box dedicated to passengers and cargo, with the powertrain and other mechanical components located under the body floor.

Each box has its own specific design and packaging requirements. Since the front box primarily features the powertrain components, it is also known as the engine compartment. It contains strong longitudinal and lateral structural members that are required to carry the engine and transmission loads, as well as road excitation loads transmitted by the front suspension.

In addition to providing space and support for the powertrain, suspension, steering, and braking systems, in the event of a frontal car crash, the front box receives significant impact. Generally, car designers attempt to absorb as much crash energy as possible through the front box structural members to ensure that the central box remains safe for passengers. As such, the

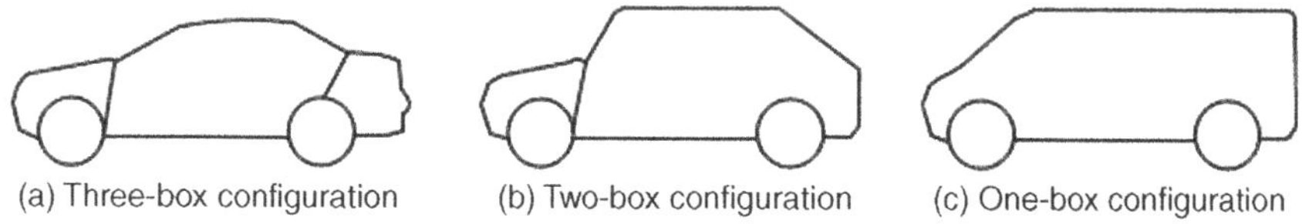

Figure 3.2 Three-, two- and one-box vehicle configurations

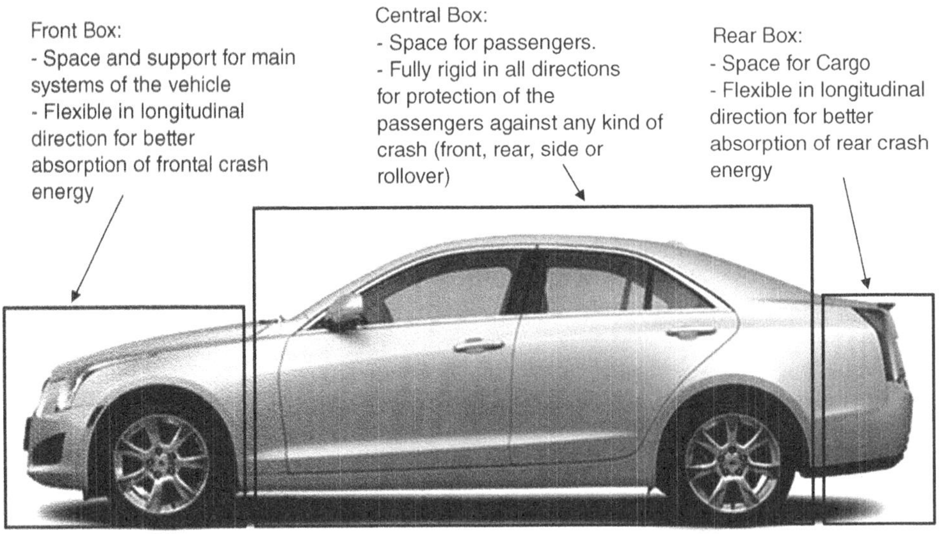

Figure 3.3 Different functions of front, rear and central boxes. *Source:* Reproduced by permission of General Motors of Canada

front box must be resilient against the bending and torsion loads generated by the powertrain and suspension. It must also be simultaneously flexible in longitudinal direction, for better absorption of crash energy. Designing a structural system that is rigid in some directions while simultaneously flexible in others can be a difficult engineering challenge for the vehicle body designer.

The design requirements of the rear box are similar to those of the front box. Like the front box, the bending and torsional strength and stiffness of the rear box should be high enough to handle loads from the rear suspension and cargo while concurrently absorbing any rear crash impact. As such, its longitudinal stiffness must be relatively low. Despite its similarities with the front box, the rear box is considerably simpler. Instead of a compact arrangement of mechanical and electrical systems such as powertrain, steering, braking, and cooling, the rear box only provides empty space for cargo (Figure 3.3).

The design and packaging requirements of the central box are fundamentally different from the front and rear boxes. The primary function of the central box is to host passengers. As such, it should be spacious and ergonomically sufficient enough to provide comfortable seating conditions for all passengers.

Traditionally, passengers occupy a large portion of the central box interior, with only a small portion of this space containing additional components, such as the cockpit and seats. Other parts, such as the exhaust pipe and fuel tank, are under the central box. The power transmission system of rear and all-wheel drive vehicles can be found under the central box floor as well.

Unlike the front and rear boxes, the central box must be stiff in all directions and capable of handling all types of load cases. During any kind of crash (front, rear, side, or rollover), the central box deformation must be kept to a minimum to protect passengers. While front and rear boxes must be flexible in longitudinal direction to absorb crash energy, the central box should not, as it protects passengers from crash impact.

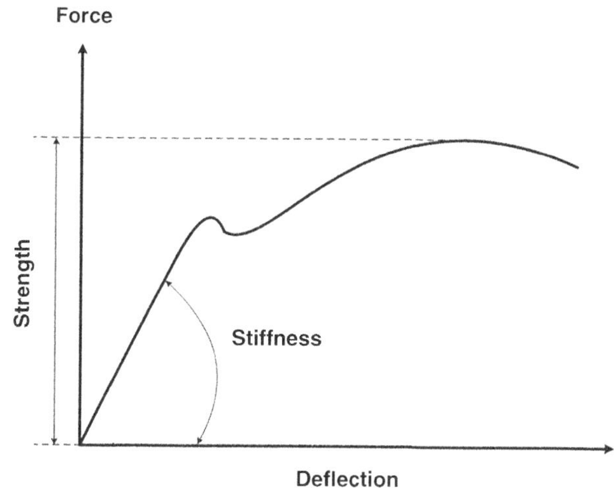

Figure 3.4 Graphical representation of strength and stiffness

3.3.2 Design Criteria

3.3.2.1 Strength

Strength is defined as the maximum load capacity of a given structure (Figure 3.4). The loads imposed on a vehicle result in stresses occurring throughout the vehicle structure. As such, under the worst possible load conditions, the stress placed on the vehicle structure must be lower than its admissible limits. Because of these limitations, preventing the vehicle structure from overstressing is a crucial design aspect of car body design. However, in most cases, factors such as buckling, fatigue failure, and joint failure determine the strength of vehicle structure, rather than the overstressing capacity of structural parts.

3.3.2.2 Stiffness

In addition to sufficient strength, the vehicle structure must also be adequately stiff. In fact, maintaining stiffness is even more important than strength in vehicle structure design. Inadequate structural rigidity can be detrimental to vehicle functionality and its other important performance aspects, such as ride comfort, handling, and noise and vibration behaviors. For example, doors cannot shut properly on a car that has low structural bending stiffness due to excessive deflections in the doorframes. It has a likewise negative impact on vehicle handling and stability due to insufficient torsional stiffness of the vehicle structure.

According to the principles of vehicle dynamics, the alignment of the wheels and the geometry of the suspension and steering must remain as is in the original design. This is to ensure predictable and safe vehicle handling. However, during instances, such as driving on uneven roads or undergoing severe maneuvers, the car body may twist excessively if the structural torsional stiffness is insufficient. This will cause the suspension geometry and wheel alignment to change unexpectedly.

In addition to the car body's bending and torsional stiffness, local stiffness of parts and panels is also important. If there is low local stiffness in the car body panels, the road and powertrain excitations will noisily resonate inside the car cabin.

3.3.2.3 Energy Absorption

As mentioned earlier, the structural elements of the front and rear boxes should be designed to absorb as much kinetic energy as possible during an accident. This improves the vehicle's crashworthiness and reduces the amount of dynamic force transferred to the vehicle occupants. Figure 3.5 can help generate a better understanding of how the structural elements can absorb crash energy. It illustrates a typical tensile stress–strain curve of brittle and ductile metals; the area under the curve is defined as the toughness. There is a direct relationship between the energy absorption on impact and the toughness.

Materials with high toughness absorb more energy when fractured. The value of toughness is directly related to the ultimate tensile stress of the material, as well as its maximum elongation. While it is important to consider improving the maximum stress of a structural element when trying to maximize energy absorption characteristics, having a greater maximum elongation is much more important. Thus, designing a strong but sufficiently flexible structural system for the front and rear boxes are the main design goals for car body engineers.

Note that because a side crash impacts the central box laterally, energy absorption is less important in terms of keeping passengers safe. Rather, the primary focus must be to design a body structure that is as rigid and nondeformable as possible.

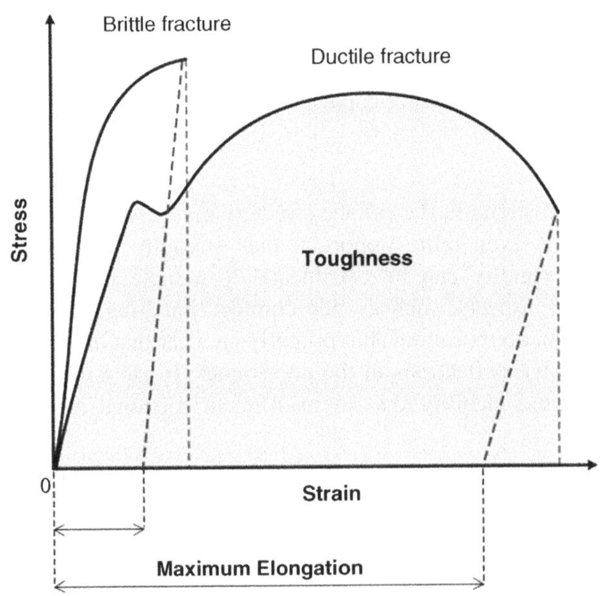

Figure 3.5 Typical tensile stress–strain curve of brittle and ductile metals

3.3.3 Design Loads

Due to the specific nature of vehicle service conditions, the load imposed on a vehicle structure is dynamic in nature. As such, to calculate the designed loads for a particular vehicle, the worst-case static load of each specific loading condition must first be determined. Following this, the load's dynamic behavior must be simulated by multiplying a specific empirical factor known as the dynamic load factor. Lastly, a safety factor is also included to compensate for unexpected overloading. The calculation formula is as follows [2]:

$$\text{Design Load} = \text{Safety Factor} \times \text{Dynamic Load factor} \times \text{Static Load}$$

The dynamic load factor and safety factor can vary, based on the type and expected functions of the vehicle. For example, off-road heavy-duty vehicles will have a greater numerical value in these factors than family sedans.

After calculating the design load, the vehicle structure can be analyzed using numerical or analytical static stress analysis techniques. The resulting maximum stress should be less than the maximum allowable stress determined by the corresponding failure theory. For low carbon steels that are widely used in the production of car bodies, the maximum allowable stress is generally 67% of yield stress [3]. Using the above procedure, designers can compensate for the dynamic nature of loads and can guarantee a sufficient fatigue life for the car body elements. However, most modern cars undergo additional fatigue failure analysis as well.

3.4 Different Types of Structural Systems

The car chassis and body can use different types of structural systems. The sections to follow discuss the most important types.

3.4.1 Body-on-Frame Construction

The body-on-frame construction is the traditional configuration of a car body structure, ever since the production of early automobiles. As Figure 3.6 illustrates, the car structure in this configuration is based on two distinct parts: the body and the chassis. The chassis is a ladder-shaped frame that carries important vehicular system components such as the fuel tank,

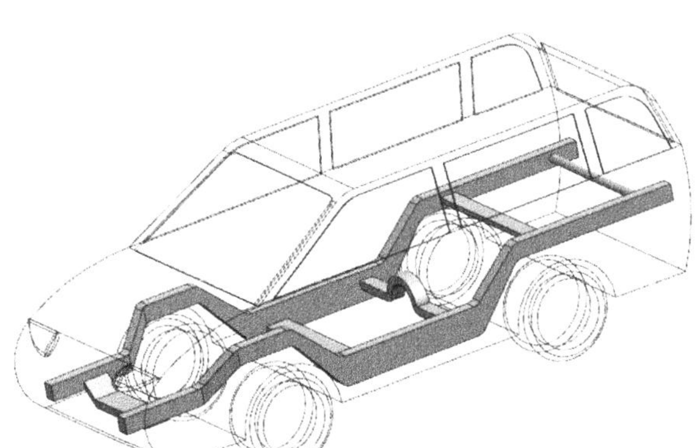

Figure 3.6 Body-on-frame construction

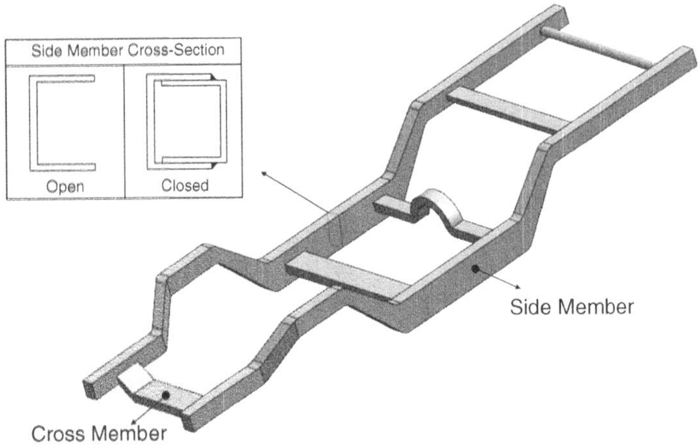

Figure 3.7 Typical ladder frame design

suspensions and axles, and the powertrain. The body includes the body shell and systems such as seats, trim parts, and other interior/exterior components. The body is mounted on top of the ladder frame.

The ladder frame is the primary structural member of the car and carries the loads of all of the aforementioned systems and parts. By contrast, the body is not theoretically considered to be a fully structural member, since its intended purpose is to provide space for passengers and cargo while forming the vehicle's exterior.

As Figure 3.7 shows, a ladder frame is constructed based on longitudinal and lateral members. Longitudinal members, also known as side members, maintain the bending stiffness and handle the bending loads applied to the chassis structure. The cross-section of the side members can be open or closed, based on the expected load-carrying capacity of the chassis. Closed section side members are stronger, but are more complicated and costly to build.

Lateral members, also known as cross-members, provide torsional stiffness for the chassis. How the cross-members attach to the side members heavily influences the torsional stiffness and strength of the chassis. Stiffer connections of cross-/side members can considerably improve the torsional performance of the chassis.

Most early automobiles adopted a ladder frame construction, but since the early 1960s, there has been a significant shift towards more modern construction (mainly unibody construction) for passenger vehicles. Nowadays, the body-on-frame construction is not used in any passenger car produced worldwide. The main reason for this shift is its heavy weight and high cost. Nevertheless, nearly all trucks, buses, and vehicles that need more strength and loading capacities still use this structural system.

3.4.2 Backbone Construction

Backbone construction can be considered as a type of body-on-frame construction. In this type of car body structure, the body and the chassis are still separate components. The distinguishing factor between the classical body-on-frame and the backbone construction is the shape of the chassis. Figure 3.8 depicts a backbone construction vehicle using a tubular backbone chassis, rather than a ladder-type frame. The body is usually composed of non-metallic materials such

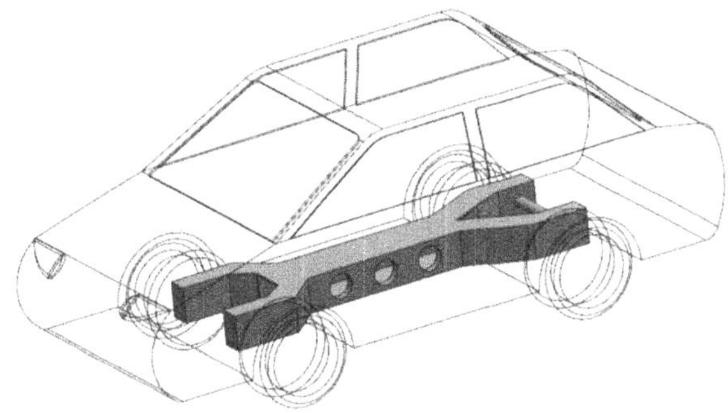

Figure 3.8 Backbone construction

as plastics and composites, and is mounted on top of the backbone chassis. The body does not have any structural performance, and the chassis provides almost all of the mechanical strength and rigidity.

The cross-section of the backbone tube is commonly rectangular, but can be circular as well. The inside space of the tubular backbone chassis can be used for the drive shaft in front-engine, rear-wheel drive vehicles. Some vehicles with backbone construction use a specific space frame, rather than the tubular member. This adjustment makes the chassis lighter and stronger.

Backbone construction is not a popular structural system, and only has limited use in low-volume productions of small sports cars. In addition to not being strong enough to support larger vehicles, the primary weaknesses of backbone construction are low torsional stiffness and strength, and ineffective crash protection. Backbone construction is especially fragile against lateral impact or off-set crashes.

Furthermore, backbone construction is not cost-effective for mass production. Due to its simple design, backbone construction is most suitable for hand-made cars and most economical for low-volume production. On the other hand, the backbone chassis is well-organized and space-efficient, with the chassis structure occupying only a small portion of the car's interior.

3.4.3 Space Frame Construction

As Figure 3.9 illustrates, a pure space frame chassis consists of several triangulated, tubular frames. It is a complex and resilient three-dimensional structure with considerably higher

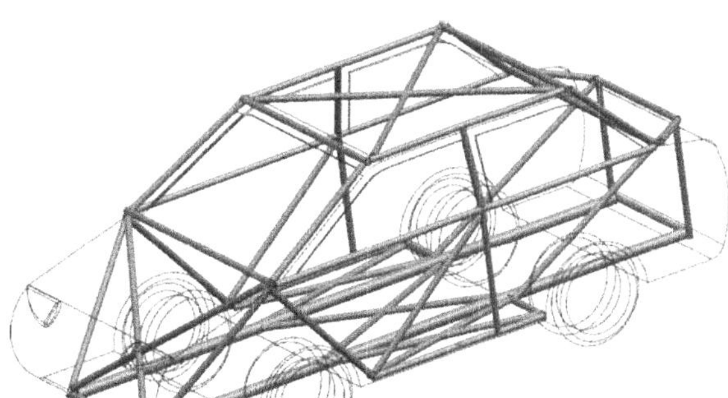

Figure 3.9 Space frame construction

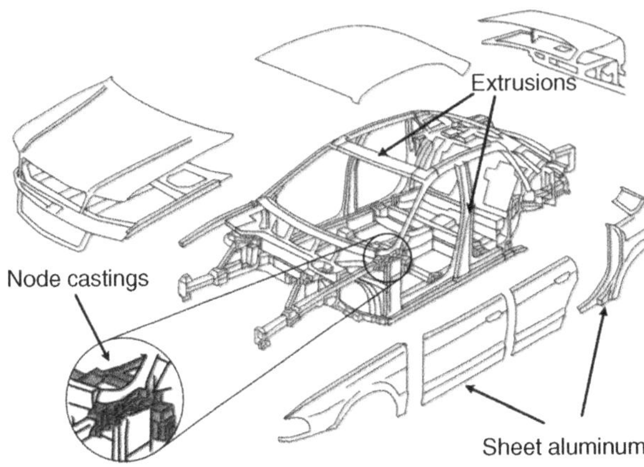

Figure 3.10 Space frame aluminum vehicle structure

rigidity than other types of vehicle constructions. In this type of design, the structural members resist compression or tension loads. Since the tubular members are much stronger against axial loads than bending loads, the design is very light and rigid.

Theoretically, the rigidity and strength of the chassis in a pure space frame construction should not be sensitive to joint stiffness, and the structural performance of the chassis should not suffer even if the joints can freely rotate. In practice, a pure space frame construction is impossible because frames, such as door openings and windshields, cannot use diagonal bracing members. As such, stiff joints are necessary in some instances, and some members should be designed to handle both bending and axial loads.

To form the vehicle outlook, a space frame chassis is usually covered by body panels. These panels are not structural members and can be made by light non-metallic materials such as fibreglass or plastic.

Ultimately, classical tubular space frames are generally strong and light, but are also complicated, costly, and time-consuming to build. As such, they are only limited to use in handmade sports and racing cars.

Recently, some auto-makers have introduced an advanced concept of the space frame chassis. In this method, extruded aluminum or steel members are connected to one another by rigid joint members made by casting in order to construct the chassis. Next, body panels that are stamped aluminum panels, non-metallic plastic or composite panels are used to cover the assembled space frame (Figure 3.10). Due to its complexity, currently only few manufacturers consider this type of space frame construction as an applicable method. However, the popularity of this method can increase in the future.

3.4.4 Unibody Construction

Nowadays, many mass-manufactured vehicles use unibody construction, in which the body and chassis form a single unit, rather than two separate units. This method uses metalwork techniques to press-make several shell elements and panels using stamping machines. These elements are subsequently joined together using spot and laser welding to form the car body, as shown in Figure 3.11. Unibody construction is also referred to as stressed-skin construction

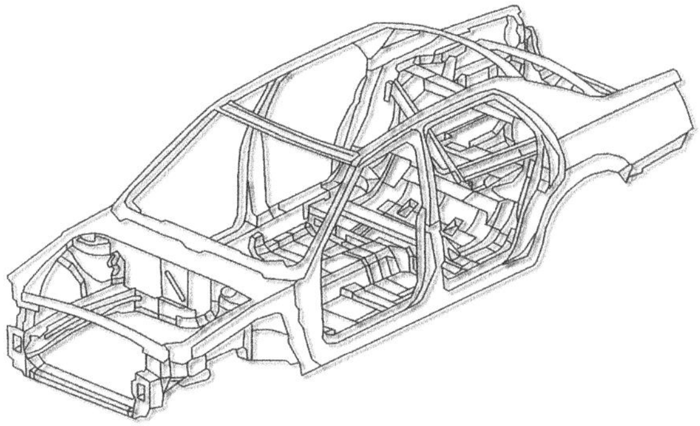

Figure 3.11 Unibody construction

because rather than ensuring overall strength and rigidity with a chassis frame, each body part provides structural support and strength to the entire vehicle.

The main advantages of unibody construction are the reductions in vehicle mass and production cost. This structure also exhibits better crash protection performance than other types of structures because of its greater ability to absorb energy during a collision.

However, there are also numerous disadvantages of unibody construction. First, while unibody vehicles are suitable for mass production, they are nearly impossible to produce on a small-scale production. The set-up cost for the tools, stamping, and molding machines are too expensive for specialized manufacturing. Furthermore, unibody construction provides enough strength and rigidity for most passenger cars, minivans, and even small sport utility and cross-over utility vehicles, but lacks the strength and rigidity for heavy-duty vehicles, such as trucks and buses.

Figure 3.12 depicts the main parts of a unibody structure, they are [4]:

- *Front rails:* These are two parallel box-section longitudinal members located alongside of the front box and extended to the central box at the floor level. The primary functions of front rails are to maintain bending stiffness for the car body, support loads of the engine and suspension mounts, and absorb crash energy in the event of a head-on collision.
- *Upper side members:* At the upper level, there are two smaller longitudinal members located between the firewall and front nose of the car, known as the upper side members. They help the front rails carry the engine and suspension mounting loads, and absorb crash energy from the front. Furthermore, upper side members also provide support for the fenders and the hood.
- *Front wheel arches:* This panel is stretched between the front rail, upper side members, and the firewall. It transmits the suspension load to the surrounding structural members.
- *Front cross-member(s):* In modern cars, a cross-member connects the end of the front rails to one another. This cross-member supports the front bumper and evenly distributes crash loads between the front rails. In some cars, other cross-members connect the upper side members.
- *Firewall:* This panel separates the front box from the central box. In modern cars, it is reinforced by open or closed box-sections. The firewall is strengthened by front pillars on the

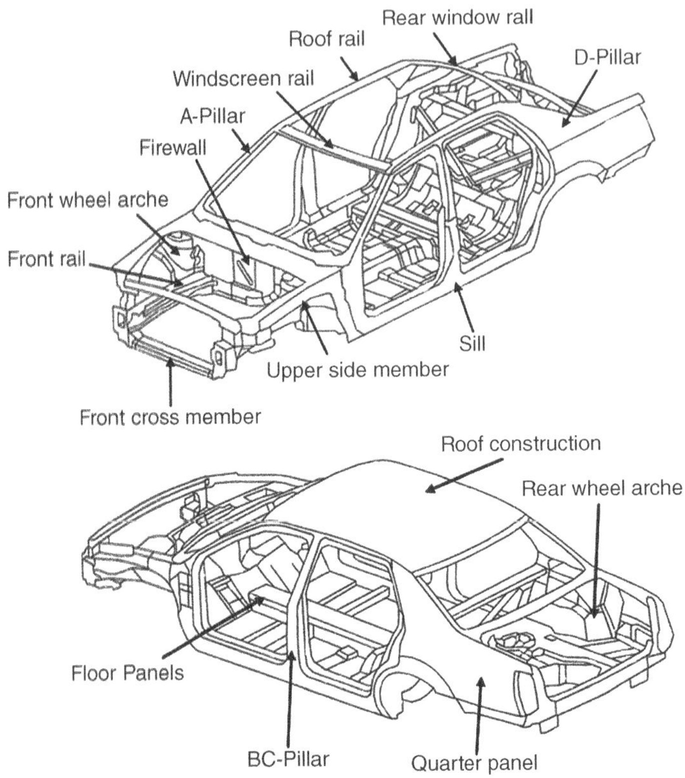

Figure 3.12 The main parts of a unibody construction

left and right sides, and behaves as a bracing member for the car body. The total torsional stiffness of the body is strongly influenced by the firewall design.

- *A-pillars:* These are also known as front pillars. Two A-pillars support the front end of the roof and enclose the windscreen. The front door hinges are also located on the lower part of the A-pillars. The joints between the A-pillar and the sill, roof rail, and windscreens rail must be strong and rigid enough to protect the passenger compartment from frontal crash impact.
- *Windshield rail:* This is a box-section lateral rail that connects the upper ends of the A-pillars. The windshield is supported by the windshield frame, which includes the upper part of the A-pillars, the firewall's upper reinforcing member, and the windshield rail. Interestingly, because the windshield is a non-metallic part, it acts as a bracing member for the windscreen frame and can considerably increase the body torsional stiffness.
- *Sills:* Sills, which help provide protection against lateral crash impacts, are longitudinal beams on each side that connect the lower ends of the roof pillar. They are laterally integrated with the floor panels, and are one of the main bending elements in the car body.
- *Roof rails:* Roof rails support the roof structure, and also connect the upper ends of the roof pillars on each side of the car body. They act as bending elements as sills to carry the bending

loads caused by the vehicle, passenger, and cargo weights. Roof rails must be rigid enough to protect passengers in the event of a rollover and roof crash.

- *BC-pillars:* The BC-pillar is a central pillar on each side of the car body that supports the roof and the rear door hinges. They are important structural members for side crash protection, and should be rigid and strong enough to withstand side crash loads with minimum deflection.
- *Floor panels:* Floor panels enclose the bottoms of the central and rear boxes. Typically, they are constructed from three parts: front floor panel, seat panel, and boot panel. The front floor panel is the first part of the floor assembly and is located between the firewall and second row seats. It usually includes a central tunnel, which was originally intended to provide space for the gearbox and drive shaft of front-engine, rear-wheel drive vehicles. However, because of its structural performance, this tunnel is also found in front-wheel drive vehicles. Using a deep central tunnel can considerably improve the bending rigidity and strength of the floor assembly. The seat panel forms a raised step to support the second row seat. Usually, the fuel tank is placed under the seat panel in the space provided by this step. The boot panel is the last part of the floor assembly; it covers the bottom of the trunk. Many narrow channels and ribs are pressed into the floor panels to improve their rigidity and their resistance to transmitted vibration. This helps decrease the intensity of the induced noise inside the car cabin.
- *Roof structure:* The roof is a simple panel that is welded to surrounding structural members including the windscreen rail, rear window rail, and roof rails. The roof structure is designed with slight longitudinal and lateral curvatures to make the roof panel stiffer. The design and rigidity of its corner joints have considerable influence on the torsional rigidity of the car body. Controlling the roof panel vibration can improve the internal noise of the car cabin. As such, one or more cross-members are used to support the roof panel in larger vehicles.
- *D-pillars:* D-pillars support the rear end of the roof.
- *Rear window rail:* This rail connects to the upper end of the D-pillars and supports the upper side of the rear window. In the case of hatchbacks, vans, and sport utility vehicles, the rear window rail also supports the hinges of the rear lift gate.
- *Quarter panel:* This side panel occupies the space between the rear door and the rear end of the car. Its upper side integrates with the D-pillar. The quarter panel and rear wheel arch form the rear wheelhouse.
- *Rear wheel arches:* These side panels form the inner portions of the rear wheelhouse on each side. If the rear suspension includes a strut, the real wheel arch supports the upper strut mounting point and carries a considerable part of the rear suspension load.
- *Rear rails:* These are two parallel box-section longitudinal members located alongside the rear box. They maintain bending stiffness and strength for the car body and absorb the rear crash energy.
- *Hinged members:* Each of the afore-mentioned members and parts is welded together to form an integrated one-piece structure known as "body in white" (BIW). Other parts, such as the side doors, hood, trunk door, and fenders, connect to the BIW with hinges or bolts. The hood, trunk door, and fenders can help absorb crash energy, while side doors can protect passengers from side crash impact. However, none of these parts have an influence on the overall torsional or bending structural characteristics of the car body. As such, they are non-structural members in a car body design.

3.5 Body and Chassis Materials

The ideal materials for car body applications require many specific properties. Some of the key requirements for ideal materials include: high yield and tensile stress for bearing loads; high modulus of elasticity for providing sufficient stiffness; high toughness for greater energy absorption; high elongation and formability; low weight; and low cost. Additionally, other desirable requirements are recyclability, paint-friendliness, and high corrosion resistance.

The subsequent section introduces and briefly discusses some car body application materials.

3.5.1 Low Carbon Steel

Manufacturers traditionally make the vehicle body and chassis from low carbon steels. Low carbon steels are ideal materials for stamping and sheet metal forming because they are deformable, with a maximum elongation of around 50%. Despite their manufacturing benefits, low carbon steels are not very strong. The yielding strength of most low carbon steel sheets is less than 150 MPa, with an ultimate tensile strength of less than 200 MPa.

As a result, designers have to use complex designs and reinforcement parts with complicated topologies to lower the stress on critical load-bearing elements, such as pillars and rails. This not only increases the weight of the chassis and car body, but also becomes more time-consuming and costly to manufacture. Nevertheless, low-carbon steel is optimal for low-stress parts, such as body panels.

3.5.2 Advanced High Strength Steels

Typically, high-strength steels have very low elongation and are not very suitable for sheet metal forming. In the last decade, there have been developments in a new generation of high strength steels for automotive applications. These steels make an ideal material for critical high-stress body parts because of their high strength and elongation.

There are two categories of advanced high strength steels: high-strength steels (HSS) and ultra-high strength steels (UHSS). The yield strength of HSS is between 210–550 MPa, with tensile strengths ranging from 270–700 MPa. Its maximum elongation is about 30%. Alternatively, UHSS possess yield strengths greater than 550 MPa, and tensile strengths greater than 700 MPa. Its maximum elongation is about 20%, which is still sufficient for the formability requirements of car body manufacturing.

The higher strength of these advanced steels allows the body and chassis parts to have simpler designs, reduced weights, and improved performance-related specifications, such as rigidity. However, advanced high strength steels are expensive and usually appear in certain high-stress parts, such as the front and rear rails, the central tunnel, or pillars and joints. On average, 50% –70% of the body and chassis structure of modern cars are made from advanced high strength steels, of which approximately 25% is UHSS. In addition to improvements in crash performance and structural rigidity, advanced steels can also reduce the vehicle structure weight by 15% –40% depending on its level of usage. This reduction in weight translates to approximately 50–100 kg for a typical passenger car.

3.5.3 Nonferrous Metals

Nonferrous metals, such as aluminum, magnesium, and titanium, have many characteristics that make them appropriate for use on car body applications. Perhaps the most suitable nonferrous metal for this purpose is aluminum, which is a formable material that is light, corrosion-free, and has excellent energy absorption and recyclability. Recently, automobile manufacturers such as Audi have considered aluminum for car body purposes and have introduced a few mass-produced vehicles with whole aluminum bodies onto the market.

Despite their advantages, the drawbacks of using aluminum car bodies include its high production costs and distinct manufacturing process. Therefore, having a full aluminum car body is a debatable matter. However, some luxury brands use or consider aluminum for specific car parts, such as the hood and doors. Magnesium and titanium are more advanced and expensive nonferrous metals. As such, their car body applications are even more limited than aluminum.

3.5.4 Nonmetallic Materials

Many modern cars use plastics and composite materials for non-structural applications, such as trim parts, bumpers, and exterior accessories. These materials are light, non-corrosive, and fatigue-free, making the production of complexly integrated parts much easier.

The mechanical properties of reinforced plastics and composites are considerably better than classic plastics. In some cases, they are even better than low carbon steels as the baseline material for the car body. Nevertheless, plastics and composites have numerous issues, such as poor recyclability, poor paintability, poor energy absorption, and sensitivity to humidity or sunlight. As such, their applications in modern car body structures are limited to closure parts, such as lift gates.

3.5.5 Multi-Material Approach in Car Body Design

The traditional use of low carbon steel sheets to create body and chassis parts has resulted in inadequate designs in terms of rigidity, weight reduction, durability, and crash performance. The body and chassis of modern cars, however, are designed by carefully selecting from various materials, such as low carbon steels, high strength steels, ultra-high strength steels, aluminum, and even plastics and composites. Typically, the composition of the body of a modern car is 50% low carbon steels, 40% high strength steels, and 10% ultra-high strength steels. In some luxury cars, a small percentage of body parts may be made using aluminum or composites.

Low carbon steels are generally used for low-stress body panels, while high strength steels are used for important parts, such as the front rails, rear rails, and upper side members. The characteristics of high-strength steels benefit the front rails, upper side members, and rear rails as these parts are strongly involved in the event of a frontal or rear crash. Ultra-high strength steels can be used to make A-pillars, central pillars, sills, and roof rails because they provide a rigid non-deformable central box for enhanced passenger protection.

As previously stated, a few high-end luxury cars use aluminum to produce hoods and lift gates.

3.6 Specific Considerations in Body and Chassis Design of Electric and Hybrid Electric Vehicles

While today's electric and hybrid electric vehicles use many car body design techniques of modern internal combustion engine vehicles, there are many other considerations that call for more specific design and packaging approaches for EVs and HEVs. The subsequent sections discuss these specific considerations and challenges.

3.6.1 Packaging

Packaging is a key challenge for chassis and body designers for EVs and HEVs. One of the major hurdles for engineers to overcome is the placement of a large battery pack(s). This issue is an even greater concern for pure electric vehicles, which need significantly larger battery packs.

Similarly, car body designers face a different challenge for HEVs powered by two independent electric and internal combustion power sources. Instead of accommodating large battery packs, they must ensure there is adequate space for both power sources without affecting the passenger and cargo spaces.

As a result, efficient packaging is essential for the body and chassis design of electric and hybrid electric vehicles. The next section further discusses the packaging problems described above.

3.6.1.1 Battery Pack Packaging

Since the battery pack is fairly heavy and sensitive, determining its placement requires consideration of the following factors:

- *Structural factors:* The weight of the battery pack should be supported by the main longitudinal and lateral structural members of the chassis. Generally, the front and rear rails, sills, and central tunnel are capable of supporting the battery pack load. As such, the battery pack should be located as close to these elements as possible. Furthermore, designers need to ensure that redesigning or reworking the car body for battery pack placement does not adversely affect the overall bending and torsional performance of the car body structure.
- *Protection against crash impact:* The high sensitivity and cost of batteries make it important to protect them against crash impact. Designers should also consider the ability of batteries to harm passengers in the event of a collision. First, battery packs are capable of penetrating the passenger compartment area due to their concentrated heavy mass. Second, batteries may explode or catch fire during a crash. Thus, it is important to place the battery pack away from the front, rear, and side crash zones. While the central box is the ideal location to host the battery pack because of its safe location, it is sometimes impossible to place the battery pack there due to other design considerations.
- *Handling and stability:* To improve the vehicle handling and rollover threshold, the vehicle's center of gravity must be kept close to the road surface. This can be achieved by installing the battery pack as low as possible. Moreover, the location of the battery pack can also strongly affect the longitudinal and lateral position of the vehicle's center of gravity. As such, engineers must carefully determine the longitudinal and lateral position of the battery pack to maintain the appropriate balance between the normal force of the front/rear and left/right wheels.

- *Protection against the road effects:* The battery pack must be able to withstand road-splashing and denting. Moreover, the location of the battery pack must be as far above the road as possible, or shielded.
- *Thermal factors:* It is important to ventilate batteries due to their propensity to generate heat. It is more efficient to place them in an area with natural ventilation, rather than one that consumes energy through forced ventilation.
- *Accessibility factors:* The battery pack should be easily accessible for replacement, inspection, or repair.
- *Ergonomic factors:* The space dedicated to the battery pack should not adversely affect the ergonomic design of the car or occupy the projected passenger and cargo space.
- *Manufacturing and economic factors:* There are two approaches for designing and manufacturing electric and hybrid electric vehicles. The first approach converts and redesigns the body and chassis of internal combustion engine vehicles to meet the requirements of electric and hybrid powertrains. However, designers cannot make extensive changes (such as redesigning the main platform parts) to the existing body and chassis design due to financial considerations.

  While this method does reduce the expenses of redesigning, it prevents the development of an optimal vehicle design. The inability to make extensive modifications inhibits designers from placing battery packs in ideal locations. Due to these numerous limitations, design goals for converted vehicles should focus on finding a satisfactory solution, rather than an optimal one. On the other hand, some EVs and HEVs are manufactured based on original designs or extensive changes made to an existing design. In such vehicles, it is possible to place the battery pack without consideration of the limitations mentioned above.

Note that some of the factors mentioned above are contradictory in nature, making it difficult to satisfy all of them. This is especially pertinent when considering that limited feasible alternatives exist. According to a survey on production and concept EVs/HEVs, feasible alternatives for battery pack locations are as follows (Figure 3.13).

- *Front box:* In early pure electric vehicles, and even in today's low-production converted vehicles, battery packs usually occupy some or all of the front box space (Figure 3.13(a)). In most cases, the electric motor is installed in the front box as well. However, this alternative is not applicable to HEVs, since the electric motor and internal combustion engine of these vehicles are usually placed in the front compartment space.

  Mounting the battery pack on the front rails of the front box offers several advantages:
  - The battery pack is easily accessible and naturally ventilated.
  - The car body structure does not need to be extensively modified.
  - There is no effect on the structural performance of the car body.
  - There is no reduction in the amount of space dedicated to passengers or cargo.

  However, despite these advantages, the front box is not the ideal location for a battery pack. In this location, the batteries are more prone to impact in the event of a frontal crash. Moreover, its installation on top of the electric motor can adversely change the vehicle's center of gravity.
- *Rear box: Inside of the trunk:* (Figure 13.3(b)). Placing a battery box inside of the car trunk has long been a common alternative method to host batteries. Battery boxes are effectively

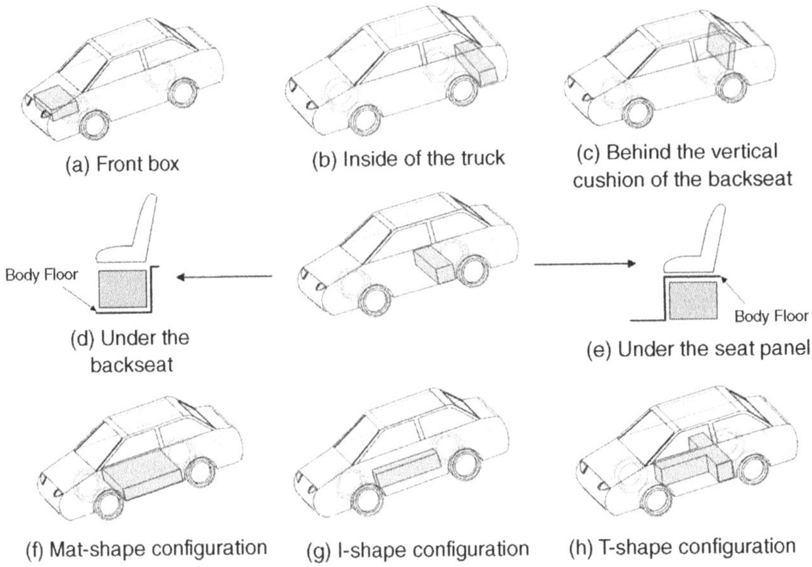

Figure 3.13 Different feasible alternatives for battery pack location

supported by rear rails and are easily accessible. The primary disadvantage of placing battery packs in the rear box is its inability to protect the batteries in rear crash events. Moreover, the mass of the battery box can shift the vehicle's center of gravity to the rear, adversely affecting the vehicle's handling and stability. Early EVs and HEVs placed the battery box under the trunk floor to maximize cargo space. These types of designs restrict the mobility and hill-climbing ability of the vehicle because of the battery box geometry. Some modern electric and HEVs, such as the Toyota Prius V, still install battery packs inside of the trunk. In such cases, batteries are usually over the trunk floor and positioned behind the back seat (Figure 3.13(b)). This keeps the batteries away from the rear crash zone. Placing the battery pack within the trunk can be an adequate choice for hybrid vehicles, because they use smaller battery packs that do not reduce cargo space. In contrast, pure electric and plug-in hybrid vehicles need a larger battery pack. Putting these batteries inside the trunk drastically reduces the amount of cargo space available, making it a poor choice for smaller vehicles. Larger hatchbacks, lift backs, station wagons, and SUVs may have marginally acceptable amounts of cargo space remaining.

- *Behind the vertical cushion of the backseat:* Figure 3.13(c) depicts a battery pack attached to the structural element located behind the vertical cushion of the backseat. This element is only in sedan cars and it divides the central and rear boxes; as such, this battery packaging method is only applicable to sedans. The Honda Civic hybrid is a well-known vehicle that uses this configuration. In general, this is a restricted area, adequate only for small battery packs of hybrid vehicles. Although this location is far from the rear crash zone, side crash impact can still affect it. Installing the battery in this location does not require extensive structural redesign. A vertical installation of the battery pack, however, may adversely elevate the height of the vehicle's center of gravity, though this is not a significant concern for small and light battery packs.

- *Central box:* As previously mentioned, the central box is the safest place for the battery pack because the batteries are well protected from both front and rear crashes. Most electric and HEVs that are produced using redesigned platforms use the central box to host their battery packs. Most designers echo the sentiment that the central box is the ideal location for a battery pack. In addition to crash protection, placing a battery pack close to the passenger compartment floor has the added benefit of not disturbing the weight distribution of the vehicle. In fact, even handling and rollover behavior would improve because of lowering the vehicle's center of gravity. Moreover, the floor structure and surrounding structural elements of the central box can adequately support the battery load. There are several areas to place the battery pack within the central box, as discussed below.
- *Under the backseat:* In some hybrid electric vehicles, such as the Toyota Prius C, the Toyota Highlander Hybrid, and the Lexus RX Hybrid, the battery pack is under the backseat of the passenger compartment. While this location cannot provide enough space for large battery packs, it can be satisfactory for the smaller battery packs of hybrid vehicles. Installing a battery pack under the backseat requires changing the shape of the seat panel. Specifically, the seat panel should be flattened or reduced in height in order to accommodate the battery pack between the seat panel and the backseat, as depicted in Figure 3.13(d). While batteries are well protected there from front and rear crashes at this location, they are still vulnerable to side crash impact. Auto-makers attempt to compensate for this insufficiency in different ways. For example, the Toyota Prius C utilizes a protection structure to shield batteries in the event of a side crash. There are still negative aspects to placing battery packs under the backseats. First, it reduces head and leg space, making the backseat uncomfortable for passengers. It can also disorganize the ergonomic design of the second row seats. Second, battery ventilation becomes a challenge because it is difficult to access fresh air from this location.
- *Under the seat panel:* It is possible to install the battery pack under the seat panel, a location typically reserved for the fuel tank (Figure 3.13(e)). The value of this location is similar to battery packs placed under the backseat in terms of crash protection. However, this area offers better ventilation and allows easier access from the underbody. The battery pack should be well shielded against road splashing and similar damages.
- *Mat-shape configuration:* The mat-shape configuration is ideal for pure electric vehicles. As shown in Figure 3.13(f), the underbody is mounted by a large box containing many batteries, which occupies the entire floor area of the passenger compartment. Such a big battery box meets the demands of pure electric vehicles because it can provide a higher energy capacity. The floor structure should be designed from scratch in this configuration. This is because the structure needs to be able to host such an unusually hefty battery pack and handle its heavy load. However, the mat-shape does not provide adequate protection against side crashes, especially when compared to other configurations. Furthermore, implementing the mat-shape configuration may disturb the ergonomic design of the car. Based on the elevation of the battery box, the height of the passenger compartment floor may be adversely increased in order to meet minimum road clearance standards. This increase in height can make it difficult for passengers to enter and exit the vehicle. It is especially challenging for small cars made for the city, which is a key production target for pure electric vehicles. Nissan Leaf uses this approach for battery pack packaging.
- *I-shape configuration:* This configuration uses a wider and deeper central tunnel to accommodate an I-shaped battery pack. This pack attaches to the underbody and is

accessible from the bottom of the vehicle (Figure 3.13(g)). The central location of the battery pack makes it well protected from front, rear, and side crashes, providing better crash protection than the available alternatives. While the space provided by the over-sized central tunnel can support a hybrid vehicle battery pack, it is not satisfactory for the battery packs of plug-in hybrids or pure electric vehicles. The advantage of the large central tunnel is its ability to strengthen the floor structure bending. It is also capable of handling battery pack load to assist the cross-members, which transmit the load to the sills. The primary disadvantage of the I-shape configuration is that the over-sized central tunnel can reduce the seating space of the front and rear passengers. Specifically, backseat passengers can only sit beside the central tunnel, reducing the total sitting capacity of the vehicle to four.

- *T-shape configuration:* This configuration places a T-shaped battery pack in a space that mirrors its shape (Figure 3.13(h)). This space is formed by combining the area under an over-sized central tunnel with the space under the seat panel; it is sufficient for large battery packs capable of powering pure electric and plug-in hybrid vehicles. The T-shape configuration was first used by the GM EV1, and is currently in the GM Volt, a plug-in hybrid vehicle. The implementation of a T-shaped configuration requires a dedicated platform and is only applicable to vehicles with an original design. The crash protection performance of the T-shape configuration is satisfactory, much like that of the I-shape. The difference between the two is that in this configuration, some of the batteries are under the seat panel, thus making them more susceptible to side crash impact. The T-shape battery pack is easily accessible from the bottom of the vehicle and should be shielded against road effects. Similar to the I-shape configuration, the T-shape also adversely affects passenger seating. Despite its disadvantages, the T-shape can meet most of the requirements of modern electric and hybrid electric vehicles.

To conclude, while most of the feasible battery packaging alternatives have been discussed above, they are not limited to just these configurations. Any combination of these as well as other innovative alternatives can be considered.

Example 3.2

Consider a conventional vehicle with the following specifications (Table 3.1).

Convert this conventional vehicle to an HEV with the given components in Table 3.2. Table 3.3 summarizes the battery cell specifications.

Table 3.1 Conventional vehicle specifications

Dimension (L × W × H)	4400 mm × 1750 mm × 1450 mm
Track	1500 mm
Wheelbase	2700 mm
Powertrain layout	Front engine, front-wheel drive
Engine power	110 kW

Table 3.2 HEV components' specifications

Powertrain layout	Front engine and electric motor, front-wheel drive
Engine power	60 kW
Electric motor power	45 kW
Battery energy capacity	2 kWh

Table 3.3 Battery cell specifications

Type	Nickel-Metal Hybrid
Capacity	6.5 Ah
Voltage	7.2 V
Energy	46.8 Wh
Dimension (L × W × H)	280 mm × 20 mm × 120 mm
Volume	0.67 L
Mass	1 kg

Table 3.4 Approximate available space in each part of the vehicle

Section	Available space (L)
Front Box	600
Rear Box, inside trunk	200
Rear Box, behind vertical cushion of backseat	80
Central Box, under backseat	80
Central Box, Mat-shape	400
Central Box, I-shape	60
Central Box, T-shape	130

Find the battery pack size and mass, and then evaluate the effect of each possible battery location on the vehicle CG location and weight distribution. Table 3.4 summarizes the approximate available space in each location to place the battery. The CG coordinates of the vehicle components as well as CG coordinates of the battery pack in possible locations are shown in Table 3.5.

Solution

$$Number\ of\ battery\ cells = \frac{Given\ total\ battery\ pack\ energy\ capacity}{Battery\ cell\ energy} = \frac{2000\ \text{Wh}}{46.8\ \text{Wh}} \simeq 43$$

In order to provide given energy for the vehicle, 43 battery cells should be connected in series.

$$Volume\ of\ batteries = 43\ \times 0.67 = \ 28.8\ \text{L}$$

$$Mass\ of\ batteries = 43\ \times 1 = \ 43\ \text{kg}$$

Table 3.5 Weight and CG coordinates of components

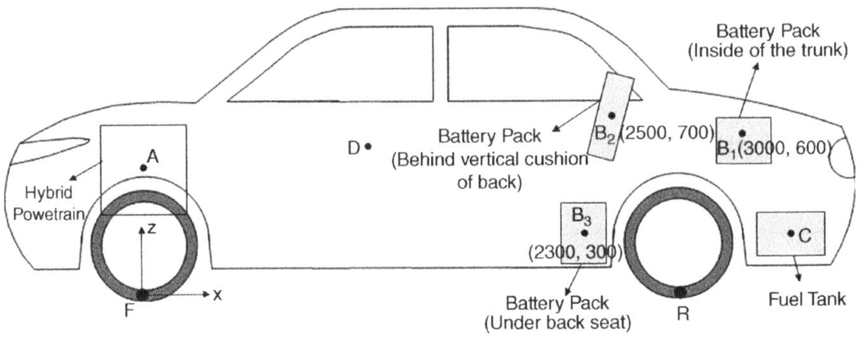

Components	Mass [kg]	CG coordinates (x_i, z_i) [mm]
Vehicle without powertrain	1100	D (1215,670)
Hybrid powertrain (Electric motor, engine, transmission, etc.)	150	A (0,600)
Engine and transaxle (Conventional vehicle)	200	A (0,600)
Full fuel tank	40	C (2900,350)

Note that the battery pack consists of some other components, such as frame, wiring, and control module. Considering that there is a 40% increase in volume and 30% increase in mass of the battery pack, the total volume and weight of a battery pack are equal:

$$Volume\ of\ beattery\ pack = 28.8 \times 1.4 = \ 40.3\,\mathrm{L}$$

$$Mass\ of\ battery\ pack = 43 \times 1.3 = \ 55.9\,\mathrm{kg}$$

In the first case where the battery pack CG is in option B_1 in Table 3.1, the CG location of the vehicle can be calculated from:

$$x_c = \frac{\sum m_i x_i}{M},\ z_c = \frac{\sum m_i z_i}{M}$$

where M is total weight of the vehicle, (x_c, z_c) are CG coordinates of the vehicle, m_i is the mass of each component, and (x_i, z_i) are their CG coordinates. Therefore,

$$x_c = \frac{(1100 \times 1215) + (150 \times 0) + (40 \times 2900) + (55.9 \times 3000)}{(1100 + 150 + 40 + 55.9)} = 1203.8\ \mathrm{mm}$$

$$z_c = \frac{(1100 \times 670) + (150 \times 600) + (40 \times 350) + (55.9 \times 600)}{(1100 + 150 + 40 + 55.9)} = 649.8\ \mathrm{mm}$$

Table 3.6 CG location of HEV with different battery configurations

Battery Locations	Battery Pack CG [mm]	Vehicle CG [mm]	Weight Distribution (Front/Rear) [%]	Description
Hybrid electric vehicle				
Front box	N/A	N/A	N/A	Available space in the front box is occupied by the engine and electric motor.
Inside of the Trunk	B_1 (3000,600)	(1204,650)	55.4/44.6	
Behind vertical Cushion of back-seat	B_2 (2500,700)	(1183,654)	56.2/43.8	
Under backseat	B_3 (2300,300)	(1175,637)	56.5/43.5	
Mat-shape	N/A	N/A	N/A	These locations need significant changes in the floor structure and are not usually chosen for mounting battery packs in HEVs especially if they do not need a large battery pack.
I-shape	N/A	N/A	N/A	
T-shape	N/A	N/A	N/A	
Conventional Vehicle				
—	—	(1084,650)	59.9/40.1	

Weight distribution, which is the percentage of weight in the front and rear axles is obtained by:

$$Mgb - m_f gl = 0, Mga - m_r gl = 0$$

$$m_f = \frac{Mb}{l}, m_r = \frac{Ma}{l}$$

$$Front\ axle\ load\ [\%] = \frac{m_f}{M} \times 100 = \frac{b}{l} \times 100$$

$$Rear\ axle\ load\ [\%] = \frac{m_r}{M} \times 100 = \frac{a}{l} \times 100$$

where a and b are the distances from CG to the front and rear wheels, respectively, l is the wheelbase of the vehicle, and m_f and m_r are portions of vehicle mass on front and rear axles. Therefore,

$$Front\ axle\ load[\%] = \frac{(2700 - 1203.8)}{2700} \times 100 = 55.4\%$$

$$Rear\ axle\ load[\%] = \frac{1203.8}{2700} \times 100 = 44.6\%$$

Table 3.6 provides details of the CG location of the HEV with different battery locations, as well as the CG of the conventional vehicle.

According to the results, different battery locations do not make a significant change in the weight distribution and CG height of the HEVs. This is because battery packs are relatively small in HEV's and do not make a significant change in the mass distribution of the vehicle.

Example 3.3

Convert the conventional vehicle in Example 3.2 to an electric vehicle with the given components in Table 3.7. Battery cell specifications are summarized in Table 3.8.

Find the battery pack size and weight, and evaluate the effect of each possible battery location on the vehicle CG location and weight distribution. Table 3.9 summarizes the CG coordinates of the vehicle components as well as the CG coordinates of the battery pack in possible locations.

Solution

$$Number\ of\ battery\ cells = \frac{Given\ total\ battery\ pack\ energy\ capacity}{Battery\ cell\ energy} = \frac{20000\ \text{Wh}}{495\ \text{Wh}} \simeq 40$$

Table 3.7 EV components' specifications

Powertrain layout	Front motor, Front wheel drive
Motor power	80 kW
Battery energy capacity	20 kWh

Table 3.8 Battery cell specifications

Type	Lithium-ion
Capacity	66 Ah
Voltage	7.5 V
Energy	495 wh
Dimension (L × W × H)	300 mm × 220 mm × 35 mm
Volume	2.31 L
Mass	3.8 kg

Table 3.9 Weight and CG coordinates of components

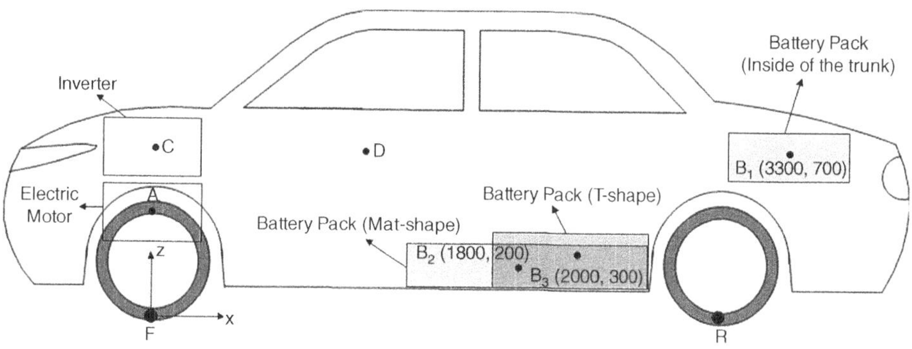

Components	Mass (kg)	CG coordinates (x_i, z_i) (mm)
The vehicle without powertrain	1100	D (1215,670)
Electric motor and transmission	70	A (0,550)
Inverter	20	C (0,700)

In order to provide given energy for the vehicle, 40 battery cells should be connected in series.

$$Volume\ of\ batteries = 40 \times 2.31 = 92.4\,\mathrm{L}$$

$$Mass\ of\ batteries = 40 \times 3.8 = 152\,\mathrm{kg}$$

With 40% increase in volume and 30% increase in mass, the battery pack volume and mass will be:

$$Volume\ of\ battery\ pack = 92.4 \times 1.4 = 129.4\,\mathrm{L}$$

$$Mass\ of\ battery\ pack = 152 \times 1.3 = 197.6\,\mathrm{kg}$$

Table 3.10 summarizes the CG location of the EV with different battery locations, as well as the CG of the conventional vehicle.

According to the results, mounting the battery pack in the trunk of the EV is unsuitable because the rear axle of the vehicle becomes heavier than the front, which can downgrade handling. On the other hand, for the mat-shape and the T-shape, the battery location has a lower CG height, which is a desirable feature for the vehicle handling and stability.

Table 3.10 CG location of EV with different battery configurations

Battery configuration	Battery pack CG [mm]	Vehicle CG [mm]	Mass distribution (front/rear) [%]	Description
Electric vehicle				
Front box	N/A	N/A	N/A	Available space in the front box is occupied by the electric motor and inverter.
Inside of the Trunk	B1 (3300,700)	(1433, 669)	46.9/53.1	
Behind vertical Cushion of backseat	N/A	N/A	N/A	Enough space is not available.
Under backseat	N/A	N/A	N/A	Enough space is not available.
Mat-shape	B2 (1800,200)	(1219, 597)	54.8/45.2	
I-shape	N/A	N/A	N/A	Enough space is not available.
T-shape	B3 (2000,300)	(1248, 612)	53.8/46.2	
Conventional vehicle				
—	—	(1214, 634)	59.9/40.1	

Example 3.4

As a simple approach to evaluate the structural performance of a vehicle, it can be modeled by a beam and the components and the body weight are respectively modeled by concentrated and distributed loads. Table 3.11 shows the beam model of a conventional and two different types of electric vehicles.

Use the information provided in Table 3.12 and draw the diagrams of applied shear force and bending moment to the vehicle structure. Compare the results of EVs with those of the conventional vehicle and discuss the results. Can we directly use the calculated loads for the vehicle's structural design?

Table 3.11 Beam models of a conventional and two different types of electric vehicles

Vehicle	Beam model
Conventional	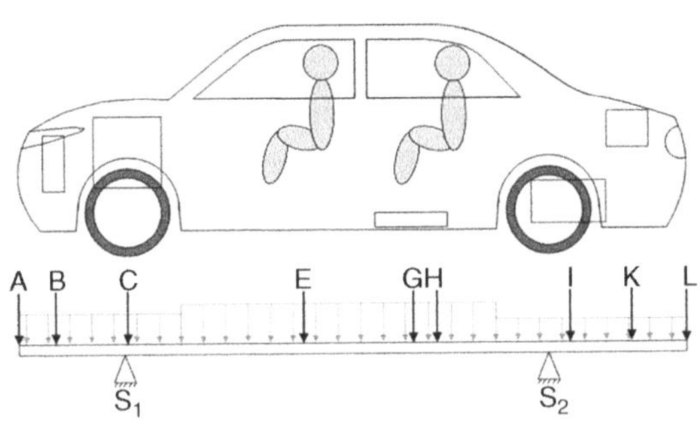

Vehicle	Beam model
EV (Battery pack in the trunk)	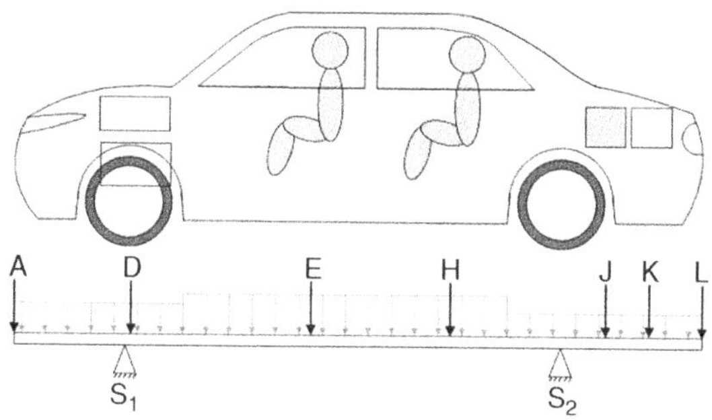
EV (Mat-shape battery pack)	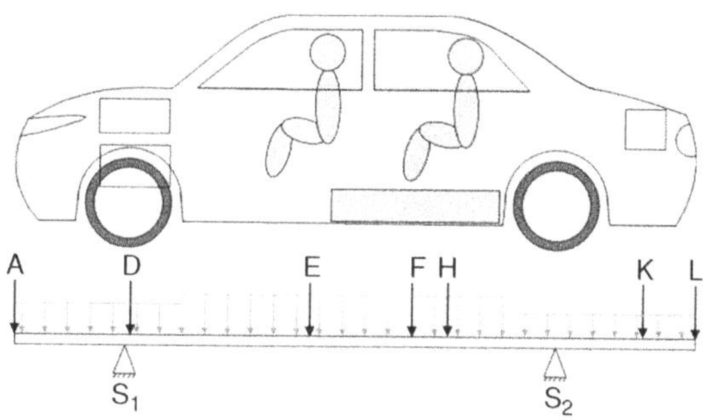

Table 3.12 Location and magnitude of applied loads to the vehicles

Components	Distance from point A (mm)	Load
Front bumper	A(0)	130 N
Radiator	B (300)	60 N
Engine and transmission	C (700)	1900 N
Motor and inverter	D (700)	900 N
Front passengers	E (2100)	1700 N
Batteries (Mat-shape)	F (2500)	1950 N
Exhaust	G (3000)	150 N
Rear passengers	H (3100)	2200 N
Fuel tank	I (3600)	400 N
Batteries (in the trunk)	J (4000)	1950 N
Luggage	K (4100)	500 N
Rear bumper	L (4400)	140 N
Body weight (Front box)	0–1500	2.4 N/mm
Body weight (Central box)	1500–3300	2.7 N/mm
Body weight (Rear box)	3300–4400	1.5 N/mm
Front axle	S_1 (700)	Front and rear axles reaction forces (can be calculated from the other loads)
Rear axle	S_2 (3400)	

Solution

Figures 3.14 and 3.15 show the applied shear force and bending moment of the vehicle structure for different types of vehicles. As can be seen, bending moment in the central box of the EV with mat-shape battery pack is increased by 32.4% compared to that of the conventional vehicle, while the value for the EV with in-the-trunk battery pack is decreased by 20%. However, when batteries are located in the trunk, there is 82.6% increase in the bending moment in rear box. More details are provided in Table 3.13.

The shear force and bending moment diagrams are found based on static loads, but, as mentioned before, we should multiply them by dynamics load factor and safety factor before using them for design. In this case, recommended values are 3 and 1.5, respectively.

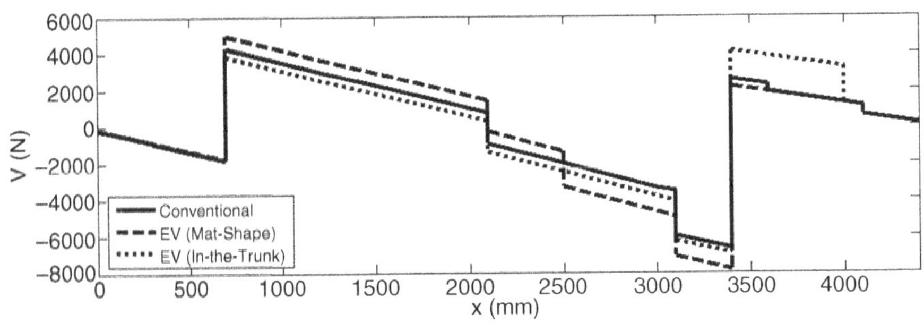

Figure 3.14 Applied shear force to the body structure

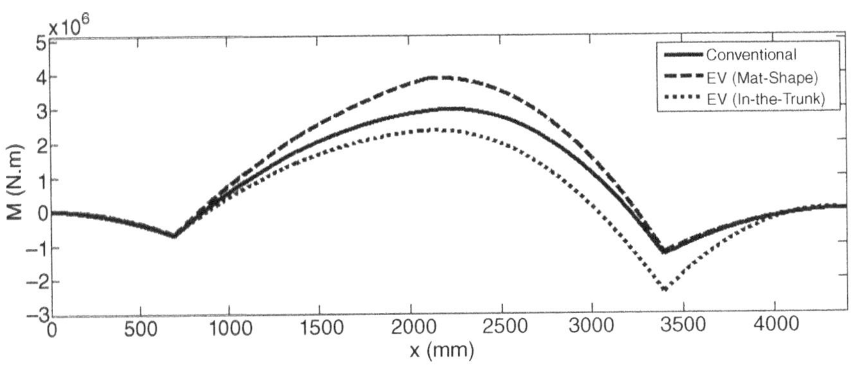

Figure 3.15 Applied bending moment to the body structure

Table 3.13 Maximum shear force and bending moment in different parts of the vehicles' body

Vehicle type	Maximum bending moment in the front box [Nm]	Maximum bending moment in the central box [Nm]	Maximum bending moment in the rear box [Nm]
Conventional	−7.03e+5	2.90e+06	−1.32e+06
EV (Mat-shape)	−6.79e+5	3.84e+06	−1.24e+06
EV (In-the-trunk)	−6.79e+5	2.32e+06	−2.41e+06

3.6.1.2 Powertrain Packaging

In modern electric vehicles, one or more electric motors replace the internal combustion engine. The gearbox is either eliminated or relegated to a small fixed-ratio gearbox. Due to the specific torque-speed performance curve of electric motors, EVs do not usually need the multi-speed gearboxes of conventional cars.

An electric powertrain includes an electric motor, its incorporated controller, and a transmission system. Generally, a typical electric powertrain system occupies less space than a typical internal combustion engine powertrain. Smaller electric powertrain systems can provide additional space for other components such as batteries.

Electric powertrain systems are even more compact in motorized wheel electric vehicles, especially those equipped with in-wheel motors. Such vehicles do not utilize a large concentrated motor; rather, two or four smaller motors are used inside or next to each wheel. In addition, it eliminates the entire transmission system, including the gearbox, differential, and drive shafts. For this reason, motorized wheel systems are occasionally known as transmission-less systems.

Powertrain packaging for hybrid electric vehicles is challenging as these vehicles contain two power units within their powertrain. As a result, the car body must accommodate the additional power unit without compromising the structural integrity of the body. Hybrid vehicles still use the basic components of conventional powertrains, including an internal combustion engine, a fuel tank, and a complex transmission system. As such, powertrain packaging becomes a difficult task, as two powertrain systems need to fit into the space of one.

A practical solution for this problem is to find more efficient down-sized engines, electric motors, controllers, and other subsystems. Most modern hybrid vehicles follow the approach suggested above and place all the powertrain components in the front box space with very tight packaging. The maintenance and inspection of such vehicles, however, are difficult.

The difficulties of powertrain packaging of hybrid vehicles can be resolved by using other powertrain layouts, such as a combined motorized wheel configuration. As Figure 3.16 depicts, it is possible to form a hybrid powertrain layout based on an internal combustion engine that drives the front wheels and two electric motors that drive the rear wheels independently. As

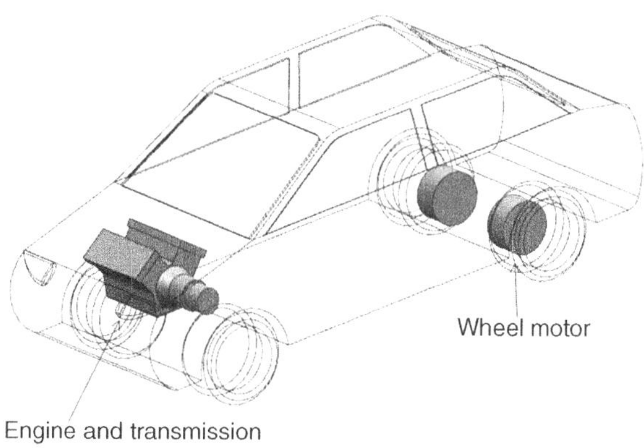

Figure 3.16 Motorized wheel hybrid configuration with considerable packaging benefits

such, placing the electric motor outside of the front box would resolve the packaging problems of the engine compartment.

The complex architecture of hybrid powertrain systems presents significant challenges in packaging, body, and chassis design requirements that must be addressed carefully.

3.6.2 Material Selection

Reducing energy consumption is an essential requirement for electric and hybrid electric vehicles because of their restricted onboard energy. Optimizing the vehicle weight is an effective way to reduce energy consumption. Unfortunately, electric and hybrid electric vehicles have additional components, such as battery packs and electric power units, making them heavier than conventional cars.

Figure 3.17 is a graphical comparison of curb weight versus vehicle length for various compact cars in the American market. The plug-in hybrid Chevrolet Volt is the heaviest vehicle in this class, with a curb weight of 3781 lbs. The Volt is 21% (688 lbs) heavier than its internal combustion engine version, the Chevrolet Cruze.

Comparing the Ford Focus Electric with the gasoline-fueled Ford Focus leads to a similar conclusion. The electric version is 704 lbs (24%) heavier than the internal combustion engine version. This discrepancy in weight is less for hybrid electric vehicles due to their smaller battery packs. For example, the Honda Civic Hybrid is only 247 lbs (9.5%) heavier than a standard Honda Civic. Reducing the weight of the body and chassis can partially compensate for the extra weight of an electric powertrain.

Furthermore, the heavier weight of electric or hybrid electric vehicles increases its kinetic energy. As a result, the vehicle structure must absorb a greater amount of energy in the event of a crash. Ultimately, the energy absorption capability of an electric or hybrid electric vehicle body structure must be greater than that of a gas-fueled vehicle.

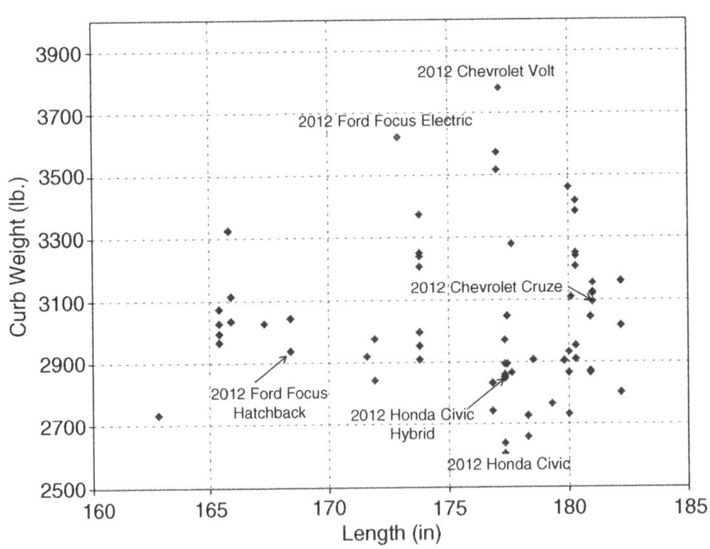

Figure 3.17 Curb weight versus vehicle length for different 2012 model compact cars in the American market

The extra weight of electric and hybrid electric vehicles are mostly in the form of lumped masses, such as a battery pack. As such, the car body structure must be locally and globally stronger and stiffer than normal, in order to handle the added dynamic loads.

Creating a body and chassis design for electric and hybrid vehicles is challenging due to the design load, demanded energy absorption capability, and expected body stiffness which are all greater than those of conventional vehicles, despite requiring lighter body structures.

Essentially, the aforementioned design goals can be achieved by using two different approaches: shape optimization and the utilization of advanced materials. The effectiveness of shape and topology optimization for the body and chassis has limitations because they are subject to the influence of other factors, such as styling, ergonomics, manufacturing requirements, and functionality.

Historically, electric and hybrid electric vehicle designers have considered advanced materials since the mass-production of early electric vehicles. The GM EV1 (pure electric) and Honda Insight (hybrid electric) pioneered the use of an aluminum body structure in their vehicle types, with the GM EV1 even using composite body panels. However, high-strength steels are mainly used in modern electric and hybrid electric vehicles.

The GM Volt is an example of a modern plug-in hybrid electric vehicle that uses advanced high-strength steel materials extensively in its body structure. According to General Motors, the main design philosophy of the Volt body structure is to use advanced steels based on an intelligent multi-material approach. Only 28% of the Volt's upper body structure has been made using low carbon steel. For lower body and chassis parts, such as floor panels, rails, and sills, the percentage of low carbon steel usage is further reduced to a mere 16%. The average use of advanced high-strength steels in GM Volt is about 80%, considerably higher than the 60% usage in modern conventional cars.

The Toyota Prius family provides another example of advanced multi-material design. The third generation of Prius vehicles expands its use of high-strength and ultra-high-strength steels, in addition to using aluminum for the hood and rear lift gate.

3.6.3 Aerodynamics

Aerodynamic forces are applied to the vehicle when it moves relative to the surrounding air. Drag force and lift forces are the most important aerodynamic forces. Drag force affects the energy consumption of the vehicle significantly, whereas lift force slightly influences vehicle handling and stability.

Until the mid-1960s, knowledge of vehicle aerodynamics was limited. As such, industrial designers and artists determined the car shape, resulting in a "boxy" car with large frontal areas. The advances in computing technologies in the 1970s and 1980s made it possible to analyze air flow around a vehicle. Since then, the primary and secondary goals of engineers are to reduce drag force and maximize lift force [5].

The early mass-produced EVs and HEVs followed the energy consumption reduction strategy and had a design based on drag force reduction. The GM EV1 and the first generation of the Honda Insight have the lowest drag coefficient among their competitors. The GM EV1 had an extremely low drag coefficient of 0.19, while the drag coefficient of the first generation of Honda Insight was 0.25. The typical drag coefficient value for conventional passenger cars in the early 2000s was around 0.3–0.35, demonstrating the superiority of EVs/HEVs in drag performance. However, the extensive improvements towards aerodynamic performance came at the expense of unconventional styling of early EVs and HEVs.

Presently, drag reduction is no longer the dominant body design goal for electric and hybrid electric vehicles because of certain legal and market-oriented factors. The body design priorities of today's electric and HEVs are style and safety, with aerodynamic performance as a lower priority. Rather than being aerodynamically effective at the expense of aesthetics, modern designers prefer to manufacture cars that are stylish and visually appealing in an effort to attract more customers.

Furthermore, modern safety standards and crashworthiness requirements do not permit the tiny body structures of the early electric and hybrid vehicles. These structures are what led to a smaller frontal area and lower drag coefficients.

The drag coefficients of the latest electric/hybrid vehicles, such as the Toyota Prius third generation, GM Volt, and Honda Insight second generation are 0.25, 0.28, and 0.28, respectively. These numbers are similar to the drag coefficients of modern conventional cars.

Unlike the early EVs and HEVs, aerodynamics no longer affects the overall shapes of modern versions of these vehicles. Nevertheless, details of the body design, such as the underbody geometry, the side mirror design, the internal flow, and grille opening area, are specifically designed to reduce the drag coefficient as much as possible.

3.7 The Chassis Systems of Electric and Hybrid Electric Vehicles

Chassis systems are defined as systems that are directly connected to the chassis, and include the suspension, steering, and brake systems. Some important car performance features, such as ride comfort, handling, and safety are strongly affected by these systems. This section briefly discusses the fundamentals of and technologies behind each system, and subsequently addresses specific considerations regarding electric and hybrid electric vehicles.

3.7.1 The Suspension System

3.7.1.1 Suspension System Functions

The suspension system connects the wheels to the vehicle chassis. An unconnected wheel has six degrees of freedom – it can freely move along longitudinal, lateral, and vertical directions, and can freely rotate about these axes. As illustrated in Figure 3.18, a suspension system restricts the wheels' degrees of freedom to three for vertical, rotational, and steering rotations. The steering motion is restricted in the case of nonsteerable wheels.

An ideal suspension system needs to meet some key requirements. First, to provide adequate ride comfort, the suspension system should be capable of managing road irregularities, while also isolating the vehicle body and passengers from road-induced vibrations and noises as much as possible. The suspension vertical compliance should be carefully optimized for this purpose.

Second, a suspension system should precisely control the wheels and vehicle body positions and motions to improve the handling and stability of the vehicle. The wheels need to be maintained in the proper positional attitude with respect to the road surface. On the other hand, wheels should not be deflected during cornering, braking, and accelerating. As such, a good suspension system should be almost non-deformable in both longitudinal and lateral directions. Moreover, the suspension should be adequately designed to resist angular motions of the

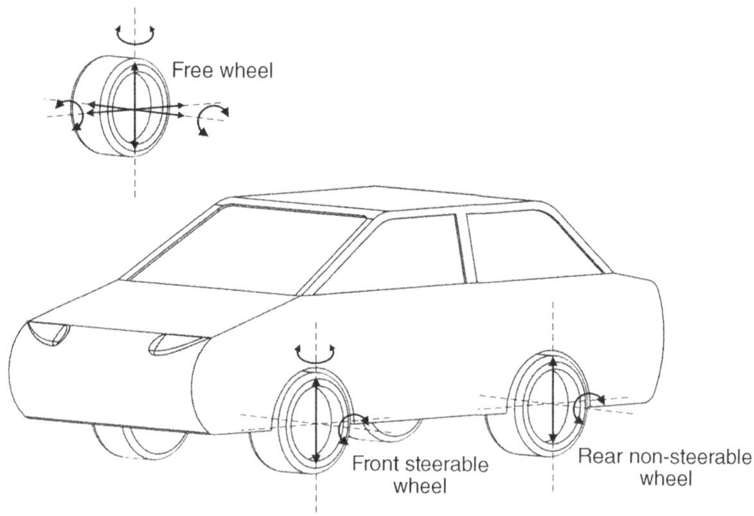

Figure 3.18 Suspension systems for restricting the wheel's degrees of freedom

vehicle body, specifically roll and pitch motions (for more information on vehicle dynamics terminologies, please see Chapter 4).

Finally, the suspension system should effectively keep the wheels in contact with the road with minimal tire normal load fluctuation. Minimizing the normal load fluctuation provides more effective longitudinal and lateral forces, which can eventually provide better braking, traction and handling performances. Since these requirements are very difficult to achieve simultaneously, designers should compromise between them, according to their design priorities.

3.7.1.2 Suspension System Components

As seen in Figure 3.19, a typical suspension system comprises four main parts:

- *Suspension mechanism:* This is a system of linkages that controls the relative motions of the wheels and the vehicle chassis.
- *Suspension spring:* The suspension spring stores the road-induced kinetic energy of the wheel and isolates the vehicle body from road irregularities. There are usually two types of springs that are used in suspension systems. The main spring controls the vehicle body bounce (vertical) motion and the anti-roll bar helps to control the body roll (angular) motion. An automobile suspension system can use different types of mechanical, pneumatic, hydraulic, and electronically-controlled springs. Of these, the most popular option is mechanical springs, including coil and leaf springs, and torsion bars. Modern cars most frequently use coil springs.
- *Shock absorber:* The shock absorber is a hydraulic device for damping the road-induced kinetic energy and controlling the wheel and body motions.
- *Bushing:* Elastic parts are used in joints and mounting points of the suspension system.

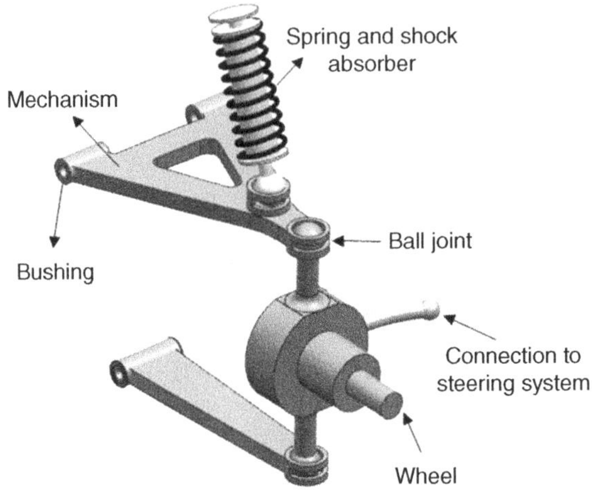

Figure 3.19 Main parts of a typical suspension system

3.7.1.3 Suspension System Types

There are several suspension systems regularly used in cars, and the type of suspension used depends on various factors, including: the location of the power unit (engine or electric motor), and whether the wheels are driven or nondriven, and steerable or nonsteerable.

Suspension systems are generally classified as dependent or independent. With dependent suspension systems, the motion of a wheel on one side of the vehicle is dependent on the motion of its partner on the opposite side; that is, when a wheel on one side of the vehicle strikes a road bump, its effect transmits directly to its partner on the other side. This has a detrimental effect on the ride and handling of the vehicle. Independent suspensions feature independent wheel motion for each wheel. As such, a disturbance on one wheel is not directly transmitted to its partner. This ensures better ride and handling capabilities.

Each dependent and independent suspension category includes different types of suspension systems that are applicable for different purposes. Table 3.14 illustrates the configuration and basic information of the most popular suspension systems.

3.7.1.4 Active Suspension Systems

As mentioned earlier, it is impossible to simultaneously achieve all key requirements by using a classical suspension system. Historically, suspensions have been passive systems, meaning they have simple mechanical elements, without any kind of energy-consuming active control process. In recent decades, there has been the emergence in a new category of suspension systems called active and semi-active suspension systems.

Figure 3.20 illustrates an active suspension system, using a controllable actuator in conjunction with, or instead of, a spring and damper set. Semi-active suspensions only use a controllable shock absorber. As expected, the performance of semi-active suspension systems is lower than that of fully active suspension systems.

Active and semi-active suspensions can considerably improve vehicle ride, handling, and road holding. These systems can change the suspension characteristics, according to the road

Table 3.14 Configuration and basic information of different suspension systems

Suspension System Name	Configuration	Type
Double wishbone	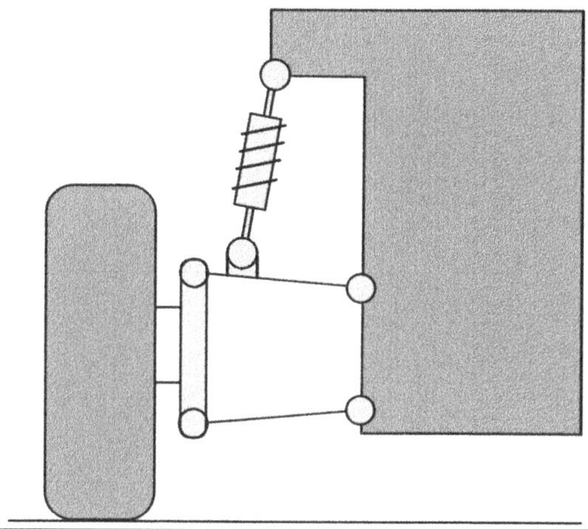	Independent
McPherson	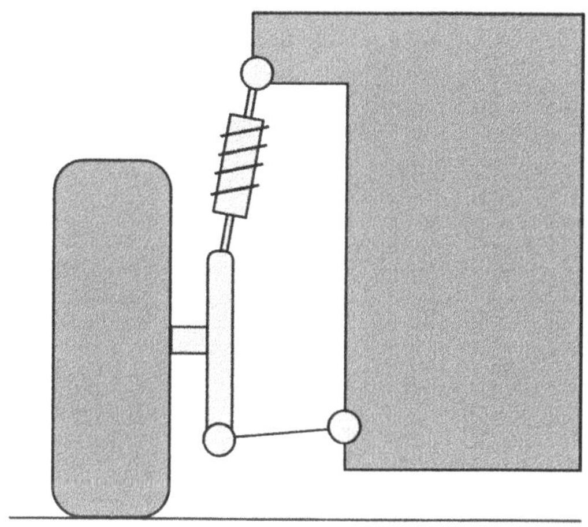	Independent
Swing axle	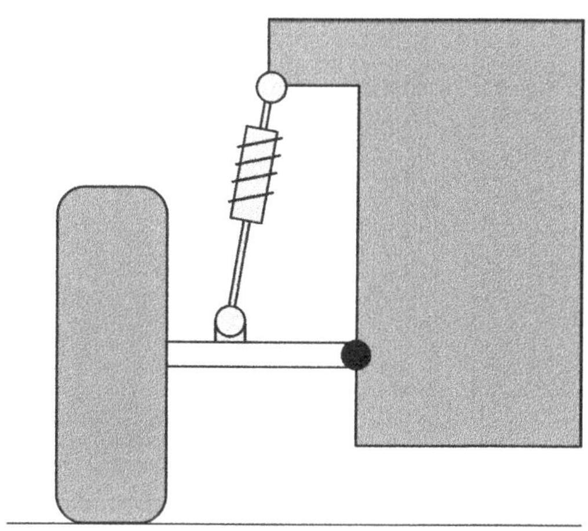	Independent

(continued)

Table 3.14 (*Continued*)

Suspension System Name	Configuration	Type
Solid axle	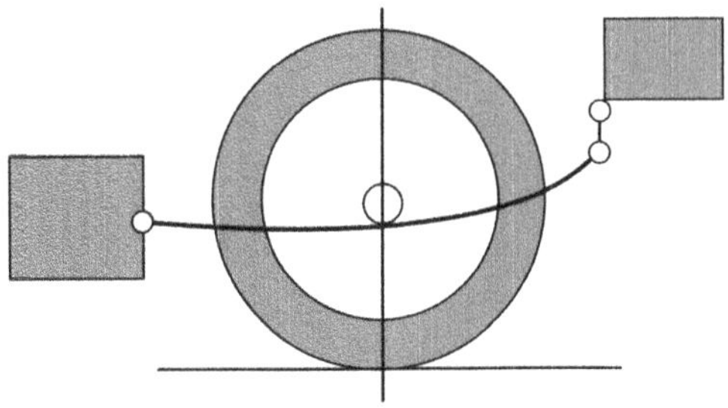	Dependent
Trailing arm	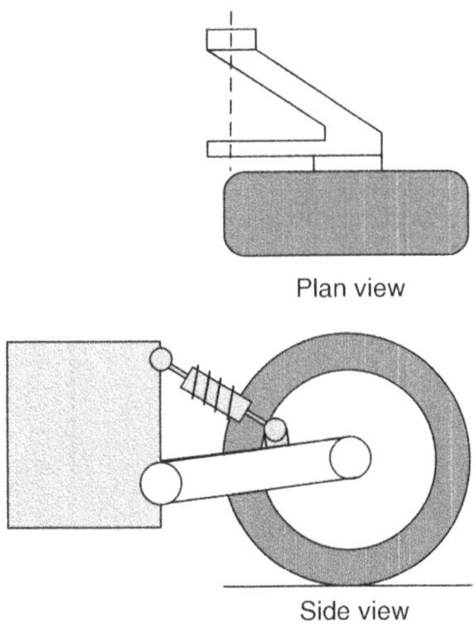	Independent
H-beam (torsion beam)	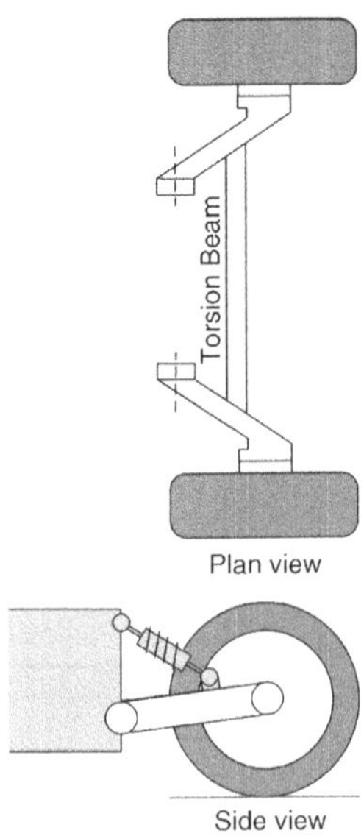	Semi-independent

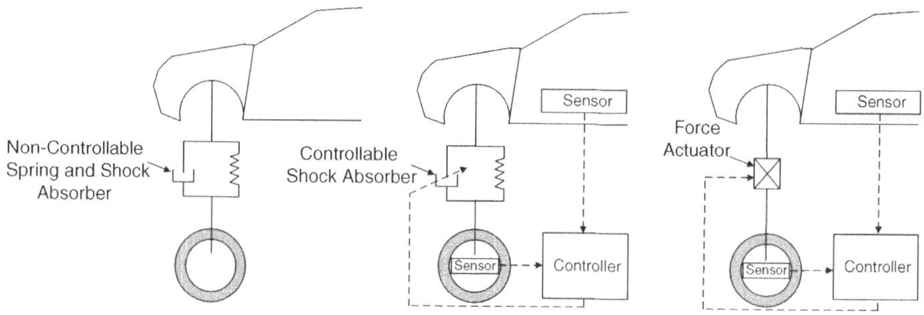

Figure 3.20 Working principle of passive, active and semi-active suspension systems

and the driving conditions at the time. For example, to improve vehicle handling and safety, an active/semi-active system can make the suspension stiffer during a harsh maneuver. Likewise, they can lower the suspension stiffness when driving in a straight line, to improve the ride comfort of passengers.

The application of active systems does have some problems. Regardless of their complexity and higher cost, the most significant drawback of active systems is their high energy consumption, a feature that has become an important cause for concern for today's consumers. As a result, consuming significant amounts of extra energy limits the widespread implementation of active suspension systems.

3.7.1.5 Regenerative Suspension System

In ordinary passive suspension systems, shock absorbers convert vibration energy into thermal energy using its viscosity, so that the vibration energy dissipates. This dissipated energy is not utilized. Regenerative suspension systems can potentially harvest the energy from suspension vibration, while also controlling the vibration.

Currently, regenerative suspension systems are still progressing through research and development. In recent years, different mechanical, hydraulic, and electromagnetic types of regenerative suspension systems have been the subjects of investigation.

Among the different variations of regenerative suspension systems, electromagnetic types are the most promising. The main part of this system is a regenerating shock absorber that recovers the vehicle's vibrational energy and converts it into usable electricity. Like regenerative braking systems, regenerative suspension systems are considered to be a secondary source of energy for recovering energy and reducing the energy consumption of the vehicle. Unfortunately, recent research predicts that a typical electromagnetic regenerative suspension system can theoretically only generate small amount of power under normal driving conditions – not an adequate amount of recovered energy. Moreover, this amount of energy can only be recovered in specific driving conditions, such as low-frequency and high-amplitude displacement, which features low-speed driving on very harsh and uneven roads. Ride, handling, and energy harvesting all have different requirements, and optimizing one of them may have an adverse effect on the others. There is on-going research seeking to extend the harvesting capacity of regenerative suspension systems.

3.7.1.6 Suspension Systems of EVs and HEVs

At a first glance, the differences between the suspension systems of current EVs/HEVs and conventional cars may seem insignificant. For example, the Toyota Prius has a MacPherson suspension system in the front axle and an H-beam suspension system in the rear axle, which is the same system as conventional cars in this class. The Nissan Leaf is another example of an electric vehicle that also uses MacPherson and H-beam suspension systems in the front and rear axles, which are the dominant suspension system configurations for compact cars.

There are still some specifics to consider when designing a suspension system for EVs and HEVs. The technological differences between the suspension systems of conceptually designed electric and hybrid vehicles for the future and a traditionally designed suspension systems are more significant. These specific considerations are:

- *Heavy-duty design:* As previously mentioned, heavy energy storage systems, such as batteries and ultra-capacitors, are on-board in pure EVs and as additional power units in HEVs. These storage systems are relatively heavier than internal combustion engines. As a general estimate, the weight of EVs and HEVs with storage systems can be up to 25% higher than that of similarly sized conventional cars. This additional weight significantly affects the ride and handling performance of the vehicle. The severity of the problem can also increase if there is an adverse change to the front/rear weight distribution and the vehicle center of gravity height.

  Aside from using appropriate tires with optimized cornering coefficients, the suspension system should also accommodate the undesirable situations mentioned above. To achieve optimized ride and handling performances, the characteristic parameters of the suspension system (in terms of natural frequencies, damping ratio, etc.) should be set to specific optimized values.

  In terms of the increased mass and inertia of the vehicle's sprung mass and the undesirable location of the vehicle's center of gravity (very close to the front or rear axles, with space above the road surface), the suspension system should be designed in a way that keeps the natural frequency, damping ratio at their optimum values. Following this guideline, optimized suspensions for EVs and HEVs generally have stiffer spring and shock absorbers with higher damping coefficients than a common suspension. In some cases, to control the roll motion of the car body, the torsional stiffness of the anti-roll bar(s) should also be increased. Considering the increased loading of the suspension, the load-carrying capacity of the suspension mechanism may require improvement. This will result in stronger suspension arms and linkages with stiffer bushings.
- *Design for in-wheel drive configuration:* As discussed in Chapter 2, in-wheel drive configuration is an innovative powertrain arrangement for EVs and HEVs that will likely become more popular. The in-wheel motor is an electric motor incorporated into the hub of a wheel and drives it directly. The in-wheel motor arrangement has control and functional merits, such as higher response rates, precise and controllable torque generation, yaw moment control capability through torque vectoring, and elimination of the adverse effects of driveshaft flexibility (Figure 3.21).

  The major disadvantage of the in-wheel motor configuration is that the weight of the electric motor increases the unsprung mass, which deteriorates the ride comfort and road handling of the vehicle. As a general rule, designers usually try to minimize unsprung mass

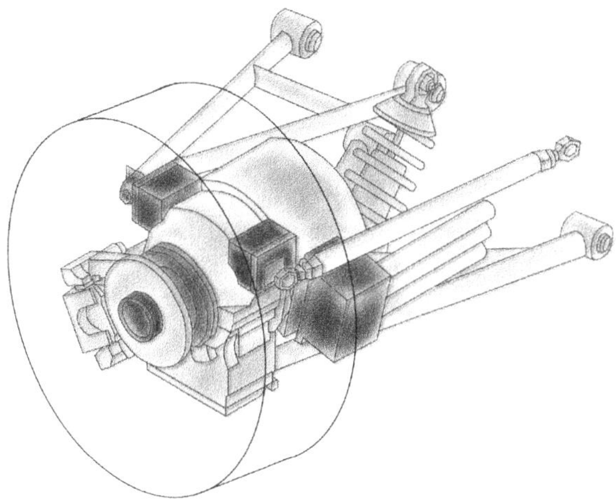

Figure 3.21 A schematic of modified suspension system for in-wheel motor drive configuration

to improve the handling and steering characteristics of the vehicle, as well as ride comfort for the vehicle's passengers. The unsprung mass of an in-wheel drive configuration is considerably higher than that of conventional vehicles. This poses a serious challenge for suspension designers. Besides minimizing the in-wheel motor weight, the most commonly suggested solution is to implement a stiffer suspension system. While a stiffer suspension can resolve some of the adverse effects of heavier unsprung mass, it cannot completely compensate for all of its negative effects. Using a stiffer spring can suppress the wheel's jounce and rebound amplitude, but it also increases the acceleration transmitted to passengers. In addition to the suspension spring, the bushings' design (specifically, their stiffness) should also be optimized based on the heavier unsprung mass.

Moreover, the use of in-wheel motor drive substantially affects the suspension and steering mechanism, as well as the chassis design and packaging. The primary concern is that suspension and steering mechanisms traditionally occupy large spaces inside and alongside of the wheel, which in this case would be dedicated to the in-wheel motor. As such, the suspension mechanism type and configuration should be carefully selected and designed in order to properly install the suspension, steering mechanisms, and in-wheel motor in a very tight space.

There are a number of factors that have an influence on the design process. First, the type of suspension system is crucial, since suspension types, such as the torsion beam and semi-trailing arm, are more likely to fit alongside an in-wheel motor than the other types, such as double wishbone. Second, the size of the motor and its requested output power are another influential factor. Since larger, more powerful motors require more space inside of the wheel, the wheel size, in terms of diameter and width should be increased accordingly. This would consequently decrease the remaining space for the suspension system and make the suspension design process more difficult. Other problems may occur if the wheel is steerable. The presence of the steering system makes it more difficult to install the motor within the wheel.

In addition to the suspension mechanism, the body and chassis need to be modified according to in-wheel drive requirements. For example, when using bigger wheels, the wheelhouse should be designed larger than usual in order to have an acceptable maximum steering angle. This design configuration has considerable effects on the body and chassis design.

In the case of the originally designed motorized wheels, for electric or hybrid electric vehicles, the aforementioned changes in the suspension, steering, chassis and body can be considered during the design phase. However, in the case of electrification of an existing platform, or the conversion of an existing conventional car, many of these problems may remain unsolved, leading to technical problems, such as limited suspension stroke and limited steering angle range. These cars may have severe problems in terms of poor ride, maneuverability and road handling.

- *Application of advanced systems:* In theory, the regeneration of energy from suspension vibration is an ideal concept for EVs and HEVs. However, in practice, the current design of regenerative suspension systems cannot recover considerable amount of energy. Nevertheless, continued research in this area will ensure the development of more efficient systems, potentially making regenerative suspension systems a standard feature of EVs and HEVs in the future.

  Active suspension systems are known as energy-consuming devices, making them unsuitable for EVs and HEVs. However, some advanced concept cars within this class have implemented these suspension systems, such as the BMW Vision Efficient Dynamic 2012. Generally, it is possible to implement active or semi-active suspension systems in these types of vehicles, if the technical limitations of electrical energy storage devices are resolved.
- *Application of advanced materials:* As mentioned above, EVs and HEVs typically suffer from their unusually heavy weights. Especially, in motorized wheel vehicles, the unsprung weights and overall vehicle weight are considerably heavier than that of conventional cars. Reducing the suspension system weight can help decrease the unsprung weight and the overall weight of the vehicle. Using advanced lightweight materials, such as aluminum alloys, is beneficial when making suspension arms and linkages. For example, some parts of suspension system used by the Toyota Prius are made of aluminum alloys. In the future, this prominent trend should continue.

3.7.2 The Steering System

The function of the steering system is to change the vehicle path in response to the driver's command input. The steerable wheels on a vehicle change in correspondence with the steering turns made by the driver. This develops a lateral force in the tires, consequently resulting in a yaw moment along the vehicle's vertical axis, eventually causing the vehicle's direction to change. A specific mechanism is used to transmit and convert the steering wheel angle to the steering angle of the wheels.

3.7.2.1 Steering System Mechanisms

Generally, steering systems use two types of mechanisms: the parallelogram mechanism and the rack and pinion mechanism. Figure 3.22 illustrates the configuration of the parallelogram steering mechanism. When the steering wheel rotates, a recirculating ball steering gear converts rotational motion to linear motion. This moves the pitman arm, which is connected to the

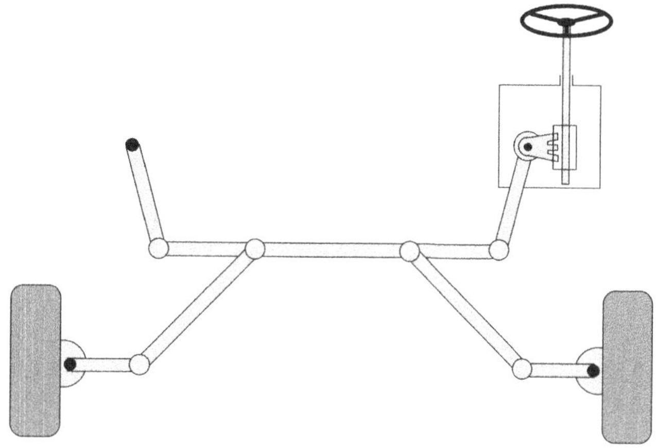

Figure 3.22 Schematic illustration of the parallelogram steering mechanism

steering linkage. The movement of the pitman arm results in the wheels being steered. The parallelogram mechanism is suitable for heavy-duty vehicles due to its high load capacity. The most significant disadvantage of this mechanism is its complexity, which produces high manufacturing costs. As alluded to above, heavy commercial vehicles typically use a parallelogram mechanism.

Figure 3.23 also depicts a rack and pinion mechanism. The pinion gear connects to the steering column and moves the rack horizontally. This movement steers the wheels through steering linkage. The rack and pinion mechanism is simple and economical, but its limited load capacity makes it a desirable choice for passenger cars.

3.7.2.2 Power-Assisted Steering

An external power source is often used to assist the driver and reduce steering effort. Power steering systems are implemented as a standard feature in most modern cars, regardless of their steering mechanism type. There are different power sources that can provide steering assistance (Figure 3.24).

Hydraulic power steering (HPS) is the most commonly used power steering system in automobiles. Figure 3.24(a) illustrates a hydraulic rack and pinion power steering system. In this system, the engine drives an oil pump to produce hydraulic pressure. When the driver turns the steering wheel, the hydraulic valve guides the flow of pressurized oil into the hydraulic cylinder and the hydraulic piston assists the movement of the rack in each direction. When the steering wheel is in the neutral position, the oil flow passes through the open control valve and returns to the oil reservoir without any significant changes in pressure. Unnecessary circulation of the steering system oil increases the fuel consumed by the vehicle.

Another type of power-assisted steering system is the electro-hydraulic power steering (EHPS) systems (Figure 3.24(b)). The working principle of an EHPS system is similar to that of

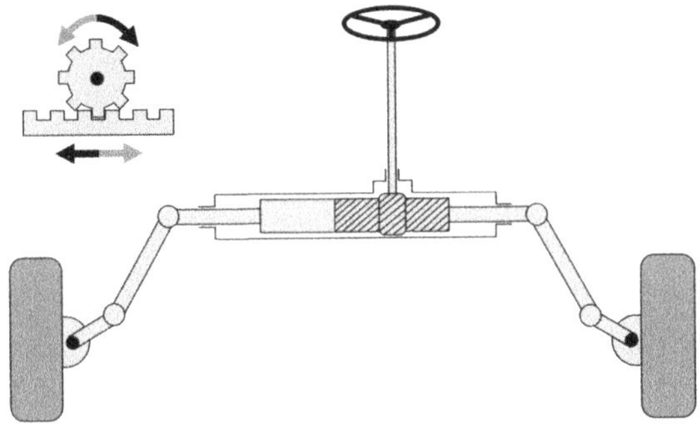

Figure 3.23 Schematic illustration of the rack and pinion steering mechanism

hydraulic power steering. The primary difference between the two is that an electric motor, rather than an engine, drives the oil pump. This system generates only the required amount of pressurized oil for each specific steering and driving condition. As such, oil does not circulate when the steering system is deactivated.

The use of energy in EHPS systems is considerably more efficient compared to hydraulic power steering systems. Another advantage of this system is that the system's power pack – which contains the electric pump, the electronic control unit, and the oil reservoir – is independent of the engine position. As a result, its position can be in any suitable position within the engine compartment.

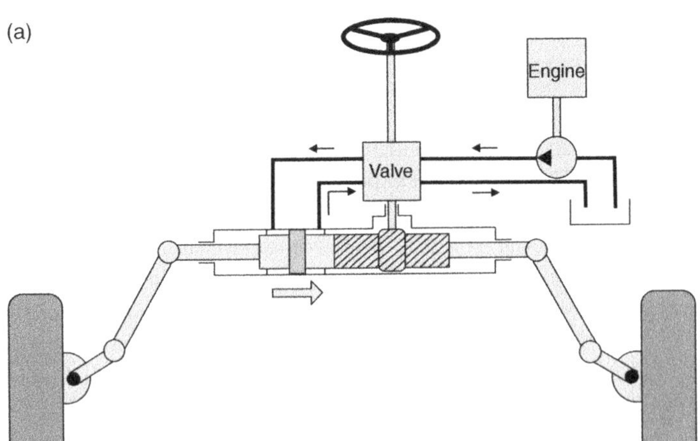

Figure 3.24 Different types of power-assisted steering systems (a) Hydraulic power-assisted steering system (b) Electro-hydraulic power-assisted steering system (c) Electric power-assisted steering system

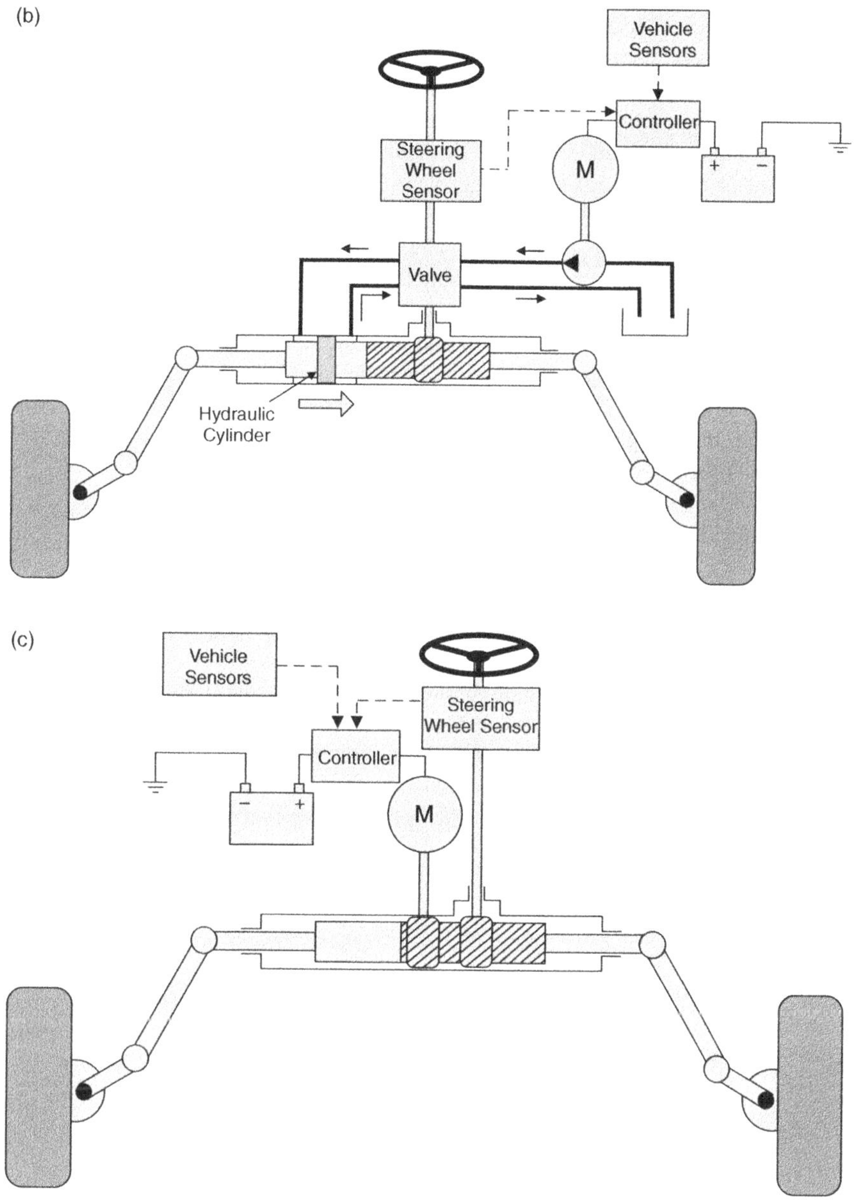

Figure 3.24 (*Continued*)

The electric power steering (EPS) system is the most advanced power steering system. In this system, an electric actuator directly generates and controls the steering assistance (Figure 3.24(c)). As a result, the system is simpler and energy-efficient, because of the elimination of all hydraulic components and circuits. In addition, the improved controllability of the electric and electronic systems (compared to the hydraulic ones) allows the system to precisely control the steering assistance and modify it based on different driving conditions. This results in enhanced vehicle safety and handling, and additionally increasing comfort and convenience for the driver. The current application of this system is limited to only small and medium-sized passenger cars. This is because providing the required amount of electric power for larger vehicles using conventional 12-volt electrical systems considerably increases the system's electric loss.

Electric actuators can be mounted on the steering column, pinion, or rack. Column drive electric power steering systems are more suitable for light vehicles that require lower amounts of steering assistance. They are more appropriate for these vehicles because applying high amounts of torque on the upper section of the steering driveline causes undesired elastic deformation and vibration within the steering system. Alternatively, it is possible to install electric actuators on the pinion; however, the amount of stress placed on the pinion increases in this configuration. Installing an electric linear motor on the gear rack provides high levels of steering assistance for heavier vehicles.

3.7.2.3 Advanced Steering Systems

- *Four wheel steering (4WS) systems:* Electronically controlled actuators steer the rear axle wheels in this system, resulting in improved vehicle maneuverability, handling, and stability. The vehicle speed determines the steering direction of the rear wheels. During lower speeds, the rear wheels steer to the opposite direction of the front wheels (Figure 3.25(b)) to reduce the vehicle turning radius and improve its maneuverability. On the other hand, higher speeds necessitate steering the rear wheels in the same direction as the front wheels to improve vehicle stability and handling (Figure 3.25(a)).
- *Active front steering (AFS):* AFS is an intelligent system that improves vehicle handling, stability, and convenience by increasing or decreasing the steering angle generated by the

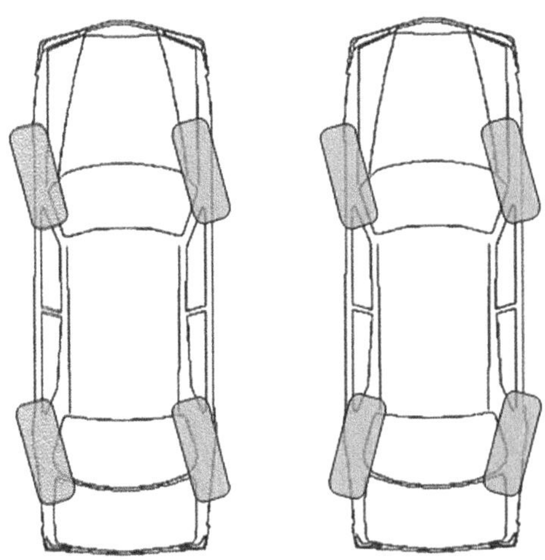

Figure 3.25 Different functions of four-wheel steering at high speeds (a) and low speeds (b)

driver. In this system, an electric motor generates a corrective steering angle to add or subtract to the steering angle of the driver. This is done by using a planetary gear set integrated into the steering column. An electronic control unit calculates the value of a corrective steering angle based on the data it collects from various sensors that measure vehicle speed, lateral acceleration, and yaw rate among others. At lower speeds, active front steering systems magnify the driver steering angle, causing a more sensitive vehicle response to the steering inputs of the driver. This increased sensitivity makes it easier to park and drive in city traffic. At higher speeds, the AFS system reduces steering system sensitivity by decreasing the steering input of the driver. This gives the driver more precise steering control, guaranteeing better stability and handling.

- *Steer-by-wire (SbW) systems:* The SbW system eliminates the mechanical connection between the steering wheel and the steered wheels. A steer-by-wire system consists of the steering wheel module, steering actuator module, and electronic control unit. There are sensors in the steering wheel module to detect the steering angle and torque applied on the steering wheel by the driver. Moreover, an electric motor is mounted upon the steering wheel to provide the driver with steering feel and to simulate road feedback. The steering actuator module consists of an electromechanical rack and pinion steering gear, as well as sensors that detect the steering angle and steering force of the wheels. An electronic control unit processes all of the data collected from the sensors. It uses this data to control the steering electric motors to steer the wheels proportional to the commands of the driver, while also providing appropriate steering feel to the driver.

  Using a SbW system increases the amount of free space in the engine compartment and allows more flexibility in the steering angle control, as well as the vehicle design, packaging, and manufacturing. Eventually, it will be possible to control the vehicle dynamics intelligently and more precisely, based on different driving conditions. However, due to a lack of reliability, this system is not yet in mass production vehicles, only in some concept cars.

3.7.2.4 The Steering System of EVs and HEVs

- *Electric power-assisted steering*: The steering system of most modern EVs and HEVs is a type of electric power-assisted rack and pinion steering systems. As mentioned above, conventional vehicles have restricted amounts of on-board electric power and limited level of electric voltage. As a result, electric power-assisted steering systems can only work on smaller conventional vehicles. However, the energy storage device of electric and hybrid electric vehicles can easily provide higher power and voltages, which are suitable for electric actuators of larger and heavier vehicles. As such, electric power-assisted steering systems can be used in any size and type of EV and HEV. For example, the Toyota Highlander Hybrid is a large SUV using the electric power-assisted rack and pinion steering system. Other EVs and HEVs that use this system are: the 2012 Hyundai Sonata Hybrid, the 2012 Ford Fusion Hybrid, the 2012 Honda Insight Hybrid, the 2012 Mitsubishi i-MiEV, the 2012 Ford Focus Electric, and the 2012 Tesla Model S.
- *Steer-by-wire system*: Modern electric and hybrid electric concept cars use more advanced steering systems, such as the steer-by-wire system. There are many concept EVs and HEVs, such as the Audi A2, that use this technology in their steering system. Concept electric and

hybrid electric vehicles work as testing platforms for advanced steering systems, such as steer-by-wire to facilitate their applications in future mass-produced vehicles.

In recent years, some suppliers have tried to develop more advanced and innovative types of steer-by-wire systems for the next-generation EVs and HEVs. For example, the eCorner system is a conceptual steer-by-wire system design developed by Siemens VDO. This system allows independent control of each wheel, and eliminates all the mechanical components of the steering system. In fact, an independent electric motor mounted upon the wheel steers each wheel independently. This arrangement provides the best maneuverability and handling performance for the vehicle.

3.7.3 The Braking System

The braking system reduces the velocity of a vehicle or completely stops its momentum. Parking brake systems are also necessary to fix the position of a parked vehicle. Traditionally, a braking system applies friction forces to the vehicle to convert its kinetic energy to useless thermal energy. Regenerative braking systems, however, are able to recover significant amounts of kinetic energy and convert it into electric energy that can be stored in the vehicle energy storage devices, such as batteries. Regenerative braking systems are widely used in electric and hybrid electric vehicles.

As illustrated in Figure 3.26, when the brake pedal is pushed, the brake booster increases the force applied by the driver. The generated force then pushes the master cylinder to produce

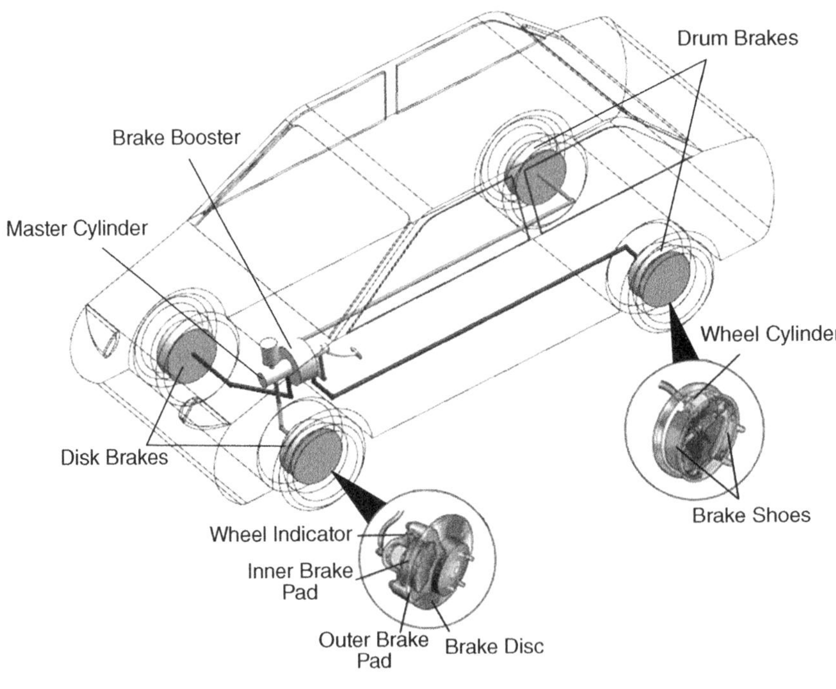

Figure 3.26 Braking system components

hydraulic pressure. The pressurized hydraulic oil flows to the wheel brake that is used to push the brake pads towards the brake disc or the brake shoes towards the drums. This push ultimately produces friction force and reduces the speed of the wheels [4].

3.7.3.1 Braking System Components

- *Vacuum brake booster:* A vacuum booster amplifies the force applied to the brake pedal in order to make braking easier and safer for the driver. This component includes a vacuum chamber, a flexible diaphragm, and a control valve. The vacuum chamber is connected to the engine's intake manifold, which functions as the source of vacuum.
- *Hydraulic transmission system:* The output force of the brake booster applies to a tandem master cylinder to produce hydraulic pressure. The tandem master cylinder has two separate hydraulic chambers, which feed two separate hydraulic braking circuits. If one circuit fails (e.g., due to leakage), the other circuit can still function to stop the vehicle. This is one of the most important safety features in vehicles' braking systems. One circuit is usually dedicated to the right front and left rear wheels, while the other circuit is for the left front and right rear wheels.
- *Wheel brake:* Wheel brakes use hydraulic pressure provided by hydraulic circuits to produce braking torque and reduce the rotational speed of the wheels. Drum and disc brakes are two common types of wheel brakes used in automotive braking systems [4].
- *Drum brakes* (Figure 3.26) consist of two brake shoes in a cylinder-shaped brake drum. During braking, these shoes press against the inner surface of the drum by a hydraulic cylinder. When braking ends, springs pull the brake shoes back. Due to their self-energizing characteristic, drum brakes need less hydraulic force than disc brakes. Furthermore, they can easily incorporate a parking brake, and are also relatively less expensive than disc brakes. However, a primary disadvantage of drum brakes is that their performance is highly sensitive to the working temperature. Under severe braking conditions, they often lose their effectiveness. The front brakes generate most of the braking force; the amount of heat generated in the rear wheel brakes is significantly less in comparison. As such, the rear wheels of passenger cars often use drum brakes.
- *Disc brakes:* Disc brakes (Figure 3.26) are the other type of wheel brakes. In these brakes, a hydraulic cylinder applies force to brake pads in the normal direction, thus pressing the brake pads against the surface of the brake disc. Disc brakes have the advantage of being less sensitive to working temperature. The front wheels of almost all passenger cars use disc brakes; for economic reasons, these brakes are found only on the rear wheels of more expensive vehicles.

3.7.3.2 Electronic Braking Control System

- *Antilock braking system (ABS):* Depending on the road and driving conditions, braking too heavily can cause excessive wheel slip or may even cause the wheels to lock. In this circumstance, the vehicle loses directional control and stability, and its stopping distance increases. The main function of ABS is to prevent the tire from locking, and to keep the slip of the tire within a desired range (Figure 3.27). This means that the tire can develop a

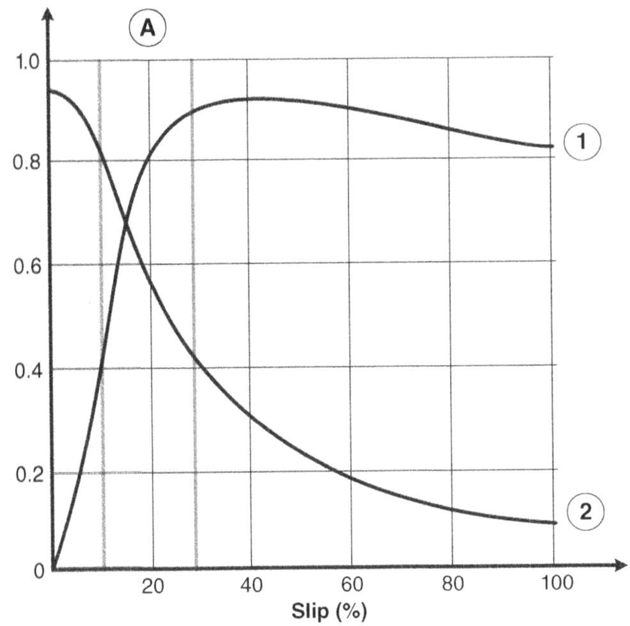

Figure 3.27 Effect of slip on braking force coefficient and lateral force coefficient. *Note*: A = desirable slip range, 1 = braking force coefficient, 2 = lateral force coefficient

sufficiently high braking force to stop the vehicle, and it can also provide an adequate lateral force for directional control and stability. For more information about a tire's slip and forces, see Chapter 4.

An ABS consists of sensors, an electronic control unit, and a brake pressure modulator (Figure 3.28). Sensors are mounted upon the wheels to detect the angular speed and the angular deceleration of the tires. This data is transmitted to the control unit for further processing. In the control unit, the measured parameters and its derivations are compared with the predetermined threshold values. When certain conditions are met, a signal is sent to the modulator to release the brake in order to prevent the wheel from locking.

- *Traction control system (TCS):* Traction control systems were developed to help improve vehicle tractive performance and to maintain directional control and stability during acceleration. This system improves the vehicle's traction and stability on road surfaces with different levels of adhesion for left and right tires, and also prevents the tire from spinning during acceleration on slippery roads. One of the techniques used in the TCS is the application of braking systems in preventing the wheels from spinning. In this technique, an extra module is used to provide the necessary braking pressure to the spinning wheels. The working principle, hardware (hydraulic system and sensors), and software of a TCS and an ABS are similar. As such, vehicles with ABS can also have a braking-based TCS with some extra modules.
- *Electronic brake force distribution (EBD):* The EBD system distributes braking force intelligently between the front and rear axles to increase the vehicle's stability during braking and to reduce the stopping distance. Parameters such as deceleration level, road condition,

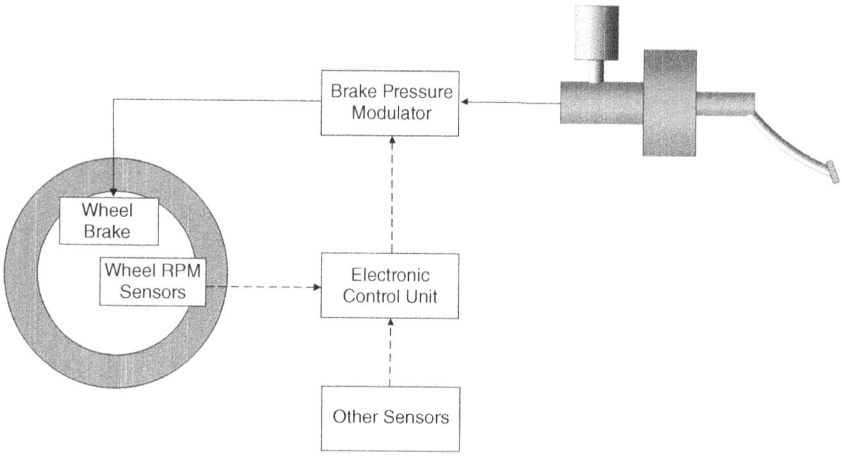

Figure 3.28 General configuration of ABS

and weight distribution of the vehicle determine the optimum braking force distribution. Electronic brake force distribution systems mainly function with ABS components. As such, considerable hardware improvement is not a requirement for this system.

- *Electronic stability control (ESC):* Electronic stability control, also known as the electronic stability program (ESP), is an advanced active safety system for improving the dynamic behavior and safety of vehicles. The working principle of this system is differential braking, which allows the system to automatically apply brakes on an individual wheel to keep the vehicle on the desired track. During cornering, ESC prevents the vehicle from spinning (oversteering) by braking outside the front wheel, and from drifting out (understeering) by braking inside the rear wheel (Figure 3.29). Appling brakes on a wheel changes its longitudinal and lateral force, consequently generating a corrective yaw moment (a moment around the vehicle's vertical axis). This yaw moment can correct the vehicle direction. The system configuration of an ESC is mainly based on ABS/TCS hardware, with some limited changes in the hydraulic system, and some added components, such as lateral acceleration and yaw rate sensors. ESC software is entirely different from ABS/TCS software, and includes more sophisticated control algorithms.

3.7.3.3 The Brake-By-Wire (BBW) System

A BBW system eliminates the need for a mechanical/hydraulic connection between the brake pedal and the wheel brakes. When the driver pushes the brake pedal, a signal is sent to the electric actuators mounted on each wheel. The electric actuators generate braking force to push the brake pad towards the brake disc. Due to a lack of reliability, this technology is still in its research and development stage and has not yet been used in mass-produced vehicles.

3.7.3.4 The Regenerative Braking System

As mentioned earlier, conventional braking systems convert the vehicle's kinetic energy of the vehicle into thermal energy. However, regenerative braking systems are able to recover

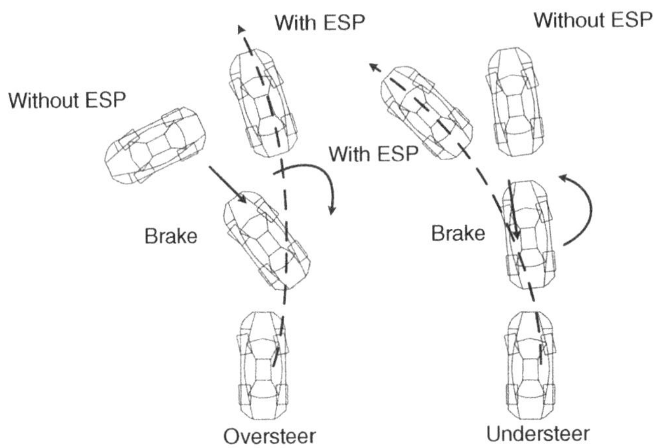

Figure 3.29 ESC working principle

significant amounts of kinetic energy and convert it into other types of energy to store in energy-storage devices.

Using electrical energy-storage devices, such as batteries and ultra-capacitors, and using mechanical energy-storage devices, such as flywheels, are two ways of storing recovered energy. In a regenerative braking system with an electrical energy storage device (Figure 3.30), an electric generator converts the kinetic energy of the vehicle into electrical energy during braking. This energy then charges the vehicle batteries or ultra-capacitors.

In a mechanical regenerative braking system, a clutch engages the wheels to a flywheel during braking. The kinetic energy of the vehicle transfers to the flywheel, making it spin gradually faster as the energy is introduced. During acceleration, the stored energy in the flywheel can be returned to the vehicle.

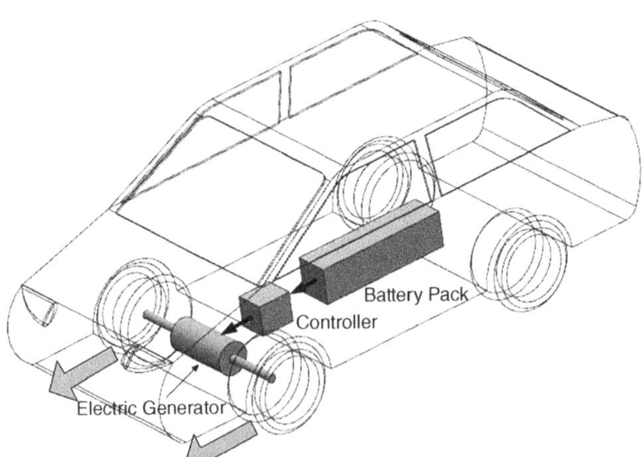

Figure 3.30 The concept of of regenerative braking systems

In an electrical regenerative braking system, energy conversion occurs four times: mechanical- to-electric, electric-to-chemical, chemical-to-electric and electric-to-mechanical. As such, its overall efficiency is very low (approximately 35% in some cases). By contrast, there is only mechanical-to-mechanical energy conversion happening in the flywheel regenerative braking systems. In this arrangement, the translational kinetic energy of the vehicle is converted into rotational kinetic energy of the flywheel during braking. However, in cases where longer-term energy storage is a necessity (e.g., more than an hour), the friction rises between the mechanical components of the flywheel, making it less efficient than electrical systems. Nevertheless, flywheels are more efficient when stored energy is needed in a shorter period of time (e.g., while driving in city traffic).

Conversely, using batteries as an electrical energy storage device can impose further limitations on the regenerative braking system. Mainly, batteries are not capable of absorbing excessive amounts of energy in a short period of time. Moreover, the operational life (maximum number of charge–discharge cycles) of batteries is limited. As such, they are not the ideal energy-storage devices for regenerative braking systems.

An alternative solution is the use of ultra-capacitors as an energy-storage device. Ultra-capacitors have an extended life cycle, and can be charged and discharged quickly. The simultaneous use of ultra-capacitors and batteries as the energy-storage devices for regenerative braking systems can greatly improve fuel efficiency in stop-and-go urban driving conditions. However, ultra-capacitors are very expensive.

As previously mentioned, regenerative braking systems have a limited rate of energy absorption. This is especially true during emergency braking, where regenerative braking cannot stop the vehicle safely within a short stopping distance. As such, regenerative braking systems should be used in conjunction with conventional friction brakes. The apportioning of the braking force between the friction and the regenerative brakes is accomplished by simultaneous control of the hydraulic brake system and the regenerative braking system. As a result, the total brake force of the hydraulic brake and the regenerative brake matches the brake force required by the driver. If the regenerative brake fails, the friction brake system operates the hydraulic brake system to supply the driver with the entire brake force required.

3.7.3.5 The Braking System of EVs and HEVs

EVs and HEVs, similar to conventional vehicles, typically have modern features of braking systems, such as ABS, EBD, TCS and ECS. Additionally, EV and HEV braking systems also contain the specific features discussed below:

- *Electric vacuum pump:* As mentioned above, conventional braking systems require a vacuum source used in the brake booster to magnify the force of the driver's pedal. In the case of EVs and HEVs, sources other than the engine intake manifold should provide the needed vacuum energy. This is because HEV engines do not permanently run, and EVs do not have an engine at all. As Figure 3.31 illustrates, some EVs and HEVs are equipped with an electronically controlled electric vacuum pump, providing vacuum energy for the brake booster. Using a vacuum reservoir is necessary to prevent the vacuum pump from constantly cycling.
- *Electro-hydraulic power booster:* Electro-hydraulic power boosters are the design alternative to electric vacuum pumps. The hydraulic pressure magnifies the driver braking force

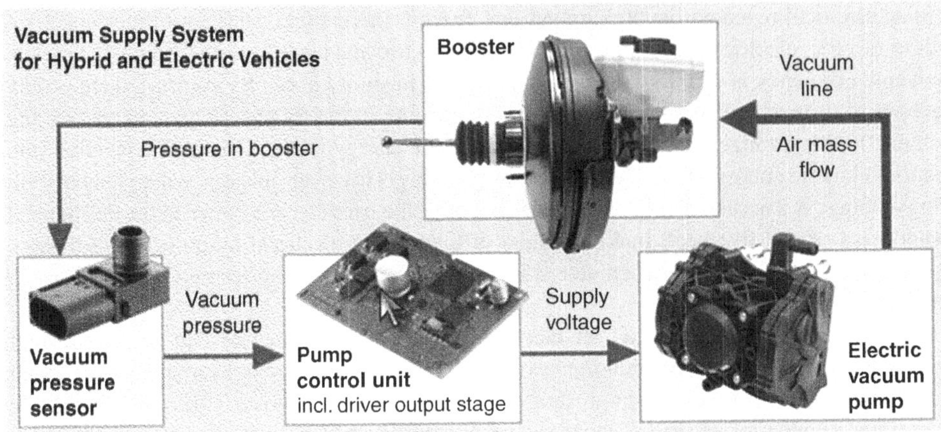

Figure 3.31 Vacuum supply system for HEVs and EVs. *Source:* Reproduced by permission from Continental

generated by an electrically driven hydraulic pump. In some cases, the steering system hydraulic pump is the source of hydraulic pressure for electro-hydraulic boosters.

- *Regenerative braking system:* Due to its lack of on-board energy, regenerative braking systems are very useful for electric and hybrid electric vehicles. Furthermore, as these vehicles already have a battery pack and an electric motor/generator, implementing regenerative braking systems is relatively easy. Note that in addition to regenerative braking; all modern EVs and HEVs also implement mechanical friction braking systems because regenerative braking systems cannot generate enough braking torque independently, especially during emergency braking.
- *Brake-by-wire system:* As previously mentioned, BBW systems are not yet found in mass-produced cars. This is because by-wire technology is not reliable enough for use in critical safety systems, such as braking. Much like the steer-by-wire systems, brake-by-wire technology can be energy-efficient, if a higher level of electric voltage is available. However, EVs and HEVs are a suitable testing platform for this revolutionary technology due to their onboard, high voltage circuits. The introduction of ongoing commercialized brake-by-wire systems in an EV or HEV is a likely possibility.

Problems

1. Explain the different functions of an automotive body and platform and classify the advantages of platform consolidation.
2. Explain the different functions and design considerations of front, central and rear automotive body boxes.
3. Describe the different design criteria of a car body.
4. Create an evaluation matrix and compare different types of automobile structural systems from different perspectives such as manufacturability, performance and cost.

Table 3.15

Powertrain layout	Front engine/electric motor, front-wheel drive
Engine power	60 kW
Electric motor power	45 kW
Battery pack energy capacity	13 kWh

5. Which structural members in a unibody construction are responsible for overall car body torsional/bending stiffness and front/rear/side crash safety?
6. Describe the multi-material approach in car body design.
7. Is it true or false? Drag coefficient of modern electric and hybrid vehicles is less than that of conventional cars? Explain.
8. Create an evaluation matrix and compare various battery pack locations in terms of the criteria discussed in this chapter.
9. The conventional vehicle in Example 3.2 is converted to a plug-in hybrid electric vehicle with the given components in Table 3.15 and by using the battery module specified in Table 3.8.

   Find possible battery pack locations using a similar methodology to Examples 3.2 and 3.3. Calculate the battery pack size and mass and evaluate the effect of each possible battery pack location on the vehicle CG location and weight distribution. CG coordinates of the vehicle components are given in Table 3.16.

Table 3.16

Components	Mass (kg)	CG coordinates (x_i, z_i) (mm)
Vehicle without powertrain	1100	(1215,670)
Hybrid powertrain (electric motor, engine, transmission, etc.)	150	(0,600)
Full fuel tank	40	(2900,350)
Inverter and motor controller	15	(0,700)
Battery back location		**coordinates of battery pack CG(x_i, z_i) [mm]**
Front box		(0,550)
Inside of the trunk		(3300,650)
Behind vertical cushion of backseat		(2500,700)
Under backseat		(2300,350)
Mat-shape		(2000,200)
I-shape		(1800,250)
T-shape		(2000,250)

10. Explain what need to be considered specifically for the electric and hybrid electric vehicle chassis and body designs and manufacturing.

11. Describe the main functions of automobiles' suspension systems.

12. What are the specific considerations in the design of suspension systems for electric and hybrid electric vehicles?

13. Describe the advantages and disadvantages of various types of power assist steering systems.

14. What are the specific considerations in the design of steering systems for electric and hybrid electric vehicles?

15. What are the specific considerations in the design of braking systems for electric and hybrid electric vehicles?

16. Describe the working principles of regenerative braking systems.

References

1. The Automotive Technology Roadmap Report (2004) *AutoBusiness*. See http://www.automotivecouncil.co.uk/what-we-do/reports/.
2. Happian-Smith, J. (2001) *An Introduction to Modern Vehicle Design*, Butterworth-Heinemann, Oxford.
3. Bodings, R., and Nisbett, K. (2010) *Shigley's Mechanical Engineering Design*, McGraw-Hill Education, New York.
4. Heisler, H. (2002) *Advanced Vehicle Technology*, Elsevier, Oxford.
5. Barnard, R.H. (2009) *Road Vehicle Aerodynamic Design*, MechAero Publishing, St Albans.

4

Vehicle Dynamics Fundamentals

4.1 Introduction

Understanding the basic concepts and terminologies of vehicle dynamics is an essential prerequisite for studying the dynamic behavior of electric and hybrid electric vehicles. As such, this chapter briefly addresses the general topics of vehicle dynamics, including the fundamentals and terminologies of vehicle dynamics, vehicle motions, forces and moments, and the equations of motion.

4.2 Concepts and Terminology

A vehicle is a device designed or used to transport people or cargo. There are different types of vehicles including ground vehicles, such as automobiles, trains, and snowmobiles; sea vehicles, such as ships, boats, and submarines; air vehicles, such as airplanes and helicopters; space vehicles, such as spacecrafts and other types of vehicles such as, hovercrafts. Ground vehicles can be further classified as wheeled vehicles, tracked vehicles, skied vehicles, and railed vehicles.

Wheeled vehicles are the most popular type of the ground vehicles. Wheeled vehicles include: single body-multi wheels vehicles, such as cars vans, and buses; single-body two wheels vehicles, such as bicycles and motorcycles, and multi-body multi-wheeled vehicles, such as articulated trucks. The concepts, terminology, and modeling presented in this chapter can be applied to different types of vehicles, but are illustrated with four-wheeled automobiles.

4.2.1 Evaluation Criteria for Vehicle Dynamics

In order to compare vehicles, it is necessary to define specific criteria to measure the qualities of a vehicle's dynamic behavior such as comfort, agility, safety, and controllability. Using these terms, vehicles can be judged and ranked based on different aspects of their behavior. The most important measures of the dynamic behavior of vehicles are detailed in the following sections.

Electric and Hybrid Vehicles: Technologies, Modeling and Control: A Mechatronic Approach, First Edition.
Amir Khajepour, Saber Fallah and Avesta Goodarzi.

4.2.1.1 Ride Comfort

Ride quality or ride comfort is concerned with passenger comfort and driver feeling in a moving vehicle. The transmitted vibration to the passengers is the main governing factor of the vehicle ride quality. This vibration originates from different sources, including road unevenness, airflow around the vehicle, powertrain, and tires; however, road irregularities are indeed the major source of the vibration.

Ride comfort is directly related to the acceleration transmitted to the passengers mainly in the vertical direction. Minimizing the peak values as well as the integral of the transmitted accelerations in the frequency domain in specific frequency ranges improves the vehicle ride quality.

4.2.1.2 Road handling

The term handling is often used to describe the maneuverability, turning and directional response of a road vehicle. Handling quality can be judged based on factors, such as good command following and controllability of the vehicle, predictability of the vehicle's behavior, the necessary effort by the driver to keep the vehicle on track, and the quality of the vehicle transient and steady state responses.

4.2.1.3 Acceleration/Braking Performance

Acceleration performance and braking performance or simply performance is a term that represents the vehicle agility and mobility as well as the effectiveness of its braking behavior. This quality of a car is related to factors such as accelerating time, top speed, and stopping distance when braking.

In automotive engineering terminology, a high-performance car is a car with the capability to accelerate quickly, reach a high velocity very quickly, and stop in a short distance.

4.2.1.4 Road Holding

For improved accelerating and braking performance, good handling and cornering quality, and for keeping the vehicle safe and functional, the tires should be kept in contact with the road during the different driving conditions. The term road holding is used to describe the quality of the tires and road contact.

Road holding is directly related to the tires' normal forces. In situations such as severe cornering or driving on rough roads, the normal forces of some tires may dramatically decrease or oscillate which can disturb the vehicle safety and functionality. Road holding is strongly affected by the suspension system's design and configuration.

4.2.2 Weights and Dimensions

Vehicle technical specifications include different weight and dimension factors, some of which are very important in vehicle dynamics. This section introduces these weights and dimensions.

4.2.2.1 Vehicle Weights

At a first glance, weight may seem to be simple, basic vehicle data; however, in vehicle dynamics analyses, it requires careful consideration. The weight of a vehicle may vary in

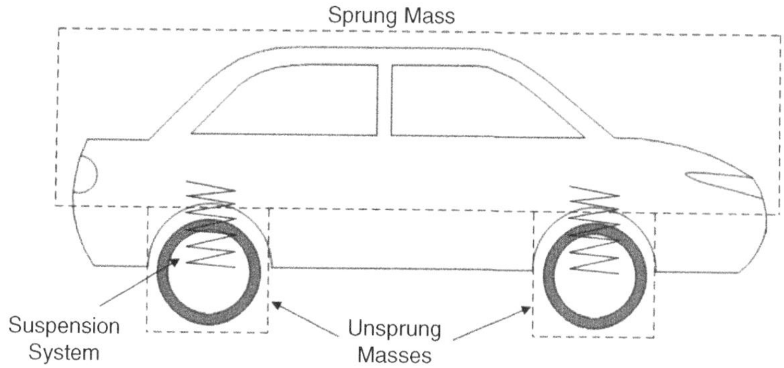

Figure 4.1 Illustration of sprung and unsprung masses

different situations. Also a vehicle is a multi-body system, formed by some distinct masses, which require independent consideration. To clarify these subjects the following standard definitions are necessary:

- *Curb weight:* The weight of the vehicle without passengers and cargo.
- *Payload:* The weight of passengers and cargo.
- *Gross vehicle weight (GVW):* The maximum admissible vehicle weight. Obviously it is the summation of the vehicle curb weight and the maximum payload.
- *Sprung/unsprung masses:* As shown in Figure 4.1, the main portion of a vehicle mass is supported by the suspension system. It includes the mass of the vehicle body and its interior/exterior parts, the passenger and cargo, the powertrain system, fuel tank, etc. In vehicle dynamics terminology it is known as the sprung mass.

The total mass of the parts and components not supported by the suspension system is the unsprung mass. Each wheel has its own unsprung mass, which includes the mass of the wheel, wheel brake components, and some parts of suspension and steering linkages. It means that it is usually possible to divide a vehicle into a sprung and four unsprung masses. However, in the case of dependent suspension systems, each axle, rather than each wheel, has a dedicated unsprung mass.

4.2.2.2 The Vehicle's Dimensions

Figure 4.2 shows the main dimensions of a vehicle. The following dimensions are specifically considered in vehicle dynamic modeling and analysis:

- *Wheelbase:* The distance between the center of the front and rear wheels.
- *Wheel track:* The transversal distance between the center lines of the wheels of an axle. Generally, front and rear axles can have different wheel tracks, but in practice, their values are close to each other.
- *Vehicle's center of gravity (CG) position:* The position of the vehicle's CG is a very important factor in vehicle dynamics. The vehicle's CG is determined by its height h and the distance of CG from the front and rear axles which are shown by a and b respectively, as

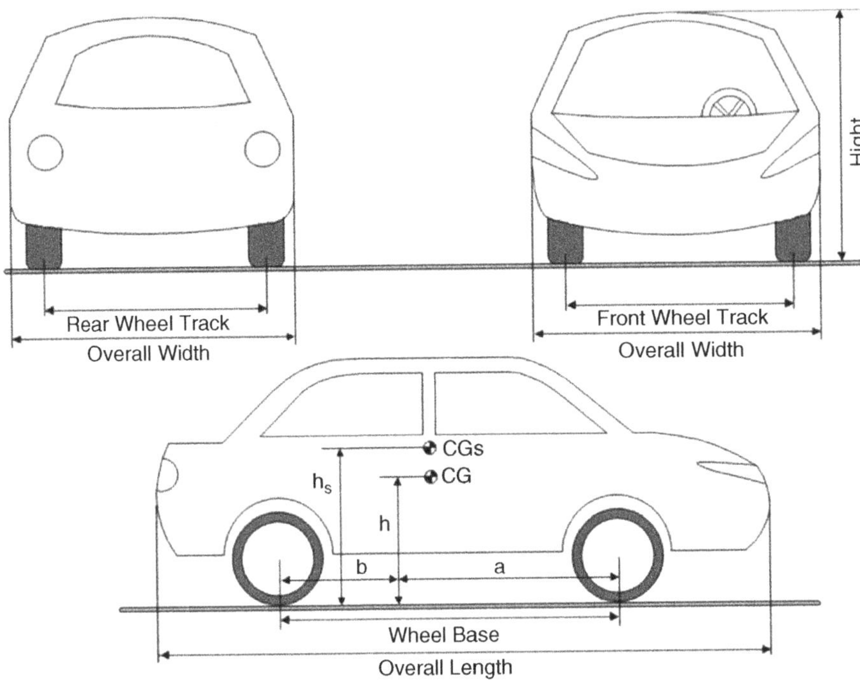

Figure 4.2 Important vehicle dimensions in vehicle dynamics

depicted in Figure 4.2. Typically, a vehicle is supposed to have a symmetric shape in transversal direction with its CG in the vehicle's middle plane.

- *Sprung/unsprung masses CG positions:* In the advanced vehicle dynamics, in addition to the vehicle CG position, the position of the sprung and each unsprung mass also needs to be known. Of the dimensions of sprung and unsprung masses' CG, the sprung mass CG height h_s is very important in vehicle dynamics and is frequently used.

4.3 Vehicle Kinematics

4.3.1 Vehicle Coordinate Systems

To study the dynamic behavior of a vehicle, it is necessary to properly define the coordinate system(s). For this purpose, two types of coordinate systems are needed: body-fixed and ground-fixed coordinate systems.

As illustrated in Figure 4.3, the body-fixed coordinate system is attached to the vehicle and all the vehicle motions are expressed in this coordinate system. According to the definition by the Society of Automotive Engineers (SAE); the body-fixed coordinate system is a right-handed coordinate system, meaning that all rotations are clockwise about the axis of rotation; the origin is at the vehicle center of gravity (CG); the z-axis is perpendicular to the ground and its positive direction is downward; the x-axis is along the vehicle body, and its positive direction is frontward and the right-handed rule determines the y-axis positive direction [1]. In vehicle dynamics terminology, each axis has a specific name. x, y and z axes are respectively known as longitudinal, lateral, and vertical axes.

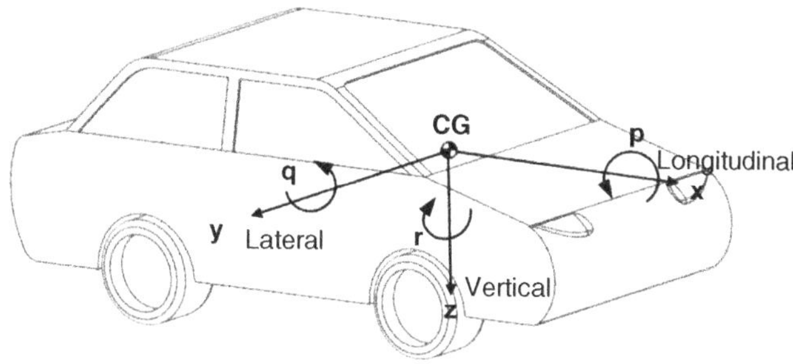

Figure 4.3 Body-fixed vehicle coordinate system

In addition to the body-fixed coordinate system, to express wheel dynamics requires a dedicated coordinate system for each wheel, as shown in Figure 4.4. Based on the SAE definition, the system is fixed to the vehicle and their origins are in each center of the tire and road contact patch. They are right-handed coordinate systems and their axes always remain parallel to the body-fixed coordinate system.

The wheel position in its local coordinate system is determined by the angles between the wheel axes of symmetry and the axes of the coordinate system. As illustrated in Figure 4.4, the angle between the wheel's longitudinal axis of symmetry and x-axis of the coordinate system is defined as the steering angle. The angle between the wheel's vertical axis of symmetry and the z-axis of the local coordinate system is known as the camber angle. In automobile wheels, the camber angle is small (up to 2–3 degrees) but for motorcycles, the camber angle can be very large, even more than 45 degrees during high speed turns.

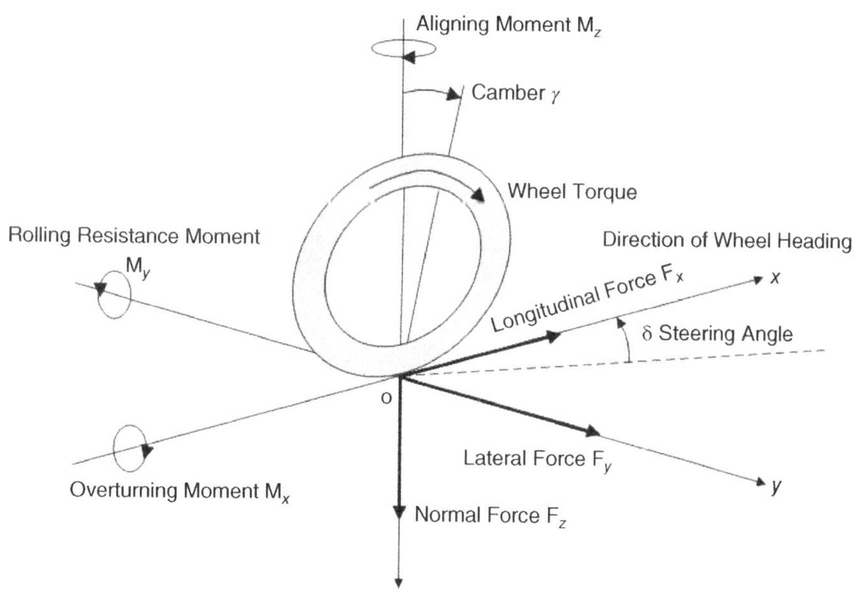

Figure 4.4 Wheel coordinate system

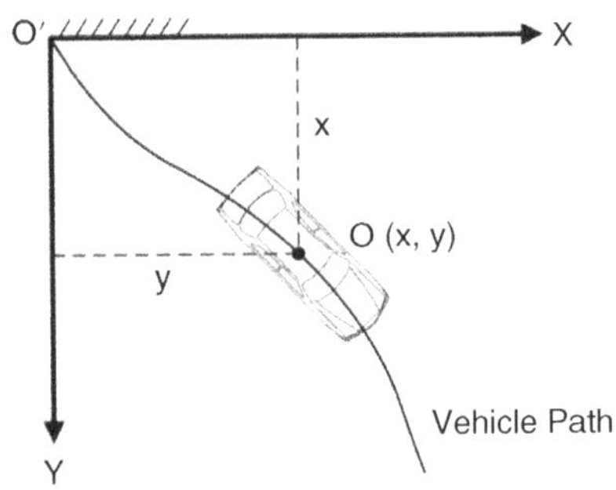

Figure 4.5 Ground-fixed coordinate system

Finally, as shown in Figure 4.5, a ground-fixed coordinate system is needed to determine the vehicle path.

4.3.2 Vehicle Motions

A vehicle has linear and angular motions along and about each axis of the body-fixed coordinate system. Figure 4.3 earlier showed the vehicle motions. In vehicle dynamics terminology, each motion has a specific name.

The linear motion of the vehicle along the x-axis is longitudinal motion, and the vehicle dynamics in this direction is longitudinal dynamics. The vehicle's velocity, acceleration, and movement along the x-axis are respectively known as the longitudinal velocity, u; the longitudinal acceleration, a_x; and traveled distance, x. There is a very close relationship between the longitudinal dynamics behavior of a vehicle and its accelerating/braking performance. Longitudinal dynamics variables are the basis of expressing and judging a vehicle's performance during accelerating and braking; these include longitudinal velocity, longitudinal acceleration, and traveled distance.

The angular motion of the vehicle about the x-axis is roll motion. It is mainly specified by the roll angle φ and the roll rate p. Roll motion has significant effects on the vehicle handling and rollover stability.

The linear motion of the vehicle along the y-axis is the lateral motion and the vehicle dynamics in this direction is lateral dynamics. Lateral velocity v and lateral acceleration a_y are the velocity and acceleration of the vehicle along the y-axis, respectively. They are very important variables in vehicle handling.

Pitch motion is the angular motion of the vehicle about the y-axis. It is specified by pitch angle θ and pitch rate q. Hard braking and accelerating as well as driving on harsh uneven roads typically excite the pitch motion.

When a vehicle is driven on uneven roads, it is excited by the road irregularities and shows oscillatory motion along the z-axis. This motion is bounce motion and is specified by z, w and a_z as vertical displacement, vertical velocity, and vertical acceleration, respectively.

The angular motion of a vehicle about the z-axis is yaw motion and the vehicle dynamics in this direction is yaw dynamics. Yaw angle ψ and yaw rate r are the angular displacement and velocity of the vehicle about z-axis, respectively. They are very important variables in vehicle handling and stability.

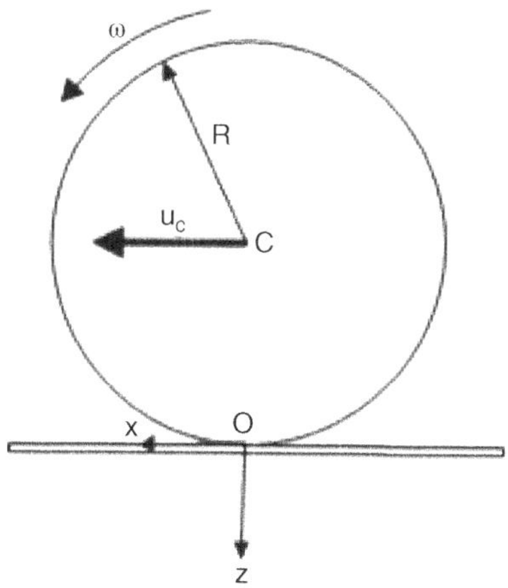

Figure 4.6 Schematic of a rotating wheel

4.3.3 Longitudinal and Lateral Slips

4.3.3.1 Longitudinal Slip of the Wheel

Figure 4.6 shows a rotating wheel. Generally, the rotation of an automobile wheel is not free rolling. When either driving or braking torque is applied to a wheel, it no longer rotates purely, and longitudinal slip occurs. The wheel longitudinal slip s can be formulated by:

$$s = \begin{cases} 1 - \dfrac{u_c}{R\omega} & \textit{if} \quad R\omega > u_c \quad \textit{Accelerating} \\ -1 + \dfrac{R\omega}{u_c} & \textit{if} \quad R\omega \le u_c \quad \textit{Braking} \end{cases} \tag{4.1}$$

where u_c, ω and R are the longitudinal velocity of the wheel's center, the wheel's angular velocity and the wheel radius, respectively. According to Equation 4.1, longitudinal slip is a parameter ranging between −1 to 1. It is negative during braking and is positive during accelerating. For a completely locked wheel during harsh braking, $u_c \neq 0$ but, $\omega = 0$ then s is -1, on the other hand, for a completely spinning wheel during harsh accelerating, $u_c = 0$ but $\omega \neq 0$, then s is 1. Sometimes the longitudinal slip is expressed in terms of percentage and its value is in [-%100, %100] range.

Example 4.1

Consider a vehicle is moving with constant speed of 20 m/s on a straight road. If the wheel radius is 0.35 m, calculate the longitudinal slip when the angular velocities of the wheels are (a) 630 rpm, (b) 460 rpm. Is the vehicle in accelerating or braking mode?

Solution

(a)

$$R\omega = 0.35 \times \left(630 \times \frac{2\pi}{60}\right) = 23.1\ m/s$$

$$u_c = 20\ m/s \quad R\omega > u_c \rightarrow Accelerating$$

$$s = 1 - \frac{u_c}{R\omega} = 1 - \frac{20}{23.1} = 13.4\%$$

(b)

$$R\omega = 0.35 \times \left(460 \times \frac{2\pi}{60}\right) = 16.9\ m/s$$

$$u_c = 20\ m/s \quad R\omega < u_o \rightarrow Braking$$

$$s = -1 + \frac{R\omega}{u_c} = -1 + \frac{16.9}{20} = -15.5\%$$

4.3.3.2 Wheel Sideslip Angle

The angle between a wheel's longitudinal axis of symmetry and the velocity vector at the tire-ground contact point V is known as wheel or tire sideslip angle. In other words, it is the angle between the tire's direction of heading and the tire's direction of travel. According to Figure 4.7, which depicts the top view of a tire, the tire sideslip angle can be expressed by the following equation:

$$\alpha = \delta - \tan^{-1}\frac{v_0}{u_0} \tag{4.2}$$

where δ is the steering angle and u_0, v_0 are the longitudinal and lateral velocity of the contact point, respectively.

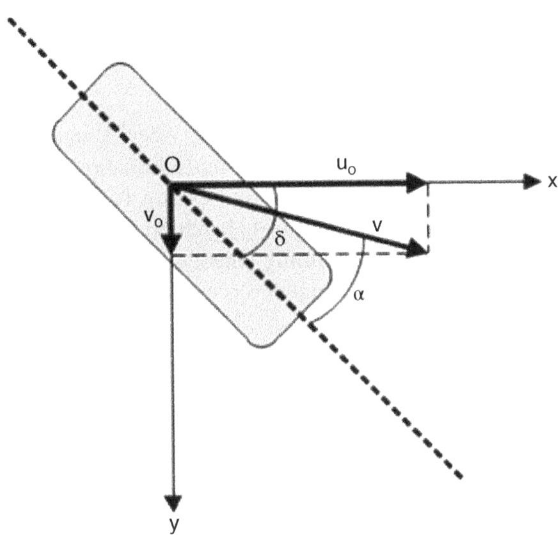

Figure 4.7 Definition of tire side slip angle

During normal driving conditions, the tire's sideslip angle is low, normally below 2–4 degrees. However, in severe conditions it can be much higher. In this book, the terms slip angle or lateral slip refer to the sideslip angle.

4.3.3.3 Vehicle Sideslip Angle

The vehicle sideslip angle is the consequence of the tire's sideslip angle. It is the angle between the longitudinal axis of the vehicle and its velocity vector. The vehicle sideslip angle is represented by:

$$\beta = -\tan^{-1}\frac{v}{u} \tag{4.3}$$

During normal driving conditions, the vehicle sideslip angle as well as the tire sideslip angle is very low, up to 2–4 degrees. In this situation, the sideslip angle is not recognizable and it seems that the vehicle always moves tangential to its path. But in the event of severe maneuvers, the sideslip angle can be dramatically increased.

Example 4.2

A vehicle is moving on a circular path. The vehicle dimensions as well as its linear and angular velocities are given in Table 4.1. Find the longitudinal and lateral velocity of the contact point of each wheel and accordingly calculate the sideslip angle of all wheels as well as the vehicle sideslip angle.

Table 4.1 The vehicle dimensions and velocities

Dimensions	Vehicle Configuration
$\delta = 0.05\ rad$ $T = 1.5\ m$ $a = 1\ m$ $b = 1.5\ m$	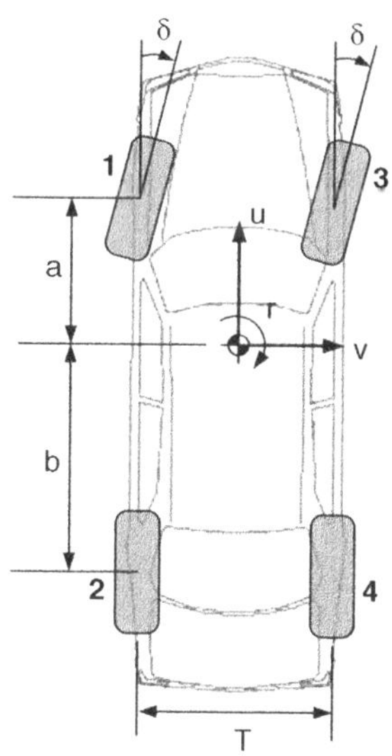
Velocities	
$u = 12\ m/s$ $v = 0.8\ m/s$ $r = 0.21\ rad/s$	

Solution

The longitudinal and lateral velocity of the contact point of each wheel can be obtained as follows:

$$u_{01} = u_{02} = u + \left(r \times \frac{T}{2}\right) = 12 + \left(0.21 \times \frac{1.5}{2}\right) = 12.16\, m/s$$

$$u_{03} = u_{04} = u - \left(r \times \frac{T}{2}\right) = 12 - \left(0.21 \times \frac{1.5}{2}\right) = 11.84\, m/s$$

$$v_{01} = v_{03} = v + (r \times a) = 0.8 + (0.21 \times 1) = 1.01\, m/s$$

$$v_{02} = v_{04} = v - (r \times b) = 0.8 - (0.21 \times 1.5) = 0.48\, m/s$$

The sideslip angle of the wheels and the vehicle sideslip angle are equal to:

$$\alpha_1 = \delta - \tan^{-1}\frac{v_{01}}{u_{01}} = 0.05 - \tan^{-1}\frac{1.01}{12.16} = -0.033\, rad = -1.89 \deg$$

$$\alpha_3 = \delta - \tan^{-1}\frac{v_{03}}{u_{03}} = 0.05 - \tan^{-1}\frac{1.01}{11.84} = -0.035\, rad = -2.01 \deg$$

$$\alpha_2 = -\tan^{-1}\frac{v_{02}}{u_{02}} = -\tan^{-1}\frac{0.48}{12.16} = -0.039\, rad = -2.23 \deg$$

$$\alpha_4 = -\tan^{-1}\frac{v_{04}}{u_{04}} = -\tan^{-1}\frac{0.48}{11.84} = -0.041\, rad = -2.35 \deg$$

$$\beta = -\tan^{-1}\frac{v}{u} = -\tan^{-1}\frac{0.8}{12} = -0.067\, rad = -3.83 \deg$$

4.3.4 *Planar Vehicle Kinematics*

According to Figure 4.8, consider a vehicle to be a rigid body moving on the XY plane of the ground coordinate system. The vehicle has its own SAE-defined xyz body-fixed coordinate system. The vehicle is in position A at $t = 0$, then moves into position B at $t = dt$. Suppose that at $t = 0$, the longitudinal and lateral velocities of point o, the origin of the xyz coordinate system is u and v respectively, the vehicle yaw rate is r, and the vehicle yaw angle is ψ. At $t = dt$, the terms u, v, r, and ψ increase incrementally *to* $u + du, v + dv, r + dr$ and $\psi + \delta\psi$.

Now suppose an arbitrary point P is somewhere in the vehicle. The velocity terms of P at $t = 0$ and $t = dt$ can be expressed as below:

$$@t = 0 \quad \begin{cases} u_p = u - y_p r \\ v_p = v + x_p r \end{cases} \qquad @t = 0 \quad \begin{cases} u_p' = (u + \delta u) - y_p(r + \delta r) \\ v_p' = (v + \delta v) + x_p(r + \delta r) \end{cases} \tag{4.4}$$

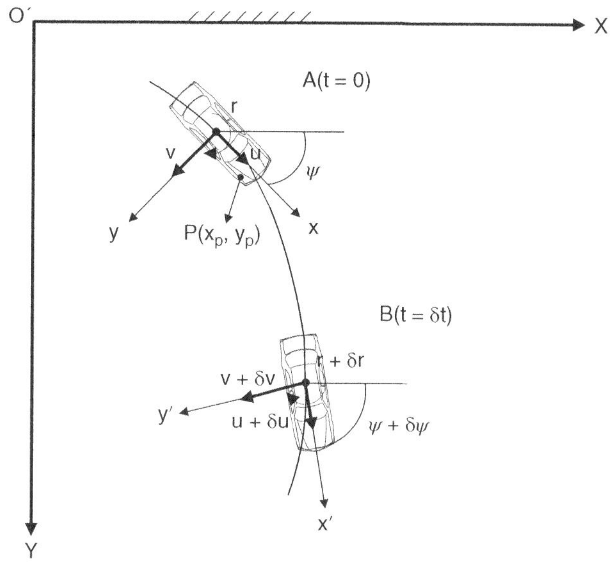

Figure 4.8 Vehicle planar kinematics

To find the incremental changes in the velocity of point P after dt, the velocity vectors at $t = dt$ should be projected into the initial position of x and y axes, then we have:

$$\begin{cases} \delta u_p = u_p^{'}\cos\delta\psi - v_p^{'}\sin\delta\psi - u_p \\ \delta v_p = u_p^{'}\sin\delta\psi + v_p^{'}\cos\delta\psi - v_p \end{cases} \tag{4.5}$$

By substituting the values of $u_p, v_p, u_p^{'}, v_p^{'}$ from Equations (4.4) into Equation (4.5), the incremental changes in longitudinal and lateral velocities of point P are found as:

$$\begin{cases} \delta u_p = [(u+\delta u) - y_p(r+\delta r)]\cos\delta\psi - [(v+\delta v) + x_p(r+\delta r)]\sin\delta\psi - (u - y_p r) \\ \delta v_p = [(u+\delta u) - y_p(r+\delta r)]\sin\delta\psi + [(v+\delta v) + x_p(r+\delta r)]\cos\delta\psi - (v + x_p r) \end{cases} \tag{4.6}$$

$\delta\psi$ and δt are very small, then the following assumptions can be considered:

$$\frac{\delta}{\delta t} = \frac{\mathrm{d}}{dt}, \quad \cos\delta\psi = 1, \quad \sin\delta\psi = \delta\psi, \quad \dot{\psi} = \frac{d\psi}{dt} = r$$

By applying the above simplifying assumptions in Equation (4.6) and performing some algebraic operations, the acceleration of point P in x and y directions can be written as:

$$\begin{aligned} a_{xp} &= \frac{\delta u_p}{\delta t}_{\delta t \to 0} \Rightarrow a_{xp} = \dot{u} - vr - y_p\dot{r} - x_p r^2 \\ a_{yp} &= \frac{\delta v_p}{\delta t}_{\delta t \to 0} \Rightarrow a_{yp} = \dot{v} + ur + x_p\dot{r} - y_p r^2 \end{aligned} \tag{4.7}$$

Now, consider the generalized form of Newton's second law for a group of masses:

$$\sum F = \sum_i m_i a_i \tag{4.8}$$

A vehicle can be considered as a group of lumped masses, thus according to Equation 4.8, the vehicle equations of motion in x, y and z directions can be listed as:

$$\begin{cases} \sum F_x = \sum_p \delta m_p a_{xp} \\ \sum F_y = \sum_p \delta m_p a_{yp} \\ \sum M_z = \sum_p \delta m_p \left(x_p a_{yp} - y_p a_{xp}\right) \end{cases} \tag{4.9}$$

Based on the standard definition for center of mass and polar moment of inertia:

$$\begin{aligned} &\sum_p \delta m_p = M \\ &\sum_p \delta m_p x_p = M\bar{x} \\ &\sum_p \delta m_p y_p = M\bar{y} \\ &\sum_p \delta m_p \left(x_p^2 + y_p^2\right) = I_{zz} \end{aligned} \tag{4.10}$$

where $\bar{x}, \bar{y}$ are the coordinates of the vehicle center of mass. By substituting the acceleration terms from Equations 4.7 in the vehicle equations of motion (4.9) and considering Equations 4.10, eventually the general form of vehicle planar equations of motion can be written as below:

$$\begin{cases} \sum F_x = M(\dot{u} - vr) - M\bar{y}\dot{r} - M\bar{x}r^2 \\ \sum F_y = M(\dot{v} + ur) + M\bar{x}\dot{r} - M\bar{y}r^2 \\ \sum M_z = I_{zz}\dot{r} + M\bar{y}(\dot{u} - vr) + M\bar{x}(\dot{v} + ur) \end{cases} \tag{4.11}$$

Remember that based on the SAE definition, the origin of the vehicle xyz coordinate system is placed at the vehicle CG, so we have:

$$\begin{cases} \bar{x} = 0 \\ \bar{y} = 0 \end{cases} \tag{4.12}$$

Finally, the planar vehicle equations of motion are simplified to:

$$\begin{cases} \sum F_x = M(\dot{u} - vr) \\ \sum F_y = M(\dot{v} + ur) \\ \sum M_z = I_{zz}\dot{r} \end{cases} \tag{4.13}$$

4.3.5 Three-Dimensional Vehicle Kinematics

The above equations are only valid when considering the vehicle as a simple rigid body which moves in the XY plane. For more accurate modeling and analysis of vehicle dynamics behavior, the vehicle should be modeled as a three-dimensional system, considering pitch and roll motions, as well as vertical motion. For this purpose, at first these motions should be studied individually.

4.3.5.1 Roll Dynamics Model

At first glance, studying roll motion requires a complex multi-body model of the vehicle, including the suspension linkages, but by using the concept of roll center and roll axis, the roll motion can be analyzed without direct engagement with the suspension system kinematics. In other words, the roll center position concisely summarizes the effect of the suspension linkages. With a known roll center height, it is easy to calculate the roll angle and the load transfer at each of the front and rear axles.

According to the SAE definition, the roll center is: "The point in the transverse vertical plane through any pair of wheel centers at which lateral forces may be applied to the sprung mass without producing body roll" [2]. In other words, the roll center is the instantaneous center of rotation of the sprung mass relative to the ground.

The position of the roll center depends on the type and geometry of the suspension system. For each type of suspension system, there is a specific geometrical approach for finding the roll center position. Table 4.2 depicts the roll center position of different types of suspension systems.

According to Figure 4.9, knowing the front and rear roll center positions, the roll axis can be found by drawing a line between them. The sprung mass rotates around this hypothetical line relative to the ground. Generally, the roll axis can be pointing upward or downward or can be parallel to the ground.

Table 4.2 Roll center position for different types of suspension systems

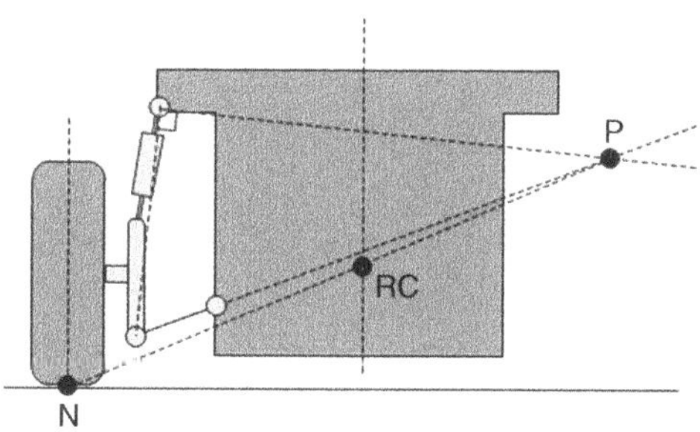 McPherson	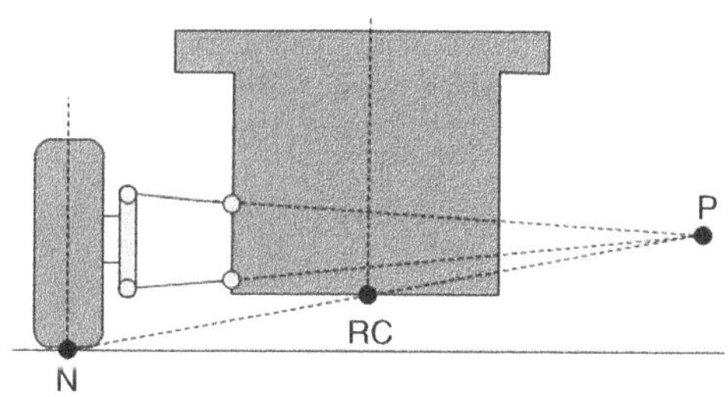 Double wishbone
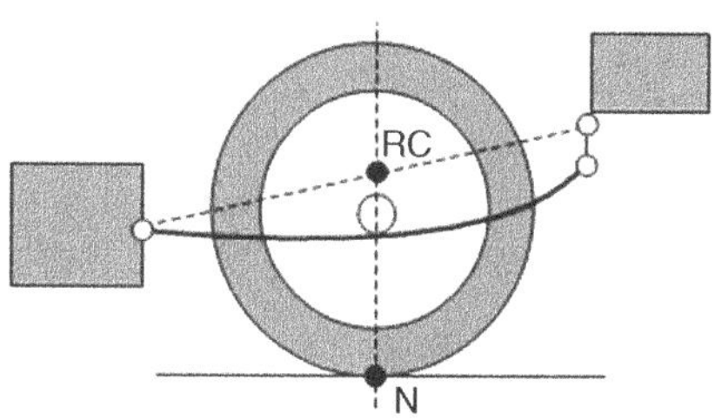 Solid axle	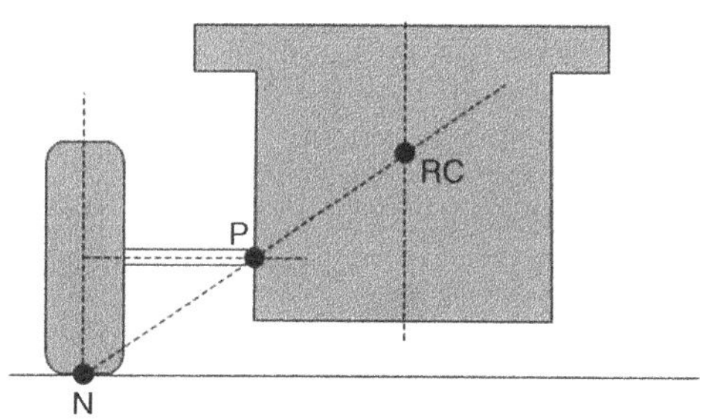 Trailing arm

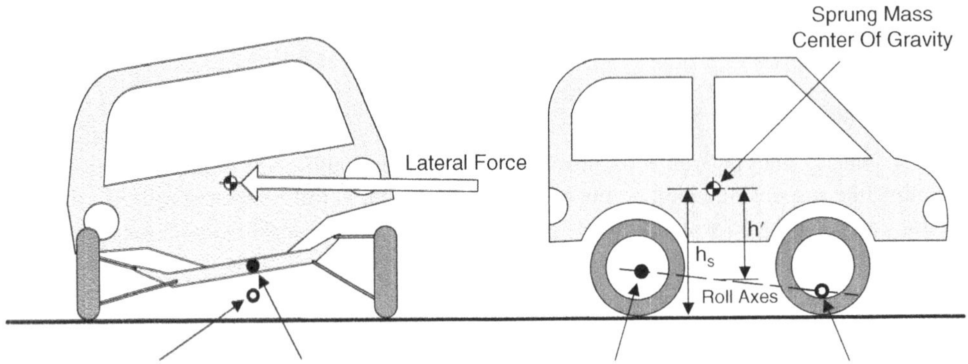

Figure 4.9 General illustration of front/rear roll centers and the roll axis

Having the roll axis position, roll dynamics can be expressed by a one degree of freedom roll plane model. For more simplicity, let's assume that the front and rear roll centers' height are equal and that the roll axis is parallel to the ground. Figure 4.10 depicts the roll plane model in detail. During cornering, the inertia force due to lateral acceleration is applied to the vehicle. As a result, the vehicle body will turn to the outside of the bend. By assuming that the roll angle is small, the moment equation about the roll axis can be written as:

$$\left(I_{xx_s} + M_s h'^2\right)\ddot{\varphi} = -C_\phi \dot{\varphi} - \left(K_\phi - M_s g h'\right)\varphi - M_s a_y h' \tag{4.14}$$

Where I_{xx_s}, C_ϕ, K_ϕ, h', and M_s are sprung mass moment of inertia around the x axis, the roll damping coefficient, the roll stiffness coefficient, the distance between the sprung mass CG and the roll axis, and the sprung mass, respectively.

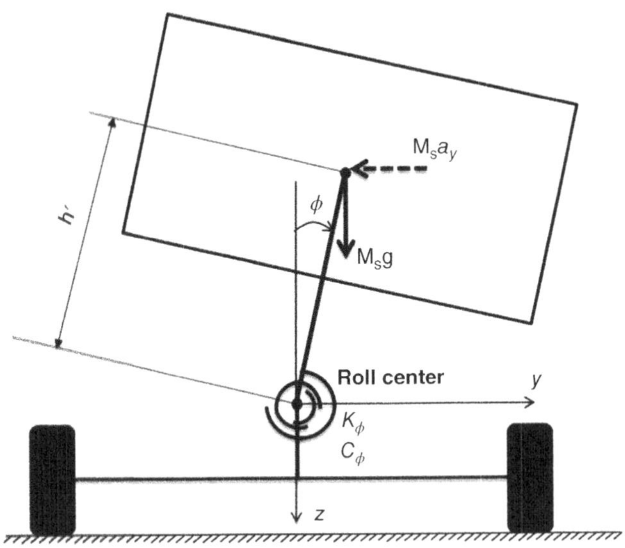

Figure 4.10 Roll plane model (rear view)

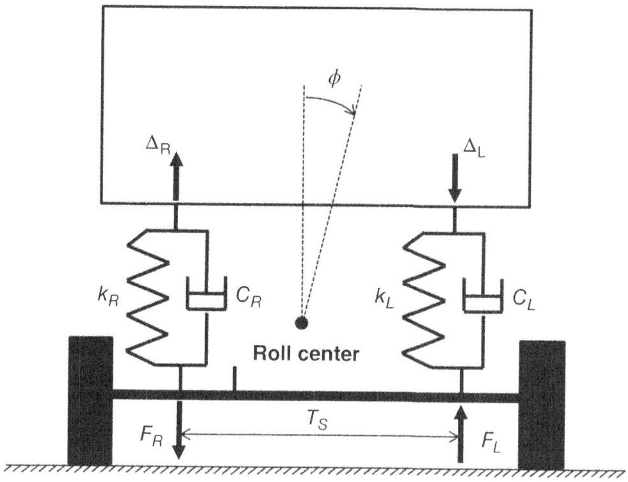

Figure 4.11 Rear axle suspensions (rear view)

The next step is to calculate the roll stiffness and roll damping coefficients. Figure 4.11 shows how the suspension is deflected by the roll motion. Assume the left and right suspensions are the same ($k_L = k_R = k$, $c_L = c_R = c$).

The absolute amount of deflection of right and left suspensions are equal and can be expressed as:

$$\Delta = \Delta_R = \Delta_L \approx \frac{T_s \phi}{2} \tag{4.15}$$

Due to the left and right spring deflections, left and right spring forces are generated as:

$$F_R = F_L = k\Delta = \frac{kT_s\phi}{2} \tag{4.16}$$

These forces induce the roll moment around the roll axis:

$$M_\phi = \frac{kT_s^2\phi}{2} \tag{4.17}$$

Therefore, the equivalent roll stiffness coefficient of each axle is:

$$k_\phi = \frac{M_\phi}{\phi} = \frac{1}{2}kT_s^2 \tag{4.18}$$

Here, the effect of the anti-roll bar has to be added because an anti-roll bar acts as a torsional spring and resists roll motion. By doing the same for the front and rear axles, the final equation for the roll stiffness coefficient is:

$$k_\phi = \frac{1}{2}k_f T_{sf}^2 + \frac{1}{2}k_r T_{sr}^2 + K_{ARB,f} + K_{ARB,r} \tag{4.19}$$

where k_f and k_r are the front and rear spring stiffness coefficients, $K_{ARB,f}$ and $k_{ARB,r}$ are the front and rear anti-roll bar stiffness coefficients, and T_{sf} and T_{sr} are the front and rear suspension installation span, respectively. By applying the same approach, the roll damping coefficient can be calculated as

$$C_\phi = \frac{1}{2} c_f T_{sf}^2 + \frac{1}{2} c_r T_{sr}^2 \tag{4.20}$$

where c_f and c_r are the front and rear suspensions damping coefficients, respectively.

4.3.5.2 Roll Effects

Roll motion affects passenger comfort and feeling, but the roll motion has more serious effects on the vehicle's handling and stability. The most important roll motion effects are:

- *Camber-by-roll:* Roll motion can change the wheels' camber angle. Camber change due to roll relates to suspension kinematic or elasto-kinematics behavior. Figure 4.12 illustrates a typical camber-by-roll characteristics curve. Changes in the tire camber angle through roll motion can affect the lateral force of the tire, and can eventually affect the vehicle yaw and lateral motions.
- *Steer-by-roll:* Roll motion can change the wheel steering angle as well as the camber angle. Again, the steering angle change due to roll relates to suspension/steering systems' kinematic or elasto-kinematic characteristics. Evidently, the roll motion, by changing the steering angles in the front and rear wheels, can affect the tires' lateral forces, and eventually affect the yaw and lateral dynamics.
- *Scrub-by-roll:* The wheel can move laterally on the road due to roll motion. This is known as scrub motion. Scrub motion is a characteristic of the suspension system and its kinematic or elasto-kinematic behavior. In addition to adverse effects of scrub motion on tire wear and its service life, it influences the yaw and lateral dynamics.

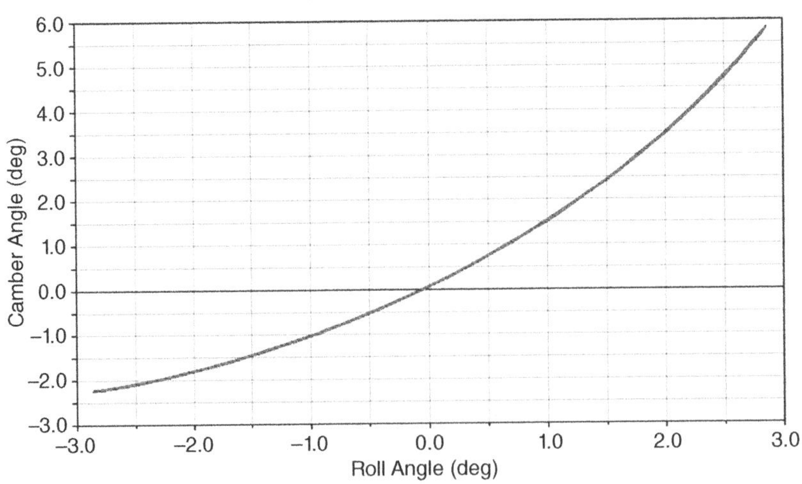

Figure 4.12 Typical camber-by-roll characteristics curve

- *Lateral load transfer:* During cornering because of lateral acceleration, lateral load transfer (LLT) occurs, and a portion of vehicle weight transfers from the inside of the bend wheels to the outside wheels. Roll motion can magnify the lateral load transfer. Lateral load transfer has significant effects on vehicle handling and overturning threshold.

4.3.5.3 Pitch Dynamics Model

In some events, such as harsh accelerating and braking, pitch dynamics can significantly affect the overall behavior of the vehicle. Figure 4.13 depicts the pitch plane model of a vehicle. The pitch moment equation around the pitch center is as follows:

$$\left(I_{yy_s} + M_s h''^2\right)\ddot{\theta} = -C_\theta\dot{\theta} - \left(K_\theta - M_s g h''\right)\theta + M_s a_x h'' \tag{4.21}$$

where I_{yy_s} is the sprung mass moment of inertia around the y axis, h'' is the distance between the sprung mass CG and the pitch center, C_θ is the pitch damping coefficient, and K_θ is the pitch stiffness coefficient. The pitch center location can be found based on suspension geometry at each instant [2]. However, Equation 4.21 is calculated based on the assumption that the vehicle pitch center is known and fixed.

Here, it is necessary to calculate the pitch stiffness and pitch damping coefficients for this model. In Figure 4.14, the vehicle's right-side view with the suspension system is shown. The same approach applied in the previous section also works to calculate these coefficients. Here, the front and rear suspension deflections are:

$$\Delta_f \approx L_f\theta \tag{4.22}$$

$$\Delta_r \approx L_r\theta \tag{4.23}$$

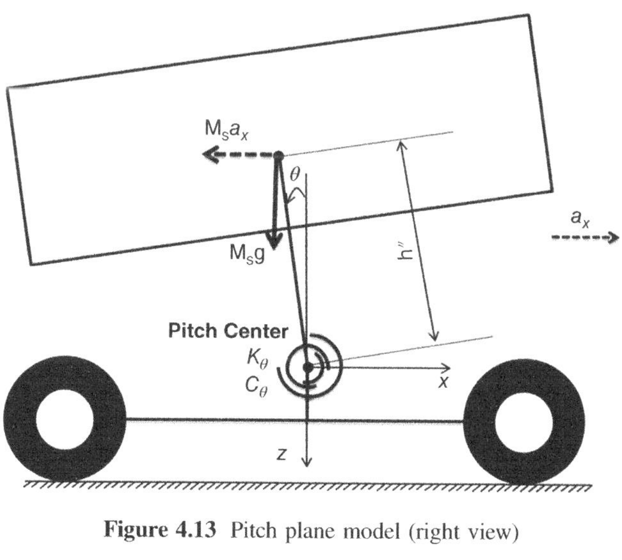

Figure 4.13 Pitch plane model (right view)

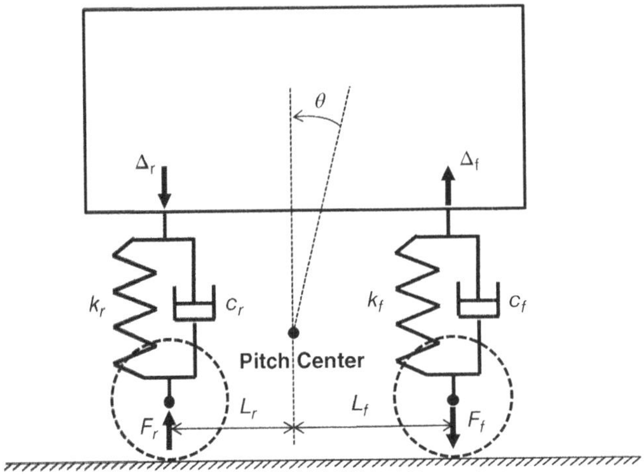

Figure 4.14 Calculation of pitch stiffness coefficient

The front and rear suspension forces and induced pitch moment are:

$$F_f = k_f\Delta_f = k_f L_f\theta \tag{4.24}$$

$$F_r = k_r\Delta_r = k_r L_r\theta \tag{4.25}$$

$$M_\theta = 2k_f L_f^2\theta + 2k_r L_r^2\theta \tag{4.26}$$

The final form of the suspension pitch stiffness is:

$$K_\theta = \frac{M_\theta}{\theta} = 2k_f L_f^2 + 2k_r L_r^2 \tag{4.27}$$

By applying the same approach, the pitch damping coefficient can be calculated as:

$$C_\theta = 2c_f L_f^2 + 2c_r L_r^2 \tag{4.28}$$

4.3.5.4 Three-Dimensional Vehicle Equations of Motion

The three-dimensional equations of motion are suitable for sophisticated and precise vehicle dynamics modeling and analysis, but they are obviously more complicated than simple planar vehicle equations.

To develop such equations, Newton's as well as Lagrange's approaches can be used. The details of three-dimensional vehicle kinematics can be found in many books in the field of vehicle dynamics such as [3]. However considering the purposes of this book, here a simplified form of the three-dimensional vehicle equations of motion is presented:

$$\begin{aligned}
&\sum F_x = M(\dot{u} - vr)\\
&\sum F_y = M(\dot{v} + ur + h\ddot{\varphi})\\
&\sum M_z = I_z\dot{r}\\
&\left(I_{xx_s} + M_s h'^2\right)\ddot{\phi} + C_\phi\dot{\phi} + \left(K_\phi - M_s g h'\right)\phi - M_s(\dot{v} + ur)h' = 0
\end{aligned} \tag{4.29}$$

The proposed equations include longitudinal, lateral, yaw, and roll dynamics. To simplify the equations further, it is assumed that the roll axis is parallel to the ground.

Note that for more accurate vehicle dynamics simulation and analysis, in addition to yaw, lateral and longitudinal motions, it is important to consider roll motion. The roll motion can influence the vehicle dynamics behavior through its effects on the wheels' angles and movement, as well as the vehicle's lateral load transfer during cornering. Usually it is acceptable to neglect the effects of vertical and pitch motions on the vehicle handling behavior during driving on even roads.

4.3.6 Vehicle Forces and Moments

According to the vehicle equations of motion in Equation 4.29, in addition to the vehicle mass, inertias, and other vehicle parameters, the external forces and moment must be known as the inputs. Generally, the applied forces and moments to the vehicle can be categorized as tires and aerodynamics forces, and moments.

4.3.6.1 Tire Forces and Moments

Tires are the primary source of the applied forces and moments on the vehicle. The forces and moments exerted on the tire by road via the small contact patch area almost entirely control vehicle motion. Consider the wheel local coordinate system, illustrated in Figure 4.4; the tire's force and moments are:

- *Longitudinal force (F_x):* A force along the x-axis that significantly affects vehicle accelerating and braking performances. Positive and negative longitudinal forces are known as the tractive and braking forces, respectively.
- *Lateral force (F_y):* A force along the y-axis that directly affects the vehicle's handling and stability.
- *Normal force (F_z):* A force along the z-axis also known as the holding force. The vehicle's road holding is directly affected and the vehicle handling and performance are indirectly affected by the tires' normal forces.
- *Rolling resistance moment (M_y):* A moment about the y-axis. Rolling resistance moment is always applied in the negative direction, therefore it is considered a source of resistance acting against the vehicle's longitudinal motion. It is produced as a result of non-uniform deformation in the tire and road contact patch. As shown in Figure 4.15, the rolling resistance moment can be formulated as:

$$M_y = eF_z \tag{4.30}$$

where e is the offset between applied normal force on the tire and its equivalent reaction acting on the contact patch. For simplicity, the effects of rolling resistance moments of all wheels acting on the vehicle are replaced by a single resistance force called the rolling resistance force F_{RR}, which is assumed to be applied to the vehicle directly. This force can be formulated as:

$$F_{RR} = \mu_R W \tag{4.31}$$

where μ_R and W respectively are the rolling resistance coefficient and the vehicle weight. Generally, the rolling resistance coefficient is a function of many factors such as road

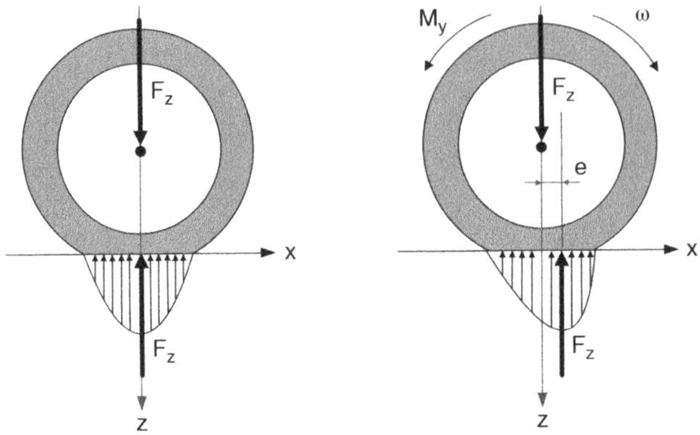

Figure 4.15 The generation of rolling resistance moment

condition, tire construction, tire inflation pressure and mainly the vehicle's speed. But for simplicity, it can be considered a constant parameter. As a rough estimation, its value can be taken as 0.015 for passenger cars and 0.01 for trucks.

- *Self-aligning moment (M_z):* A moment about the z-axis. This moment can affect the yaw motion and eventually the vehicle handling and stability. Its magnitude, however, is relatively small and vehicle handling is not considerably affected by the self-aligning moment.
- *Overturning moment (M_x):* This moment is about the x-axis. For automobile tires, its value is very small and can be ignored but for motorcycle tires it has considerable value and should be taken into the account in the motorcycle dynamics analysis.

The above-mentioned tire forces and moments will be formulated in Section 4.4.

4.3.6.2 Aerodynamics Forces and Moments

The surrounding air of a moving vehicle is the other source of external forces and moments to the vehicle. Unlike tire forces and moments, the aerodynamics forces and moments are directly applied to the vehicle body. The nature of vehicle aerodynamics is a complicated matter; analysis of the airflow around a car has only been possible in recent decades after considerable advances in computing technologies. As a result, in recent decades the importance of aerodynamics in vehicle performances, handling, and stability has been recognized more than before.

As shown in Figure 4.16, the aerodynamic forces and moments that act upon a vehicle are: drag force in the longitudinal direction, lift force in the vertical direction, side force and pitch, roll and yaw moments. Lift and side forces can have significant effects on the stability of vehicles. Aerodynamic side forces when driving in a cross-wind can be considerable for buses, trailers, vans and SUVs. In addition, the lift force can positively or negatively affect the tires' holding force and can indirectly affect the vehicle's performance, stability, and handling. However, the effect of lift force on the vehicle dynamics is important at high speed driving.

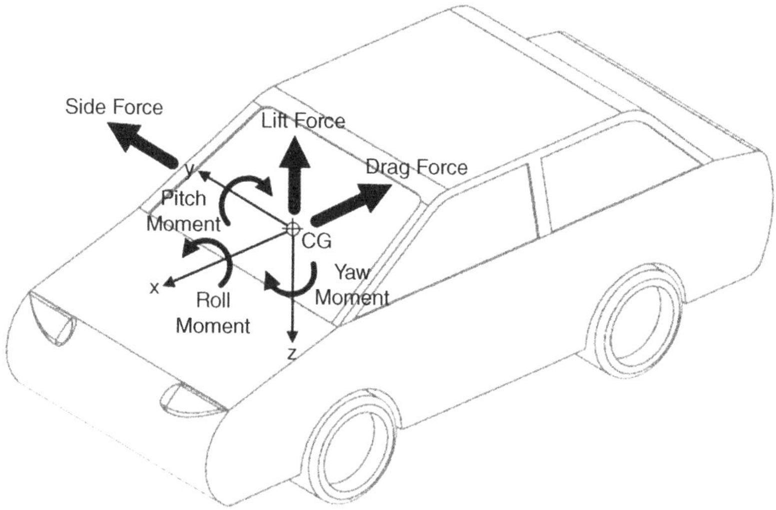

Figure 4.16 The aerodynamic forces and moments

Aerodynamic drag force is the most important aerodynamic effect that acts on a vehicle. Drag force has a strong influence on the vehicle acceleration, top speed, and fuel consumption. It can be formulated as:

$$F_{RA} = \frac{1}{2}\rho_a C_d A_f u^2 \tag{4.32}$$

where ρ_a is the air density, C_d is the drag coefficient, A_f is the vehicle frontal projection area and u is the vehicle speed. Drag coefficient is mainly determined based on the vehicle general shape. It is also affected by the details of the body design including the tail geometry, the side mirror design, the underbody geometry, the wheel cover design, the internal flow, the grille opening area, etc. The drag coefficients of different vehicle types vary greatly. For example, the typical value of a modern passenger car drag coefficient is about 0.35, for a modern van, buses, and articulated trucks, they are considerably higher at 0.45, 0.6, and 0.7, respectively.

Example 4.3

Consider a vehicle moving along a flat road. Draw graphs of the aerodynamics force, rolling resistance force, and total resistance force versus longitudinal speed. Use the following vehicle parameters in Table 4.3. At the speeds of 10 m/s and 35 m/s, find the percentage of aerodynamics force from the total resistance force.

Table 4.3 Vehicle parameters

$\rho_a = 1.2\frac{kg}{m^3}$	$C_d = 0.3$	$A_f = 2.1\,m^2$	$M = 1200\,kg$	$\mu_R = 0.015$

Solution

$$F_{RA} = \frac{1}{2}\rho_a C_d A_f u^2 = \frac{1}{2}\times 1.2\times 0.3\times 2.1\times u^2 = 0.378u^2$$

$$F_{RR} = \mu_R W = 0.015\times 1200\times 9.81 = 176.58N$$

$$F_R = \frac{1}{2}\rho_a C_d A_f u^2 + \mu_R W = 0.378u^2 + 176.58$$

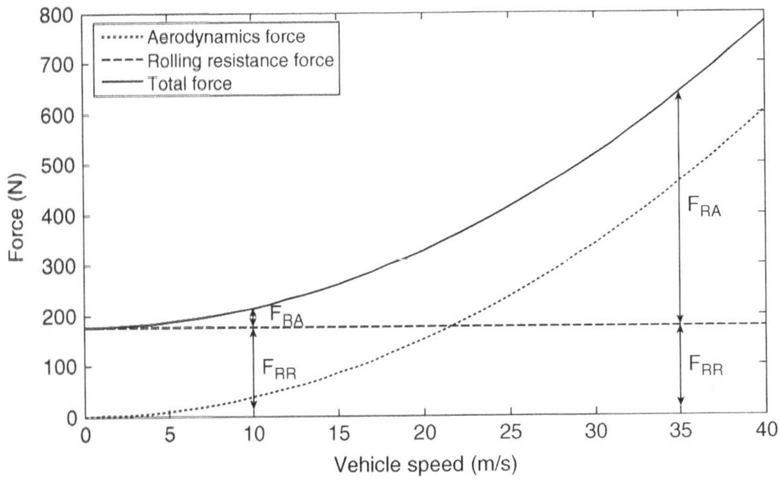

$$u = 10\,m/s$$

$$F_{RA} = 0.378\times 10^2 = 37.8\,N$$

$$F_R = 214.38$$

$$\frac{F_{RA}}{F_R} = \frac{37.8}{214.38} = 17.63\%$$

$$u = 35\,m/s$$

$$F_{RA} = 0.378\times 35^2 = 463.05\,N$$

$$F_R = 639.63$$

$$\frac{F_{RA}}{F_R} = \frac{463.05}{639.63} = 72.39\%$$

The results show that the effect of aerodynamics force at higher speeds is more significant.

4.4 Tire Mechanics and Modeling

Tire forces and moments are the most important forces in a ground vehicle, and thus the accurate calculation of these forces is an essential part of vehicle dynamics simulation and analysis. Generally, due to the nonlinear viscoelastic behavior of the tire materials, the mechanism of force and the moment generation by tires are very complicated.

The longitudinal and lateral tire forces, as well as self-aligning moment, are the results of longitudinal and lateral shear stresses at the tire and road contact patch. The shear stresses are formed due to deformations and distortions of the tire in and around the contact patch.

Experimental investigations, as well as analytical studies, have shown that the deformations, the stresses and eventually the forces and moments of a tire are functions of the following input variables:

- *Normal force:* As illustrated in Figure 4.15, the applied normal force on a rotating tire generates a non-uniform asymmetric distribution of normal stress in the contact patch. Due to the non-uniform and asymmetric shape of the normal stress distribution, in addition to the reaction force, it produces moments about the longitudinal and lateral axes. As mentioned earlier, these moments are overturning and rolling resistance moments. Moreover, considering the road friction, normal stress can affect the shear stresses, and eventually, the lateral and longitudinal forces and self-aligning moments.
- *Longitudinal and lateral slips:* The tire distortions and deformations in the contact patch are the result of longitudinal and lateral slips.
- *Camber angle:* The camber angle can change both the normal and shear stress distribution on the contact patch.
- *Time:* In fact, tires do not generate forces and moments instantaneously. However, the tire response is very fast, and in most cases, its transient response can be ignored.

4.4.1 Tire Characteristic Curves

Tire characteristic curves are formed based on experimental data show the relationship between the tire forces and moments and the above-mentioned input variables in steady state conditions. They are very important for understanding the tire mechanics, and eventually the vehicle's dynamics behavior.

Generally, according to different possible road and vehicle driving situations, the tire working conditions can be categorized as pure slip and combined slip conditions, and each has dedicated characteristic curves, discussed below.

4.4.1.1 Pure Slip Characteristic Curves

Pure slip is the situation where only lateral or longitudinal slip occurs in the tires. In the majority of normal vehicle maneuvers, tires work in some kind of pure slip condition. For instance, during straight line braking, the lateral slip is almost zero, but longitudinal slip exists. On the other hand, during normal cornering with constant speed, longitudinal slip does not exist, but a sideslip angle occurs in the tires. Both of these conditions are considered pure slip.

Figure 4.17 illustrates a typical pure slip longitudinal force characteristic curve of a tire. The longitudinal force can be either braking or tractive force. The curves illustrate the variation of the longitudinal force F_x versus the longitudinal slip s for various normal forces F_z. As the curves represent a pure slip condition, the sideslip angle α does not exist. The camber angle γ does not have a significant effect on the longitudinal force so it is not necessary to include it in the calculation.

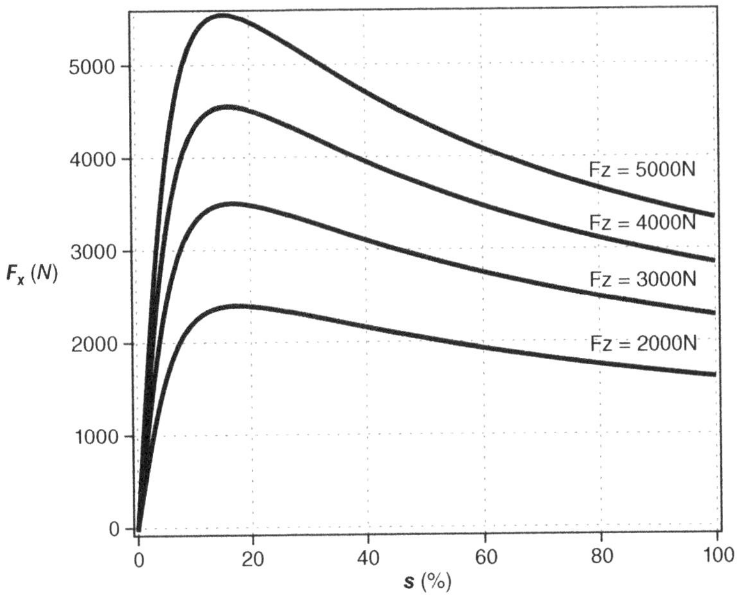

Figure 4.17 A typical pure slip tire longitudinal force characteristics curve

The curves have three different recognizable zones: linear, peak, and saturation zones. In the linear zone, the longitudinal force increases linearly by the longitudinal slip. Thus, the relationship between longitudinal force and longitudinal slip in the linear zone is:

$$F_x = c_s s \tag{4.33}$$

where c_s is the slope of the curve at the origin and is known as the tire longitudinal stiffness. In the peak zone, the longitudinal force reaches its peak value when the longitudinal slip is around 15–20%. The peak value is a function of the road coefficient of friction μ and the normal force F_z, represented as:

$$F_{x\max} = \mu F_z \tag{4.34}$$

After the peak point in the saturation zone, the force slightly decreases until the slip reaches 100%.

The lateral force characteristics curve is the other important tire pure slip characteristic curve and its typical illustration is shown in Figure 4.18. The family of curves represents the relationship of the lateral force F_y and the sideslip angle α for different normal forces F_z. As expected, the longitudinal slip should be zero. First, assume that the tire is upright to the road and the camber angle does not exist. A discussion of the effects of camber angle on the lateral force appears later.

As the curves show once again, there are three easily recognizable zones of linear, peak and saturation. In the linear zone, the lateral force is a linear function of sideslip angle:

$$F_y = c_\alpha \alpha \tag{4.35}$$

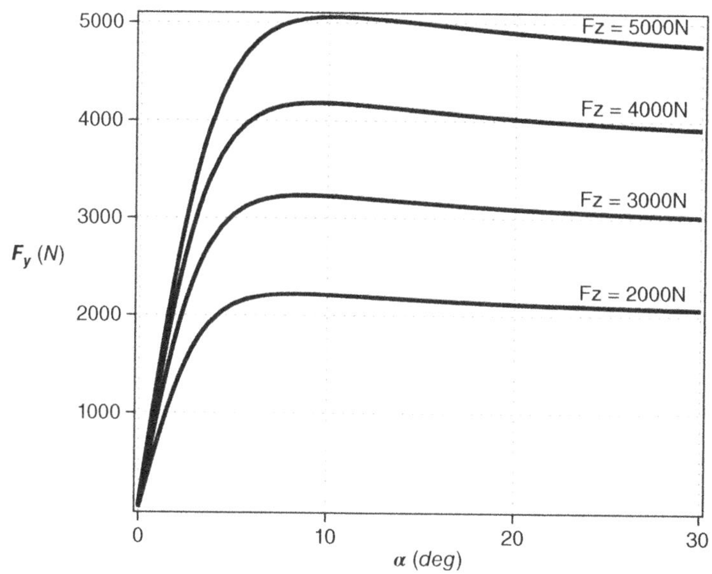

Figure 4.18 A typical pure slip tire lateral force characteristics curve

where c_α is the slope of the curve at the origin, it is known as the tire cornering stiffness. During regular handling and normal driving conditions, the tire working point is in the linear zone. The lateral force reaches its peak value around 4-8 degrees of side slip angle. The peak value can be formulated as:

$$F_{y\max} = \mu F_z \tag{4.36}$$

Similar to the longitudinal force, the lateral force is saturated after the peak point. In the saturation zone the value of lateral force slightly decreases, but practically it can be supposed to be a constant value.

Figure 4.19 illustrates the effect of the camber angle on the lateral force for a constant value of normal force. The relationship between the lateral force and camber angle is linear and in the absence of side slip angle, has the following formula:

$$F_y = c_\gamma \gamma \tag{4.37}$$

where c_γ is the camber stiffness. If both the camber and side slip angle exist and the tire working point is in the linear zone, the total lateral force becomes:

$$F_y = c_\alpha \alpha + c_\gamma \gamma \tag{4.38}$$

4.4.1.2 Combined Slip Characteristics Curves

Combined slip is the situation when lateral and longitudinal slips simultaneously occur in the tires. In other words, the tire concurrently generates both longitudinal and lateral forces.

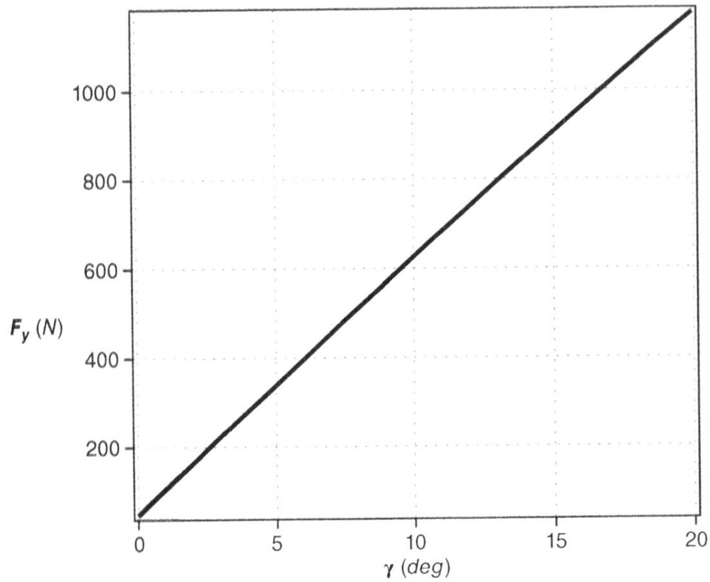

Figure 4.19 Effect of camber angle on the lateral force

This happens in specific driving conditions, such as accelerating and braking while turning, and driving on slippery roads. The tire behavior in combined slip situations is almost nonlinear and characteristic curves are more complex.

To understand the nature of the tire force and moment generation in combined slip situations, it is necessary to study friction circle theory. It is a simple theory that is not a seamless match with reality, but it does describe the concept of combined slip situation with satisfactory accuracy.

As illustrated in Figure 4.20a, consider a rotating tire under the normal force of F_z that simultaneously generates longitudinal and lateral forces. The resultant tire contact force is:

$$F = \sqrt{F_x^2 + F_y^2} \tag{4.39}$$

By using simple friction theory, the value of resultant forces F is equal to:

$$F = \mu F_z \tag{4.40}$$

By substituting Equation 4.40 into Equation 4.39 and with some algebraic operation, we finally have:

$$F_x^2 + F_y^2 = (\mu F_z)^2 \tag{4.41}$$

Figure 4.20(a) is a graphical representation of Equation 4.41. As depicted, a circle indicates the relationship between the tires' longitudinal and lateral force capacity in combined slip situations. As such, in combined slip situations, the capacities of the tire force generation in longitudinal and lateral directions are interconnected. In other words, generating force in each direction can reduce the capacity of the tire force in the other direction. This is a key

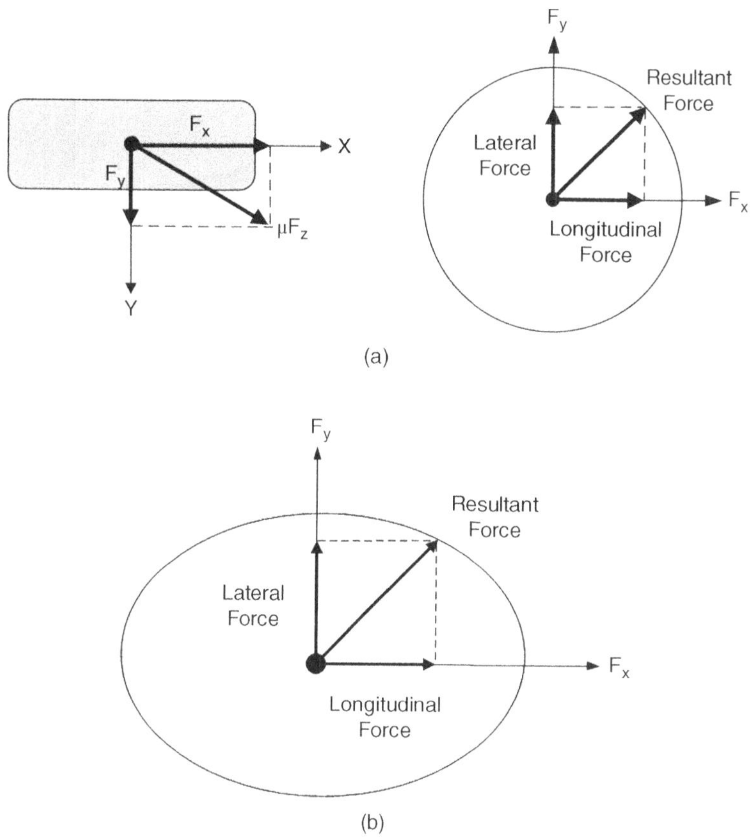

Figure 4.20 (a) Friction circle theory, (b) friction ellipse theory

understanding of the tire behavior in combined slip situations. However, in practice, due to the different coefficient of friction in the longitudinal and lateral directions, the friction circle is replaced by an ellipse. This is a more accurate theory which can precisely represent the tire behavior in the combined slip situations (Figure 4.20(b)).

Figure 4.21 shows a typical combined slip longitudinal force characteristics curve. This is a family of curves depicting the variation of longitudinal force versus longitudinal slip for various slip angles assuming a constant normal force. As predicted by the friction circle/ellipse theories, in the presence of the side slip angle, the tire longitudinal force capacity decreases. The deterioration of the tire capacity is not significant in the linear and saturation zones, but the side slip angle limits the peak capacity of the tire.

Figure 4.22 is a typical combined slip lateral force characteristics curve. Again, the normal force is assumed to be constant and the curves depict the variation of lateral force versus side slip angles for various values of longitudinal slip. As illustrated, the tire lateral force dramatically decreases as the result of longitudinal slip. In this case, reduction in the lateral force capacity is visible in the entire span of the tire working zones. This curve completely supports the predictions of the friction circle/ellipse theory.

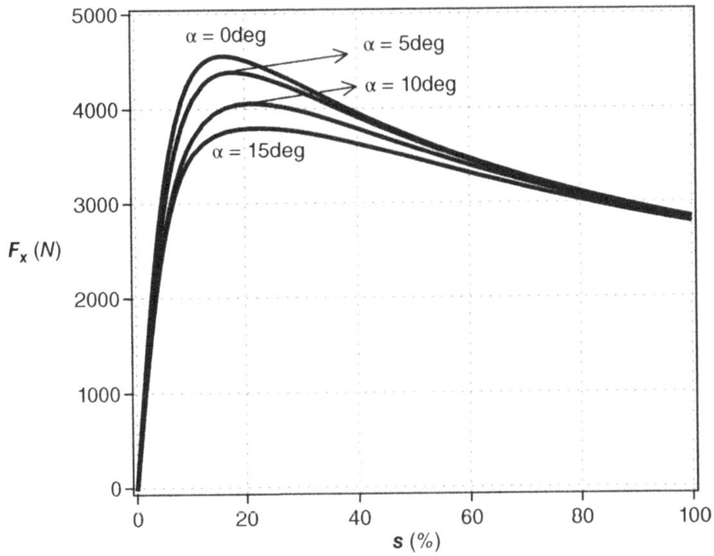

Figure 4.21 Combined slip longitudinal force characteristics curve

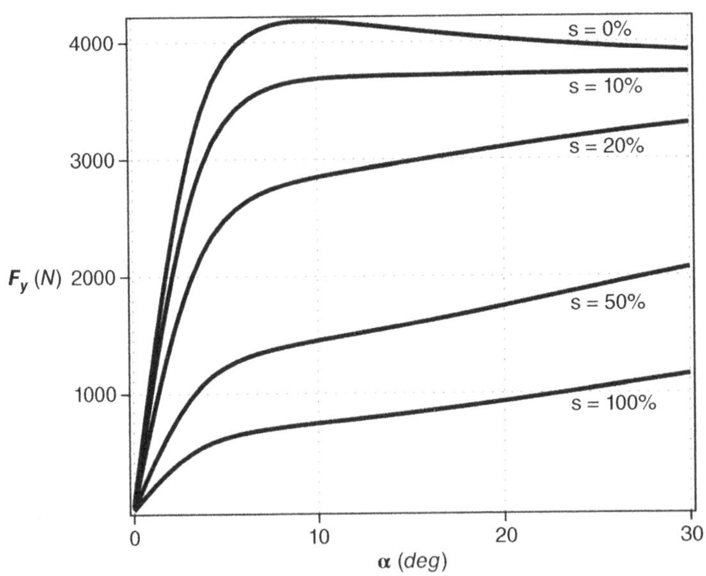

Figure 4.22 Combined slip lateral force characteristics curve

4.4.2 Tire Models

The tire forces and moments in a steady state situation can be represented by the following general forms:

$$\begin{aligned} F_x &= F_x(\alpha, s, F_z, \gamma) \\ F_y &= F_y(\alpha, s, F_z, \gamma) \\ M_z &= M_z(\alpha, s, F_z, \gamma) \\ M_y &= M_y(\alpha, s, F_z, \gamma) \\ M_x &= M_x(\alpha, s, F_z, \gamma) \end{aligned} \tag{4.42}$$

The above forms are known as the tire model in vehicle dynamics terminology. They can be either explicit mathematical functions, look-up tables, or even finite element models. However, regardless of their structure; they should be able to accurately calculate the tire forces and moments based on the input variables. In the other words, a desired tire model should be able to exactly reproduce the above-mentioned experimental characteristics curves. A tire model is an essential part of vehicle dynamics simulation and analysis.

In recent decades, developing simple, accurate, and cost-effective tire models has been an important research area, and numerous tire models have been presented by different researchers. For instance, Equations 4.33, 4.35 and 4.38 represent very simple tire models that are only valid in linear zone of the tire working zones. This is not a perfect tire model but it is very useful in preliminary analytical vehicle dynamics analysis. For accurate vehicle dynamics analysis, more sophisticated models are needed.

Tire models can be categorized into three basic groups as:

- *Analytical tire models:* In this type of tire models, an attempt is made to find a closed form mathematical function to calculate the tire forces and moments based on a theoretical approach. In the other words, by making some assumptions about the tire mechanical behavior as well as the shape of stress distributions, the tire model is formed. Brush and Fiala tire models are some examples of analytical tire models. These types of tire models are relatively simple and the effects of the tire's physical parameters are explicitly addressed in the model but they are not accurate enough and do not exactly match the experimental data.
- *Numerical tire models:* These are numerical computer codes instead of explicit mathematical equations. Finite element analysis ("FEA") methods are used to model tire dynamics with parameters obtained from experimental data or tire material properties. These models are capable of generating accurate results, however, their applications in vehicle dynamics simulation or analysis are limited due to their computational time.
- *Empirical models:* These types of tire models are formed based on physical tests. Curve fitting is usually used to fit the date with equations. Look-up tables are also used to present the tire behavior for different inputs. Empirical models are very accurate and are widely used for accurate vehicle dynamics simulation. However, they require extensive experiments and are expensive and their development is time-consuming. The Magic Formula (MF) Pacejka tire model is the most famous empirical tire model that is briefly introduced in the next section.

Table 4.4 Pacejka tire model main parameters

Tire model parameters	Tire parameter description
X	Input Variable – Slip Angle or Longitudinal Slip
Y	Output Variable – longitudinal Force, Lateral Force or Aligning Moment
S_h	Horizontal Shift
S_v	Vertical Shift
B	Stiffness Factor
C	Shape Factor
D	Peak Value
E	Curvature Factor

4.4.3 The Magic Formula (FM) Tire Model

The Magic Formula is a very accurate empirical tire model that now is the standard model in most vehicle dynamics simulation and analysis software packages. It takes its name from its main developer, Hans B. Pacejka [4].

Below is a simple explanation of the Pacejka tire model. The Pacejka tire model calculates longitudinal and lateral forces, as well as the self-aligning moment for pure and combined slip conditions. The Pacejka tire force equations for fitting physical tire test data are:

$$\begin{aligned} x &= X - S_h \\ Y(x) &= y(x) + S_v \\ y(x) &= D\sin\left\{C\tan^{-1}\left(Bx - E\left(Bx - \tan^{-1}Bx\right)\right)\right\} \end{aligned} \tag{4.43}$$

Table 4.4 defines the main parameters of the Pacejka tire force equations. The full description of this model can be found in [4].

Problems

1. At a wheel speed of 500 rpm, determine the vehicle speed during acceleration and braking if the slip ratio is equal to 20%. ($R = 0.35\,m$).

2. Consider a city bus with the following parameters:

$$T = 2\,m, a = 2.9\,m, b = 1.5\,m$$

In a maneuver with $u = 12\,m/s, v = 0.8\,m/s, r = 0.21\,rad/s, \delta = 0.05\,rad$, determine the side sideslip angle of all the wheels as well as the vehicle sideslip angle.

3. Consider a heavy-duty vehicle moving on a flat road. Draw graphs of the aerodynamics force, rolling resistance force, and total resistance force versus the longitudinal speed of the vehicle with the following parameters.

$$\rho_a = 1.2\frac{kg}{m^3}, C_d = 0.7, A_f = 7.1\,m^2, M = 4500\,kg, \mu_R = 0.008$$

Compare the results with those of Example 4.3, and discuss the differences between the resistance forces of conventional and heavy-duty vehicles.

References

1. Gillespie, T.D. (1992) *Fundamentals of Vehicle Dynamics*, SAE International, Warrendale, PA.
2. Dixon, J.C. (1996) *Tires, Suspension and Handling*, SAE International, Warrendale, PA.
3. Jazar, R.N. (2008) *Vehicle Dynamics: Theory and Application*, Springer, New York.
4. Pacejka, H.B. (2012) *Tire and Vehicle Dynamics*, Butterworth-Heinemann, Oxford.

5

Modelling and Characteristics of EV/HEV Powertrains Components

5.1 Introduction

Generally, an EV/HEV powertrain includes distinct components such as power unit(s), an internal combustion engine and/or electric motor(s), a transmission system, and batteries. Performance analysis of an EV/HEV powertrain system requires the modeling of each integrated component. Depending on the powertrain topology, component models should be combined to form the overall powertrain model. Components of a powertrain system can be mathematically modeled by using the laws of physics to arrive at a set of differential equations representing powertrain dynamics. Physics-based models are very powerful tools for the design and optimization of powertrain systems. However, they are complicated and sometimes do not match the experimental results exactly.

Alternatively, output power and torque, as well as the energy consumption and efficiency of a powertrain system can be calculated through characteristic curves instead of complicated physics-based models. Characteristic curves illustrate the results of standard experiments with sufficient accuracy.

Driving cycles also are an important tool in the design and offline evaluation of powertrain systems. There are different driving cycles representing urban and highway driving conditions for different geographical regions.

This chapter discusses the performance characteristics, mathematical models, and efficiencies of EV and HEV powertrain components, as well as the features of some important driving cycles.

Electric and Hybrid Vehicles: Technologies, Modeling and Control: A Mechatronic Approach, First Edition.
Amir Khajepour, Saber Fallah and Avesta Goodarzi.

5.2 ICE Performance Characteristics

It is important to properly characterize the performance of an ICE to evaluate its operation in terms of the engine's output torque, power, and fuel economy and pollutant emissions over its operating range. The main characteristics necessary to evaluate the performance of an engine are:

- power and torque generation
- specific fuel consumption
- specific emissions
- efficiencies (fuel conversion efficiency, mechanical efficiency, volumetric efficiency)
- others (air/fuel ratio, mean effective pressure, etc.).

The following subsections further discuss these quantities and their importance to the operating conditions and efficiencies of ICEs.

5.2.1 Power and Torque Generation

The output power and torque of an ICE can be obtained through its characteristic curves data, which are usually presented in a two-dimensional look-up table. For analytical studies, it is also possible to fit a closed form algebraic function to the data. Depending on the type and operating function of an ICE, the following performance characteristic curves can be developed:

- *Constant torque:* Figure 5.1 shows a constant torque characteristic curve of an ICE. As can be seen, for a constant accelerator pedal position, the torque remains constant when the engine's rotational speed changes. Assuming a linear relationship between the torque and accelerator pedal, the output torque can be formulated as:

$$T_e(\omega_e, x_{\theta e}) = x_{\theta e} T^* \tag{5.1}$$

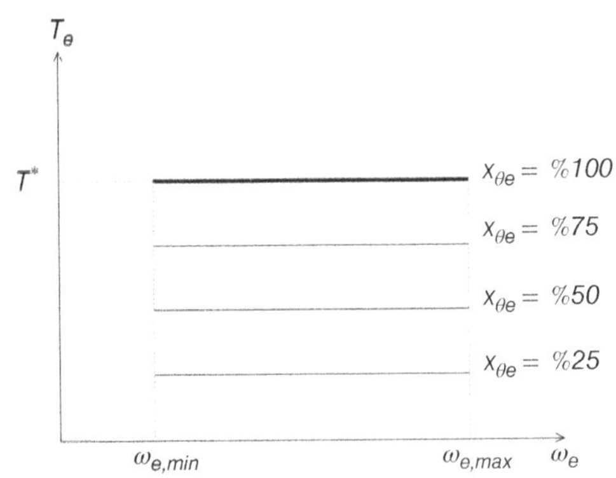

Figure 5.1 Constant torque performance characteristic curve

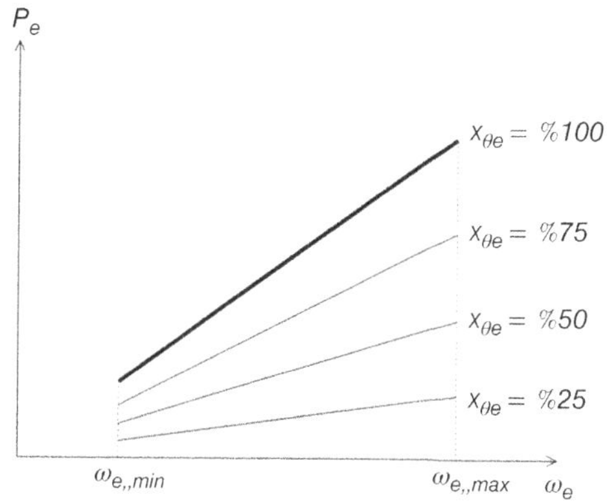

Figure 5.2 Power characteristic curve of an ICE with constant output torque

where T_e and ω_e are the engine's output torque and rotational speed, respectively. $x_{\theta e}$ has a value between 0 to 1 (0%–100%) and is considered as the ICE control signal, which can be the throttle valve opening, or the accelerator pedal position. T^* is the maximum ICE's torque. In Figure 5.1, $\omega_{e,min}$ and $\omega_{e,max}$ represent the minimum and maximum admissible values of ICE rotational speeds, respectively.

Considering the definition of mechanical power, the ICE output power can be calculated as:

$$P_e(\omega_e, x_{\theta e}) = \omega_e T_e(\omega_e, x_{\theta e}) = x_{\theta e}\omega_e T^* \tag{5.2}$$

Figure 5.2 shows a family of ICE power lines for different values of $x_{\theta e}$.

It should be noted that although diesel engine output torque typically can be approximated by a constant torque characteristic curve, its output torque is not completely constant. Alternatively, in order to have a more accurate characteristic curve, it is possible to use parabolic approximation for diesel engines.

- *Parabolic torque:* Figure 5.3 illustrates a parabolic type torque characteristic curve. It can be formulated as:

$$T_e(\omega_e, x_{\theta e}) = x_{\theta e}(A_0 + A_1\omega_e + A_2\omega_e^2) \tag{5.3}$$

where A_0, A_1, and A_2 are constant coefficients. In Figure 5.3, T_{max} and $\omega_{e,T_{max}}$ are the engine's maximum torque and its corresponding rotational speed, respectively. Also as shown by Figure 5.4, the output power is:

$$P_e(\omega_e, x_{\theta e}) = x_{\theta e}\omega_e(A_0 + A_1\omega_e + A_2\omega_e^2) \tag{5.4}$$

In Figure 5.4, $P_{\max}$ and $\omega_{e,P_{max}}$ are the engine's maximum power and its corresponding rotational speed, respectively.

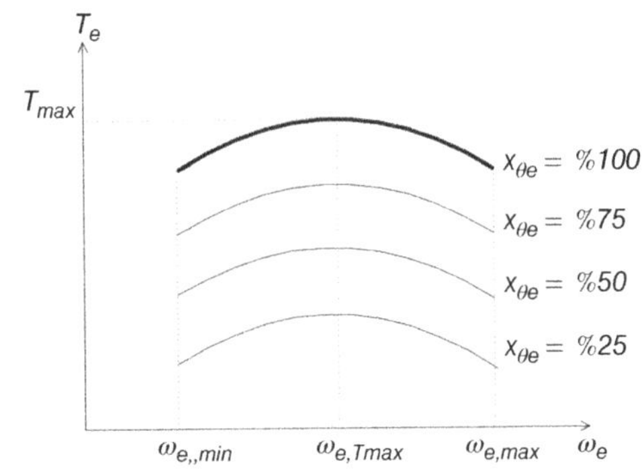

Figure 5.3 Parabolic torque performance characteristic curve for ICEs

The output torque of internal combustion engines, mainly gasoline ICEs can be modeled by a parabolic curve. As mentioned above, for more accuracy, diesel engines' torque characteristics can be modeled by parabolic curves instead of constant lines.

5.2.2 Mean Effective Pressure

The torque performance of an ICE is an important measure when examining engine capability. In order to simplify the process of comparing the torque performance of different engines, a useful relative engine performance parameter is the ratio of the work per cycle over the cylinder volume displaced per cycle. The parameter has a unit of force per area called the mean effective pressure (MEP). In fact, MEP is a fictitious pressure that, if it acted on the piston during the entire power stroke, would produce the same amount of work as produced during the actual cycle.

$$\mathrm{MEP}_e = \frac{\textit{work per cycle}}{\textit{displacement of cylinder}} \tag{5.5}$$

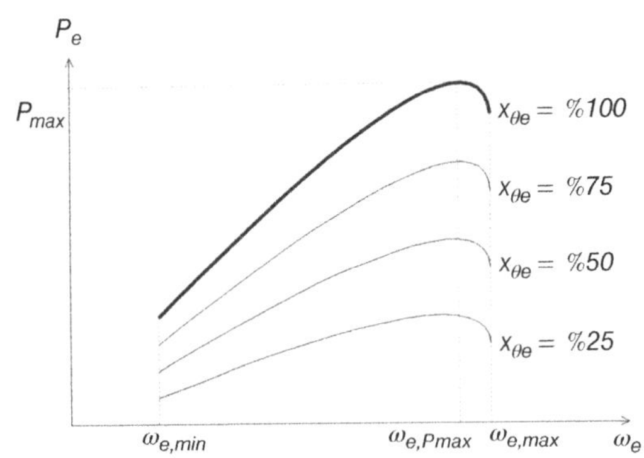

Figure 5.4 Output power of parabolic torque characteristic

The power produced by the engine is equal to the work completed per operating cycle multiplied by the number of operating cycles per second. If N_e is the number of revolutions per second, and N_r is the number of revolutions per cycle, the number of cycles per second is just their ratio. Thus,

$$W_e = \frac{P_e N_r}{N_e} \tag{5.6}$$

Note that $N_r = 2$ for a four-stroke engine and $N_r = 1$ for a two-stroke engine. According to the definition of MEP_e, it is possible to rewrite W_e as:

$$W_e = \text{MEP}_e V_d \tag{5.7}$$

where V_d is the engine displacement (e.g., total volume capacity of engine):

$$V_d = \frac{\pi}{4} D^2 h_d n \tag{5.8}$$

where D is the size of bore (the diameter measurement of the cylinders in a piston engine), h_d is the stroke, and n is the number of cylinders. By equating Equations 5.5 and 5.6, one has:

$$\text{MEP}_e = \frac{P_e N_r}{V_d N_e} \tag{5.9}$$

From the other side, the engine's rotational speed in terms of (rad/s) is:

$$\omega_e = 2\pi N_e \tag{5.10}$$

Thus, the power produced by the engine is:

$$P_e = 2\pi T_e N_e \tag{5.11}$$

Therefore, the relation for MEP_e in terms of torque becomes:

$$\text{MEP}_e = \frac{2\pi T_e N_r}{V_d} \tag{5.12}$$

- *Types of mean effective pressures:* In the context of engine design, the terms "indicated," "brake," and "friction" specify the location of measurement and method of calculation for each parameter. For example, the term "indicated" work (power) specifies the work (power) delivered to the piston through the combustion procedure. Conversely, the term "brake" work (power) specifies the work (power) that is measured at the crankshaft; it is the usable power the load receives from the engine. The term "brake" refers to the method of measurement where a brake attaches to the crankshaft to measure this quantity. Accordingly, MEPs are classified as indicated MEP, brake MEP, and friction MEP.

Measured within the cylinder, *Indicated* MEP (IMEP) uses the work (power) delivered to the piston through the combustion procedure.

Measured on the crankshaft, *Brake* MEP (BMEP) finds its basis in the useable work (power) delivered by the engine to the load.

Friction MEP (FMEP) relates to the work due to friction between moving parts of the engine. It is the difference between IMEP and BMEP. In other words,

$$\text{FMEP} = \text{IMEP} - \text{BMEP} \tag{5.13}$$

5.2.3 *Specific Fuel Consumption*

The fuel consumption of an ICE is the function of different factors, such as the engine performance characteristics (e.g., air–fuel ratio and power-to-volume-ratio), the powertrain components (e.g., gear number and ratios), vehicle speed, traffic congestion, and amount of resistance forces (e.g., rolling resistance and drag resistance). Typically, the ratio of the amount of fuel consumption per traveling distance is the measurement for ICE fuel consumption. Liter/100 km and miles/gallon are two common dimensions for measuring fuel consumption. There are different ways to evaluate the ICE's fuel consumption, using various quantities and maps.

As a measure of fuel efficiency of ICEs, Specific Fuel Consumption (SFC) is the ratio of the rate of fuel consumption and generated power. In other words, it quantifies how efficiently an engine consumes fuel to generate mechanical work at a certain operating point. The terms brake specific fuel consumption (BSFC) and specific fuel consumption typically work interchangeably. Like BMEP, the term "brake" refers to the location of measurement at the crankshaft. The engine output power measured at the crankshaft, known as the brake power, is the product of the measured engine brake torque and speed. Therefore, to calculate BSFC:

$$\text{BSFC} = \frac{\dot{m}_f}{P_{e,b}} = \frac{\dot{m}_f}{T_{e,b} 2\pi N_e} \tag{5.14}$$

By using BSFC, it is possible to compare engines with different sizes in terms of fuel efficiency. For power management of an HEV powertrain, it is important to understand the operating conditions necessary to minimize BSFC to improve fuel efficiency. Typically, BSFC is in SI units and its dimension is g/kWh, where g represents mass in grams and kWh is kilowatt multiplied by hour. Low values of BSFCs are desirable since it represents a lower rate of fuel consumption. For spark ignition ICEs, the best values of BSFCs are between 250–270 (g/kWh).

BSFC values are the function of vehicle speed and loads. For example, as displayed on the right in Figure 5.5, a gasoline-engine provides its best fuel efficiency when the engine is operating under full load conditions (i.e., when the throttle is fully opened) and close to its maximum torque.

As the plots indicate, BSFC suffers at low and high-engine speeds. This is because at low engine speeds, the heat generated by combustion has sufficient time to transfer through the walls of the cylinders from inside of the chambers to the outside, thus resulting in less useful

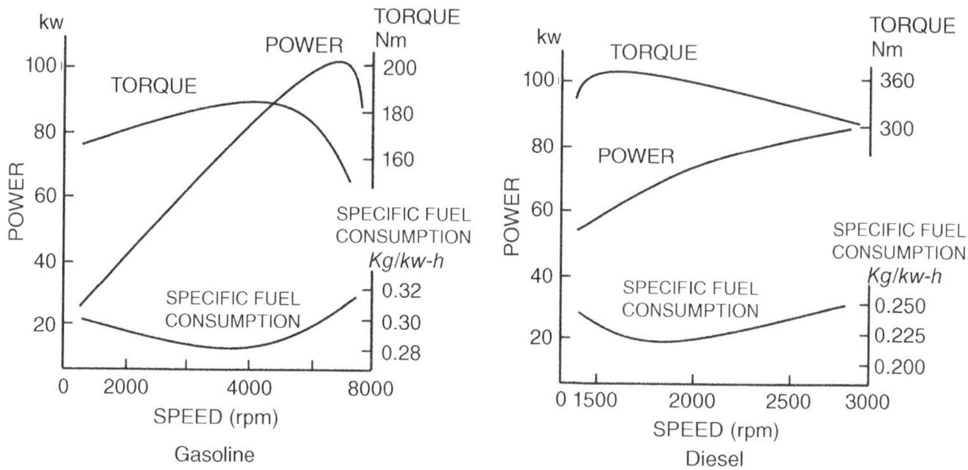

Figure 5.5 Performance characteristics of gasoline and diesel engines

mechanical work. At higher engine speeds, the frictional power losses of the engine increase drastically, thereby wasting the heat energy. Another reason that the gasoline-engine has the lowest BSFC in the middle of its speed range is because the highest air/fuel efficiency can be achieved at this range of speeds. Note that when the engine produces its maximum torque, it does not guarantee the best BSFC value.

Additionally, at some engine speeds, the BSFC suffers when the throttle opens partially (low loads). The right-hand side plot of Figure 5.5 depicts the performance of diesel engines. As the plot indicates, the BSFC of diesel engines is similar to that of gasoline engines. However, they are more efficient at low loads because the combustion heat adjusts through the amount of injected fuels and is independent of the throttle.

To clarify, the plots in Figure 5.5 are subject to full loads; in reality, however, there are different ranges of loads. In order to be more precise and represent BSFC values more realistically, Figure 5.6 represents BSFC as contours.

Such figures show BSFCs more accurately for each engine operating point. For example, for a given operating point at a power such as 50 (kW), and with the engine speed at 2500 (rpm), the best fuel efficiency possible is 280 (g/kWh). Notwithstanding the engine output, torque, and speed, the engine operating point is a function of the gear ratio of the transmission system. Manual and automatic transmission systems have a certain number of gear ratios which significantly limits their efficient operating points. For instance, Figure 5.7 represents the operating points of the engine at a constant speed with the highest gear and the second highest gear. As indicated by the plots, the engine has lower fuel efficiency in low gear compared to higher gears. In contrast, continuous transmission systems can ideally provide the gear ratio required for any given efficient operating point of the engine. This justifies the effectiveness of CVTs in regards to vehicle fuel efficiency improvement [1].

Note that the real-world performance of an ICE is complicated and is a function of different factors such as load, speed, traffic congestion, gear ratio, etc. As such, the diagrams discussed above only represent engine fuel efficiencies subject to certain and limited circumstances.

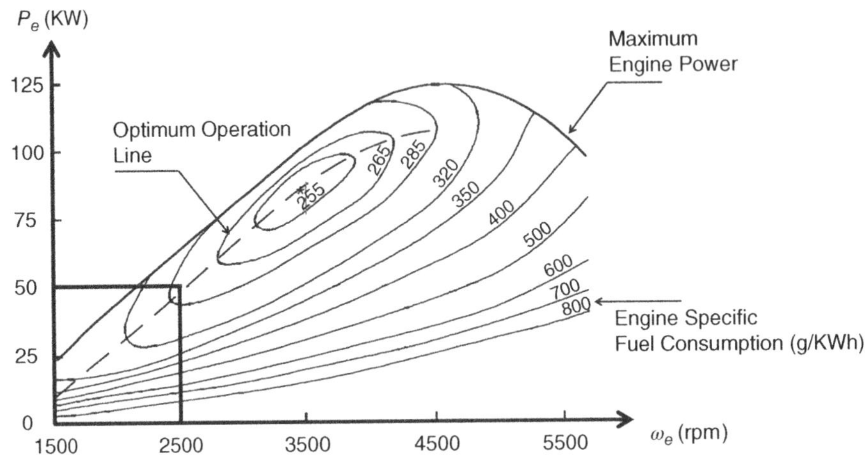

Figure 5.6 BSFC map of a typical gasoline engine

Thus, evaluating engine performance subject to different driving cycles yields an accurate evaluation regarding the fuel efficiency of an ICE after analysis and design.

Fuel consumption calculation: Figure 5.7 depicts the calculation of vehicle fuel consumption. The first step requires power and engine speed calculations. The engine speed as a function of vehicle speed and gear ratio is:

$$\omega_e = \frac{30u\beta_d\beta_i}{\pi R} \text{(rpm)} \tag{5.15}$$

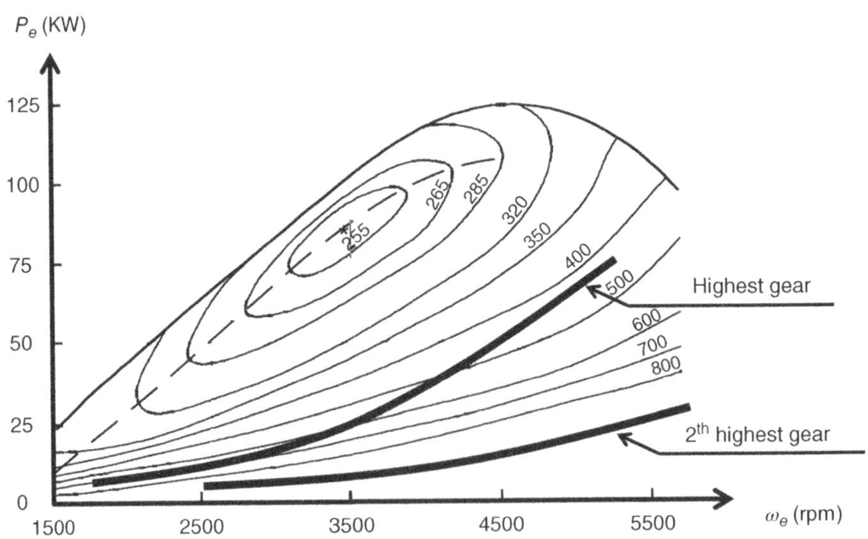

Figure 5.7 Operating point of the engine at a constant speed with highest gear and second highest gear

where β_i is the gear ratio of the transmission system, and β_d is the gear ratio of the final drive – usually the differential system. u and R are the vehicle longitudinal speed and tire radius, respectively. The engine's power can be calculated as:

$$P_e = \frac{u}{\eta_t} F_x \tag{5.16}$$

In which F_x is the summation of the vehicle longitudinal forces, and η_t is the total drivetrain efficiency (for more information about the formulation of F_x, please see Chapter 6). With the calculation of engine speed from Equation 5.15 and engine power from Equation 5.16, it is possible to find the related fuel consumption value, α_{BFSC}, from Figure 5.6. According to the definition of BSFC, α_{BFSC} is

$$\alpha_{BFSC} = \frac{\dot{m}_f}{P_e} \tag{5.17}$$

As mentioned before, the dimension of α_{BSFC} is g/kWh. Thus,

$$\dot{m}_f = \alpha_{BSFC} P_e, \tag{5.18}$$

Hence, the volumetric time rate of fuel consumption is:

$$Q_f = \frac{\alpha_{BDFC} P_e}{1000\gamma_f} \tag{5.19}$$

where γ_f is the mass density of fuel, with a dimension of kg/L. The total fuel consumption over a travel distance, x, with a cruise speed of u can be calculated as:

$$v_{fuel} = \frac{Q_f x}{u} \tag{5.20}$$

where v_{fuel} is total fuel consumption in liters.

5.2.4 Fuel Conversion Efficiency

Fuel conversion efficiency, η_f, is a measure of the effectiveness of the fuel transformation into usable energy. Each type of fuel has the potential to produce a specific amount of energy per its unit of weight. However, the combustion procedure can waste part of this energy. The fuel conversion efficiency, η_f, can be calculated with the relation below:

$$\eta_f = \frac{W_e}{m_f Q_{HV}} = \frac{P_e}{\dot{m}_f Q_{HV}} \tag{5.21}$$

where m_f is the mass of the fuel, and W_e is the work done by the engine in one cycle. $\dot{m}_f Q_{HV}$, is the fuel power potential, and represents the potential power that fuel combustion generates inside the engine. Q_{HV} is the heating value of the fuel.

The heating value is the amount of heat produced by the combustion of a unit quantity of a fuel. It defines the thermal energy released through the complete combustion of a certain fuel mass with air. A calorimeter records the released thermal energy through the combustion process. The heating value is in units of energy per unit of fuel mass (kJ/kg), and there is a higher and lower value for fuels containing hydrogen. It is popular to use the lower value because water vapor formed during combustion passes out of the system and takes the latent energy with it. Typical heating values for commercial hydrocarbon fuels for engines are in the range 42–44 (MJ/kg). The specific fuel consumption is inversely proportional to the fuel conversion efficiency for normal hydrocarbon fuels. Thus,

$$\eta_f = \frac{1}{\alpha_{BSFC} Q_{HV}} \tag{5.22}$$

Since η_f is dimensionless, it is possible to compare different ICEs in terms of power and torque generation, regardless of their size.

5.2.5 Mechanical Efficiency

Mechanical efficiency is the measure of how much the power generated through the combustion process (indicated power) is delivered to the crankshaft. The friction between the moving parts of an engine adversely affects the mechanical efficiency of an engine as well.

Friction in various moving parts, such as between the piston and piston rings, as well as between the piston rings and the cylinder walls, can reduce the engine mechanical efficiency. The valve and injection mechanisms waste power, as do various auxiliaries such as the lubricating oil, and the water-circulating pumps, among others. As a result, the actual amount of power delivered to the crankshaft that is available as mechanical power is less than the indicated power. Friction power, $P_{e,f}$ includes a combination of the mechanical losses due to friction and drive engine accessories. Thus,

$$P_{e,i} = P_{e,b} + P_{e,f} \tag{5.23}$$

where, $P_{e,i}$ is the indicated power of the ICE and $P_{e,b}$ is the brake power, the useful power on the crankshaft. The accurate calculation of friction power is difficult. In practice, the amount of friction power is measured by driving the engine with a dynamometer without firing the engine. The dynamometer measures the power supplied, which is the friction power.

The ratio of the brake power to the indicated power is the mechanical efficiency:

$$\eta_m = \frac{P_{e,b}}{P_{e,i}} = \frac{P_{e,b}}{\eta_f \dot{m}_f Q_{HV}} \tag{5.24}$$

Since the friction power includes the power required to pump gas into and out of the engine, mechanical efficiency depends on the throttle position, as well as engine speed and design. Note that in practice, the friction inside the cylinder is independent of the speed and load of the engine. Thus, to achieve maximum mechanical efficiency, the engine speed must be at a level which delivers peak power.

The mechanical efficiency of an engine in partial load conditions is lower than when the engine is operating at full load. This is because most friction power is almost independent of the load, and indicated power decreases relatively less than brake power when the load decreases. Mechanical efficiency becomes zero when an engine operates at no load (idle condition) since brake power is zero, but indicated power is not.

5.2.6 Air–Fuel Ratio

Section 1.2.3.2 comprehensively discussed the concept of the air–fuel ratio and its importance in fuel consumption and emissions; this section provides its mathematical calculation. Air–fuel ratio is the ratio between the mass of air and the mass of fuel in the fuel–air mix at any given moment. Therefore,

$$(A/F) = \frac{m_a}{m_f} = \frac{\dot{m}_a}{\dot{m}_f} \tag{5.25}$$

Engine testing typically measures the air mass flow rate, $\dot{m}_a$, and fuel mass flow rate, $\dot{m}_f$. As such, it is more convenient to calculate the air–fuel ratio using mass rate values.

Another measure of the amount of air and fuel in the air–fuel mixture is the air–fuel equivalence ratio, λ, which is the ratio of actual (A/F) to stoichiometry for a given mixture.

$$\lambda = \frac{\left(\frac{A}{F}\right)}{\left(\frac{A}{F}\right)_s} \tag{5.26}$$

Since most modern vehicles are capable of using different fuels when tuning, and since it is possible for fuel composition to change due to reasons such as seasonal variation, it is more relevant to use the air–fuel equivalence ratio than to use air–fuel ratio. $\lambda = 1.0$ is at stoichiometry; rich mixtures have $\lambda < 1.0$; and lean mixtures have $\lambda > 1.0$.

5.2.7 Volumetric Efficiency

As mentioned in Section 1.2.3.4, having an adequate amount of air ensures that the fuel will be completely combusted; as such, more energy can be converted into usable power. However, there are limits to the amount of air that an engine with a given displacement can induct. These limitations are a result of the short cycle time available, as well as the intake system, including the air filter, intake manifold, throttle plate, etc. As a result of these limitations, the available amount of air is less than the required amount of air that needs to be entered into the cylinder.

Volume efficiency is a measure that identifies the efficiency of an engine in taking the amount of air into the cylinder during a distinct induction process. The volumetric efficiency, η_v, is the ratio of volume flow rate of air into the intake system over the displaced volume rate by piston. In other words, the volumetric efficiency is the actual volumetric flow rate divided by the theoretical volumetric flow rate. For a four-stroke engine, it can be formulated as:

$$\eta_v = \frac{2\dot{m}_a}{\rho_a V_d N_e} \tag{5.27}$$

where ρ_a is the inlet air density and N_e is the engine speed.

Another expression of the volumetric efficiency is:

$$\eta_v = \frac{m_a}{\rho_a V_d} \tag{5.28}$$

where m_a, is the mass of air inducted into the cylinder per cycle. It is noted that the air density can be evaluated either at atmospheric conditions (the overall volumetric efficiency) or at inlet manifold conditions, measuring only the pumping performance of the cylinder, inlet port, and valve. Fuel type, air–fuel ratio, air–fuel mixture temperature, compression ratio, engine speed, and other engine design specifications affect the volumetric efficiency.

5.2.8 Compression Ratio

The compression ratio is the measure indicating the amount that the piston compresses in each intake charge. In other words, the compression ratio of an ICE is a measure that indicates the ratio of the volume between the piston and cylinder head before and after a compression stroke. An ICE with a higher compression ratio is capable of producing more usable work from a given mass of air–fuel ratio. The ratio may be calculated by:

$$CR = \frac{\pi D^2 h_d + 4V_c}{4V_c} \tag{5.29}$$

where D is the cylinder diameter, h_d is the piston stroke length, and V_c is the clearance volume, which is the minimum volume of the space at the end of the compression stroke. The compression ratio affects the fuel conversion efficiency of an engine. Theoretically, a higher compression ratio results in higher fuel conversion efficiency.

5.2.9 Specific Emissions

The amount of emissions produced by an ICE is an important characteristic of the engine performance. As mentioned in Section 1.2.4, oxides of nitrogen (NO_x), unburned hydrocarbons (HC), carbon monoxide (CO), and particulate matter are the main emissions of ICEs. The specific emissions are a measure of the concentrations of emissions in engine exhaust gases. Specific emissions are the mass flow rate of pollutant per unit power output [1]:

$$s_{NO_x} = \frac{\dot{m}_{NO_x}}{P_e} = \frac{\dot{m}_{NO_x}}{T_e \omega_e} \tag{5.30}$$

$$S_{CO} = \frac{\dot{m}_{CO}}{P_e} = \frac{\dot{m}_{CO}}{T_e\omega_e} \tag{5.31}$$

$$S_{HC} = \frac{\dot{m}_{HC}}{P_e} = \frac{\dot{m}_{HC}}{T_e\omega_e} \tag{5.32}$$

$$S_{PM} = \frac{\dot{m}_{PM}}{P_e} = \frac{\dot{m}_{PM}}{T_e\omega_e} \tag{5.33}$$

By collecting appropriate and sufficient data such as engine torque load and speed, fuel flow rate, mass air rate, and volumetric exhaust content through dynamometers in emission test laboratories, it is possible to develop brake-specific emission maps in order to consider emission minimization during HEV powertrain control design.

5.2.10 Relationships between ICE Performance Characteristics

It is important to use the afore-mentioned efficiencies and parameters to calculate engine performance characteristics, such as power and torque [1]. For power in a four-stroke ICE:

$$P_e = \frac{\eta_f m_a N_e Q_{HV}}{N_r(A/F)} \tag{5.34}$$

By integrating the volumetric efficiency into Equation 5.34, this equation can be reshaped to:

$$P_e = \frac{\eta_f \eta_v N_e V_d Q_{HV} \rho_a}{N_r(A/F)} \tag{5.35}$$

the torque generated by the engine becomes:

$$T_e = \frac{\eta_f \eta_v V_d Q_{HV} \rho_a}{2\pi N_r(A/F)} \tag{5.36}$$

Furthermore, the mean effective pressure can be calculated from:

$$\text{MEP} = \frac{\eta_f \eta_v Q_{HV} \rho_a}{A/F} \tag{5.37}$$

Another important parameter in the context of engine design is specific power – the power per unit piston area – which helps use the available piston area, regardless of cylinder size. From Equation 5.35, the specific power is:

$$\frac{P_e}{A_p} = \frac{\eta_f \eta_v N_e h_d Q_{HV} \rho_a}{N_r(A/F)} \tag{5.38}$$

where A_P is the area of piston head. Moreover, Equation 5.38 introduces mean piston speed as:

$$\frac{P_e}{A_p} = \frac{\eta_f \eta_v \overline{S}_P Q_{HV} \rho_a}{2N_r(A/F)} \tag{5.39}$$

where, $\overline{S}_P$ is the mean piston speed.

Another formulation of a four-stroke engine's power and torque as a function of MEP is:

$$P_e = \frac{\text{MEP}A_P\overline{S}_p}{4} \tag{5.40}$$

and

$$T_e = \frac{\text{MEP}V_d}{4\pi} \tag{5.41}$$

The parameters discussed above have both brake and indicated values. The effectiveness of these parameters is subject to variations such as speed and load range. Maximum-rated or normal-rated brake power, BMEP, and BSFC determine the engine potentiality. The maximum brake torque over the whole speed range illustrates the capability of the engine to obtain the required air–fuel ratio over the whole speed range, and to use the air effectively. Moreover, engine fuel consumption and emissions are important in the operating range of a full engine. They require careful consideration during the design process of the powertrain, powertrain control, and power management control systems.

Example 5.1

A four-cylinder four-stroke SI engine has the following specifications:

BMEP:	840 kPa
Total capacity:	1.3 liters
Engine speed:	4200 rev/min
Friction loss:	10 kW
Volumetric efficiency, η_v:	85%
Brake fuel conversion efficiency, η_f:	28%
Fuel heating value:	43.5 MJ/kg
Inlet air density, ρ_a	1.29 kg/m^3

(a) Calculate the brake power.
(b) Calculate the indicated mean effective pressure (IMEP).
(c) Calculate the air–fuel ratio.

Solution

From Equation 5.12, BMEP is:

$$\text{BMEP} = \frac{2\pi T_{e,b} N_r}{V_d} \Rightarrow T_{e,b} = \frac{840 \times 1.3}{2 \times \pi \times 2} = 86.89\ \text{Nm}$$

Then, using Equation 5.11, the brake power is:

$$P_{e,b} = 2\pi T_{e,b} N_e = \frac{2 \times \pi \times 86.89 \times 4200}{60} = 38219(\text{W}) \approx 38.22\ \text{kW}$$

The indicated power can be calculated using Equation 5.23 as:

$$P_{e,i} = P_{e,b} + P_{e,f} = 38.22 + 10 = 48.22\ \text{kW}$$

Therefore:

$$\text{IMEP} = \frac{P_{e,i} N_r}{V_d N_e} = \frac{48.22 \times 2}{1.3 \times \dfrac{4200}{60}} = 1.06\ \text{kPa}$$

The air–fuel ratio using Equation 5.35 becomes:

$$P_{e,b} = \frac{\eta_f \eta_v N_e V_d Q_{HV} \rho_a}{N_r \left(\dfrac{A}{F}\right)} \Rightarrow \left(\frac{A}{F}\right) = \frac{0.28 \times 0.85 \times \left(\dfrac{4200}{60}\right) \times 1.3 \times 43.5 \times 1.29}{38.22 \times 2} = 15.9$$

5.3 Electric Motor Performance Characteristics

Typically, electric motors used in industrial applications operate at predefined distinct operating modes. Thus, it is easy for these motors to achieve optimized performance. However, for vehicle applications, the electric motor operates at all of its operating (torque/speed) ranges. In other words, the motor needs to satisfy different driving conditions such as frequent stop–go, high acceleration/deceleration rates, etc., thereby requiring the motor to operate at its high torque density and efficiency over all speed ranges. As such, it is important to study the motor performance and efficiency of all possible torque-speed combinations within the motor's operating envelope [2].

5.3.1 Power and Torque Generation

As shown by Figure 5.8, the typical output curve of modern electric motors includes a constant torque/constant power characteristic. Constant power is the ideal characteristic of automotive powertrains, because it provides high traction in low speeds and low traction in high speeds which exactly match the vehicle's needs to overcome inertial and resistance forces. It is noted that in ICE vehicles, the function of a multi-gear transmission system is to convert the parabolic

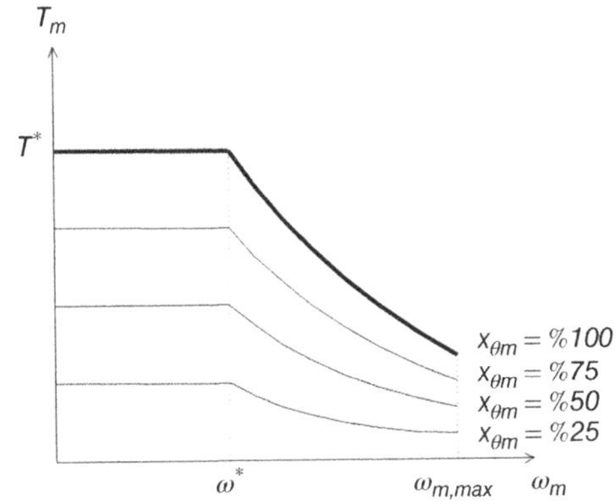

Figure 5.8 Typical torque characteristics curves of modern electric motors

or constant torque characteristic of the engine to an approximately constant power characteristic. However, for modern electric motors, constant power characteristics can be achieved by motor controllers and their power electronics. Providing such a desired output characteristic is one of the main advantages of electric drive systems, because they no longer need a complex multi-gear transmission system and they can operate with a fixed-ratio gearbox or even without any transmission.

To formulate the output torque of modern electric motors, suppose that for a constant value of motor control signal, $x_{\theta m}$ the motor always generates a constant power of P_m^* for motor speeds from base speed ω_m^* to maximum motor speed $\omega_{m,max}$. Then the motor power at any speed in this range is:

$$P_m(\omega_m,\ x_{\theta m}) = \omega_m T_m(\omega_m, x_{\theta m}) = x_{\theta m} P_m^* \tag{5.42}$$

Thus, the motor's torque characteristic can be formulated as:

$$T_m(\omega_m,\ x_{\theta m}) = x_{\theta m} \frac{P_m^*}{\omega_m} \tag{5.43}$$

As shown by Figure 5.8, the torque characteristic is a family of hyperbolic curves versus rotational speed for any value of $x_{\theta m}$.

It should be noted that according to Equation 5.43, in low speeds the motor torque gradually approaches infinity which is practically impossible. To resolve this problem, usually the low speed part of the curve (below the base speed) is changed to a constant torque characteristic by the motor controller.

Finally, the output torque characteristic of modern electric motors can be formulated as:

$$T_m(\omega_m, x_{\theta m}) = \begin{cases} \omega_m < \omega^* & x_{\theta m} T_m^* \\ \omega^* \le \omega_m \le \omega_{\max} & x_{\theta m} \dfrac{P_m^*}{\omega_m} \end{cases} \qquad \omega^* = \frac{P_m^*}{T_m^*} \tag{5.44}$$

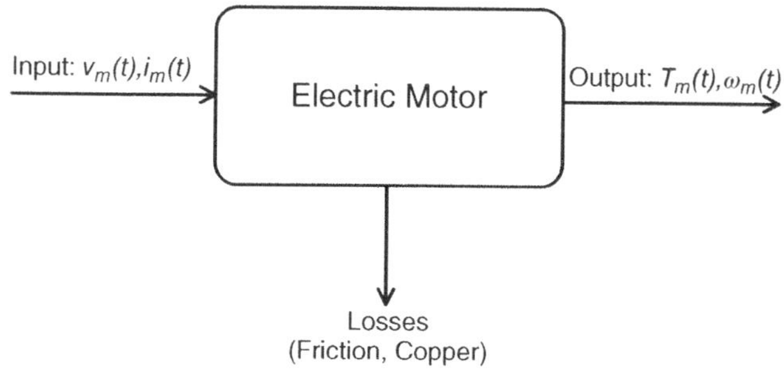

Figure 5.9 Power flow diagram of an electric motor

5.3.2 *Efficiency*

The mechanical efficiency (η_m) of electric motors is defined as the ratio of motor output power, $P_{m,out}$ to input power, $P_{m,in}$:

$$\eta_m = \frac{P_{m,out}}{P_{m,in}} \tag{5.45}$$

The inputs of a motor are the voltage and current, both of which come from storage systems, such as the battery or the ultracapacitor (Figure 5.9). Thus, the output power of an electric motor is:

$$P_{m,out} = \eta_m P_{m,in} \tag{5.46}$$

In order to calculate the output power of an electric motor, η_m should be known. This parameter can be determined from a motor efficiency map if the output torque and speed are known. From a mechanical point of view, the output power can be also calculated as:

$$P_{m,out} = T_m \omega_m \tag{5.47}$$

where, T_m is the motor torque and ω_m is the motor speed.

Efficiency maps are a convenient way to represent motor performance in complicated applications such as EVs/HEVs. By using efficiency maps, a motor is represented as a "black box" that provides a known output when a certain input is applied. Figure 5.10 illustrates a typical efficiency map of an electric motor.

Each electric motor has its own efficiency map due to its internal mechanical and electrical structure. Figure 5.11 illustrates the efficiency maps of different AC motors. As shown, depending on the motor structure and source of losses (e.g., friction, copper), the efficiency maps are different.

In order to determine the efficiency of an electric motor, it is important to understand how power losses change with motor speed. The important losses affecting motor efficiency include

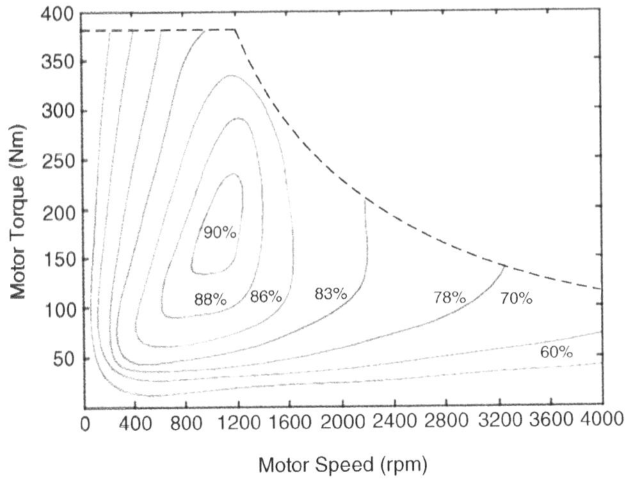

Figure 5.10 Typical electric motor efficiency map characteristics

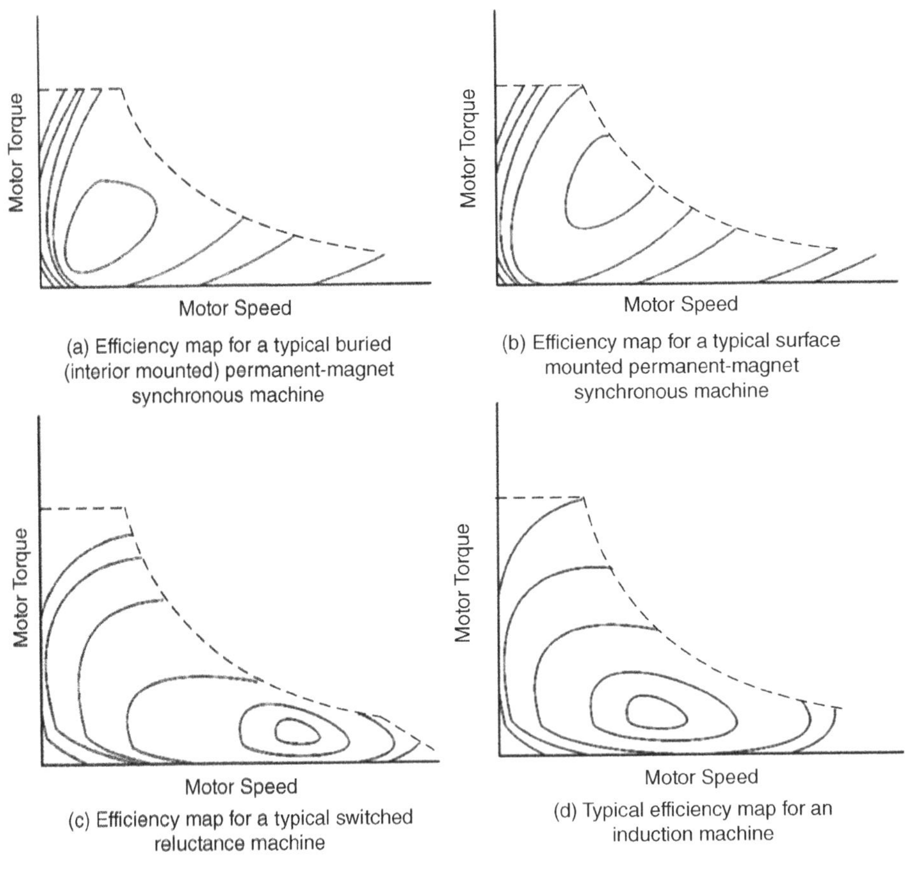

Figure 5.11 Efficiency maps of different electric motor drives

resistance losses, iron losses, stray losses, and mechanical losses. Resistance losses, also called copper losses, refer to the electrical power lost in the rotor and stator windings. These losses vary with the load in proportion to the current squared as:

$$P_{m,L} = R_m i_m^2 \tag{5.48}$$

where:

$P_{m,L}$ = Motor power losses
R_m = Motor equivalent internal resistance (Ω)
i_m = Motor equivalent current (Amp)

Iron losses are the result of dissipated magnet energy in the motor's stator and rotor magnetic cores. When the motor operates under load, it generates harmonics, which are the main cause of stray losses. The energy in electric motors dissipates due to currents in the copper windings, harmonic flux components in the iron parts, and due to leakage in the laminate core. Mechanical losses include friction in the motor bearings, as well as the fan for air cooling.

Example 5.2

A parallel HEV has a downsized engine, motor/generator, and a battery. The battery supplies energy to the motor/generator for propulsion assistance. The vehicle is working in cruise condition for 40 mins at a speed of 30 (m/s) with a combined engine and electric motor propulsion. The required traction power is 14.2 (kW). In this span, the engine's BSFC is 280 g/kWh and the battery has 2500 kJ energy, of which travel consumes 60%. Calculate the fuel efficiency while assuming that motor mechanical efficiency and transmission efficiency are 88% and 85%, respectively. Assume the engine mechanical efficiency is 100%. Fuel heating value is 45.8 (MJ/kg) and fuel density is 737 (kg/m^3).

Solution

The motor output power in the 40-min time span is:

$$P_m = P_b \eta_m = \frac{0.6 \times 2500(\text{kJ})}{40\ \text{min} \times 60\left(\frac{\text{s}}{\text{min}}\right)} \times 0.88 = 0.55\ \text{kW}$$

The required traction power is provided by the engine and electric motor simultaneously, thus:

$$P_{traction} = (P_e + P_m) \times \eta_t$$

Therefore, the engine power is:

$$P_e = \frac{P_{traction}}{\eta_t} - P_m = \frac{14.2}{0.85} - 0.55 = 16.15\,\text{kW}$$

Moreover, the fuel efficiency can be calculated using Equation 5.25:

$$\eta_f = \frac{1}{\frac{280}{3600}\left(\frac{\text{g}}{\text{kWs}}\right) \times 45.8\left(\frac{\text{MJ}}{\text{kg}}\right)} = 0.28$$

Thus, using Equation 5.27, the fuel mass rate is

$$\dot{m}_f = \frac{P_e}{\eta_f Q_{HV}} = \frac{16.15(\text{kW})}{0.28 \times 45.8\left(\frac{\text{MJ}}{\text{kg}}\right)} = 1.26\ \frac{\text{g}}{\text{s}}$$

The total vehicle travel distance is:

$$x = u \times t = 30\left(\frac{\text{m}}{\text{s}}\right) \times 40 \times 60\ (s) = 72\,\text{km}$$

And, the total mass fuel consumption is:

$$fuel\ consumption = \frac{\frac{1.26}{1000}\left(\frac{\text{kg}}{\text{s}}\right) \times 40 \times 60\ (\text{s})}{72\ (\text{km}) \times \frac{737}{1000}\left(\frac{\text{kg}}{\text{L}}\right)} = 0.06\left(\frac{\text{L}}{\text{km}}\right)\ \text{or}\ 6\left(\frac{\text{L}}{100\,\text{km}}\right)$$

5.3.3 DC Motors

The concept, structure, and classifications of DC motors were discussed in Section 2.2.2.1. This section is devoted to the mathematical modeling of DC motors. Figure 5.12 shows the model of a DC motor including an equivalent circuit for the motor armature and field excitation circuits.

The voltage equation for the armature circuit can be written based on Kirchhoff's law as:

$$v_a(t) = R_a i_a(t) + L_a \frac{di_a(t)}{dt} + v_i(t) \tag{5.49}$$

where $i_a(t)$ is the armature current, $v_i(t)$ is the induced voltage, R_a is the armature resistance, L_a is the armature inductance, and $v_a(t)$ is the armature voltage. The induced voltage, $v_i(t)$, is proportional to the field current, $i_f(t)$, and the speed of electric motor, $\omega_m(t)$, as:

$$v_i(t) = L_{af} i_f(t) \omega_m(t) \tag{5.50}$$

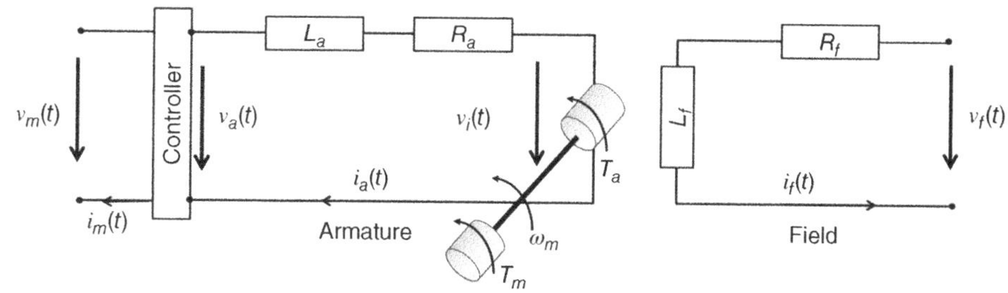

Figure 5.12 Equivalent circuit of a DC electric motor

where L_{af} is the field-armature mutual inductance. The torque of armature, $T_a(t)$, is also proportional to the field current and the armature current and can be written as:

$$T_a(t) = L_{af} i_f(t) i_a(t) \tag{5.51}$$

In separately excited DC motors, for speeds below the base speed, the field voltage is usually kept constant while for permanent magnet DC motors, usually a constant magnetic flux (e.g., a constant field current) is applied [2]. Thus, with assumption of constant field current (e.g., $i_f(t) = I_f$), Equations 5.50 and 5.51 can be rewritten as:

$$\omega_m = k_{mV} v_i(t) \tag{5.52}$$

$$T_a(t) = k_{mT} i_a(t) \tag{5.53}$$

where $k_{mV} = \left(L_{af} I_f\right)^{-1}$ and $k_{mT} = L_{af} I_f$ are speed and torque constants. It is noted that in the case of separately excited DC motors, for the speeds above the base speed, a time-variable voltage is applied, revoking the assumption made for Equations 5.52 and 5.53.

On the other hand, the armature voltage $v_a(t)$ is a linear function of motor control voltage signal $v_m(t)$ as:

$$v_a(t) = k_{mc} v_m(t) \tag{5.54}$$

where k_{mc} is the motor controller gain. By substituting Equations 5.52 and 5.53 into Equation 5.49, one has:

$$\left(\frac{L_a}{k_{mT}}\right)\frac{dT_a(t)}{dt} + \left(\frac{R_a}{k_{mT}}\right)T_a(t) + \frac{1}{k_{mV}}\omega_m(t) - k_{mc} v_m(t) = 0 \tag{5.55}$$

By applying Newton's second law to the armature shaft, it is possible to establish a relation between the torque generated by the armature and the output load on electric motor by:

$$T_a(t) = J\frac{d\omega_m(t)}{dt} + T_m(t) \tag{5.56}$$

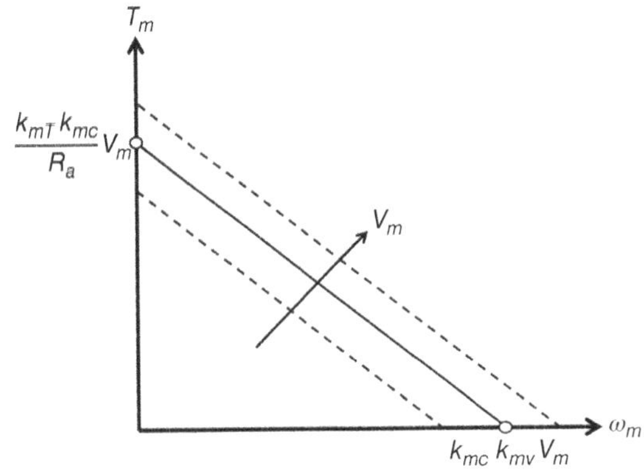

Figure 5.13 Torque-speed characteristic curve of a DC motor

where $T_m(t)$ and J are load torque on electric motor and inertia of the load, respectively. Now, by substituting Equation 5.56 into 5.55, the following differential equation can be obtained:

$$\left(\frac{L_a}{k_{mT}}J\right)\frac{d^2\omega_m}{dt^2}+\left(\frac{R_a}{k_{mT}}J\right)\frac{d\omega_m}{dt}+\frac{1}{k_{mV}}\omega_m=k_{mc}v_m(t)-\left(\frac{L_a}{k_{mT}}\right)\frac{dT_m}{dt}-\left(\frac{R_a}{k_{mT}}\right)T_m \quad (5.57)$$

If motor control voltage signal, $v_m(t)$, and the load torque (motor output torque), T_m are known, the rotational speed of the motor can be calculated by solving Equation 5.57. In steady-state conditions in which the variation of the system states goes to zero, from Equation 5.57, the relation between the torque and speed of the electric motor becomes:

$$T_m=\frac{k_{mT}k_{mc}}{R_a}V_m-\frac{k_{mT}}{k_{mV}R_a}\omega_m \quad (5.58)$$

where, V_m is the steady-state value of $v_m(t)$. Figure 5.13 illustrates the characteristic curve $T_m=f(\omega_m)$ of an electric DC motor [2]. The curve is determined by the parameters k_{mT}, k_{mc}, k_{mV}, R_a, and V_m. At stall conditions (i.e., $\omega_m=0$), the starting torque is $T_m=\frac{k_{mT}k_{mc}}{R_a}V_m$ while at the idling (no-load) conditions the motor speed is $\omega_m=k_{mc}k_{mV}V_m$. It is noted that in an ICE vehicle, the engine must be above its idle speed to provide full torque. However, Figure 5.13 indicates that in an electrified vehicle, the electric motor can provide full torque at low speeds, thereby giving the vehicle excellent acceleration from a standing position.

One can see that the calculated characteristic curve in Figure 5.13 is different from the constant torque/constant power curve that was introduced in Section 5.2.1 as a typical torque characteristic of electric motors for vehicular applications. The reason for this difference is the simplification assumptions made above to arrive at the motor equations. In practice, neither the field voltage nor the motor controller gain is constant and by controlling them, the constant torque/constant power characteristic is achieved.

Example 5.3

In steady-state conditions, find the efficiency of a DC electric motor and its internal power loss in terms of its output torque and speed.

Solution

In steady-state conditions the current signals shown in Figure 5.12 are constant. Thus, $i_a(t) = I_a$ and $i_m(t) = I_m$. Also from Equations 5.49, 5.52, 5.53 and 5.56, one has $V_a = R_a I_a + V_i$, $V_i = \frac{\omega_m}{k_{mV}}$, $T_a = k_{mT} I_a$, and $T_a = T_m$ respectively. On the other hand, the input and output powers of the electric motor are:

$$P_{m,out} = T_m \omega_m$$

$$P_{m,in} = V_m I_m = I_a V_a + P_{L,c} = \frac{T_m}{k_{mT}}\left(\frac{\omega_m}{k_{mV}} + \frac{R_a T_m}{k_{mT}}\right) + P_{L,c}$$

where $P_{L,c}$ is the power lost in the controller. According to Equation 5.46, the efficiency of the electric motor is:

$$\eta_m = \frac{P_{m,out}}{P_{m,in}} = \frac{T_m \omega_m}{\frac{T_m}{k_{mT}}\left(\frac{\omega_m}{k_{mV}} + \frac{R_a T_m}{k_{mT}}\right) + P_{L,c}} = \frac{1}{\frac{1}{k_{mT} k_{mV}} + \frac{R_a T_m}{k_{mT}^2 \omega_m} + \frac{P_{L,c}}{\omega_m T_m}}$$

The internal power losses $P_L = P_{m,in} - P_{m,out}$ is

$$P_L = \omega_m T_m \left(\frac{1}{k_{mV} k_{mT}} - 1\right) + \frac{R_a}{k_{mT}^2} T_m^2 + P_{L,c}$$

The above relation indicates that the overall internal power losses are because of ohmic resistance and losses in the controller.

5.3.4 Induction AC Motors

As mentioned in Section 2.2.2.2, three-phase squirrel-cage AC induction (asynchronous) motors have a potential for EV/HEV applications due to their advantages such as low cost and reliability. Like the DC machine, an induction motor is composed of a stator, or stationary windings, and a rotor. But, unlike the DC machines, induction machines carry alternating currents in the stator and rotor windings. The stator windings are placed in the slots cut in the inner surface of its frame. AC current is provided to the stator windings from an external AC source. The rotor windings are positioned on the slots in the outer surface of the rotating segment. The principal operation of an induction machine is based on electromagnetic voltage induction. In an induction machine, an electromagnetic field (EMF) rotating synchronously with the stator magnetic field is created when an AC voltage is supplied to the stator winding terminals. In a three-phase induction machine, the stator and rotor windings are physically distributed windings which create the sinusoidal space distribution of magnetic fields in three

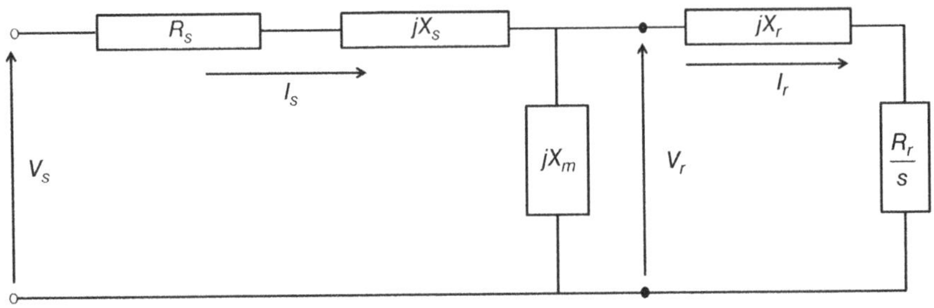

Figure 5.14 Equivalent circuit of an induction electric motor

corresponding 120-degree phase-shifted phases of the stator and rotor. The rotating magnetic flux induces current in the windings of the rotor by electromagnetic induction. The rotor and stator are separated by a uniform air gap which allows free rotation of the rotor. It is noted that an induction motor with six stator windings is called a two-pole motor. If it includes 12 windings, it is called a four-pole motor, and so on.

5.3.5 *Steady-State Performance Analysis*

A simplified equivalent circuit of an induction machine is shown in Figure 5.14. In this equivalent circuit, it is assumed that the magnetic circuit is separated by an air gap between the stator winding and the rotor winding. This circuit is a single-phase representation of a multiphase induction motor that is valid in steady-state balanced-load conditions. Here, R_s and X_S represent the stator resistance and leakage reactance, respectively. Also rotor resistance and leakage reactance, and slip are shown by R_r, X_r and s_m, respectively. X_m is the magnetizing (air-gap) reactance.

It is worth outlining the following definitions before moving on to the modeling section.

- *Leakage reactance* is a self-induced back EMF in the respective windings due to the leakage of magnetic flux caused by imperfect coupling between the stator and rotor. In general, reactance is the opposition of a circuit element to a change of current, caused by the build-up of electric or magnetic fields in the element.
- *Impedance* is the effective resistance of an electric circuit or component to alternating current (AC), arising from the combined effects of ohmic resistance and reactance. Impedance generalizes the concept of resistance to AC circuits, and possesses both magnitude and phase. In fact, it is

$$z_m = R_m + jX_m \tag{5.59}$$

where z_m is the electrical impedance, R_m is the resistance and X_m is the reactance. Impedance is measured in ohms. In quantitative terms, it is the complex ratio of the voltage to the current in an alternating current circuit. It is noted that when a circuit is driven with direct current (DC), there is no distinction between impedance and resistance; the latter can be considered as impedance with a zero phase angle.

- *Magnetizing reactance* is an equivalent reactance representation of the air gap reluctance in electric circuits.
- *Magnetomotive force (MMF) or magnetomotance* is any physical driving (motive) force that produces a magnetic flux.
- *Synchronous speed* is the rotating speed of the stator magnetic field which determines the speed of machine and is a function of the power source frequency and the number of poles in the motor. An AC motor's synchronous speed, n_{ms}, is expressed in revolutions per minute (rpm) as

$$n_{ms} = 120\frac{f_s}{p_m} \tag{5.60}$$

where f_s is the supply's frequency in Hertz and p_m is the number of magnetic poles. That is, for a two-pole three-phase machine, p_m equals 2 and n_{ms} equals 3600 (rpm) for 60 (Hz) supply frequency.

The stator magnetic field causes the rotor shaft rotation due to induced current. When the induction machine operates as a motor, the rotor speed is always slower than the synchronous speed inducing an EMF in the rotor. In other words, if the rotor rotates synchronously at the magnetic field speed, the rotor will be stationary with respect to the rotating field. This operating condition forces the induced current flow in the rotor to cease, thereby creating a zero average torque and stalling the motor. The ratio between the rotational speed of EMF induced in the rotor circuit and the speed of stator's rotating field is called the slip and is expressed as a percentage of the synchronous speed:

$$s_m = \frac{n_{ms} - n_{mr}}{n_{ms}} \tag{5.61}$$

where s_m and n_{mr} are slip and the rotor speed, respectively. Here, $s_m = 0$ means that the rotor runs at synchronous speed while $s_m = 1$ indicates that the motor is stationary. For a motoring condition, the slip is between 0 and 1 while during generating mode (regeneration), it is a negative value. It is noted that under load conditions, the rotor's speed decreases while the slip increases enough to create sufficient torque to turn the load. Accordingly, the slip speed in rpm can be defined as

$$n_{slip} = s_m n_{ms} \tag{5.62}$$

The frequency of the induced voltage in the rotor winding is calculated as:

$$f_r = \frac{p n_{slip}}{120} \tag{5.63}$$

where f_r is the rotor frequency in Hz. From Equation 5.61, f_r may be expressed as:

$$f_r = s_m f_s \tag{5.64}$$

Thus, the frequency of the signals in the rotor circuit is a slip portion of the frequency of the corresponding signals in the stator winding.

In the equivalent circuit shown in Figure 5.14, the resistance and reactance of the rotor are transferred to the stator side as:

$$R_r = \left(\frac{N_s}{N_r}\right)^2 R_{r0} \text{ and } X_r = \left(\frac{N_s}{N_r}\right)^2 X_{r0} \tag{5.65}$$

where $\frac{N_s}{N_r}$ is the stator-to-rotor turn ratio. It is a fixed ratio as a manufacturing parameter. R_{r0} and X_{r0} are resistance and reactance of the rotor at the supply frequency, respectively. Thus, if the frequency of the induced voltage in the rotor changes, the resistance and reactance of the rotor circuit will also change. Based on the equivalent circuit, the electrical parameters of an induction motor in the steady-state condition can be expressed as in [3]:

$$z_m = (R_s + jX_s) + \frac{\left(\frac{R_r}{s} + jX_r\right)(jX_m)}{\frac{R_r}{s} + j(X_r + X_m)} \tag{5.66}$$

where z_{EM} is the equivalent impedance of the induction machine. Using this impedance, the stator current is evaluated as:

$$I_s = \frac{V_s}{z_m} = \frac{V_s}{(R_s + jX_s) + \frac{\left(\frac{R_r}{s} + jX_r\right)(jX_m)}{\frac{R_r}{s} + j(X_r + X_m)}} \tag{5.67}$$

where V_s is the voltage phasor of the stator terminals. Using Kirchhoff's current law, the rotor current is:

$$I_r = \frac{jX_m}{\frac{R_r}{s} + j(X_r + X_m)} I_s \tag{5.68}$$

It is noted that an induction motor can be wired in the form of either wye or delta connections. The input power to the induction motor is

$$P_{m,in} = 3Re[V_s I_s] = 3V_s I_s \cos(\theta_v - \theta_i) = 3V_s I_s(\text{PF}) \tag{5.69}$$

where θ_v and θ_i are the phase angles of input voltage and current phasors, and PF is the power factor of the induction motor which equals $\cos(\theta_v - \theta_i)$. The developed mechanical power of the motor is

$$P_m = 3\left(\frac{1-s}{s}\right) I_r^2 R_r \tag{5.70}$$

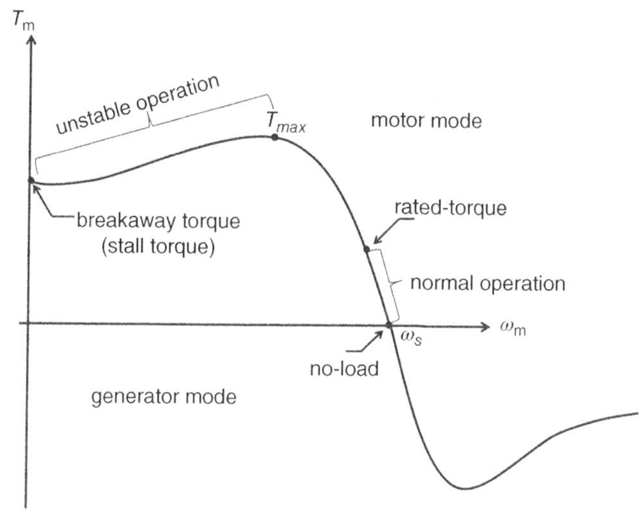

Figure 5.15 Torque-speed characteristic curve of an induction machine

The rotational speed and torque developed by the motor can be determined by

$$\omega_m = \omega_r = \frac{2\pi n_{mr}}{60} \text{ and } T_m = \frac{P_m}{\omega_m} \tag{5.71}$$

The efficiency of the motor is

$$\eta_m = \frac{P_{m,out}}{P_{m,in}} = \frac{P_m - P_L}{P_{m,in}} \tag{5.72}$$

where P_L is the power losses due to friction and windage.

Figure 5.15 illustrates torque-speed characteristic curve of an induction machine for a fixed supply voltage and frequency. As Figure 5.15 indicates, at no mechanical load conditions, the induction motor rotates at synchronous speed. However, in practice, internal friction and windage losses create some load on the motor shaft and as the load increases, the motor speed decreases till the load torque reaches its rated value. The range between the synchronous speed and the one corresponding to the maximum generated torque (T_{max}) is the stable region of operation for the induction motor. If the load on the motor shaft is more than the rated torque, the motor is overloaded. As load increases, the generated torque by the motor will reach T_{max}. If the motor is loaded beyond T_{max}, the rotor speed drops and the motor enters its unstable operation region. In this region, the slope of the machine-generated torque is positive with respect to the rotor shaft, indicating that as the motor speeds drop, it generates less torque. This process continues until the motor speed reaches zero when the motor will stall at its stall torque. For speeds higher than the synchronous speed, the induction machine operates as generator $T_m < 0$.

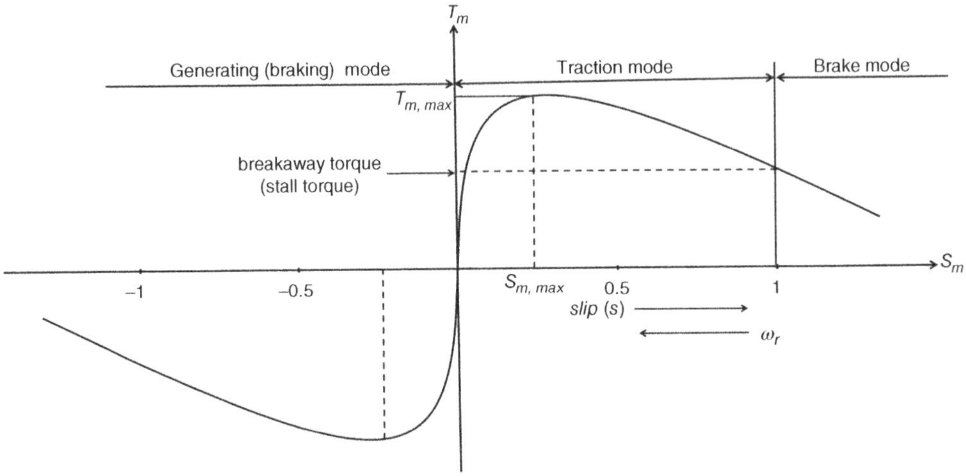

Figure 5.16 Torque-slip characteristics curve of an induction machine

Another characteristic curve of an induction motor is its torque-slip curve. A schematic torque-slip curve is shown in Figure 5.16. For low values of slip, $\frac{R_r}{s} \gg X_r$, thus from Equations 5.70 and 5.71,

$$T_m = \frac{3\left(\frac{1-s_m}{s_m}\right) I_r^2 R_r}{\omega_r} = \frac{3\left(\frac{1-s_m}{s_m}\right) I_r^2 (z_r)^2 R_r}{\omega_r (z_r)^2}$$

$$= \frac{3\left(\frac{1-s_m}{s_m}\right)(V_r)^2 R_r\, 3\left(\frac{1-s_m}{s_m}\right)(V_r)^2 R_r}{\omega_r (Z_r)^2 \qquad \omega_r (Z_r)^2} \approx \frac{3\left(\frac{1-s_m}{s_m}\right)(V_r)^2 R_r}{\omega_r \left(\frac{R_r}{s_m}\right)^2} \tag{5.73}$$

$$\Rightarrow T_m \approx \frac{3V_r^2 s_m}{\omega_s R_r}$$

Since in the steady-state condition all variables are constant, the above equation indicates that T_m changes linearly with respect to slip (s_m). Similarly, for large values of slip $\frac{R_r}{s_m} \ll X_r$, it can be shown that,

$$T_m = \frac{3V_r^2 R_r}{\omega_s X_r^2 s_m} \tag{5.74}$$

The above equation indicates that for large values of slip, the torque characteristic is hyperbolic. Generally speaking, for a traction mode in which $0 < s_m < 1$, the rotor is driven by the faster rotating magnetic field where electrical power is transformed into mechanical power. In the region of $0 < s_m < s_{m,max}$, where $s_{m,max}$ is the slip at the motor maximum torque ($T_{m,max}$), the torque almost increases linearly with the increase of slip. After reaching the operating point of $\left(s_{m,max}, T_{m,max}\right)$, the torque starts decreasing as the slip increases. At $s_m = 1$, the motor is at a standstill and the corresponding torque is the stall or starting torque. For values of s greater than one ($s_m > 1$), the rotor is forced to turn against the magnetic flux (brake range). For slip values less than zero ($s_m < 0$), the rotor is forced to turn faster than the magnetic field. In this case, mechanical power is transformed into electrical power and the induction machine acts as a generator.

The characteristic curves of an induction motor indicate that, for EV/HEV applications, the induction motor must be properly controlled to provide demanded torque-speed characteristics so as to improve the motor's limitations such as the low starting torque, limited speed range and existing of unstable operation range.

Example 5.4

A three-phase induction motor with a delta connection and with 5 pairs of poles is connected to a 60 Hz voltage source of 440 V. The slip is 5% and the windage and friction losses are 1.8 kW. The equivalent circuit parameters are $R_s = 1.6\,\Omega$, $X_s = 0.6\,\Omega$, $R_r = 1\,\Omega$, $X_r = 2\,\Omega$ and $X_m = 26\,\Omega$. Calculate (a) the input power; (b) the speed of the rotor; (c) the mechanical power; (d) the output torque; and (e) the motor efficiency.

Solution

(a)

$$V_s = 440\angle 0^\circ$$

$$z_m = (R_s + jX_s) + \frac{\left(\frac{R_r}{s_m} + jX_r\right)(jX_m)}{\frac{R_r}{s_m} + j(X_r + X_m)} = (1.6 + 0.6j) + \frac{\left(\frac{1}{0.05} + 2j\right)26j}{\frac{1}{0.05} + j(2 + 26)} = 13.02 + \mathrm{j}10.61$$

$$I_s = \frac{V_s}{z_m} = \frac{440\angle 0^\circ}{13.02 + \mathrm{j}10.61} = 20.31 - j16.55 = 26.2\angle -39.18^\circ$$

$$P_{m,in} = 3V_s I_s \cos(\theta_v - \theta_i) = 3 \times 440 \times 17 \times \cos(61.8^\circ) = 26.808\ \mathrm{kW}$$

(b)

$$n_{ms} = \frac{120f}{p} = \frac{120 \times 60}{10} = 720\ (\mathrm{rpm})$$

$$s_m = \frac{n_{ms} - n_{mr}}{n_{ms}} \Rightarrow n_r = n_s(1 - s) = 720 \times (1 - 0.05) = 684\ \mathrm{rpm}$$

(c)

$$V_r = 440 - I_s z_s = 440 - (20.31 - j16.55)(1.6 + j0.6) = 397.57 + j14.29$$

$$I_r = \frac{V_r}{z_r} = \frac{397.57 + j14.29}{\dfrac{1}{0.05} + j2} = 19.65 - j1.25 = 19.69\angle - 3.64°\text{A}$$

$$P_m = 3\left(\frac{1 - s_m}{s_m}\right) I_r^2 R_r = 3\left(\frac{1 - 0.05}{0.05}\right)(19.69)^2(1) = 22.099\ \text{kW}$$

(d)

$$\omega_r = \frac{2\pi n_{mr}}{60} = \frac{2\pi \times 684}{60} = 71.63 \frac{\text{rad}}{\text{s}}$$

$$T_m = \frac{P_m}{\omega_r} = \frac{22.099 \times 1000}{71.63} = 308.52\ \text{Nm}$$

(e)

$$\eta_m = \frac{P_{m,out}}{P_{m,in}} = \frac{P_m - P_L}{P_{m,in}} = \frac{22.099 - 1.8}{26.808} = \%76$$

5.3.6 *Permanent-Magnet AC Motors*

Synchronous motors with a permanent magnet in the rotor are rapidly replacing induction motors in automotive applications due to their advantages such as compact size, high efficiency, low cost, and high power density. Permanent magnet synchronous motors (PMSM) and permanent magnet brushless DC motors (BLDC) belong to the permanent magnet AC family of motors. Similar to induction AC motors, the major components of permanent magnet AC motors include the stator and the rotor where the external three-phase AC voltage is applied to the armature windings of the stator. In terms of the torque generation principle, these two electric motors are identical. Both motor types are synchronous machines in which the operation is through the realization of the excitation on the rotor with permanent magnets. However, the main difference is that PMSMs develop a sinusoidal back EMF, as compared to a rectangular, or trapezoidal, back EMF for BLDC motors. It is noted that the term brushless DC indicates that this motor behaves similar to brushed DC motors where the mechanical commutators/brushes system are replaced by power electronic or electronic commutators. On the other hand, the term permanent magnet synchronous motor indicates that the machine uses permanent magnets to create the field excitation [2]. Assuming that the back EMF is sinusoidal for both types of motors, the performance analysis is identical.

5.3.6.1 Steady State Performance Analysis of Permanent Magnet Motors

A simplified equivalent circuit of one-phase BLDC in quasi steady-state conditions is shown in Figure 5.17 [5]. In the quasi steady state condition, all signals are sinusoidal. Here I_F is the

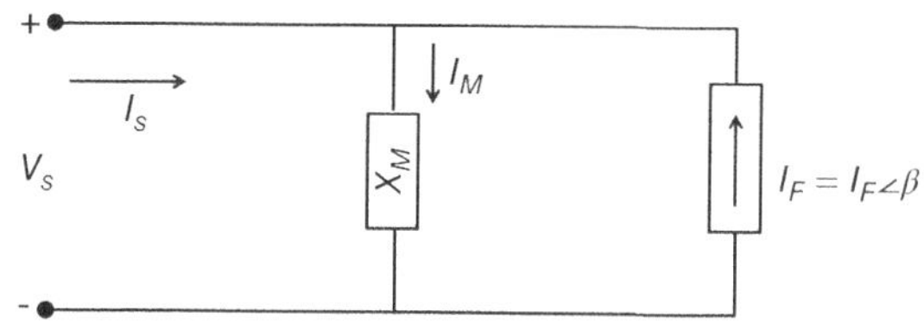

Figure 5.17 Simplified equivalent circuit of a PMSM

equivalent field source current. V_s and I_s are voltage and current of the stator, respectively. β_m is the angle between V_s and I_F. Two other angles which need to be defined are the angle between I_s and V_s (θ_m) and the angle between I_s and I_F (γ_m). Figure 5.18 shows the corresponding phasor diagram of the PMSM.

Using the phasors in Figure 5.18, it can be shown that the input power and the electromagnetic torque of the PMSM are [5]:

$$P_{m,in} = 3X_M I_s I_F \sin(\gamma_m) \tag{5.75}$$

$$T_m = \frac{3p}{2} L_M I_s I_F \sin(\gamma_m) \tag{5.76}$$

For a fixed stator voltage and power (and torque) level, in order to have minimum stator losses, I_s and V_s needs to be in zero phase ($\theta_m = 0$) as shown in Figure 5.19. If this condition holds true, the relation among I_s, I_M, and I_F is as [5]:

$$I_F^2 = I_M^2 + I_s^2 \tag{5.77}$$

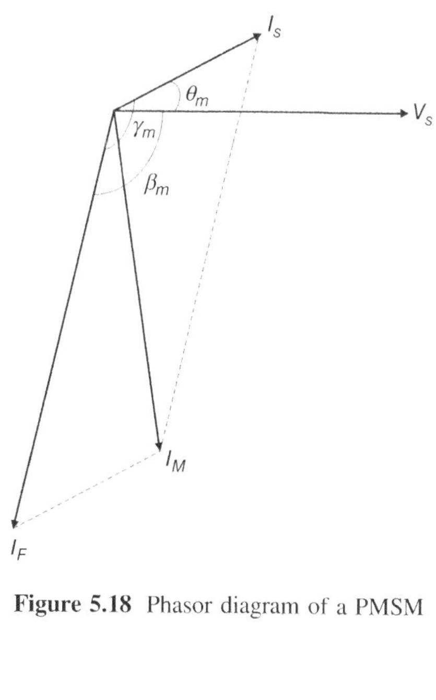

Figure 5.18 Phasor diagram of a PMSM

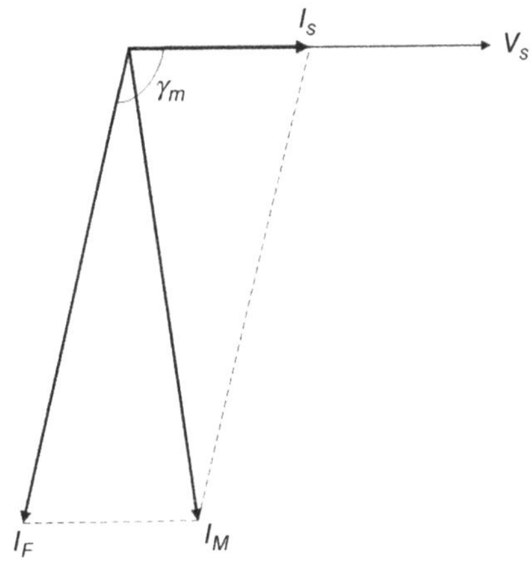

Figure 5.19 Phasor diagram of the PMSM when $\theta = 0$

For a given torque, the minimum losses require a minimum value of the stator current. To minimize the value of I_s with constant power and fixed I_F, γ_m should equal to 90°. Figure 5.20 depicts the corresponding phasor diagram.

While the rotor speed increases, the stator voltage $V_s = \omega_s L_M I_s$ increases till the stator voltage reaches ω_{sb}, the base speed. It is noted that for speeds beyond the base speed, the required voltage exceeds the maximum voltage that the power source can provide. Thus, to increase the speed beyond the base speed, it is not possible to keep $\gamma_m = 90°$. Since at and above the base speed, the voltage has reached its upper limit $V_s = V_{s,\max}$, the value of $I_M = V_{s,\max}/X_M$ is known. In this case,

$$T_{EM} = \frac{3p}{2\omega_s} V_s I_s \cos(\theta_m) = \frac{3p}{2} L_M I_M I_s \cos(\theta_m) \tag{5.78}$$

$$I_F^2 = I_M^2 + I_s^2 + 2 I_M I_s \sin(\theta_m) \tag{5.79}$$

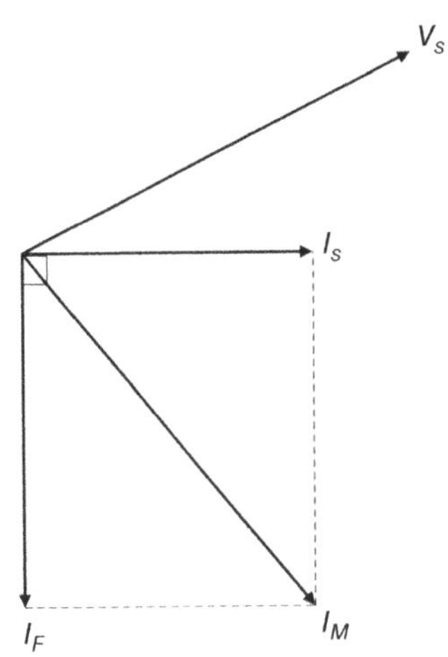

Figure 5.20 Phasor diagram of the PMSM when $\gamma = 90$

Example 5.5

A three-phase, four-pole, wye-connected, permanent magnet synchronous motor is rated at 400 (V), 50 (Hz), and 50 (kVA). Its magnetizing inductance is 2.5 (mH) and its equivalent field source current, I_F, is 310 (A). Assume the stator resistance is negligible and a unity power factor.

(a) If the motor is operated as a generator at rated frequency, determine the maximum and minimum values of the stator phase voltage as the load current is varied from zero to rated value.
(b) If the motor is operated by a variable speed drive motor from a variable voltage, variable frequency source, what should be the voltage and frequency in order to provide torque of 300 (Nm) at 600 (rpm)?

Solution

(a) The rated phase voltage is $V_S = \frac{400}{\sqrt{3}} = 231$ V and the rated stator current is $I_s = \frac{50 \times 10^3}{3 \times 231} = 72.2$A. With no load and at the rated frequency, the phase voltage is:

$$V_s = \omega_s L_M I_F = 2\pi \times 50 \times 2.5 \times 10^{-3} \times 310 = 243.5\text{V}$$

If the motor is operated at unity factor, the stator current is:

$$I_M^2 = I_F^2 - I_s^2 \Rightarrow I_M = \sqrt{310^2 - 72.2^2} = 301.5\text{A}$$

and the stator voltage is:

$$V_s = \omega_s L_M I_M = 236.8\text{V}$$

(b) The motor has four poles, thus

$$\omega_s = \frac{p}{2} 600 \frac{2\pi}{60} = 125.6 \text{ rad/s}$$

and the output torque is

$$T_m = 3\frac{p}{2\omega_s} V_s I_s pf = 3\frac{p}{2\omega_s}(\omega_s L_M I_M) I_s = 300 \text{ Nm}$$

Moreover, the equivalent field current for unity *pf* is:

$$I_F^2 = I_M^2 + I_s^2 = 310^2$$

Thus,

$$I_M = 303\text{ A } and\ I_s = 66\text{ A or}$$

$$I_M = 66\text{ A } and\ I_s = 303\text{ A}$$

As the stator current should remain in the safe range, the first set of solutions is acceptable. So, with $I_M = 303$ A, the phase voltage and output power are:

$$V_s = \omega_s L_M I_M = 95.1\text{ V}$$

$$P_{m,out} = \frac{2}{p}\omega_s T_m = 18.85\text{ kW}$$

5.4 Battery Performance Characteristics

Section 2.2.3 explained the concept of energy source/storage systems, including batteries and their parameters. This section describes the mathematical representation of battery performance characteristics and its related parameters, including capacity and state of charge (SoC) [3].

5.4.1 Battery Capacity

Generally, the amount of electric charge that a battery can store is the battery's capacity. The battery capacity size directly relates to the amount of electrolyte and electrode material inside the battery. Hence, greater capacity can be achieved if more electrolyte and electrode material are provided. The battery capacity is also a function of other battery parameters such as the magnitude of the current, the allowable terminal voltage of the battery, the temperature, and other factors. The measurement unit of a battery's capacity is Ah (1Ah = 3600C or coulomb). In vehicle applications, it is preferable to measure energy stored in the battery as watt-hour (Wh). The energy capacity of a battery measured in Wh can be converted to Ah using Ohm's rule that states battery power $P_b = v_b i_b$, where v_b and i_b are the voltage and current of the battery. Thus:

$$E_b = \text{power} \times \text{time} = v_b i_b \times \text{time} \tag{5.80}$$

Therefore:

$$\text{Wh} = \text{Ah} \times v_b \tag{5.81}$$

Theoretically, the battery's capacity can be calculated using Faraday's law of electrolysis. The law states that for a given quantity of electric charge, the mass of an elemental material altered at an electrode is directly proportional to the element's equivalent weight. The equivalent weight of a substance is its molar mass divided by an integer that depends on the reaction undergone by the material. Faraday's law can be expressed by:

$$m_s = \left(\frac{Q}{F_b}\right)\left(\frac{M_m}{z_b}\right) \tag{5.82}$$

where:

m_s is the mass of the substance altered at an electrode in kg
Q is the total electric charge passed through the substance
F_b= 96,485 C/mol is the Faraday constant
M_m is the molar mass of the substance in g/mol
z_b is the valency number of ions of the substance (electrons transferred per ion).

Note that the equivalent weight of the substance altered is $\frac{M_m}{z_b}$. In Equation 5.82, Q, F_b, and z_b are constants; thus, the larger equivalent weight $\frac{M_m}{z_b}$ results in the larger m_s. Therefore, the theoretical capacity of a battery, $C_{T,b}$, can be calculated as:

$$C_{T,b} = n_b F_b z_b \tag{5.83}$$

where the dimension of capacity is in coulombs. Here, n_b is the amount of substance ("number of moles") altered: $n_b = m_s/M_m$. In the above equation, symbol $C_{T,b}$ is used for the battery capacity since typically symbol C is used for battery capacity. The theoretical capacity in Ah is:

$$C_{T,b} = 0.278 n_b F_b z_b \tag{5.84}$$

The capacity can be expressed in energy unit using Equation 5.84 if the voltage of the battery is known. It is noted that the theoretical capacity of a battery in Equation 5.84 is derived with assumption of constant current while, in practice, a variable electrical current is the case. Thus, the usable capacity, $C_{U,b}$, of a battery is the electric current $i(t)$ integrated over time:

$$C_{U,b} = \int_{t_0}^{t_{cut}} i(t)dt \tag{5.85}$$

where, t_0 is the time when a battery is at a full charge and t_{cut} is the time when a battery terminal voltage is at the voltage cut, v_{cut}. As mentioned in Section 2.2.3, battery capacities are limited to the voltage cut in order to prevent sustaining permanent damage. Hence, the practical capacity is always less than the theoretical one because of practical limitations.

5.4.2 Open Circuit and Terminal Voltages

The voltage of a battery is one of its most important characteristics. It is a function of the chemical reactions in the battery, the concentration of the battery components, and the polarization of the battery. The simplest battery model comprises of a series connection of an internal voltage, V_b, and internal resistance, R_{bi}, as Figure 5.21a depicts. Open circuit voltage, also known as internal voltage, is the voltage between the battery terminals when not connected to an external load or no electric current flows between the terminals. The open-circuit voltage is a function of other battery parameters such as the state of charge, temperature, and past discharge/charge history. For example, the variation of open circuit voltage in terms of state of charge is plotted in Figure 5.21b. As shown, when the battery gradually discharges, the

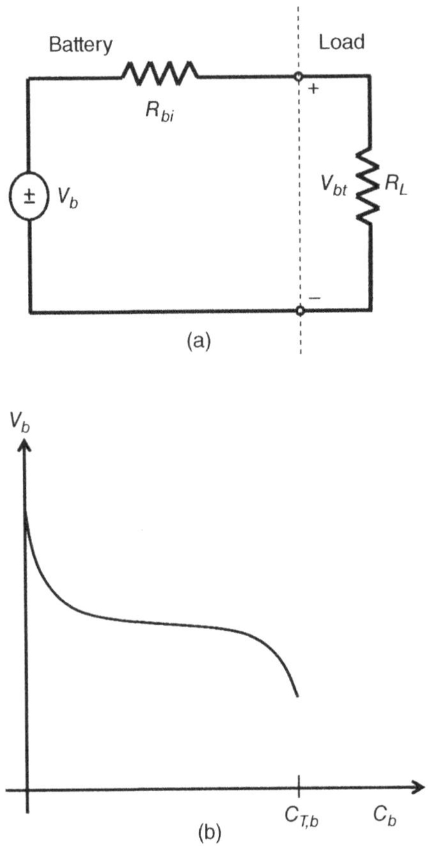

Figure 5.21 (a) Steady-state battery equivalent circuit, (b) battery open circuit voltage characteristics with constant current

internal voltage decreases and the internal resistance increases. It is noted that as internal resistance increases, the battery efficiency decreases, since more of the charging energy is converted into heat. By measuring the open circuit voltage, it is possible to roughly estimate the residual capacity of the battery. However, accurate estimation of the battery capacity needs a careful consideration of the discharge's current characteristics, as well as battery chemistry, temperature effects, and other factors. When a load is connected to the battery, the voltage measured at the terminals is the battery terminal voltage, V_{bt}.

5.4.3 Charge/Discharge Rate

As mentioned in Section 2.2.3, measurement of the charge and discharge current of a battery is in C-rate or C/h-rate, where C is the rate of battery capacity and h is the time in hours. A discharge of 1C draws a current equal to the rated capacity. For instance, a battery rated at 100 Ah provides 100 A for one hour if discharged at a rate of 1C. The same battery discharged at 0.5C provides 50 A for 2 h. At 2C, the same battery delivers 200 A for 30 min. 1C is often referred to as a 1 h discharge; 0.5C would be a 2 h discharge, and 0.1C is a 10 h discharge.

The maximum current at which the battery can be continuously discharged is called the *maximum continuous discharge current*, which is usually defined by the battery manufacturer in order to prevent excessive discharge rates that would damage the battery or reduce its capacity.

Charge and discharge rates indicate constant current loads representing many applications with constant resistance. In this case, the current relates to the voltage in relation to Ohm's law. Most battery systems, such as nickel-cadmium, lead-acid, and lithium-sulfur, have a fairly constant voltage during discharge, which means that under a purely resistive load, the current is relatively constant.

5.4.4 State of Charge/Discharge

State of the charge (SoC) is a measure of residual capacity of a battery and is the equivalent of a fuel gauge for the battery pack in EVs/HEVs. In other words, it is the amount of capacity that remains after the discharge from the fully charged condition. The units of SoC are percentage points (0% = empty; 100% = full). Direct determination of SoC is not usually possible. However, it can be theoretically calculated using battery voltage and current.

In the voltage method, the battery voltage is converted to SoC by a given discharge curve (voltage vs. SoC). However, as mentioned in Section 5.4.2, the voltage is significantly affected by the battery current and temperatures. Therefore, the discharge curves are subject to variation under different operating conditions, thereby making this method unreliable.

On the other hand, SoC can be theoretically calculated using the battery current and integrating it in time. The current is the rate of charge given by:

$$i(t) = C_{T,b}\frac{dq}{dt} \tag{5.86}$$

where q is the per-unit charge (charged divided by the capacity) flowing thorough the circuit. For a time interval, dt, the theoretical battery state of charge, $SoC_{T,b}$ is:

$$dSoC_{T,b} = -dq = -\frac{1}{C_{T,b}}i(t)dt \tag{5.87}$$

Integrating from the initial time, t_0, to the final time, t, and with consideration of $SoC_{T,b}$ typically measured as the percentage of battery capacity, the instantaneous battery SoC is:

$$SoC_{T,b}(t) = SoC_{T,b}(t_0) - \left(\frac{1}{C_{T,b}}\int_{t_0}^{t} i(\tau)d\tau\right) \tag{5.88}$$

Discharging the battery results in an decrease of the $SoC_{T,b}$. If the state of the charge is 100% at initial time, then the $SoC_{T,b}$ is:

$$SoC_{T,b} = 1 - \frac{\int_{t}^{t_0} i(\tau)d\tau}{C_{T,b}} \tag{5.89}$$

If $i(t)$ represents the charging current and the state of the charge is zero at initial time, the formula for $SoC_{T,b}$ is

$$SoC_{T,b} = \frac{\int_{t}^{t_0} i(\tau)d\tau}{C_{T,b}} \tag{5.90}$$

It is noted that calculation of *SoC* using the afore-mentioned equations requires integration of the current signal, which can suffer from long-term drift and lack of a reference point. A more accurate estimation of *SoC* can be obtained by more advanced estimation algorithms such as Kalman filters.

The dynamic (rate) of $SoC_{T,b}$ can be expressed as a function of open-circuit voltage, internal resistance, and battery power. By applying Kirchhoff's voltage law, according to Figure 5.18a, one has:

$$V_{bt} + R_{bi}i - V_b = 0 \tag{5.91}$$

Considering the terminal battery power $P_b = V_{bt}i$, by multiplying above equation with V_t, the following quadratic equation can be achieved.

$$V_{bt}^2 + R_{bi}V_{bt}i - V_{bt}V_b = o \Rightarrow V_{bt} = \frac{E_v \pm \sqrt{V_b{}^2 - 4R_{bi}P_b}}{2} \tag{5.92}$$

Using Ohm's law and assuming constant internal resistance, the voltage V_{bt} can be rewritten as:

$$V_{bt} = V_b - R_{bi}i = V_b + R_{bi}C_{T,b}\frac{dSoC_{T,b}}{dt} \tag{5.93}$$

Inserting Equation 5.93 in Equation 5.92 results in:

$$\frac{dSoC_{T,b}}{dt} = -\frac{V_b \pm \sqrt{V_b{}^2 - 4R_{bi}P_b}}{2R_{bi}C_{T,b}} \tag{5.94}$$

5.4.5 *Depth of Discharge*

As mentioned in Section 2.2.3, depth of discharge, DoD, is a measure of the amount of discharged energy capacity from the battery, typically expressed as a percentage of maximum capacity. The state of discharge can be given as:

$$DoD_{T,b} = \frac{1}{C_{T,b}}\int_{t_0}^{t} i(\tau)d\tau - DoD_{T,b}(t_0) \tag{5.95}$$

Deep discharging beyond the cut-off voltage must be avoided, especially under heavy loads, to prevent serious damage to the batteries. The withdrawal of at least 80% of battery (rated) capacity is referred to as deep discharge.

Using the battery depth of discharge and Peukert's law, it is possible to estimate the travel range of an EV/HEV. Peukert's law, originally proposed by the German scientist, W. Peukert, in 1897 for a lead-acid battery, states that, subject to constant current discharge characteristics, the battery capacity relates to the battery's current rate at which it discharges.

$$I_{cdc}{}^{k_{b1}} t_{cut} = k_{b2} \tag{5.96}$$

where:

I_{cdc} is the constant discharge current.
t_{cut} is the time (h) when the terminal voltage reaches to the cut-off voltage limit.
k_{b1} and k_{b2} are constant parameters of a particular battery. k_{b1} is dimensionless and k_{b2} has Amp-hour (Ah) unit.

t_{cut} can be calculated from Equation 5.85. Assuming $t_0 = 0$ and constant discharge current $i(t) = I$, the usable capacity of a battery is:

$$C_{U,b} = I_{cdc} t_{cut} \Rightarrow t_{cut} = \frac{C_{U,b}}{I_{cdc}} \tag{5.97}$$

Substituting t_{cut} into Peukert's relation, one has:

$$I_{cdc}{}^{k_{b1}} \left(\frac{C_{U,b}}{I_{cdc}} \right) = k_{b2} \Rightarrow C_{U,b} = \frac{k_{b2}}{I_{cdc}{}^{k_{b1}-1}} \tag{5.98}$$

Since $0 < (k_{b1} - 1) < 1$, for $I > 1$ the battery capacity decreases as I increases.

Moreover, from Equation 5.95, the small variation of DoD is:

$$d(DoD) = \frac{i(t)dt}{C_{U,b}} = \frac{I_{cdc} dt}{\dfrac{k_{b2}}{I_{cdc}{}^{k_{b1}-1}}}$$
$$\Rightarrow \frac{d(DoD)}{dt} = \frac{I^{k_{b1}}}{k_{b2}} \tag{5.99}$$

The above equation is subject to constant discharge current. However, in reality, the discharge current is a time-varying parameter. For the purposes of demonstration, let us substitute the time-varying current into the above equation to estimate the battery's range. Thus:

$$\frac{d(DoD)}{dt} = \frac{i(t)^{k_{b1}}}{k_{b2}} \tag{5.100}$$

Integrating from both sides of this equation, one can obtain:

$$DoD(t) - DoD(t_o) = \int_{t_0}^{t_1} \frac{i(t)^{k_{b1}}}{k_{b2}} dt \tag{5.101}$$

$DoD(t_0) = 0$, if the battery is fully charged at $t = t_0$.

Example 5.6

An electric vehicle has a battery pack with the maximum 80% depth of discharge. If the vehicle starts cruising at 90 km/h while the batteries are at full charge, calculate the vehicle range. The given k_{b1} and k_{b2} of the battery from the manufacturer are 1.2 and 645 Ah. The discharge current is varying as:

$$i(t) = \begin{cases} 5t & 0 < t < 50\,s \\ 100 & 50s < t \end{cases}$$

Solution

$$(0.8 - 0) = \frac{1}{645 \times 3600}\left[\int_0^{50} (5t)^{1.2}dt + \int_{50}^{t} (100)^{1.2}dt\right] \Rightarrow$$

$$\Rightarrow 0.8 \times 645 \times 3600 = \frac{(5)^{1.2} \times (50)^{2.2}}{2.2} + \frac{(100)^{1.2}}{1}[t - 50]$$

$$\Rightarrow t = 7377\,s = 2.05\,h$$

Thus,

$$x = 90 \times 2.05 = 184.5\,\text{km}$$

5.4.6 *Battery Energy Density and Specific Energy*

According to the definition in Section 2.2.3, energy density is the amount of energy capacity per unit volume, while specific energy is the amount of usable energy capacity per unit mass in Wh/kg. The theoretical energy stored in a battery directly relates to the battery voltage and its theoretical capacity. Typically, the theoretical stored energy is:

$$E_{T,b} = V_b C_{T,b} \tag{5.102}$$

where V_b is the nominal no load terminal voltage (open circuit voltage) and $C_{T,b}$ is the theoretical battery capacity in coulombs. It is noted that 1 Ah is equivalent to 3600 coulombs (C). $C_{T,b}$ can be calculated using Equation 5.84. The calculation of available energy is more complex than the theoretical one as it depends on the manner in which the battery discharges. The available energy can be formulated using Equation 5.80 as:

$$E_b = \int_{t_0}^{t_{cut}} V_{bt}(t)i(t)dt\ \text{Wh} \tag{5.103}$$

where:

t_0 is the time at which the battery is fully charged
t_{cut} is the time at which battery terminal voltage reach to cut-off voltage
V_{bt} (t) is the battery terminal voltage
$i(t)$ is the battery discharge current.

Accordingly, the battery specific energy is

$$\mathrm{SE}_b = \frac{E_b}{M_b} \tag{5.104}$$

where M_b is the total mass of battery.

5.4.7 *Battery Power Density and Specific Power*

As defined in Section 2.2.3.1, power density refers to the energy rate (in Watts) that can be delivered per unit volume, while specific power is the energy rate (in Watts) available to deliver per unit mass. According to the definition of power, the instantaneous battery terminal power is:

$$P_b(t) = V_{bt}(t)i(t) \tag{5.105}$$

where $V_{bt}(t)$ is the battery terminal voltage and $i(t)$ is the battery discharge current. Using the battery equivalent circuit shown in Figure 5.18a, the battery terminal voltage can be calculated based on Kirchhoff's voltage as:

$$V_{bt}(t) = V_b(t) - R_{bi}i(t) \tag{5.106}$$

Substituting Equation 5.106 into Equation 5.105 results in:

$$P_b(t) = V_b(t)i(t) - R_{bi}i(t)^2 \tag{5.107}$$

As shown by the equation above, battery power is a quadratic function of current. Thus, it has a maximum value at a certain current value. The maximum power is:

$$P_{\max,b} = \frac{V_b{}^2}{4R_{bi}} \tag{5.108}$$

Note that since V_b and R_{bi} are a function of the battery state of charge, $P_{max,b}$ varies as the state of charge changes.

During heavy load operating conditions such as rapid acceleration, the maximum power output is needed. In such situations, the electric motor draws a significant amount of current to provide the maximum power required for traction.

The rated power specifications, based on the ability of the battery to dissipate heat, can assess the performance of the batteries in fulfilling heavy duty operating conditions, such as rapid acceleration and hill climbing. The definition of rated continuous power is the maximum power at which the battery can provide prolonged discharge intervals without damaging the battery, whereas the rated instantaneous power is the maximum power that the battery can deliver over very short discharge intervals without damaging the battery. The rated power specifications, along with the maximum continuous discharge current, the top sustainable speed and acceleration of the vehicle can be defined.

According to the definition, the specific power of a battery is:

$$\mathrm{SP}_b = \frac{P_b}{M_b} \tag{5.109}$$

Example 5.7

A given battery for an HEV has a nominal voltage of 200 V, and a nominal capacity of 8 Ah.

(a) Calculate the energy rating of the battery.
(b) If the maximum discharged rate is 100 A while it is fully charged, and if 50% can be discharged, calculate the discharge time.
(c) If the maximum charge rate is 80 A and the current SoC is 40%, calculate the charging time that takes the battery SoC to 90%.

Solution

$$E_{T,b} = 200\ V \times 8\ \mathrm{Ah} = 1.6\,\mathrm{kWh}$$

$$t_{dis} = \frac{50\% \times 8\mathrm{Ah}}{100\ \mathrm{A}} = 0.04\,\mathrm{h} = 144\,\mathrm{s}$$

$$t_{chg} = \frac{(90\% - 40\%)8\ \mathrm{Ah}}{80\ \mathrm{A}} = 0.05\,\mathrm{h} = 180\,\mathrm{s}$$

Example 5.8

An electric vehicle is cruising at 120 km/h and has a battery with a nominal voltage of 200 V, specific energy of 55 Wh/kg, and a mass of 150 kg. If the discharged rate is 80 A while it is fully charged, and if the minimum SoC is 40%, calculate the distance that the vehicle can travel. Assume the voltage is constant.

Solution

$$E_b = \mathrm{SE}_b \times M_b = 55\frac{\mathrm{Wh}}{\mathrm{Kg}} \times 150\ \mathrm{Kg} = 8250\ \mathrm{Wh}$$

$$C_{T,b} = \frac{E_b}{V_b} = \frac{8250}{200} = 41.25\ \mathrm{Ah}$$

$$t_{dis} = (100\% - 40\%) \times \frac{41.25\ \mathrm{Ah}}{80\ \mathrm{A}} = 0.3\ \mathrm{h}$$

$$x = 120 \times 0.3 = 36\ \mathrm{km}$$

5.4.8 Battery Efficiency

Similar to other systems, batteries are not 100% efficient. Some battery input power is lost due to internal resistance. This can be presented by:

$$P_{b,out} = P_{b,in} - P_{b,L} \tag{5.110}$$

where, $P_{b,out}$ and $P_{b,in}$ are output and input battery powers, while $P_{b,L}$ is the internal lost power. Based on Figure 5.18a, the internal power loss of a battery can be obtained as:

$$P_L = R_{bi} i^2 \tag{5.111}$$

As well, the battery discharging power is:

$$P_{b,out} = V_{bt} i \tag{5.112}$$

By defining the load resistance as $R_l = \frac{V_{bt}}{i}$ and noting that the discharge efficiency is the ratio of output power to input power, the battery discharge efficiency is:

$$\eta_{b,dis} = \frac{R_l}{R_{bi} + R_l} \tag{5.113}$$

R_l can be obtained by measuring the terminal voltage and current values.

During the charging phase, the direction of the current is reverse of that in the discharging phase, and the charging (terminal) voltage, V_c, is higher than the open circuit voltage. With the same procedure for calculating the efficiency of the battery during the discharging phase, the battery charging efficiency is:

$$\eta_{b,chg} = 1 - \frac{R_{bi}}{R_c} \tag{5.114}$$

with $R_c = \frac{V_c}{i}$

As mentioned before, the internal resistance varies during charging and discharging phases and depends on the battery state-of-charge, temperature, etc.

5.5 Transmission and Drivetrain Characteristics

Section 1.6 featured a comprehensive discussion of the concept of different types of transmission systems, such as manual, automatic, CVT, and semi-automatic. The following subsections will discuss the fundamental relations governing these systems.

5.5.1 Gearboxes

The way that power transmits from the engine to the wheels governs the vehicle motion in different driving conditions. Thus, gearboxes typically are designed to meet vehicle torque requirements under different driving conditions. For example, low gears provide high tractions to meet the torque requirements of accelerating and hill-climbing, while high gears are used in

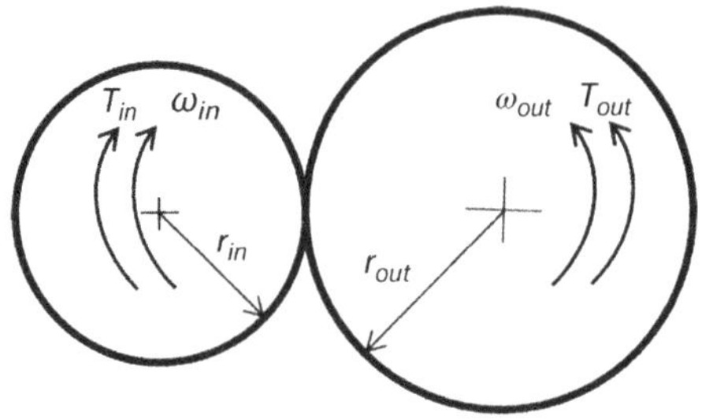

Figure 5.22 Gear mechanism

vehicle cruising at higher speeds. The torque and speed adjustments in the gearbox are carried out through the gears.

Figure 5.22 shows a basic gear mechanism. The gear ratio is the ratio of the torque of driven (output) gear to the torque of driving (input) gear, which itself is the inverse of the ratio of output and input speeds.

$$\beta_g = \frac{T_{out}}{T_{in}} = \frac{\omega_{in}}{\omega_{out}} \tag{5.115}$$

In fact, the above relation shows the principle of the conservation of energy, since torque multiplied by speed is the power that remains constant in the transmission process, assuming there is no power loss in power transfer.

Since the contact force is the same for both engaged gears, and since it is known that torque is equal to the contact force multiplied by gear radius, there is also a relation between the gear ratio and the radius of the gears. Conversely, the diameter of the gear is proportional to the number of teeth. Thus, the general gear law can be expressed as:

$$-\frac{T_{out}}{T_{in}} = \frac{r_{out}}{r_{in}} = \frac{N_{out}}{N_{in}} = -\frac{\omega_{in}}{\omega_{out}} \tag{5.116}$$

where N_{out} and N_{in} are the number of teeth of output and input gears, respectively. r_{out} and r_{in} are the radius of the output and input gears, respectively.

The negative sign indicates the opposite direction of the torques and rotations in engaged gears. The gear ratio for gear systems with more than two gears in mesh can be calculated using:

$$\beta_g = \frac{\prod_{i=1}^{p} N_{driving_i}}{\prod_{j=1}^{q} N_{driven_j}} \tag{5.117}$$

where $N_{driving_i}$ is the tooth number of ith driving gear, N_{driven_j} is the tooth number of jth driven gear, and p and q are the number of driving and driven gears in the gear set.

Gear sets can work as torque-coupling devices in HEVs. Figure 5.23 conceptually represents a three-port torque-coupling device. Port 1 is a unidirectional power input port, while ports 2 and 3 are bidirectional power input or output. However, both are not power inputs at the same

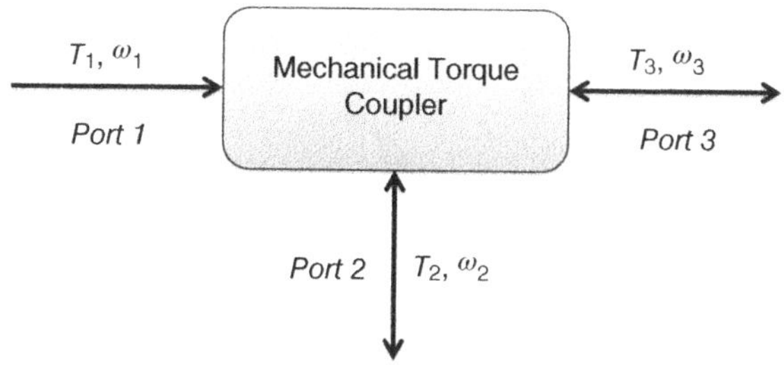

Figure 5.23 Torque-coupling device

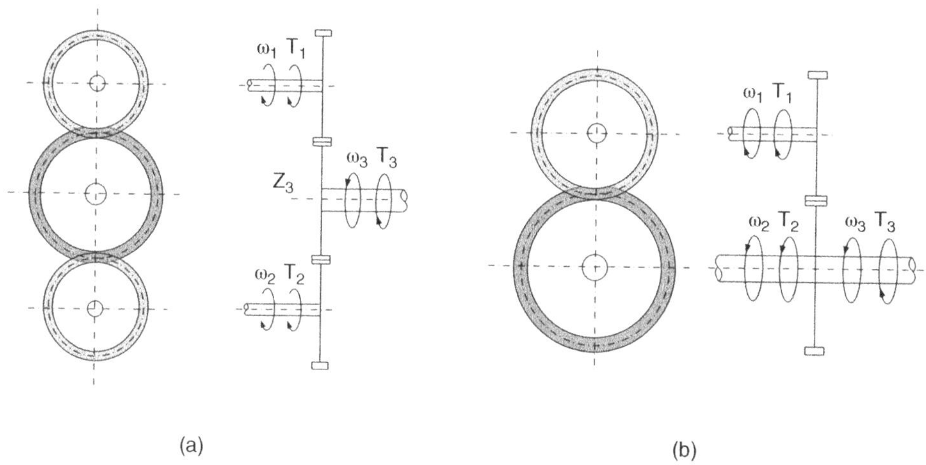

Figure 5.24 Gearbox mechanical torque-coupling devices

time. Typically, an IC engine is directly or through a transmission system is connected to port 1, while the shaft of the electric motor is directly or through a transmission is connected to port 2. Port 3 is connected to the driven wheels through a mechanical transmission system. Figure 5.24 shows two possible gearbox configurations that work as a mechanical torque-coupling device. The torque relation for the mechanical torque-coupling device of a HEV is:

$$T_3 = \beta_1 T_1 + \beta_2 T_2 \tag{5.118}$$

The constants for configuration (a) are $\beta_1 = \dfrac{N_{G_3}}{N_{G_1}}$ and $\beta_2 = \dfrac{N_{G_3}}{N_{G_2}}$, while for the configuration (b) are $\beta_1 = \dfrac{N_{G_2}}{N_{G_1}}$ and $\beta_2 = 1$

5.5.2 *Planetary Gear Set*

5.5.2.1 Gear Ratio

Figure 5.25 shows a typical planetary gear set that includes a ring gear, carrier and the sun gear. As discussed in Chapter 1, gears of a planetary gear set can be connected in various

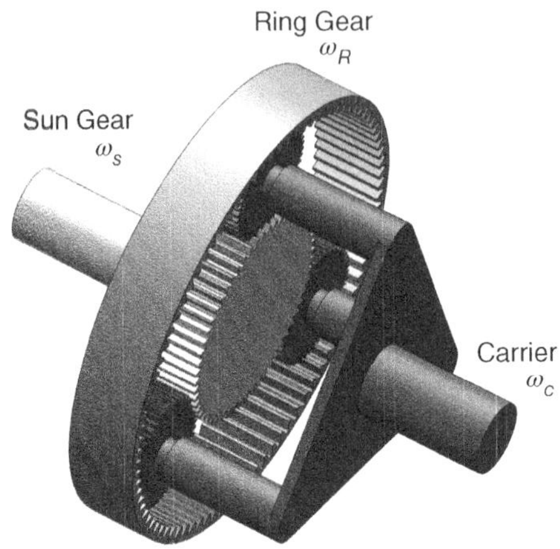

Figure 5.25 A typical planetary gear set configuration

combinations, thereby providing different velocity reduction ratios. For example, the ring gear and carrier can be the drive inputs and the sun gear can be the output. Alternatively, the ring gear can be fixed, the carrier can be the drive input and the sun gear can be the output. In these two different configurations, the reduction ratios are different.

In a planetary gear set, the planet gears can rotate around a moving axis which makes the velocity relationships between its gears different from Equation 5.116. Generally, due to the relative motion of planetary gear set gears, the angular velocity of a gear is a function of the angular velocities of the other gears controlled independently. Here, independent control means that any velocity can apply freely to the gear [8].

Generally, for a given planetary gear set with N number of independently controlled gears, the governing kinematic equation is:

$$\omega_{G_1} = k_{g1}\omega_{G_2} + k_{g2}\omega_{G_3} \ldots + k_{gN}\omega_{G_{(N+1)}} \tag{5.119}$$

where ω_{G_1} is the angular velocity of a given gear, G_1,while $\omega_{G_2}, \omega_{G_3}, \ldots, \omega_{G_{(N+1)}}$ are angular velocities of the N number of gears. $k_{g1}, k_{g2}, \ldots, k_{gN}$ are constants that depend upon the planetary gear set geometry. For each certain configuration of gears, the constants can be determined by fixing components at zero angular velocity. Without loss of generality, the determination of the constants can be explained by considering the case of a simple planetary gear set shown in Figure 5.25.

Let us consider the ring gear and carrier be the drive inputs and the sun gear is selected as the output. Equation 5.119 is then:

$$\omega_S = k_{g1}\omega_R + k_{g2}\omega_C \tag{5.120}$$

where ω_S, ω_R, and ω_C are the absolute angular velocities of the sun gear, ring gear, and carrier, respectively. Constant k_{g1} can be identified by considering a drive connection with zero carrier velocity ($\omega_C = 0$). Hence, from Equation 5.120:

$$k_{g1} = \frac{\omega_S}{\omega_R} \tag{5.121}$$

Table 5.1 Relative angular velocities for a planetary gear set with fixed ring gear

Procedure	Planetary gear set components			
	Ring gear	Planet gear	Sun gear	Carrier
Assuming all gears locked together, rotate carrier one revolution clockwise	1	1	1	1
Assuming a fixed carrier, rotate the ring gear one revolution counter clockwise	-1	$-\dfrac{N_R}{N_P}$	$\dfrac{N_R}{N_S}$	0
Using superposition, add the resulting revolutions from the above steps	0	$1-\dfrac{N_R}{N_P}$	$1+\dfrac{N_R}{N_S}$	1

and from gear geometry, one has:

$$\omega_S = \frac{N_P}{N_S}(-\omega_P) \tag{5.122}$$

$$\omega_R = \frac{N_P}{N_R}(\omega_P) \tag{5.123}$$

where ω_P is the planet angular velocity, and N_P, N_S, and N_R are the gear radius or number of the teeth on the sun, ring, and planet. Thus,

$$k_{g1} = \frac{\omega_S}{\omega_R} = \frac{\left(\dfrac{N_P}{N_S}\right)(-\omega_P)}{\left(\dfrac{N_P}{N_R}\right)(\omega_P)} = -\frac{N_R}{N_S} \tag{5.124}$$

Similarly, constant k_{g2} can be determined by considering a drive connection in which the ring is fixed ($\omega_R = 0$). Thus, from Equation 5.120, one has:

$$k_{g2} = \frac{\omega_S}{\omega_C} \tag{5.125}$$

In this case, since the carrier is moving, it is impossible to directly determine constant k_{g2} from the gear geometry. As an alternative, it is possible to use Table 5.1, as well as the well-known superposition method to calculate constant k_{g2}. Essentially, this method converts the system into a fixed geartrain for the purpose of analysis. According to Table 5.1, with one revolution of the carrier, the sun gear rotates $1+\dfrac{N_R}{N_S}$. Thus, there is a relation between angular velocities:

$$\frac{\omega_S}{\omega_C} = \frac{\left[1+\dfrac{N_R}{N_S}\right]}{1} \tag{5.126}$$

thereby resulting in:

$$k_{g2} = \frac{\omega_S}{\omega_C} = 1+\frac{N_R}{N_S} \tag{5.127}$$

Table 5.2 Different configurations of a planetary gear set and their related velocity-reduction ratios

Input	Fixed input	Output	Velocity-reduction ratio*
S	C	R	$-\beta_{RS}$
R	C	S	$-1/\beta_{RS}$
C	S	R	$\beta_{RS}/(1+\beta_{RS})$
R	S	C	$(1+\beta_{RS})/\beta_{RS}$
S	R	C	$1+\beta_{RS}$
C	R	S	$1/(1+\beta_{RS})$

Note: $*\beta_{RS} = N_R/N_S$ = Number of ring gear teeth/Number of sun gear teeth

Defining $\beta_{RS} = \frac{N_R}{N_S}$, the governing kinematic equation for a simple planetary gearset is:

$$\omega_S = k_{g1}\omega_R + k_{g2}\omega_C = \left(-\frac{N_R}{N_S}\right)\omega_R + \left(1 + \frac{N_R}{N_S}\right)\omega_C = -\beta_{RS}\omega_R + (1+\beta_{RS})\omega_C \tag{5.128}$$

For a simple planetary gear set composed of a single carrier, sun gear, and ring gear, six different configurations are possible with a unique velocity reduction gear. As mentioned previously, a simple gear set is a two-degrees-of-freedom device with two inputs (e.g., two drive components controlled independently) and one output which its velocity is a function of the input velocities. Usually, one drive component is stationary for speed-reduction applications. Table 5.2 shows the possible configurations and their velocity-reduction ratios.

As mentioned in Section 2.4.2, a simple planetary gear set can work as a mechanical speed coupling device in HEVs to combine the power generated by two power plants through their speeds. The governing kinematic relations are similar to the above-mentioned relations.

Example 5.9

Determine the velocity-reduction ratios for the two-stage planetary gear set given in Figure 5.26.

Solution

Similar to a simple planetary gear set, we can develop a comparable expression for more complex planetary configurations, such as this example. To solve this example, treat the drive configuration as two separate planetary gear sets coupled on the shaft. This makes developing the governing kinematic equations easier. Figure 5.26 shows that components C_1 and R_2 are coupled, which is why their angular velocities are identical. Assuming β_1 and β_2 are the ratios of ring to gear radii for the two planetary gear sets, the governing kinematic equation for each planetary gear set is:

$$\begin{aligned}\omega_{S_1} &= -\beta_1\omega_{R_1} + \left[(1+\beta_1)\omega_{C_1}\right]\\ \omega_{S_2} &= -\beta_2\omega_{R_2} + \left[(1+\beta_2)\omega_{C_2}\right] = -\beta_2\omega_{C_1} + \left[(1+\beta_2)\omega_{C_2}\right]\end{aligned}$$

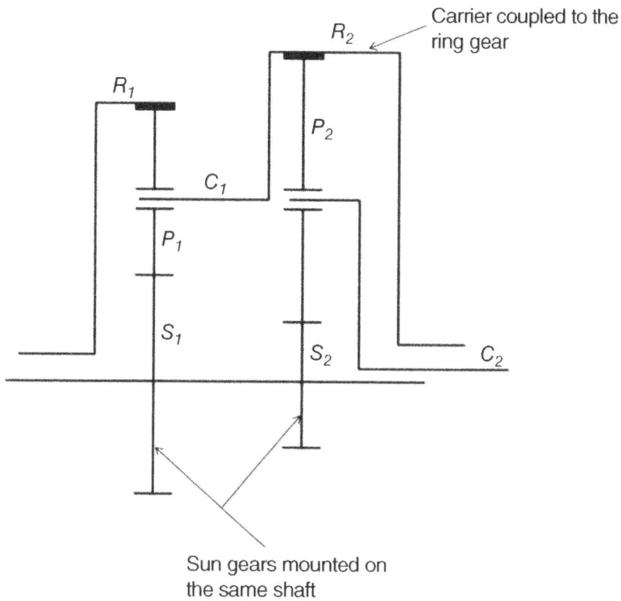

Figure 5.26 Two-stage planetary gear set

By solving the above equations for ω_{C_1} and setting them equal to one another, it is possible to obtain the following relation (which is also the governing kinematic relation for the overall system):

$$(1+\beta_1+\beta_2)\omega_{S_1} = (-\beta_1\beta_2)\omega_{R_1} + \left[(1+\beta_1)(1+\beta_2)\omega_{C_2}\right]$$

5.5.2.2 Torque

Developing the governing relations for torques applied on the components of a planetary gear set is not a trivial task. However, there are some basic rules by which it is possible to analyze the torques. For example, the ratio of ring torque to sun torque in a simple planetary is equal to the ratio of the ring and sun gear radius. In addition, the torque ratio for rotating input and output drive components with external torque is equal to the reciprocal of the velocity ratio. Generally, the algebraic sign of the velocity ratio also applies to the torque ratio, if the drive input and output shafts are parallel and point in opposite directions. Considering the correct sign for the input and output torques and ignoring the losses, the following equation is imposed on the torques due to the principle of power conservation:

$$\sum T_i\omega_i = 0 \tag{5.129}$$

Example 5.10

Determine the torque ratio for a simple planetary gear set in which the sun gear and ring are the input drives and the carrier is the output drive. Ignore the losses.

Solution

From Equation 5.132, it is possible to establish the following relation:

$$T_C\omega_C + T_S\omega_S + T_R\omega_R = 0 \Rightarrow T_C = -T_S\frac{\omega_S}{\omega_C} - T_R\frac{\omega_R}{\omega_C}$$

From Equation 5.129, one has:

$$\frac{\omega_S}{\omega_C} = 1 + \beta_{RS}, \quad \text{and} \quad \frac{\omega_R}{\omega_C} = \frac{1 + \beta_{RS}}{\beta_{RS}}$$

Thus,

$$T_C = -(1 + \beta_{RS})T_S - \frac{1 + \beta_{RS}}{\beta_{RS}}T_R$$

5.5.2.3 Efficiency

Gear mesh losses are the most important losses in a planetary gear set, which occur at a constant percentage of the power transferred across each mesh. Unfortunately, the application of the power flow concept to planetary drives usually requires difficult-to-drive mathematical expressions. However, if the data of velocities and torques of planetary gear set components are given, it is possible to evaluate the efficiency of the system. Generally, it is most convenient to just add the assumed losses at each mesh and then calculate the efficiency. Therefore, the procedure described below can help estimate the efficiency, with only negligible impact on accuracy.

First, determine the angular velocity and torque of one gear in each mesh to calculate the transmitted power. Next, attribute a certain percentage of transmitted power to mesh losses. For example, a 1% loss can be assumed for internal meshes, and a 2% loss for external meshes. Finally, sum the calculated losses and subtract from the input power to estimate the overall efficiency. Note that when calculating, use absolute values for the torque and velocity quantities to avoid the possibility of negative power losses, resulting from algebra inherent in the power loss equations.

Example 5.11

Calculate the efficiency of a fixed-ring simple planetary gear set. Assume there is a 1% loss at the internal mesh between the ring and the planet, and a 2% loss at the external mesh between the planet and sun.

Solution

Estimate the drive power losses as:

$$P_L = 0.01|(\omega_R - \omega_C)T_R| + 0.02|(\omega_S - \omega_C)T_S|$$

Example 5.10 shows the following torque ratios for a fixed-ring planetary:

$$\frac{T_C}{T_S} = 1 + \beta_{RS}, \ \frac{T_R}{T_S} = -\beta_{RS}$$

Thus,

$$P_L = 0.03\left(\frac{\beta_{RS}}{1+\beta_{RS}}\right)\omega_C T_C$$

Hence, the driving efficiency η_G is:

$$\eta_G = \frac{\omega_C T_C - P_L}{\omega_C T_C} = \frac{1 + 0.97\beta_{RS}}{1 + \beta_{RS}}$$

where $\omega_C T_C$ is the input power.

5.5.3 V-Belt CVTs

Figure 5.27 illustrates the geometry of a typical V-belt CVT, where both pulleys rotate around their fixed axes. The sides of each pulley are controlled so as to laterally move towards or opposite of each other. Assuming no slip between the belt and pulleys, the speed ratio between the drive and driven pulleys is [7]:

$$\beta_{CVT} = \frac{\omega_{driven}}{\omega_{drive}} = \frac{r_{p1}}{r_{p2}} = \frac{A - x_p}{B + x_p} \tag{5.130}$$

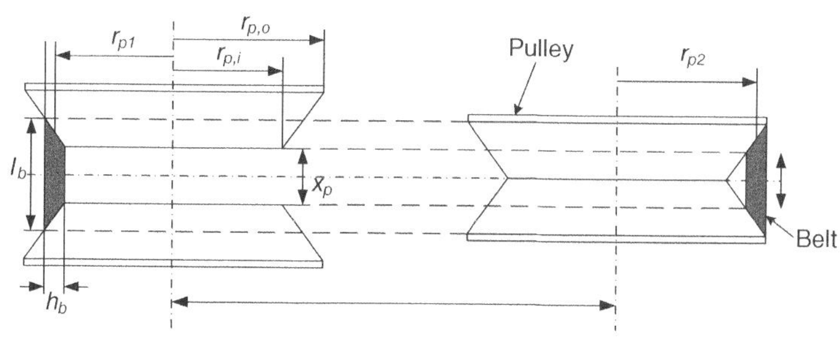

Figure 5.27 Geometry of a typical belt CVT

where x_p is the distance between the two sides of the drive pulley and:

$$A = \frac{2r_{p,o} - h_b}{2(r_{p,o} - r_{p,i})} l_b \tag{5.131}$$

$$B = \frac{2r_{p,i} + h_b}{2(r_{p,o} - r_{p,i})} l_b \tag{5.132}$$

Variable x_p has limits between the two extremes as:

$$0 \le x_p \le x_{p,max} = l_b \left\{ 1 - \frac{h_b}{r_{p,o} - r_{p,i}} \right\} \tag{5.133}$$

The speed ratio at these extremes is:

$$\frac{\omega_{driven}}{\omega_{drive}} (x_p = 0) = \frac{2r_{p,o} - h_b}{2r_{p,i} + h_b} \tag{5.134}$$

$$\frac{\omega_{driven}}{\omega_{drive}} (x_p = x_{p,max}) = \frac{2r_{p,i} + h_b}{2r_{p,o} - h_b} \tag{5.135}$$

Accordingly, the speed ratio equal to 1 can be reached when $x_p = \frac{1}{2} x_{p,max}$. By increasing the difference between the outer and inner radius, it is possible to increase the overall ratio of the CVT.

5.5.4 *Driveline Losses*

The internal frictions between driveline components cause losses. Consider a drivetrain system as shown in Figure 5.28. The input and output powers are:

$$P_i = T_i \omega_i \qquad P_o = T_o \omega_o \tag{5.136}$$

Assuming rigid connections, the input and output speeds are related as:

$$\omega_o = \frac{1}{\beta_t} \omega_i \tag{5.137}$$

where β_t is the drivetrain ratio.

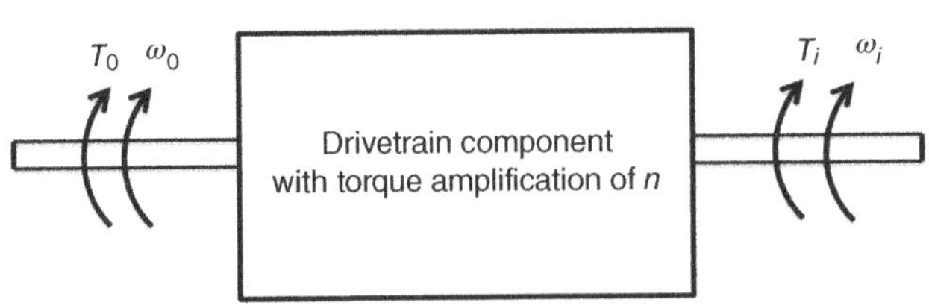

Figure 5.28 A drivetrain system

Thus, the power efficiency is:

$$\eta_P = \frac{T_o\omega_o}{T_i\omega_i} = \frac{T_o}{\beta_t T_i} \tag{5.138}$$

and accordingly, the energy efficiency of the torque-transfer is:

$$\eta_E = \frac{\int T_o\omega_o dt}{\int T_i\omega_i dt} = \frac{\int T_o\omega_o dt}{\beta_t \int T_i\omega_i dt} \tag{5.139}$$

For a drivetrain with different components, such as clutch, transmission, joint, differential, and axles, the overall efficiency of the driveline is:

$$\eta_t = \prod_{i=1}^{s} \eta_{P_i} \tag{5.140}$$

where η_t is the total drivetrain efficiency.

Example 5.12

At a certain engine speed during a vehicle test, the engine generated a torque of 185 Nm whereas the measured wheel torque was 1650 Nm. If the total gear ratio is 12, calculate the drivetrain efficiency and the amount of torque loss.

Solution

Calculate the drivetrain efficiency as:

$$\eta_t = \frac{T_w}{\beta_t T_e} = \frac{1650}{12 \times 185} = 0.74$$

The torque loss is:

$$P_w = P_e + P_L \Rightarrow P_L = \beta_t T_e - T_w = 1650 - (12 \times 185) = 570 \text{ Nm}$$

5.6 Regenerative Braking Characteristics

As discussed in Section 2.2.6, one of the main features of EV/HEVs is their ability to recuperate vehicle kinetic energy typically wasted during braking or deceleration by reversing the electric motor as a generator. However, the efficiency of powertrain components limits the amount of recuperated energy. Two primary components affecting the regenerative braking system efficiency are the electric motor and the storage system, which mainly consists of the battery, though it can also include ultra-capacitors or a flywheel. Regenerative braking power and hence torque are dependent on vehicle speed. It is evident that at very low speeds or at a speed of zero, the generator is not capable of braking action. Conversely, at high speeds, the generated torque during braking is low due to the electric motor performance, as shown in Figure 2.9.

The battery's state of charge is another limiting factor. For instance, when the battery is full or the state of the charge is high, charging the battery might not be possible. Additionally, the charging current of the battery is a major concern in absorbing the energy during regenerative braking. Generally, the efficiencies of the generator and battery systems significantly influence the effectiveness of the regenerative braking system.

The deliverable mechanical power to the generator input during braking is [7]:

$$P_{gen,in} = \eta_t(Ma_x - F_R)u = \eta_t(Ma_x u - P_R) \tag{5.141}$$

Where η_t is the total drivetrain efficiency, F_R is the total vehicle resistance force (i.e., rolling resistance and aerodynamic forces, for more information, see Chapter 6) $P_R = uF_R$ is the power loss due to resistance forces, and a_x is the vehicle acceleration. Thus, the power deliverable to the battery from the generator is:

$$P_{gen,out} = \eta_{gen} P_{gen,in} \tag{5.142}$$

where η_{gne} is the generator efficiency. Finally, the power recuperated inside the battery is:

$$P_b = \eta_{gen}\eta_{b,chg}\eta_t(Ma_x u - P_R) \tag{5.143}$$

Note that with the assumption of constant voltage,v_b, and constant current, i_b, for the battery, the maximum absorbable power by the battery is:

$$P_{b,\max} = v_b i_b \tag{5.144}$$

Accordingly, if the recuperated kinetic energy is higher than $\dfrac{P_{b,max}}{\eta_t\eta_{gen}\eta_{b,chg}}$, the power absorbed by the battery will be $P_{b,max}$. The efficiency of the regenerative braking system can be formulated as:

$$\eta_r = \frac{P_b}{P_{gen_{in}}} = \eta_t\eta_{gen}\eta_{b,chg} \tag{5.145}$$

Likewise, the efficiency of the regenerative braking system can be defined as the ratio of the available braking energy to the energy absorbed by the battery as:

$$\eta_r = \frac{E_b}{E_{gen,in}} = \frac{\int_{t_f}^{t_0} P_b(t)dt}{\int_{t_f}^{t_0} P_{gen,in}(t)dt} \tag{5.146}$$

Example 5.13

A hybrid electric vehicle with a mass of 1200 Kg is cruising at the speed of 120 km/h. The driver encounters an obstacle and needs to brake. To completely stop, the vehicle decelerates with a speed of $-5t\ m/s$ and travels 110 m. The vehicle is equipped with a battery that has a voltage of 280 V, with a maximum current of 100 A. Using this information, calculate the percentage of the energy absorbed by the battery using the regenerative braking system. The overall efficiency from the wheel to the battery is 0.80. Ignore the resistance power losses.

Solution

The total kinetic energy of the vehicle is:

$$\frac{1}{2}M(u_1)^2 - \frac{1}{2}M(u_2)^2 = \frac{1}{2}1200\left(\frac{120}{3.6}\right)^2 - 0 = 666.7\ \text{kJ}$$

The deceleration is $a_x = -5\ m/s^2$, while the time of stop is:

$$t_b = \frac{u_2 - u_1}{a_x} = \frac{0 - \frac{120}{3.6}}{-5} = 6.66\ \text{s}$$

On the other hand, the maximum energy absorbed by the battery is:

$$P_{b,\max} = v_b i_b = 280 \times 100 = 28\ \text{kW}$$

It is then necessary to calculate the time at which the available mechanical power is less than the maximum energy absorbed by the battery. Thus,

$$P_{b,max} = \eta_r\ Ma_x u = \eta_r M a_x^2 (t_b - t^*) \Rightarrow t^* = t_b - \frac{P_{b,max}}{\eta m_v a_v^2} = 6.66 - \frac{28000}{0.8 \times 1200 \times (-5)^2}$$

$$\Rightarrow t^* = 5.49\text{s}$$

Hence, the energy stored in the battery can be calculated as:

$$E_b = P_{b,max} t^* + \int_{t^*}^{t_b} \eta_r\ Ma_x u dt$$

$$= (28000) \times 5.49 + 0.8 \times 1200 \times (-5) \times \left(-\frac{5}{2}\right) \times \left[(6.66)^2 - (5.49)^2\right] = 324.3\text{kJ}$$

Therefore, the percentage of the recuperated energy is:

$$\text{percentage of the recuperated energy} = \frac{324.6}{666.7} \times 100 = 48.6\%$$

5.7 Driving Cycles

A driving cycle is a collection of data points for the speed versus time of a vehicle traveling on a certain route. These routes vary in different countries but they represent driving speeds and average driver's habits in how to accelerate and decelerate when traveling in a city or highway. Driving cycles can be used to estimate the amount of energy needed for a vehicle to complete the trip and also the potential energy for regenerative braking. Driving cycles can also be used for offline computation to determine the power-splitting logic between multiple power sources. These calculations help determine the proportioning ratios of power extracted from the power-producing units. Additionally, driving cycles play an important role in the design, optimization, and sizing of hybrid drivetrains.

A driving cycle is also important in assessing a vehicle performance in various ways, such as: fuel consumption, emissions, and the potential for hybridization. In fact, the primary purpose of a driving cycle is to simulate actual driving patterns to test vehicle exhaust emissions and fuel consumption. Typically, formation of driving cycles is a function of time in terms of vehicle speed and gear selection. Driving cycles can either be constructed from a number of constant acceleration and constant speed phases, or based on data from a vehicle in real traffic conditions. Moreover, driving cycles can be classified as steady-state (cruise) cycles and transient cycles based on the variation in speed and engine load. A steady-state cycle is formed by a sequence of constant engine speed and load modes. Usually, steady-state driving cycles are used for the evaluation of heavy-duty diesel engine vehicles [9].

Driving patterns vary based on vehicle usage and geography. Driving cycles could be specific to a category of vehicles in a specific environment and are not easily extrapolated to other types of vehicles or driving environments. Moreover, fuel consumption and emissions depend on many geographical, cultural, and legislative factors. For this reason, researchers focus on developing driving cycles based on real-world driving tests (for transient conditions) and steady state (cruising) conditions encountered during road driving.

In order to estimate fuel efficiency (consumption) of an ICE over a driving cycle, driving cycles need to be divided into specific time intervals. The summation of fuel consumption at each time interval estimates total engine fuel consumption. Thus, the fuel consumption over a given driving cycle can be calculated as:

$$v_{fuel} = \sum_i Q_{f_i} \Delta t_i \tag{5.147}$$

where Q_{f_i} is the average fuel consumption of the engine in the time interval Δt_i. The average fuel and time interval are in liters and hours, respectively.

Drive cycles have been developed in different countries for the purpose of emission certification and testing fuel economy. However, currently, the major driving cycles used by automotive testing labs include US EPA driving cycles, European NEDC and Japanese 10–15 mode.

5.7.1 EPA Driving Cycles

The US Environment Protection Agency designed EPA driving cycles, a series of tests that evaluate the tailpipe emissions and fuel economy of passenger cars. The procedure consists of four tests, including city driving (FTP-75), highway driving (HWEFT), aggressive driving (SFTP US06), and a conditioning test (SFTP SC03).

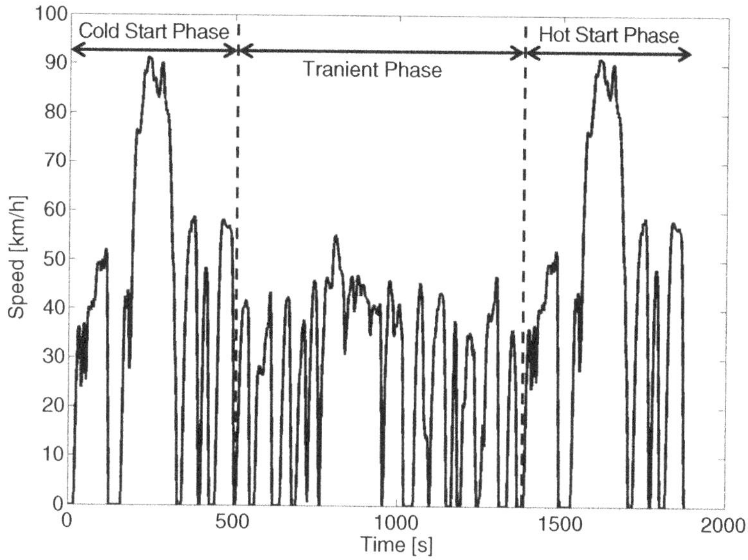

Figure 5.29 The FTP75 city driving cycle

- *City Cycle (FTP-75):* FTP (Federal Test Procedure)-75 consists of three phases including cold start, transient, and hot start phases. The first phase begins with the cold start of the engine while the second phase starts after 10 min of stopping the engine. The hot phase begins after the engine is stopped for 10 min. The cold and hot start phases are identical in terms of driving patterns except in starting conditions of engine. FTP-75 represents an urban route of 12.07 km with stop-and-go driving conditions. The average speed is 31.5 km/h with the maximum speed of 91.2 km/h. The FTP-75 cycle is formed by conducting a test procedure in which the distance traveled is 17.77 km during 1874 s. Figure 5.29 demonstrates the cycle pattern of FTP-75.
- *Highway Cycle (HWEFT):* The design of the Highway Fuel Economy Driving Schedule (HWFET) uses a warmed-up engine without stops and an average speed of 77 km/h, along with a peak speed of 97 km/h over a 16 km distance. This cycle has a relatively constant speed, meaning that there is little braking, and as a result, regenerative energy recovery is less than 1%. Figure 5.30 shows the HWEFT pattern.
- *SC03 Supplemental Federal Test Procedure:* The development of SC03 Supplemental Federal Test Procedure (SFTP) considers the engine load and emissions associated with the use of air conditioning units in vehicles certified over the FTP-75 test cycle. The cycle includes a 5.8 km route with an average speed of 34.8 km/h, peak speed of 88.2 km/h, and duration of 596 s. The driving pattern of SC03 SFTP is shown in Figure 5.31.
- *US06 Supplemental Federal Test Procedure:* The FTP-75 has some shortcomings in the representation of common human behaviors, such as aggressive, high-speed, and/or high-acceleration driving, rapid speed fluctuations, and how driving occurs after vehicle start-up. The design of US06 Supplemental Federal Test Procedure (SFTP) covers these shortcomings with the FTP-75 test cycle. The cycle includes a 12.8 km route with an average speed of 75.9 km/h, peak speed 129.2 km/h, and duration of 596 s. Figure 5.32 illustrates the US06 SFTP driving cycle.

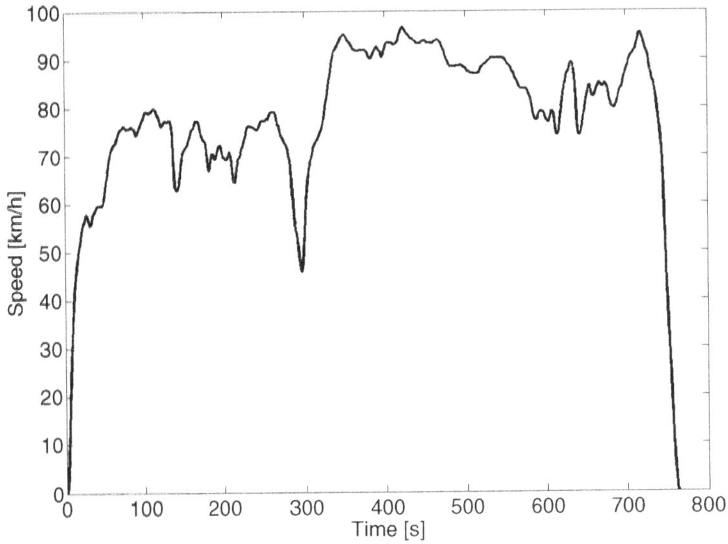

Figure 5.30 The HWFET highway driving cycle

5.7.2 *The European NEDC*

The New European Drive Cycle (NEDC) comprises of four segments of the European urban driving cycle, followed by one segment of the European extra-urban (highway) driving cycle (Figure 5.33). The urban driving cycle represents typical driving conditions of busy European cities, characterized by low engine load, low exhaust gas temperature, and a maximum speed of

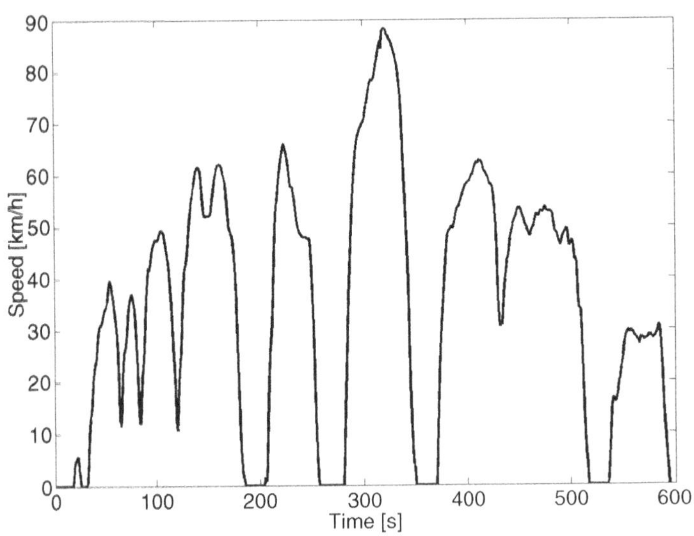

Figure 5.31 The SC03 FSTP driving cycle

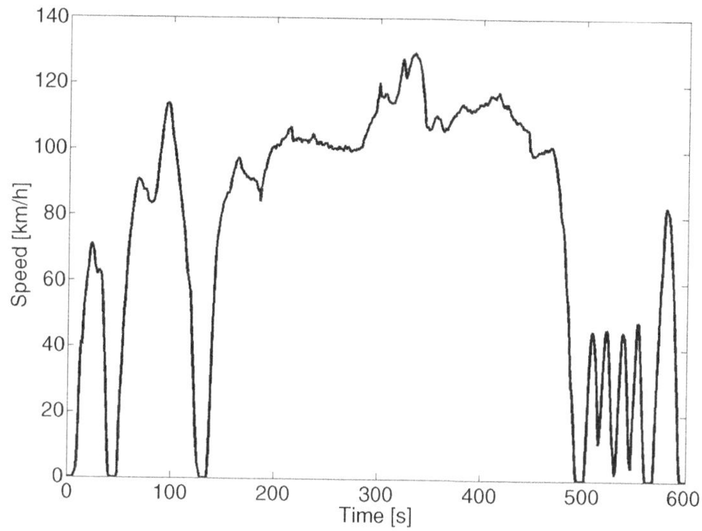

Figure 5.32 The US06 FSTP driving cycle

50 km/h, while the extra-urban driving cycle represents more high speed driving modes. The maximum speed of the extra-urban (highway) driving cycle is 120 km/h; low-powered vehicles have a limit of 90 km/h. In this combined driving cycle, the average speed is 33.6 km/h, while the total test time reaches 1180 s.

The NEDC mainly functions in light-duty models in the European Union, with a combined fuel economy calculated by the total consumption of urban and extra-urban cycles. Typically,

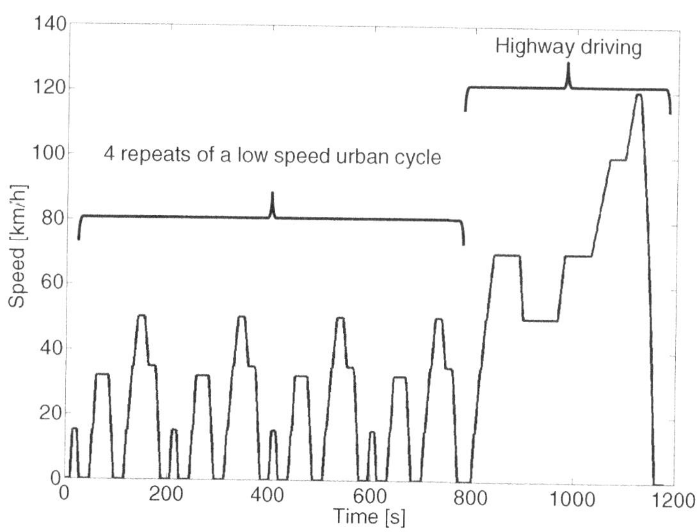

Figure 5.33 The NEDC European driving cycle

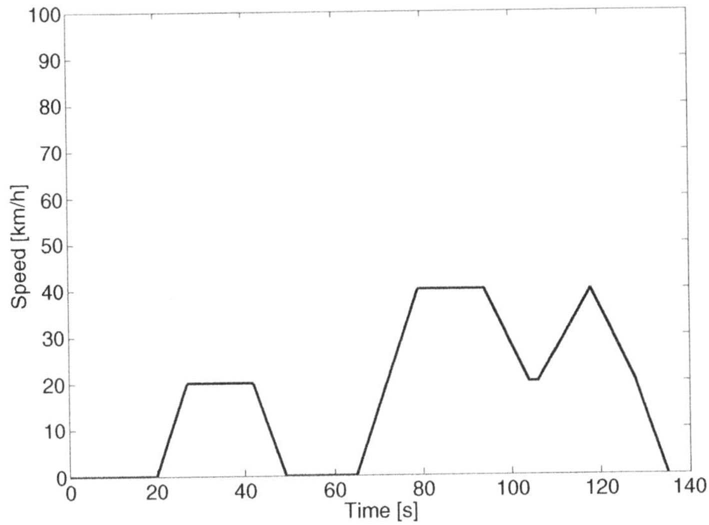

Figure 5.34 The 10-mode Japanese driving cycle

the data are obtained through tests, while all auxiliary loads (e.g., the air conditioning compressor and the fan, lights, heated rear window, etc.) are off. The NEDC constitutes periods of low acceleration, deceleration, constant speed cruises, and several idling stops. These periods are not complete representations of real driving patterns on the road, wherein transient accelerations are much steeper and more dynamic in modern engines [10].

5.7.3 The Japan 10–15 Mode

Representing urban driving conditions, the design of a 10-mode cycle (Figure 5.34) is for light-duty vehicles in Japan. One 10-mode cycle simulates a journey of 0.664 km of distance, with an average speed of 15.7 km/h and a peak speed of 40 km/h during a 135-second period of time. The newer version of the 10-mode cycle is the 10–15 mode cycle (Figure 5.35). This cycle comprises a 15.mode segment with a peak speed of 70 km/h and an average speed of 33.88 km/h over a period of 231 s plus three consecutive 10-mode segments. The 15.mode simulates higher speed driving in open road situations and works to decelerate at higher speeds than the 10-mode cycle. Simulating typical urban and highway driving patterns, the 10–15 mode cycle includes idling, acceleration, steady running, and deceleration. The 10–15 mode cycle mostly represents the driving patterns in Japan, where average speeds are low and traffic laws are very strictly enforced.

In order to cover a wider range of vehicles, as well as more driving patterns, the Japanese government designed the JC08 cycle. The cycle includes frequent stop-and-go driving conditions and long idle times, simulating driving patterns of congested city traffic. The test applies to both gasoline and diesel vehicles to measure emission, and to determine fuel economy. The overall cycle simulates a journey of 8.171 km total distance, with an average speed of 24.4 km/h, and with peak speed of 81.6 km/h in 1204 s.

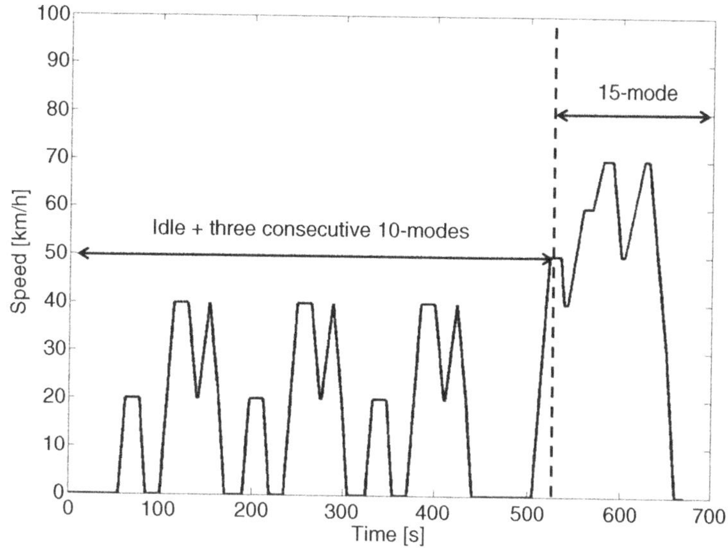

Figure 5.35 The 10–15 mode Japanese driving cycle

Problems

1. A four-stroke spark-ignition (SI) engine has the following specifications:

Number of cylinders	6
Bore (area) of cylinder	9 cm
Stroke	8 cm
Fuel consumption rate	0.3 L/min
Fuel density	750 kg/m^3
Heating value	44 MJ/kg
Brake load	180 N
Torque arm	0.5 m

Calculate the following:

(a) The brake power
(b) The mean effective pressure
(c) The indicated power
(d) The mechanical efficiency.

2. The specifications of a four-stroke combustion engine and its test results are:

Number of cylinders	4
Bore (area) of cylinder	9 cm
Stroke	8 cm
Speed	5000 rev/min
Fuel consumption rate	0.09 kg/min
Heating value	44 MJ/kg

Brake load	60 N
Length of engine shaft	0.5 m
MEP	280 kPa
Total capacity	1.3 L

Calculate the following:

(a) The mechanical efficiency
(b) The fuel conversion efficiency
(c) The air–fuel ratio
(d) The volumetric efficiency
(e) The mean piston speed

3. An electric vehicle cruises at 100 km/h, and has a battery with the voltage and current performance, shown in Figure 5.36. If the maximum discharged rate is 100 A while fully charged, and if the minimum SoC is 30%, calculate the distance that the vehicle can travel.

4. A parallel HEV has a downsize engine, motor/generator, and a battery. The battery has a specific energy of 65 Wh/kg, with a mass of 125 kg, and a nominal voltage of 200 V. The vehicle is cruising at a speed of 120 km/h with combined engine and electric motor propulsion, while the required traction power is 15 kW. During the trip, the battery SoC drops by 50%. In this span, the engine's BSFC is 285 g/kWh. Given this information, calculate what percentage improvement of fuel efficiency is in comparison to an ICE vehicle with an engine BSFC of 265 g/kWh. Assume that the motor mechanical efficiency and transmission efficiency are 90% and 95%, the fuel heating value is 45.8 MJ/kg, and the fuel density is 737 kg/m^3.

5. A three-phase induction motor with wye connection is connected to a 220 V, 60 Hz source. The slip is 5% and rotor speed is 885 rpm. The equivalent circuit parameters are: $R_s = 0.4\Omega$, $X_s = 1\Omega$, $R_r = 0.8\Omega$, $X_r = 3.5\Omega$, and $X_m = 10\Omega$. Calculate:

(a) Number of poles
(b) Input power
(c) Mechanical power
(d) Motor torque
(e) Efficiency.

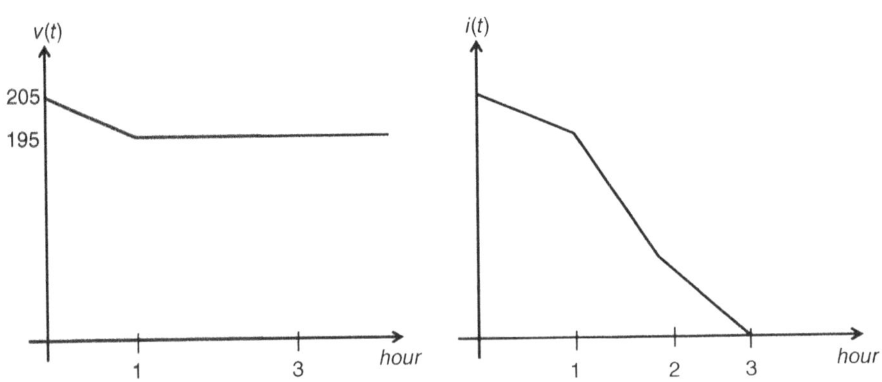

Figure 5.36

6. A given battery for an EV has an internal resistance that changes subject to the battery's state of the charge during charging and discharging modes. The variation of battery internal resistance with the battery state of charge is given (in Ohm) as:

$$R_{b,chg} = 0.456x^2 - 0.539x + 0.933$$

$$R_{b,dis} = -x^3 + 2.705x^2 - 1.965x + 1.498$$

where x is the battery state of the charge. Plot the efficiency map of the battery in a power-SoC diagram by assuming a constant terminal voltage of 280 V.

References

1. Ehsani, M., Gao, Y., and Emadi, A. (2010) *Modern Electric, Hybrid Electric, and Fuel Cell Vehicles: Fundamentals, Theory, and Design*, 2nd edn, CRC Press, Boca Raton, FL.
2. Williamson, S., Lukic, M., and Emadi, A. (2006) Comprehensive drive train efficiency analysis of hybrid electric and fuel cell vehicles based on motor-controller efficiency modeling. *IEEE Transactions on Power Electronics*, **21** (3), 730–740.
3. Guzzella, L., and Sciarretta, A. (2007) *Vehicle Propulsion Systems: Introduction to Modeling and Optimization*, 2nd edn, New York: Springer.
4. Liang, X., and Ilochonwu, O. (2011) Induction motor starting in practical industrial applications. *IEEE Transactions on Industry Applications*, **47** (1).
5. Strangas, E.G. (n.d.) Notes for an introductory course on electrical machines and drives. MSU Electrical Machines and Drives Laboratory.
6. Hossein, I. (2010) *Electric and Hybrid Vehicles: Design Fundamentals*, 2nd edn, CRC Press, Boca Raton, FL.
7. Crolla, D., and Mashadi, B. (2012) *Vehicle Powertrain Systems: Integration and Optimization*, Wiley-Blackwell, Chichester.
8. Ferguson, R.J. (1983) Short cuts for analyzing planetary gears. *Machine Design*, May 26.
9. Barlow, T.J., Latham, S., McCrae, I.S., and Boulter, P.G. (2009) *A Reference Book of Driving Cycles for Use in the Measurement of Road Vehicle Emissions*, Project Report PPR354, version 3, TRL Limited.
10. Mahlia, T.M.I., Tohno, S., and Tezuka, T. (2012) A review on fuel economy test procedure for automobiles: implementation possibilities in Malaysia and lessons for other countries. *Journal of Renewable and Sustainable Energy Reviews*, **16**.

6

Modeling and Analysis of Electric and Hybrid Electric Vehicles' Propulsion and Braking

6.1 Introduction

Propulsion and braking performances, such as maximum speed, accelerating time, gradeability, and stopping distance, are among the most important factors to judge in a car. During designing, analyzing and sizing of vehicle powertrain and braking systems, the calculation of these performances at the early stages of the vehicle development process is essential. For this purpose, it is necessary to develop vehicle dynamics models to address vehicle longitudinal dynamics behavior. Such models are based on the following assumptions:

- The model should be simple. Simplicity is beneficial as it helps us solve the model analytically, and makes the model an effective design and optimization tool. In this regard, only those degrees of freedom that significantly affect the vehicle's longitudinal dynamics are considered. In addition, it assumed that the vehicle accelerates or brakes during straight line driving.
- During straight line accelerating or braking, the longitudinal velocity and the wheels' angular velocities are the most effective degrees of freedom. As such, the model's degrees of freedom are selected from these.
- The model does not consider the lateral motion, yaw, roll, pitch, and vertical motions; as their effects on longitudinal motion are negligible.
- While accelerating on dry roads, longitudinal wheel slips are small, however, during braking; the wheel slip is no longer small. Therefore, under different conditions, there are different models for accelerating and braking events.

As discussed in the previous chapters, electric and hybrid electric vehicles differ from conventional internal combustion engine vehicles in their propulsion, and braking systems.

Electric and Hybrid Vehicles: Technologies, Modeling and Control: A Mechatronic Approach, First Edition.
Amir Khajepour, Saber Fallah and Avesta Goodarzi.

In addition, the component specifications, in terms of torque, power, and efficiency curves are significantly different from those of internal combustion engine vehicles. Therefore, to model and analyze electric and hybrid electric vehicles' longitudinal dynamics behavior, it is necessary to develop tailor-made models. However, these models are formed based on common approaches used in conventional vehicles considering the above-mentioned assumptions.

In this chapter, first, the general approach in vehicle longitudinal dynamics modeling and analysis is discussed. Then, specific models for EVs and HEVs to analyze the vehicle's accelerating and braking are developed. These models are used to calculate the vehicle's propulsion and braking performances.

6.2 The Longitudinal Dynamics Equation of Motion

By considering the modeling assumptions specified in the previous section, it is possible to simplify the general form of vehicle planar kinematics, Equation 4.13, into a single equation of vehicle longitudinal dynamics:

$$\sum F_x = M\frac{du}{dt} \tag{6.1}$$

where $\sum F_x$ is the summation of all forces applied to the vehicle, and M and u are the vehicle mass and its longitudinal speed, respectively. According to Figure 6.1, one can expand the summation of applied external forces to the vehicle by:

$$\sum F_x = F_{xf} + F_{xr} - F_R \tag{6.2}$$

where F_{xf} and F_{xr} are the longitudinal forces of the front and rear axles, respectively; and F_R is the total resistance force acting upon the vehicle, defined as:

$$F_R = F_{RA} + F_{RR} + F_{RG} \tag{6.3}$$

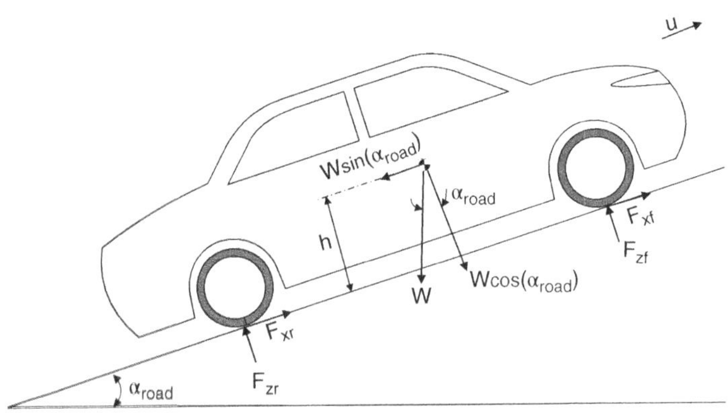

Figure 6.1 Vehicle longitudinal dynamics configuration

where F_{RA}, F_{RR}, and F_{RG} are aerodynamic resistance, rolling resistance, and gradient resistance forces. Chapter 4 introduces and discusses aerodynamic and rolling resistance forces in Equations 4.31 and 4.32.

As illustrated by Figure 6.1, gradient resistance is the tangential component of vehicle weight vector when the vehicle goes up or down a slope that and can be simply formulated as:

$$F_{RG} = W\sin(\alpha_{road}) \tag{6.4}$$

where W and α_{road} are the vehicle weight and the road slope angle, respectively. Finally, by substituting Equations 4.31, 4.32 and 6.4 into Equation 6.3, the total vehicle resistance force is formulated as:

$$F_R = \frac{1}{2}\rho_a C_d A_f u^2 + \mu_R W + W\sin(\alpha_{road}) \tag{6.5}$$

Returning to Equation 6.2, the longitudinal force of a front or rear axle can be positive or negative during accelerating and braking. During accelerating, the powertrain system produces positive longitudinal force by applying positive driving torque to the wheels. On the other hand, in the event of braking, the friction brake system and/or electric motor(s) apply negative braking torque to the wheels and produce negative longitudinal force. As such, in order to formulate the longitudinal dynamics equation of motion, the powertrain and braking systems must be mathematically modeled first and, then, by integrating them based on the vehicle configuration, the equation of motion for each specific type of vehicle is achieved. Following this approach, the vehicle dynamics models during accelerating and braking for different types of vehicles are individually addressed and discussed in the next sections.

6.3 Vehicle Propulsion Modeling and Analysis

The purpose of this model is to calculate the vehicle propulsion performances, such as maximum speed, accelerating time, and maximum gradeability under standard dry road conditions. As previously discussed, during accelerating on dry roads, the wheels' longitudinal slip is very small and can be ignored. Hence, longitudinal velocity is the only degree of freedom of the model. Additionally, considering the purpose of modeling, some other effects such as clutch engagement/disengagement, and torsional vibration of driveline system can also be ignored.

In the following, considering the above-mentioned assumptions, the models of the different types of vehicles, including conventional internal combustion engine vehicles, and the most common types of electric and hybrid electric vehicles are developed. The same methodology can be used for any other type of vehicle.

6.3.1 Internal Combustion Engine Vehicles

Figure 6.2 shows the typical configuration of a two-wheel drive vehicle, equipped with an internal combustion engine. The engine is modeled by a torque characteristic curve and a rotary inertia that represents the inertia of rotational and moving parts of the engine. The engine

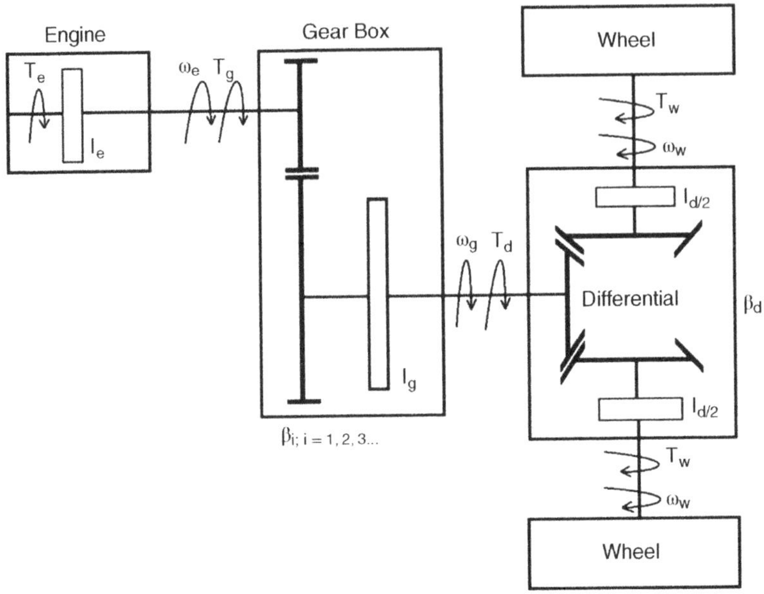

Figure 6.2 Powertrain model of a two-wheel drive internal combustion engine vehicle

generates driving torque of $T_e(\omega_e, x_{\theta e})$ and the gearbox applies T_g as the load to the engine. The equation of the engine's rotational motion can then be formulated as:

$$T_e(\omega_e, x_{\theta e}) - T_g = I_e \frac{d\omega_e}{dt} \tag{6.6}$$

where $\omega_e, x_{\theta e}$ and I_e are the engine rotational speed, the throttle position, and equivalent rotational inertia of engine, respectively. As discussed in Chapter 5, $x_{\theta e}$ is assumed as a variable in the range of [0%–100%].

The gearbox is the next part of the powertrain. It is modeled by a pair of gears with ratio of β_i that changes for different gearbox gear ratios, and an inertia representing the inertia of the gearbox rotational parts. Assuming an efficiency of η_g for the gearbox, the equation of motion can be written as:

$$\eta_g \beta_i T_g - T_d = I_g \frac{d\omega_g}{dt} \tag{6.7}$$

where, T_d is the applied load torque from the differential to the gearbox. The gearbox rotational speed, ω_g, is related to the engine speed by:

$$\omega_g = \frac{\omega_e}{\beta_i} \tag{6.8}$$

The differential receives the engine torque after magnification in the gearbox. Suppose the differential is a common open differential, then the differential distributes the torque equally between the two driving wheels, yet it enables each wheel to have different speeds. The governing equation for each differential output is:

$$0.5\eta_d\beta_d T_g - T_w = 0.5 I_d \frac{d\omega_w}{dt} \tag{6.9}$$

where η_d, β_d and I_d represent the differential efficiency, the reduction ratio, and the equivalent rotational inertia, respectively, and T_w is the delivered torque to each wheel. During straight line driving, the differential output speeds are equal and related to the input speed, ω_g, by:

$$\omega_{wR} = \omega_{wL} = \omega_w = \frac{\omega_g}{\beta_d} \tag{6.10}$$

Finally, the wheel's rotational dynamics equation can be written as:

$$T_w - 0.5F_x R = I_w \frac{d\omega_w}{dt} \tag{6.11}$$

where, F_x is the total axle's tractive force, R is the wheel's effective radius and I_w is the wheel inertia. Assuming the longitudinal slip is very small; the vehicle's velocity according to the wheel speed is:

$$u = R\omega_w \tag{6.12}$$

Now, by combining Equations 6.6 to 6.12, the total tractive force can be given as:

$$F_x = \frac{\eta_g\eta_d\beta_i\beta_d}{R} T_e(\omega_e, x_{\theta e}) - \frac{1}{R^2}\left(\eta_g\eta_d\beta_i^2\beta_d^2 I_e + \eta_d\beta_d^2 I_g + I_d + 2I_w\right)\frac{du}{dt} \tag{6.13}$$

This tractive force can be applied to the vehicle in the front or rear axle depending on the vehicle powertrain configuration. By substituting the tractive force into the vehicle longitudinal dynamics equation, Equation 6.1, the final form of the vehicle equation of motion is given as:

$$\frac{\eta_g\eta_d\beta_i\beta_d}{R} T_e(\omega_e, x_{\theta e}) - F_R = M_d \frac{du}{dt} \tag{6.14}$$

where M_d is the vehicle dynamic mass, defined as:

$$M_d = M + \frac{1}{R^2}\left(\eta_g\eta_d\beta_i^2\beta_d^2 I_e + \eta_d\beta_d^2 I_g + I_d + 2I_w\right) \tag{6.15}$$

As can be seen in Equation 6.15, the effect of the moving and rotational parts of the powertrain appears as an additional mass to the vehicle's actual mass.

By combining Equations 6.8, 6.10 and 6.12, the vehicle speed is found as:

$$u = R\frac{\omega_e}{\beta_i\beta_d} \tag{6.16}$$

At last, the engine torque in Equation 6.14, should be substituted by a parabolic characteristic curve as discussed in Chapter 5, which is an acceptable approximation for internal combustion engine torque map.

Example 6.1

Consider an internal combustion engine vehicle with the specifications in Table 6.1. The engine torque map is approximated by a parabolic characteristic curve as:

$$T(\omega_e, x_{\theta e}) = x_{\theta e}\left(A_2\omega_e^2 + A_1\omega_e + A_0\right) = x_{\theta e}\left(-0.0018\omega_e^2 + 1.7537\omega_e - 65.096\right)$$

(a) Calculate the dynamic mass in all gears by considering that the wheels, differential, and transmission rotational inertias are negligible.
(b) Plot the tractive force vs. the longitudinal speed diagram in all gears.
(c) Simulate the vehicle's response to the driver's inputs (throttle position and gear changing) shown in Figure 6.3. The initial speed is 10 m/s. Calculate the distance traveled, the vehicle speed, the acceleration, and the engine speed.

Solution

(a) According to Equation 6.15, the dynamic mass for the first gear is:

$$M_d = 1705 + \frac{1}{(0.33)^2}(0.85 \times 1 \times (3.842)^2 \times \left(3.538\right)^2 \times 0.2\right) = 1993.44\ \text{kg}$$

The dynamic mass for all gears is summarized in Table 6.2.

Table 6.1 Vehicle specifications

Symbol	Parameter	Value
M_{min}	Minimum vehicle mass (kg)	1705
M_{max}	Maximum vehicle mass (kg)	2055
R	Tire radius (m)	0.33
C_d	Drag coefficient	0.28
A	Frontal area (m^2)	2.31
μ_R	Rolling resistance coefficient	0.015
β_1	First gear ratio	3.842
β_2	Second gear ratio	2.353
β_3	Third gear ratio	1.529
β_4	Fourth gear ratio	1
β_5	Fifth gear ratio	0.839
β_d	Differential ratio	3.538
η_g	Gearbox efficiency	0.85
η_d	Differential efficiency	1
I_e	Engine inertia (kgm^2)	0.2
$\omega_{e,max}$	Minimum engine speed (rpm)	1000
$\omega_{e,min}$	Maximum engine speed (rpm)	6600
l	Wheelbase (mm)	2900
a	Distance of CG from front axle (mm)	1350
b	Distance of CG from rear axle (mm)	1550
h	C.G height (mm)	360
μ	Road friction coefficient	0.98

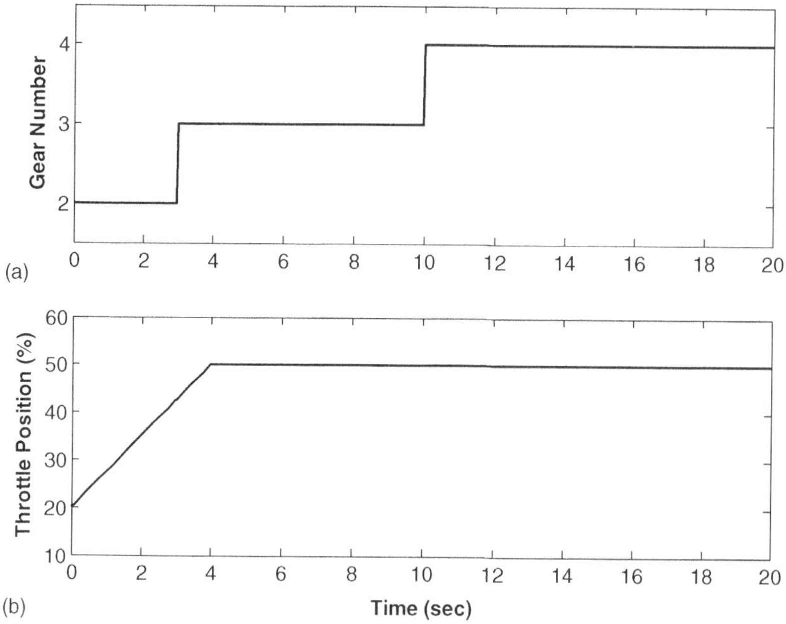

Figure 6.3 Driver inputs: gear changing pattern (a), and throttle position (b)

Table 6.2 Vehicle dynamic mass in all gears

Gear	Vehicle dynamic mass (kg)	Change compared to static mass (%)
1	1993.44	16.92
2	1813.19	6.34
3	1750.68	2.68
4	1724.54	1.15
5	1718.75	0.81

(b)

$$F_x = \frac{\eta_g \eta_d \beta_i \beta_d}{R} T_e(\omega_e, x_\theta) = \frac{0.85 \times 1 \times 3.538}{0.33} \left(-0.0018\omega_e^2 + 1.7537\omega_e - 65.096\right)\beta_i$$

$$u = \frac{R\omega_e}{\beta_i \beta_d} = \frac{0.33}{3.538}\frac{\omega_e}{\beta_i}$$

As shown in Figure 6.4, the vehicle's tractive force in each gear can be plotted versus the vehicle speed.

(c) The vehicle's acceleration, speed, and traveled distance (Figure 6.5) are obtained by substituting the parabolic engine characteristic curve in Equation 6.14 and solving it for the vehicle's speed either numerically or analytically. The vehicle's acceleration and distance traveled can then be calculated by taking the derivative and integral of the speed, respectively.

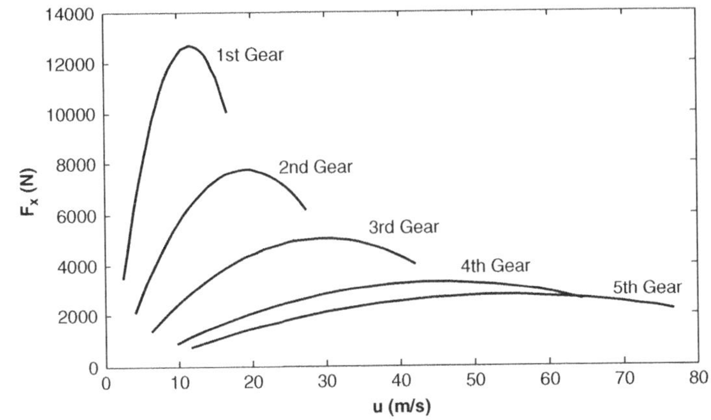

Figure 6.4 Vehicle tractive force vs. speed

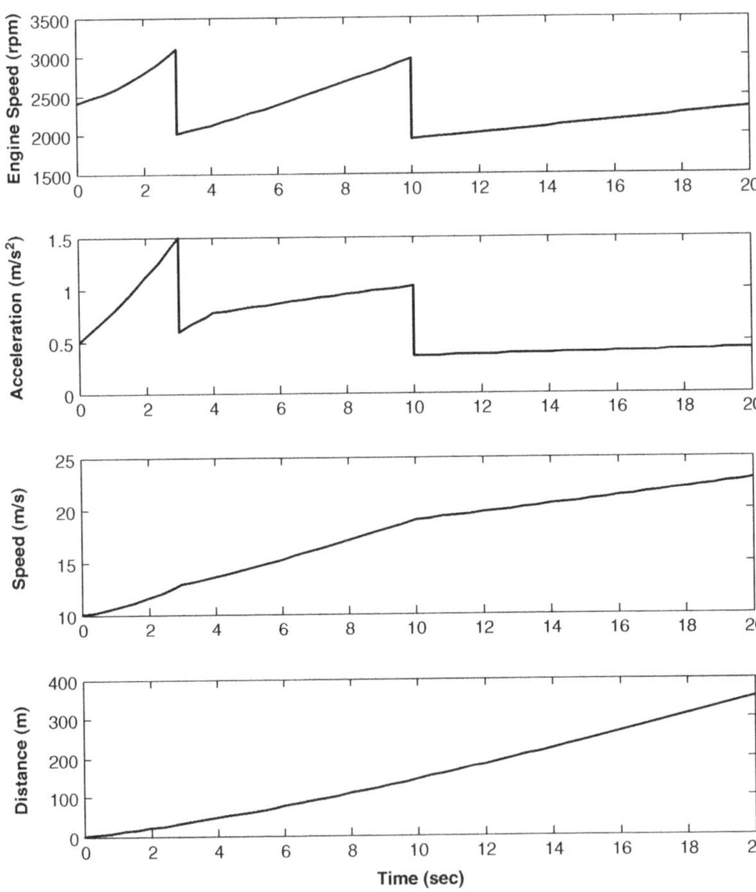

Figure 6.5 Engine speed and the vehicle's acceleration, speed, and distance traveled

In addition to a numerical simulation of the vehicle's longitudinal dynamics discussed in Example 6.1, other important vehicle performances, such as maximum speed, accelerating time, and maximum gradeability can be calculated as discussed below.

- *Maximum speed*: let us begin by calculating the vehicle's maximum speed in standard testing conditions that are: a flat road, a fully depressed accelerator pedal, the lowest gear ratio, and minimum vehicle weight. The following equations express these conditions:

$$\alpha_{road} = 0 \quad x_{\theta e} = 100\% \quad \beta_i = \beta_{min} \quad M = M_{min} \tag{6.17}$$

When the vehicle reaches its maximum speed, the speed becomes constant and the vehicle acceleration is zero:

$$\frac{du}{dt} = 0 \tag{6.18}$$

By substituting Equations 6.17 and 6.18 into Equation 6.14 and assuming a parabolic torque characteristic curve for the engine, the governing equation of the vehicle's maximum speed becomes:

$$\left(A_2 \frac{\eta_g \eta_d \beta_{min}^3 \beta_d^3}{R^3} - \frac{1}{2}\rho C_d A_f\right) u_{max}^2 + \left(A_1 \frac{\eta_g \eta_d \beta_{min}^2 \beta_d^2}{R^2}\right) u_{max} + \left(A_0 \frac{\eta_g \eta_d \beta_{min} \beta_d}{R} - \mu_R M_{min} g\right) = 0 \tag{6.19}$$

One can solve this equation either analytically or numerically to find the vehicle's maximum speed.

Example 6.2

Use the vehicle specifications in Example 6.1 to calculate the vehicle's maximum speed.

Solution

$$\left(-0.0018 \frac{0.85 \times 1 \times (0.839)^3 (3.538)^3}{(0.33)^3} - \frac{1}{2} \times 1.2 \times 0.28 \times 2.31\right) u_{max}^2$$

$$+\left(1.7537 \frac{0.85 \times 1 \times (0.839)^2 (3.538)^2}{(0.33)^2}\right) u_{max} - \left(65.096 \frac{0.85 \times 1 \times 0.839 \times 3.538}{0.33}\right.$$

$$\left. + (0.015 \times 1705 \times 9.81)\right) = 0$$

$$-1.5016 u_{max}^2 + 120.6107 u_{max} - 748.6039 = 0$$

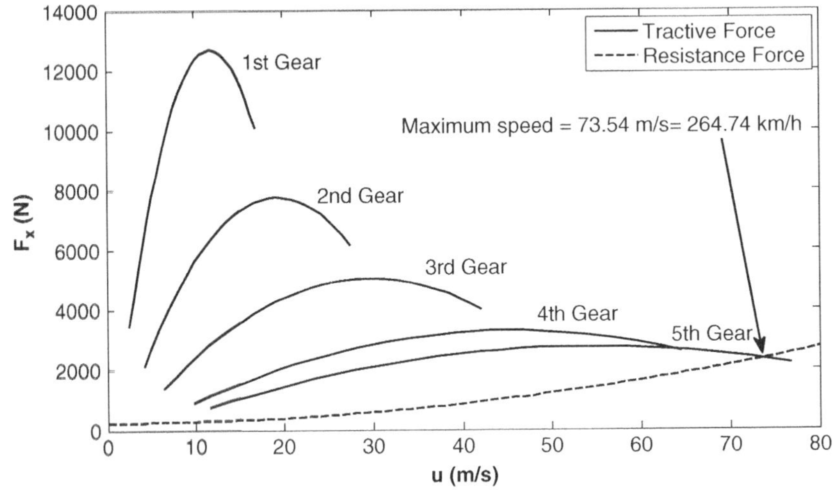

Figure 6.6 Graphical illustration of finding the maximum speed of a conventional vehicle

By solving the above equation, the vehicle's maximum speed is:

$$u_{max} = 73.54\,\text{m/s} = 264.74\,\text{km/h}$$

As Figure 6.6 shows, the vehicle's maximum speed occurs at the intersection of the tractive and resistance force curves.

- *Maximum vehicle gradeability*: This is the steepest slope that the vehicle can climb. Like the maximum speed, the maximum gradeability is calculated based on the standard conditions that are: the lowest admissible velocity, the maximum vehicle weight, and the highest gear ratio. These assumptions result in:

$$u, a_x \approx 0 \quad \beta_i = \beta_{max} \quad M = M_{max} \tag{6.20}$$

Moreover, the engine generates its maximum torque:

$$T_e(\omega_e, x_\theta) = T_{e,max} \tag{6.21}$$

By substituting Equations 6.20 and 6.21 into Equation 6.14 and solving it for A_{Road}, the maximum vehicle gradeability can be found as:

$$\alpha_{road,max} = \text{Sin}^{-1}\left(\frac{\eta_d \eta_g \beta_{max} \beta_d T_{e,max}}{R M_{max} g} - \mu_R\right) \tag{6.22}$$

However, the calculated gradeability is valid if the maximum generated tractive force is less than the road frictional capacity. In some cases, such as driving on low friction roads, and for a low-weight car with a high-power engine, the vehicle's tractive force can be higher than the road capacity. In such situations, the road capacity determines the maximum gradeability, rather than the powertrain. The following example helps explain such cases.

Example 6.3

Use the vehicle's specifications in Example 6.1 to calculate the maximum vehicle gradeability. Ensure that the tractive force is not greater than the maximum road frictional capacity. Assume the vehicle is front wheel drive.

Solution

To calculate the maximum vehicle graddability, the maximum engine torque is required. It can be obtained by finding the maximum point of a given parabolic characteristics curve which is equal to $T_{e,max} = 362$ Nm. Then, the maximum vehicle gradeability can be calculated as:

$$\alpha_{road,max} = \text{Sin}^{-1}\left(\frac{0.85 \times 1 \times 3.842 \times 3.538 \times 362}{0.33 \times 2055 \times 9.81} - 0.015\right) = 0.66\,\text{rad} = 37.86\,\text{deg}$$

Now we should ensure that the tractive force is not greater than the maximum road capacity. As discussed in Chapter 4, the maximum longitudinal force of a tire is μF_z. Then to calculate the road maximum capacity, one should know the tire normal force. According to Figure 6.1, the normal force of each axle when the vehicle is on an inclined road is determined by:

$$F_{zf} = M_{max}g\left(\frac{b}{l}\cos(\alpha_{road}) - \frac{h}{l}\sin(\alpha_{road})\right)$$

$$F_{zr} = M_{max}g\left(\frac{h}{l}\sin(\alpha_{road}) + \frac{a}{l}\cos(\alpha_{road})\right)$$

In this case, the front axle is the drive axle, and then the maximum longitudinal force capacity is:

$$F_{xf,max} = \mu M_{max}g\left(\frac{b}{l}\cos(\alpha_{road}) - \frac{h}{l}\sin(\alpha_{road})\right)$$

By substituting the above equation in the vehicle's longitudinal dynamic equation and solving it for the road slope, the maximum vehicle gradeability due to tire capacity is given as:

$$\alpha_{road,max} = \tan^{-1}\left(\frac{\mu\frac{b}{l}}{1 + \mu\frac{h}{l}}\right) = \tan^{-1}\left(\frac{0.98 \times \frac{1550}{2900}}{1 + 0.98 \times \frac{360}{2900}}\right) = 25.03\,\text{deg}$$

The calculated maximum gradeability due to the road frictional capacity (25.03 deg) is less than what was calculated according to the engine capacity (37.86 deg). As the result, 25.03 deg is the acceptable value for the vehicle maximum gradeability.

Accelerating time: There are different standard methods for evaluating a vehicle's acceleration performance. The most popular methods are based on the time of accelerating from zero to a specific speed. The most commonly used test is 0–100 km/h or its equivalent test, 0–60 miles/h. The test conditions are: a flat road, a fully depressed accelerator pedal, the minimum vehicle weight, and smart gear changing. These assumptions are:

$$\alpha_{road} = 0 \quad x_{\theta e} = 100\% \quad \beta_i = \beta_{opt} \quad M = M_{min} \tag{6.23}$$

In the above equation, β_{opt} is the optimum gear ratio. In order to maximize the vehicle's tractive force at each specific speed, an expert driver or a transmission control unit should perform optimum gear shifting.

To calculate the vehicle's accelerating time, substitute the parabolic engine torque characteristic curve and the above assumptions into Equation 6.14. After some algebraic operations, the longitudinal dynamics equation can be represented as:

$$C_0 + C_1 u + C_2 u^2 = M_{d,min}\frac{du}{dt} \tag{6.24}$$

where:

$$\begin{aligned} C_2 &= A_2\frac{\eta_g\eta_d\beta_i^3\beta_d^3}{R^3} - \frac{1}{2}\rho C_d A_f \\ C_1 &= A_1\frac{\eta_g\eta_d\beta_i^2\beta_d^2}{R^2} \\ C_0 &= A_0\frac{\eta_g\eta_d\beta_i\beta_d}{R} - \mu_R M_{min} g \\ M_{d,min} &= M_{min} + \frac{1}{R^2}\left(\eta_g\eta_d\beta_i^2\beta_d^2 I_e + \eta_d\beta_d^2 I_g + I_d + 2I_w\right) \end{aligned} \tag{6.25}$$

In Equation 6.25 C_0, C_1, C_2 and $M_{d,min}$ are functions of vehicle parameters and the transmission ratio. Alternatively, one can write Equation 6.25 as:

$$dt = \frac{M_{d,min}}{C_0 + C_1 u + C_2 u^2}du \tag{6.26}$$

By integrating Equation 6.26, the vehicle's accelerating time from an initial speed of u_1 to a final speed of u_2 will yield:

$$t_{u_1\to u_2} = \int dt = \int_{u_1}^{u_2}\frac{M_{d,min}}{C_0 + C_1 u + C_2 u^2}du \tag{6.27}$$

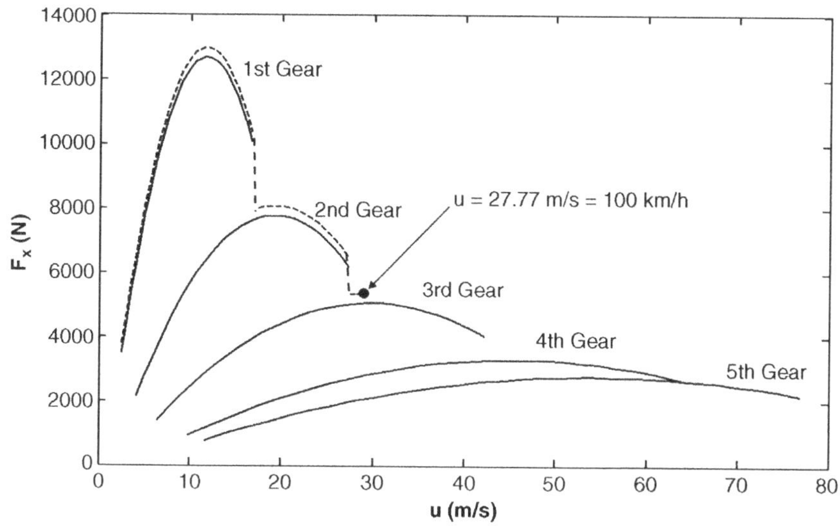

Figure 6.7 Gear shifting pattern during 0 to 100 km/h acceleration test

As mentioned above, C_0, C_1, C_2 and $M_{d,min}$ should remain constant in the integration interval of u_1 to u_2. Then, if the gear changes, the interval must be divided into constant transmission ratio sub-intervals as illustrated in Figure 6.7. Therefore, the final form of the accelerating time is:

$$t_{u_1 \to u_2} = t_{delay} + \sum_i \int_{u_i}^{u_{i+1}} \frac{M_{d,min}}{C_0 + C_1 u + C_2 u^2} du \tag{6.28}$$

where t_{delay} is the summation of gear-changing times.

Example 6.4

Use the vehicle specifications of Example 6.1, to calculate the accelerating time from 0 to 100 km/h.

Solution

As illustrated by Figure 6.7, in the first gear, the vehicle minimum and maximum speeds are obtained from minimum and maximum speeds of the engine:

$$\text{First Gear:} \begin{cases} \omega_{e,min} = 1000\, rpm \Rightarrow u_1 = \dfrac{R\omega_{e,min}}{\beta_1 \beta_d} = \dfrac{0.33 \times 104.72}{3.842 \times 3.538} = 2.54\,\text{m/s} = 9.14\,\text{km/h} \\ \omega_{e,max} = 6600\, rpm \Rightarrow u_2 = \dfrac{R\omega_{e,max}}{\beta_1 \beta_d} = \dfrac{0.33 \times 691.15}{3.842 \times 3.538} = 16.78\,\text{m/s} = 60.41\,\text{km/h} \end{cases}$$

To find the vehicle's accelerating time in first gear, first, we should find the constants of integral:

$$M_d = 1705 + \frac{1}{(0.33)^2}(0.85 \times 1 \times (3.842)^2 \times (3.538)^2 \times 0.2) = 1993.44$$

$$C_0 = -65.096\frac{0.85 \times 1 \times 3.842 \times 3.538}{0.33} - (0.015 \times 1705 \times 9.81) = -2530.05$$

$$C_1 = 1.7537\frac{0.85 \times 1 \times (3.842)^2(3.538)^2}{(0.33)^2} = 2529.16$$

$$C_2 = -0.0018\frac{0.85 \times 1 \times (3.842)^3(3.538)^3}{(0.33)^3} - \frac{1}{2} \times 1.2 \times 0.28 \times 2.31 = -107.32$$

Now, the integration of the first interval results in:

$$t_1 = \int_{2.54}^{16.78} \frac{1993.4}{-2530.05 + 2529.16u - 107.32u^2} du = 3.11\, sec$$

From the second gear, the minimum speed is equal to the maximum speed of the first gear and the maximum speed is obtained using the maximum speed of engine:

$$\text{Second Gear:} \begin{cases} u_2 = 16.78\ \text{m/s} = 60.41\ \text{km/h} \\ \omega_{e,max} = 6600\ \text{rpm} \Rightarrow u_3 = \dfrac{R\omega_{e,max}}{\beta_2\beta_d} = \dfrac{0.33 \times 691.15}{2.353 \times 3.538} = 27.39\ \text{m/s} = 98.6\ \text{km/h} \end{cases}$$

For the third gear, the minimum speed is the same as the second gear maximum speed. In this gear, the vehicle speed reaches 100 km/h, then the upper limit of integral is 100 km/h:

$$\text{Third Gear:} \begin{cases} u_3 = 27.39\ \text{m/s} = 98.6\ \text{km/h} \\ u_4 = 27.77\ \text{m/s} = 100\ \text{km/h} \end{cases}$$

With the same calculation as before, we can find accelerating time in the second and third gears:

$$t_2 = \int_{16.78}^{27.39} \frac{1813.19}{-1646.74 + 948.65u - 24.95u^2} du = 2.80\ \text{sec}$$

$$t_3 = \int_{27.39}^{27.77} \frac{1750.68}{-1157.93 + 400.57u - 7.13u^2} du = 0.12\ \text{sec}$$

The vehicle can reach 100 km/h in the third gear with two gear shifts. If each gear shift takes 0.2 s, the total acceleration time will be:

$$t_{u_1 \to u_6} = 2 \times (0.2) + 3.11 + 2.80 + 0.12 = 6.43 \text{ sec}$$

6.3.2 Electric Vehicles

There are different powertrain layouts for electric vehicles. The powertrain can have a classical single power unit, two/four-wheel drive; or an independent motorized wheel configuration that will be discussed in the following sections.

Figure 6.8 illustrates a single power unit, two-wheel drive electric powertrain. It is almost the same as conventional vehicle powertrain systems, except it uses an electric motor instead of an internal combustion engine and a fixed ratio gearbox instead of a multi-gear transmission. A fixed ratio gearbox in electric vehicles can be used because of their power unit constant power characteristic as opposed to a parabolic characteristic in combustion engines.

First, recall from Chapter 5 that the motor torque can be molded by constant torque/constant power characteristics as:

$$T_m(\omega_m, x_{\theta m}) = \begin{cases} \omega_m < \omega^* & x_{\theta m} T_m^* \\ \omega^* \leq \omega_m \leq \omega_{max} & x_{\theta m} \dfrac{P_m^*}{\omega_m} \end{cases} \qquad \omega^* = \frac{P_m^*}{T_m^*} \tag{6.29}$$

where ω_m and $x_{\theta m}$ are the motor rotational speed and the control signal. The motor control signal is a number in (−100%, 100%) interval indicating the maximum negative and positive torque.

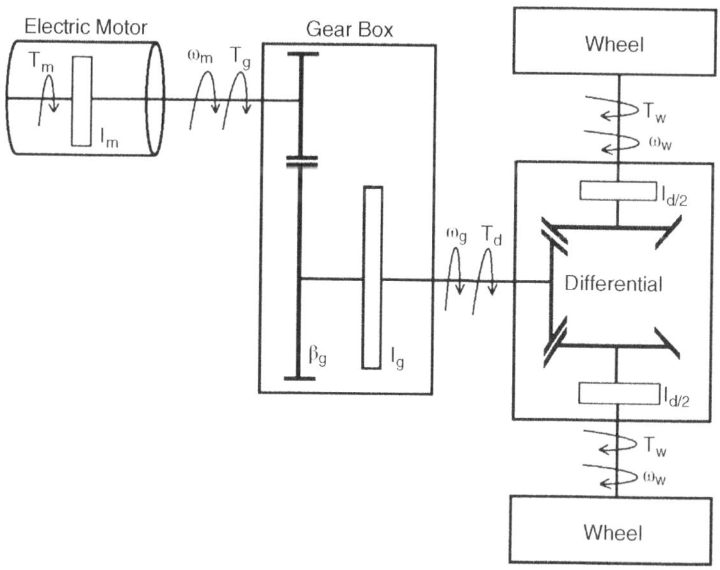

Figure 6.8 Powertrain model of a single motor, two-wheel drive electric vehicle

By substituting Equation 6.29 into Equation 6.14, the longitudinal dynamics equation of a single power unit, two-wheel drive electric vehicle is:

$$\begin{cases} u < u^* & \dfrac{\eta_g \eta_d \beta_g \beta_d}{R} x_{\theta m} T_m^* - F_R = M_d \dfrac{du}{dt} \\ u \geq u^* & x_{\theta m} \eta_g \eta_d \dfrac{P_m^*}{u} - F_R = M_d \dfrac{du}{dt} \end{cases} \qquad u^* = \frac{R\omega^*}{\beta_g \beta_d} \tag{6.30}$$

According to Equation 6.30 and following the same approach discussed in the previous section, the maximum speed of an electric vehicle can be calculated. Consider that the maximum speed of the vehicle is achieved in the constant power part of the motor characteristic curve. Then the maximum speed can be found by solving this equation for u_{max}:

$$\eta_g \eta_d \frac{P_m^*}{u_{max}} - \frac{1}{2}\rho C_d A_f u_{max}^2 - \mu_R M_{min} g = 0 \tag{6.31}$$

Also, the maximum gradeability of the vehicle becomes:

$$\alpha_{road,max} = \text{Sin}^{-1}\left(\frac{\eta_d \eta_g \beta_g \beta_d T_m^*}{R M_{max} g} - \mu_R\right) \tag{6.32}$$

To calculate the acceleration time for the vehicle from u_1 to u_2 and considering that the gearbox ratio is fixed, the integration needs to be calculated only in constant torque and constant power regions of the electric motor:

$$\begin{aligned} t_{u_1 \to u_2} = & \int_{u_1}^{u^*} \frac{M_{d,min}}{\dfrac{\eta_g \eta_d \beta_g \beta_d}{R} T_m^* - \dfrac{1}{2}\rho C_d A_f u^2 - \mu_R M_{min} g} du \\ & + \int_{u^*}^{u_2} \frac{M_{d,min}}{\eta_g \eta_d \dfrac{P_m^*}{u} - \dfrac{1}{2}\rho C_d A_f u^2 - \mu_R M_{min} g} du \end{aligned} \tag{6.33}$$

Example 6.5

Consider a single motor electric vehicle where the front wheels are driven. The electric motor torque characteristic is:

$$T_m(\omega_m, x_{\theta m}) = \begin{cases} 500 x_{\theta m} \text{ Nm} & \omega_m \leq 209 \text{ rad/s} \\ \dfrac{104500}{\omega_m} x_{\theta m} \text{ Nm} & 209 \leq \omega_m \leq 1047 \text{ rad/s} \end{cases}$$

Table 6.3 Specifications of the electric vehicle

Symbol	Parameter	Value
M_{min}	Minimum vehicle mass (kg)	2050
M_{max}	Maximum vehicle mass (kg)	2400
β_g	Gear box ratio	1.685
β_d	Final drive ratio	3.538
I_m	Motor inertia (kg.m^2)	0.15

Find the maximum speed, the maximum vehicle gradeability, and the time of accelerating from zero to 100 km/h. The vehicle specifications are summarized in Table 6.3. For other required specifications, use the information in Example 6.1.

Solution

To find the maximum speed, Equation 6.31 should be used:

$$0.85 \times 1 \times \frac{104500}{u_{max}} - \frac{1}{2} \times 1.2 \times 0.28 \times 2.31 \times u_{max}^2 - 0.15 \times 2050 \times 9.81 = 0$$
$$u_{max} = 56.99 \text{ m/s} = 205.16 \text{ km/h}$$

As Figure 6.9 shows, the maximum speed of the vehicle is found at the intersection point of the tractive and resistance forces.

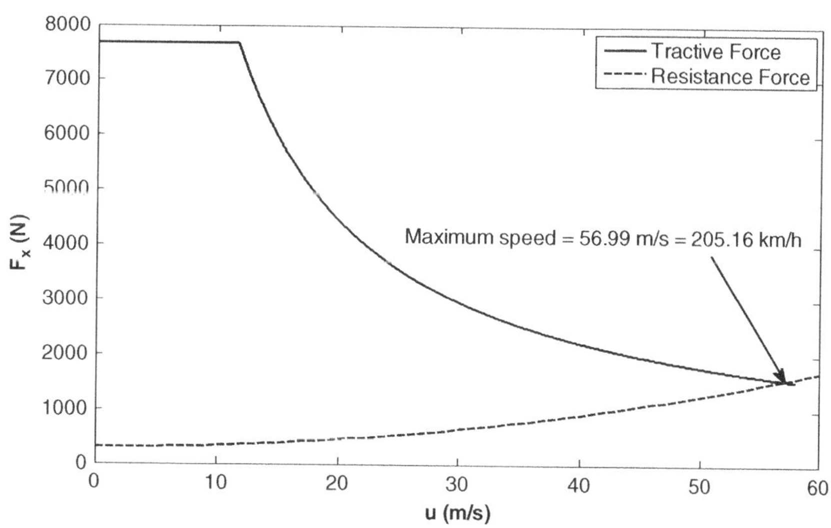

Figure 6.9 Graphical illustration of finding the maximum speed of the single motor electric vehicle

To find the maximum gradeability of the vehicle, by using Equation 6.32 we have:

$$\alpha_{road,max} = \mathrm{Sin}^{-1}\left(\frac{0.85 \times 1 \times 1.685 \times 3.538 \times 500}{0.33 \times 2400 \times 9.81} - 0.015\right) = 18.10\,\text{deg}$$

The calculated road slope is less than what was calculated based on the maximum tire capacity in Example 6.3. Therefore, the maximum gradeability of the vehicle is 18.10 deg.

To find the accelerating time of the vehicle, first u^* and $M_{\text{d,min}}$ are needed:

$$u^* = \frac{R\omega^*}{\beta_g \beta_d} = \frac{0.33 \times 209}{1.685 \times 3.538} = 11.57\,\text{m/s}$$

$$M_{d,min} = 2050 + \frac{1}{(0.33)^2}\left(0.85 \times 1 \times (1.685)^2 \times (3.538)^2 \times 0.15\right) = 2091.61\,\text{kg}$$

The accelerating time of vehicle from zero to 100 km/h is calculated by using Equation 6.33:

$$t_{0 \to 27.77\,m/s} = \int_0^{11.57} \frac{2091.61}{\frac{0.85 \times 1 \times 1.685 \times 3.538}{0.33} \times 500 - \frac{1}{2} \times 1.2 \times 0.28 \times 2.31 \times u^2 - 0.015 \times 2050 \times 9.81}\,du$$

$$+ \int_{11.57}^{27.77} \frac{2091.61}{0.85 \times 1 \times \frac{104500}{u} - \frac{1}{2} \times 1.2 \times 0.28 \times 2.31 \times u^2 - 0.015 \times 2050 \times 9.81}\,du = 11.68\,\text{s}$$

As illustrated in Figure 6.10, in motorized electric vehicles, each wheel is driven by a separate electric motor through a fixed ratio gearbox, and the powertrain does not include differentials. Assume that for all wheels, the same motors and same transmission systems are used.

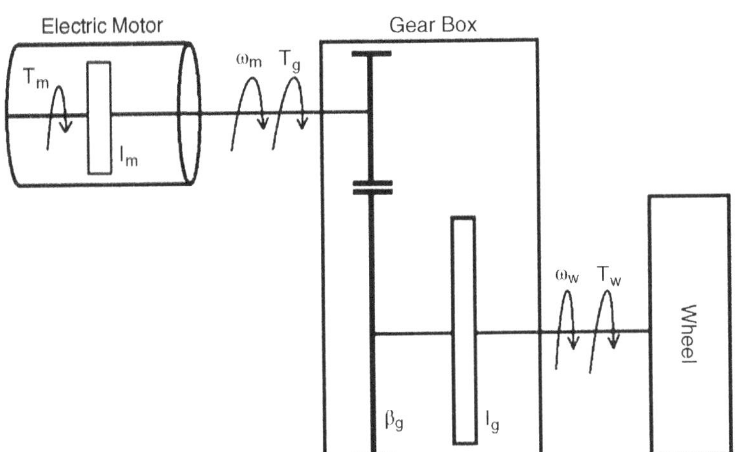

Figure 6.10 Powertrain model of each wheel for a four motorized wheels electric vehicle

The total vehicle tractive force is the summation of four wheels' traction generated by electric motors, consequently the vehicle's longitudinal dynamics equation is:

$$\begin{cases} u < u^* & \dfrac{\eta_g \beta_g}{R} \displaystyle\sum_{i=1}^{4} x_{\theta mi} T_m^* - F_R = M_d \dfrac{du}{dt} \\ u \geq u^* & \dfrac{\eta_g}{u} \displaystyle\sum_{i=1}^{4} x_{\theta mi} P_m^* - F_R = M_d \dfrac{du}{dt} \end{cases} \qquad u^* = \frac{R\omega^*}{\beta_g} \tag{6.34}$$

where $x_{\theta mi}$ is the control signal of motor i. For this vehicle, the dynamic mass is:

$$M_d = M + \frac{4}{R^2}\left(I_w + I_g\right) + \frac{4\eta_g \beta_g^2}{R^2} I_m \tag{6.35}$$

where I_m is the inertia of the motor's rotor. Also, by using the same approach used in single power vehicles, the governing equations of maximum speed, maximum gradeability, and accelerating time of a four motorized wheels electric vehicle can be given as:

$$\eta_g \frac{4P_m^*}{u_{max}} - \frac{1}{2}\rho C_d A u_{max}^2 - \mu_R M_{min} g = 0 \tag{6.36}$$

$$\alpha_{Road,max} = \mathrm{Sin}^{-1}\left(\frac{4\eta_g \beta_g T_m^*}{RM_{max} g} - \mu_R\right) \tag{6.37}$$

$$\begin{aligned} t_{u_1 \to u_2} = & \int_{u_1}^{u^*} \frac{M_d}{\dfrac{4\eta_g \beta_g}{R} T_m^* - \dfrac{1}{2}\rho C_d A_f u^2 - \mu_R M_{min} g} du \\ & + \int_{u^*}^{u_2} \frac{M_d}{4\eta_g \dfrac{P_m^*}{u} - \dfrac{1}{2}\rho C_d A_f u^2 - \mu_R M_{min} g} du \end{aligned} \tag{6.38}$$

6.3.3 *Hybrid Electric Vehicles*

As discussed in Chapter 2, there are numerous types of hybrid electric powertrains. However, their longitudinal dynamics model is formed based on similar techniques. In this section, we develop the corresponding model of a torque coupling parallel hybrid vehicle. To start with, we first model a simple four-wheel drive hybrid vehicle.

Theoretically, a hybrid vehicle has more than two power units. In the case of hybrid electric vehicles, an internal combustion engine and one or more electric motors produce the motive power. In the simplest form of a hybrid electric powertrain system, a separate drive system drives each axle. For example, an internal combustion engine drives the front wheels and the rear wheels are driven by a single or two separate electric motors. In this case, it is supposed that the rear wheels are individually powered by two separate motors. The total tractive force of the vehicle is the sum of the front and rear axles' tractive forces. Following the approach discussed in the previous sections, the vehicle's longitudinal dynamics equation is given as:

$$\frac{\eta_{ge}\eta_d\beta_i\beta_d}{R}T_e(\omega_e, x_{\theta e}) + \frac{2\eta_{gm}\beta_g}{R}T_m(\omega_m, x_{\theta m}) - F_R = M_d\frac{du}{dt} \tag{6.39}$$

$$M_d = M + \frac{1}{R^2}\left(\eta_{ge}\eta_d\beta_i^2\beta_d^2 I_e + \eta_d\beta_d^2 I_{ge} + I_d + 2I_w\right) + \frac{2\eta_{gm}\beta_g^2}{R^2}I_m + \frac{2}{R^2}\left(I_w + I_{gm}\right) \tag{6.40}$$

$$\omega_e = \frac{\beta_i\beta_d u}{R} \qquad \omega_m = \frac{\beta_g u}{R} \tag{6.41}$$

where $x_{\theta e}$ and $x_{\theta m}$ are the control signals of the engine and the electric motors, as determined by the powertrain control system, according to the accelerator pedal position x_θ and the other system's variables.

Figure 6.11 shows the configuration of a parallel-torque coupling hybrid electric powertrain. Similarly, the corresponding longitudinal dynamics equation is:

$$\frac{\eta_{ge}\eta_{ce}\eta_d\beta_i\beta_{ce}\beta_d}{R}T_e(\omega_e, x_{\theta e}) + \frac{\eta_{gm}\eta_{cm}\eta_d\beta_g\beta_{cm}\beta_d}{R}T_m(\omega_m, x_{\theta m}) - F_R = M_d\frac{du}{dt} \tag{6.42}$$

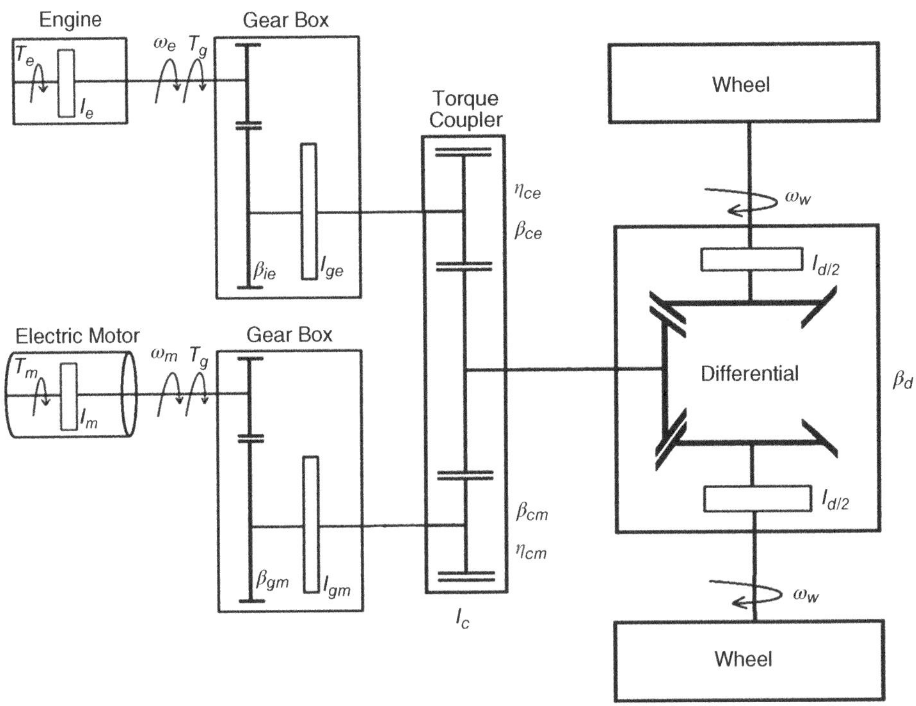

Figure 6.11 Powertrain model of a parallel-torque coupling hybrid electric vehicle

$$M_d = M + \frac{1}{R^2}\Big(\eta_{ge}\eta_{gc}\eta_d\beta_i^2\beta_{ce}^2\beta_d^2 I_e + \eta_{gm}\eta_{gm}\eta_d\beta_{gm}^2\beta_{cm}^2\beta_d^2 I_m + \eta_{ce}\eta_d\beta_{ce}^2\beta_d^2 I_{ge}$$
$$+ \eta_{cm}\eta_d\beta_{cm}^2\beta_d^2 I_{gm} + \eta_d\beta_d^2 I_c + I_d + 2I_w\Big) \tag{6.43}$$

$$\omega_e = \frac{\beta_i\beta_{ce}\beta_d}{R}u \; \omega_m = \frac{\beta_g\beta_{cm}\beta}{R}u \tag{6.44}$$

where $\eta_{ge}, \eta_{gm}, \eta_{ce}$ and η_{cm} represent the efficiencies of the engine gearbox, the motor gearbox, the torque coupler-engine input and the torque coupler-motor input. Also $\beta_i, \beta_g, \beta_{ce}$ and β_{cm} respectively, are the gear ratios of the engine gearbox, the motor gearbox, the torque coupler-engine input and the torque coupler-motor input.

According to Equations 6.39 and 6.42, the general form of hybrid electric vehicles' longitudinal dynamics equation is:

$$\begin{cases} a_e T_e(\omega_e, x_{\theta e}) + a_m T_m(\omega_m, x_{\theta m}) - F_R = M_d \dfrac{du}{dt} \\ \omega_e = \beta_e u \qquad \omega_m = \beta_m u \end{cases} \tag{6.45}$$

where $a_e, a_m, \beta_e, \beta_m$ and M_d are functions of powertrain parameters, such as: gear ratios, efficiencies, inertia, and wheel radius.

Having the general form of the vehicle dynamics equation and by using the same techniques used for conventional and electric vehicles, the following equation is found for the maximum speed of a hybrid electric vehicle:

$$a_e T_e(\beta_e u_{max}, x_{\theta e} = 100\%) + a_m T_m(\beta_m u_{max}, x_{\theta m} = 100\%) - F_R(u_{max}) = 0 \tag{6.46}$$

Similarly for maximum gradeability:

$$\alpha_{road,max} = \mathrm{Sin}^{-1}\left(\frac{F_{x,max}}{gM_{max}} - \mu_R\right) \tag{6.47}$$

where $F_{x,max}$ is the maximum vehicle traction, found as:

$$F_{x,max} = \mathrm{Max}\left[a_e T_e(\beta_e u, x_{\theta e} = 100\%) + a_m T_m(\beta_m u, x_{\theta m} = 100\%)\right] \tag{6.48}$$

And the accelerating time is:

$$t_{u_1 \to u_2} = t_{delay} + \int_{u_1}^{u_2} \frac{M_{d,min}}{a_e T_e(\beta_e u, x_{\theta e} = 100\%) + a_m T_m(\beta_m u, x_{\theta m} = 100\%) - F_R(u)} du \tag{6.49}$$

To calculate the above integral, one may divide the integral's interval into some sub-intervals according to the electric motor and engine characteristic curves, as well as changing in the transmission ratios.

Example 6.6

Consider a parallel hybrid electric vehicle which has an engine and electric motor with the following torque characteristics:

$$T_e(\omega_e, x_{\theta e}) = x_{\theta e}\left(-0.001\omega_e^2 + 0.912\omega_e - 34.456\right) \qquad 104.72 < \omega_e < 691.15\,\text{rad/s}$$

$$T_m(\omega_m, x_{\theta m}) = \begin{cases} 500x_{\theta m}\,\text{Nm} & \omega_m \leq 209\,\text{rad/s} \\ \dfrac{104500}{\omega_m}x_{\theta m}\,\text{Nm} & 209 < \omega_m \leq 1047\,\text{rad/s} \end{cases}$$

Find the maximum speed, the maximum gradeability, and acceleration time from 0 to 100 km/h. The vehicle specifications are summarized in Table 6.4. For other required specifications, use the information provided in Example 6.1.

Solution

To determine the maximum speed, first, a_e, and a_m are calculated:

$$a_e = \frac{0.85 \times 0.85 \times 1 \times 0.925 \times 1 \times 3.538}{0.33} = 7.17$$

$$a_m = \frac{0.85 \times 0.85 \times 1 \times 1.435 \times 1 \times 3.538}{0.33} = 11.12$$

Table 6.4 Specifications of the HEV

Symbol	Parameter	Value
M_{min}	Minimum vehicle mass (kg)	1900
M_{max}	Maximum vehicle mass (kg)	2250
β_{ge}	Gear ratio for engine	0.925
β_{gm}	Gear ratio for electric motor	1.435
β_{ce}, β_{cm}	Torque coupler gear ratio	1
η	Efficiency of all gears	0.85
β_d	Final drive ratio	3.538
η_d	Differential efficiency	1
I_e	Engine inertia (kgm^2)	0.1
I_m	Motor inertia (kgm^2)	0.15

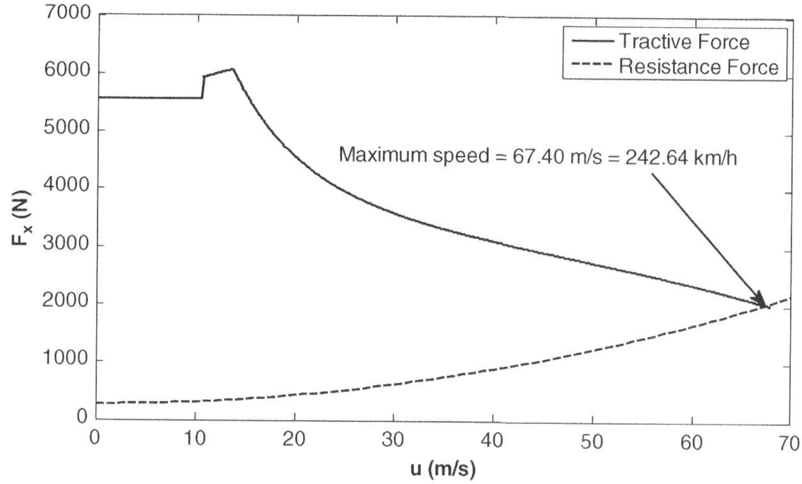

Figure 6.12 Graphical illustration of finding the maximum speed of the hybrid electric vehicle

Now, Equation 6.46 should be used:

$$7.17\times\left\{-0.001\left(\frac{0.925\times1\times3.538}{0.33}u_{max}\right)^2+0.912\left(\frac{0.925\times1\times3.538}{0.33}u_{max}\right)-34.456\right\}$$

$$+11.12\times\left\{\frac{104500}{\left(\frac{1.435\times1\times3.538}{0.33}u_{max}\right)}\right\}-\left\{\frac{1}{2}\times1.2\times0.28\times2.31\times u_{max}^2+0.015\times1900\times9.81\right\}=0$$

$$u_{max}=67.40\frac{\text{m}}{\text{s}}=242.64\ \text{km/h}$$

According to Figure 6.12, the maximum speed of the vehicle can also be obtained by determining the intersection of the tractive and resistance forces.

The maximum gradeability of the vehicle is calculated through Equation 6.47. In this Equation $F_{x,max}$ is obtained by taking the derivative of F_x with respect to u:

$$F_{x,max}=6063\ \text{N}$$

Then, the maximum gradeability of the vehicle is:

$$\alpha_{road,max}=\text{Sin}^{-1}\left(\frac{6063}{2250\times9.81}-0.015\right)=15.05\ \text{deg}$$

This calculated road slope is less than the corresponding value of the maximum tire capacity (see Example 6.3), therefore, the maximum gradeability is 15.05 deg.

Finally, Equation 6.49 is used to determine the accelerating time of the vehicle:

$$M_{d,min} = 1900 + \frac{1}{(0.33)^2}(0.85 \times 0.85 \times 1 \times (0.925)^2 \times (1)^2 \times (3.538)^2 \times 0.1)$$
$$+ \frac{1}{(0.33)^2}\left((0.85 \times 0.85 \times 1 \times (1.435)^2 \times (1)^2 \times (3.538)^2 \times 0.15)\right) = 1932.76\ \text{kg}$$

$$u^* = \frac{R\omega^*}{\beta_{gm}\beta_{cm}\beta_d} = \frac{0.33 \times 209}{1.435 \times 1 \times 3.538} = 13.58\ \text{m/s}$$

$$u_1 = \frac{R\omega_{e,min}}{\beta_i\beta_{ce}\beta_d} = \frac{0.33 \times 104.72}{0.925 \times 1 \times 3.538} = 10.56\ \text{m/s}$$

$$t_{0\to 27.77m/s} = \int_0^{10.56} \frac{1932.76}{11.12 \times 500 - \frac{1}{2} \times 1.2 \times 0.28 \times 2.31 \times u^2 - 0.015 \times 1900 \times 9.18} du$$

$$+ \int_{10.56}^{13.58} \frac{1932.76}{A} du + \int_{13.58}^{27.77} \frac{1932.76}{B} du = 9.51\ \text{sec}$$

$$A = 7.17 \times \left(-0.001\left(\frac{0.925 \times 1 \times 3.538}{0.33}u\right)^2 + 0.912\left(\frac{0.925 \times 1 \times 3.538}{0.33}u\right) - 34.456\right)$$

$$+ 11.12 \times 500 - \frac{1}{2} \times 1.2 \times 0.28 \times 2.31 \times u^2 - 0.015 \times 1900 \times 9.81$$

$$B = 7.17 \times \left(-0.001\left(\frac{0.925 \times 1 \times 3.538}{0.33}u\right)^2 + 0.912\left(\frac{0.925 \times 1 \times 3.538}{0.33}u\right) - 34.456\right)$$

$$+ 11.12 \times \frac{104500}{\frac{1.435 \times 1 \times 3.538}{0.33}u} - \frac{1}{2} \times 1.2 \times 0.28 \times 2.31 \times u^2 - 0.015 \times 1900 \times 9.81$$

6.4 Vehicle Braking Modeling and Analysis

Conventional vehicles are equipped with only friction brakes. In other words, by using friction the vehicle's kinetic energy converts to unusable thermal energy. On the other hand, in the case of electric and hybrid electric vehicles, in addition to friction braking, electric machines can provide braking torque. They can switch to the generator mode, absorb the vehicle kinetic energy, convert it to electrical energy and restore the majority into batteries or ultra-capacitors. As discussed in Chapter 3, this type of braking is known as regenerative braking. Considering the above-mentioned technological difference, the mathematical model of vehicle braking for an electric or hybrid electric vehicle will be different from that of a conventional car.

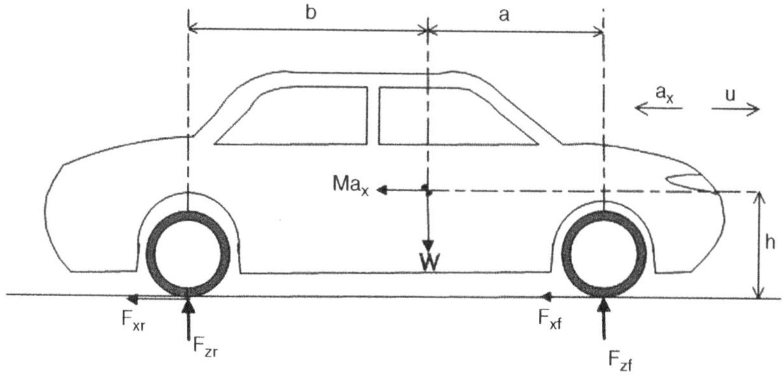

Figure 6.13 The vehicle model during braking

In this section, a mathematical model that can express the vehicle braking for electric and hybrid vehicles is developed. However, the model is a generalization, and after minor changes, can also function for conventional vehicles. The main purpose of this model is to predict the braking performances, primarily the stopping distance in various road conditions. In developing the model, all assumptions from the introduction section are valid. In addition, in the case of braking, the wheels' longitudinal slip is no longer small and must be considered in the modeling process.

First, rewrite the vehicle longitudinal dynamics equation in the case of braking:

$$-\sum_{i=1}^{4} F_{xi} - F_R = M\frac{du}{dt} \tag{6.50}$$

where F_{xi} as shown by Figure 6.13 is the braking force of each wheel. Next, considering a vehicle's rotating wheel during braking as shown in Figure 6.14, the wheel's dynamics equation becomes:

$$I_{wi}\frac{d\omega_{wi}}{dt} = T_{wi} + T_{bi} - F_{xi}R \quad i = 1..4 \tag{6.51}$$

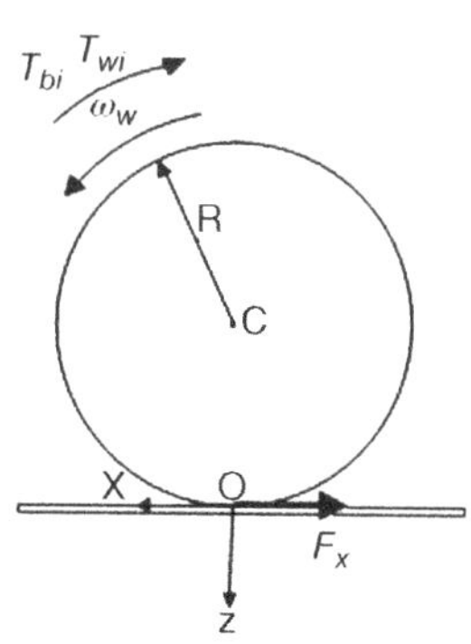

Figure 6.14 Applied forces and moments to a wheel during braking

where T_{wi} and T_{bi} respectively, are the motor and friction brake torques of each wheel and both are negative, I_{wi} is the wheel rotational inertia and ω_{wi} is the wheel angular velocity. Obviously, friction brake torque can be applied to all wheels, but availability of electric motor braking torque depends on the powertrain configuration. For instance, in the case of a four motorized electric vehicle, it is available in all-wheel, but in a two-wheel drive electric or hybrid-electric vehicle, regenerative braking can occur only in the drive axle, either front or rear axle.

To link the vehicle longitudinal dynamics and the wheels' dynamics equations, consider the wheel slip equations in braking mode as defined in Chapter 4:

$$s_i = -1 + \frac{R\omega_{w_i}}{u} \quad i = 1..4 \tag{6.52}$$

In the next step, the braking force of each wheel should be determined. According to the terminology introduced in Chapter 4, straight line braking is a pure-slip condition, then the braking force of each wheel is calculated by using a pure-slip longitudinal force tire model. The general form of such a model is:

$$F_{xi} = F_{xi}(s_i, F_{zi}) \quad i = 1..4 \tag{6.53}$$

where F_{z_i} is the normal force acting on each wheel. The normal forces of wheels change while braking. The load transfer from the rear to the front axle occurs as a result of braking. The value of this longitudinal load transfer is proportional to the vehicle's deceleration. As such, for determining the normal force of each wheel, one should formulate the longitudinal load transfer. As shown in Figure 6.13, according to the D'Alembert principle, by considering the inertial force of the vehicle as a vector in the opposite direction, it is possible to reformulate the load transfer problem as a simple static equilibrium problem. Then, the normal force of each wheel, considering longitudinal load transfer due to braking is found as:

$$\begin{aligned} &\textit{Rear Wheels:} \quad F_{z2} = F_{z4} = \frac{W}{2l}\left(a + \frac{a_x}{g}h\right) \\ &\textit{Front Wheels:} \quad F_{z1} = F_{z3} = \frac{W}{2l}\left(b - \frac{a_x}{g}h\right) \end{aligned} \tag{6.54}$$

where a_x is the vehicle deceleration, defined as:

$$a_x = \frac{du}{dt} \tag{6.55}$$

Finally, the vehicle's stopping distance is found by integrating the vehicle velocity:

$$X = \int u dt \tag{6.56}$$

Considering Equations 6.50 to 6.56, they contain 19 equations and 19 unknown variables, and these equations form a system of equations in which the applied braking torques in wheels are the inputs, and the vehicle velocity and the stopping distance are the outputs. It is easy to solve this system using numerical methods. Figure 6.15 illustrates the system's block diagram. With this braking model, one can calculate the vehicle's braking performances, such as: vehicle deceleration, velocity, and stopping distance.

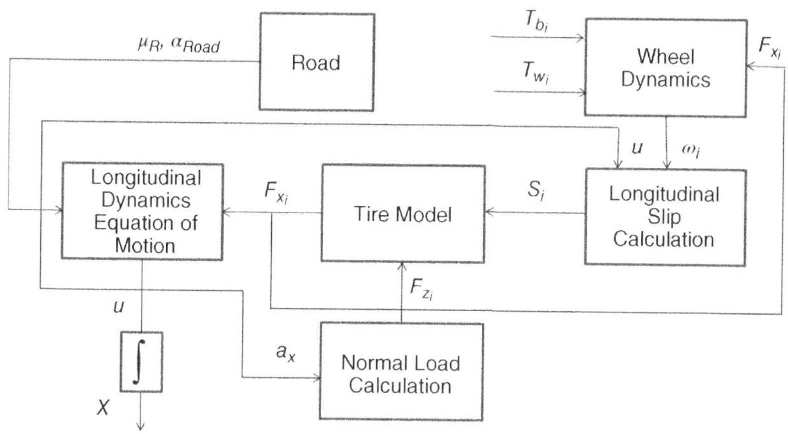

Figure 6.15 Block diagram of electric and hybrid electric vehicles' braking model

Example 6.7

Consider a four motorized wheel electric vehicle. The torque of each electric motor during braking is calculated by using the subsequent equation:

$$T_{m,max} = \begin{cases} -320\,\text{Nm if} \quad \omega \leq 70\,\text{rad/s} \\ -\dfrac{22400}{\omega}\,\text{Nm if} \;\; 70 \leq \omega \leq 210\,\text{rad/s} \end{cases}$$

Also the tire longitudinal force is modeled as:

$$F_x = \begin{cases} \dfrac{\mu F_z s}{0.2} & S \leq 0.2 \\ \dfrac{-0.3\mu F_z(s-0.2)}{0.8} + \mu F_z & S > 0.2 \end{cases}$$

Using the vehicle specifications of Example 6.1, simulate a braking maneuver with an initial speed of 100 km/h, and find the speed, deceleration, and stopping distance in subsequent cases:

(a) Without applying friction braking only, all electric motors generate the maximum possible braking torque.
(b) The electric motors of the front axle work with maximum capacity and the friction brake of rear axle applies a braking torque of 1100 Nm on each wheel.
(c) The braking torque of 1750 Nm on each wheel of the front axle and 950 Nm on each wheel of the rear axle are applied by electric motors and the friction brake simultaneously, i.e., the electric motors work with a maximum capacity and the remaining braking torque is applied by the friction brake.

Table 6.5 Stopping distance and stopping time of the vehicle

Case	Stopping distance (m)	Stopping time (s)
(a)	203.96	14.17
(b)	83.02	5.90
(c)	42.29	3.02

Solution

The braking model can be solved by using different available software such as Matlab or Maple. Here, we use Matlab SIMULINK. The stopping distance and stopping time in all cases are summarized in Table 6.5. Also, the vehicle's deceleration, speed, and the distance traveled versus time are shown in Figure 6.16.

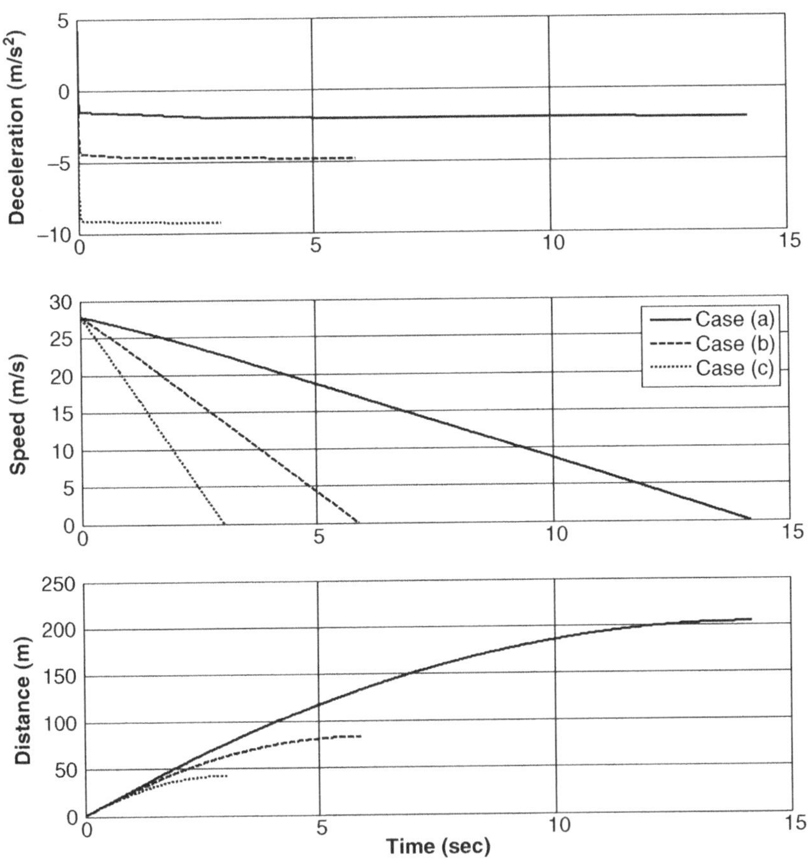

Figure 6.16 The vehicle's deceleration, speed, and distance traveled during the braking manuver

Since electric motors are connected to the wheels directly, the initial speed of the electric motor can be calculated from:

$$\omega_m = \frac{u}{R} = \frac{27.77}{0.33} = 84.15\,\text{rad/s}$$

At this speed, each electric motor can generate a braking torque of −191.32 Nm. When the speed of the vehicle decreases, the braking torque of each electric motor increases until it reaches maximum braking torque of −320 Nm. Figure 6.17 shows the braking torque on each wheel generated by the electric motors and the friction brake in all cases.

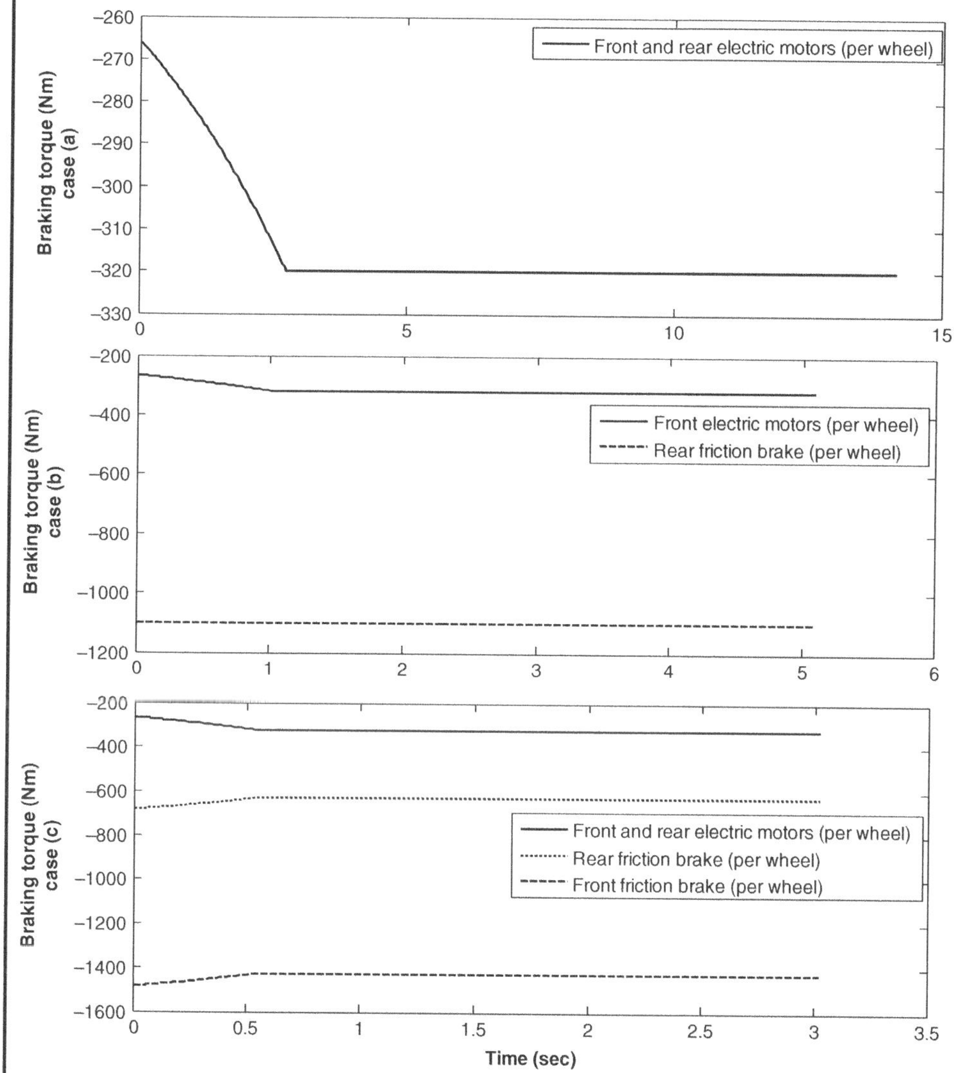

Figure 6.17 Braking torque versus time in all cases

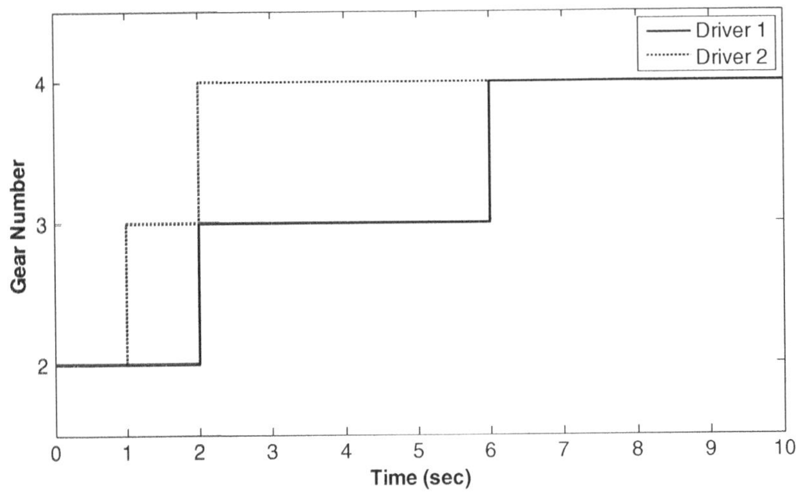

Figure 6.18 Gear shifting map for drivers 1 and 2

Problems

1. A maneuver with initial speed of 10 m/s is done by two different drivers. Both of them keep the throttle wide open and shift the gears as shown in Figure 6.18. Simulate this maneuver for both cases, and discuss how gear shifting can affect distance, speed, acceleration, and engine speed. For the required vehicle specifications, refer to the information provided in Example 6.1.

2. Using the vehicle specifications of Example 6.1, find the vehicle's maximum gradeability on a wet road ($\mu = 0.7$). Compare the results with those of Example 6.3 and show how the road friction coefficient can affect the vehicle's gradeability.

3. Consider that the engine of the vehicle in Example 6.1 is replaced by a smaller one with the following torque characteristic:

$$T(\omega_e, x_{\theta e}) = x_{\theta e}\left(-0.001\omega_e^2 + 0.912\omega_e - 34.456\right)$$

   How does the performance of the vehicle (maximum speed, maximum vehicle gradeability, and time of accelerating from zero to 100 km/h) change?

4. Consider a motorized wheel electric vehicle which has a separate electric motor for each wheel. Electric motors are connected to the wheels through a fixed ratio gearbox ($\beta_g = 1.531$). The torque of each electric motor can be calculated using:

$$T_m = \begin{cases} 170x_{\theta m}\ \text{Nm}\ \textit{if}\quad \omega \le 160\ \text{rad/s} \\ \dfrac{27200}{\omega}x_{\theta m}\ \text{Nm}\ \textit{if}\ 160 \le \omega \le 320\ \text{rad/s} \end{cases}$$

Table 6.6 Braking torque of each axle in different cases

Case	Braking torque of front axle	Braking torque of rear axle
a	Maximum possible braking torque of electric motors	Maximum possible braking torque of electric motors
b	Maximum possible braking torque of electric motors	−800 Nm on each wheel by friction brake
c	−1200 on each wheel	−750 on each wheel

Determine the maximum speed, maximum vehicle gradeability, and time of accelerating from 0 to 100 km/h. For any other required vehicle specifications, refer to the information provided in Example 6.1.

5. Consider a hybrid vehicle which has one engine and two electric motors. The front axle is driven by the engine with the following torque characteristic:

$$T_e(\omega_e, x_{\theta e}) = x_{\theta e}(-0.001\,\omega_e^2 + 0.912\,\omega_e - 34.456)\;104.72 < \omega_e < 691.15\,\text{rad/s}$$

Moreover, each wheel of the rear axle is driven by a separate electric motor through a fixed ratio gearbox ($\beta_g = 1.531$). The torque characteristic of the electric motors is:

$$T_m = \begin{cases} 170 x_{\theta m}\ \text{Nm}\ \ \textit{if}\ \ \omega \leq 160\,\text{rad/s} \\ \dfrac{27200}{\omega} x_{\theta m}\ \text{Nm}\ \textit{if}\ 160 \leq \omega \leq 320\,\text{rad/s} \end{cases}$$

Determine the vehicle's maximum speed, maximum gradeability, and time of accelerating from 0 to 100 km/h. For any other required vehicle specifications, refer to the information provided in Example 6.1.

6. In Example 6.7, the braking performance of an electric vehicle is studied in three cases. Determine the braking performance of the same vehicle when braking is done on a wet road ($\mu - 0.7$). Compare the results with those of Example 6.7 and show how much friction coefficient can affect the stopping distance and stopping time. Table 6.6 shows the braking torque for each case. In case (c), electric motors work with maximum capacity and the remaining braking torque is applied by the friction brake.

7

Handling Analysis of Electric and Hybrid Electric Vehicles

7.1 Introduction

Researchers use models with different levels of accuracy and complexity, in order to analyze the handling performance of vehicles. While some simple linear models may consider only lateral and yaw dynamics, there are other simple models for roll or pitch dynamics that ignore all other aspects of vehicle dynamics. In general, simple models are unable to express vehicle handling behavior accurately and are only useful for basic handling analyses. More complex handing models consider the lateral, longitudinal, and roll dynamics, as well as the non-linearities in the system. In the case of modern electric and hybrid electric vehicles, comprehensive handling models are necessary to consider the effect of the strong coupling between longitudinal and lateral dynamics.

This chapter focuses on developing a comprehensive handling model for electric and hybrid vehicles suitable for simulation purposes. To provide a better understanding of the basic concepts of vehicle handing, the chapter begins with linear handing models such as bicycle, roll and pitch models used for the analytical analysis of vehicle handling performance. Finally, a comprehensive handling model for electric and hybrid vehicles is introduced, discussed and simulated.

7.2 Simplified Handling Models

Simplified linear handling models are only accurate when the vehicle lateral acceleration is relatively low and the tire behavior is linear. These models are represented with simple mathematical equations allowing for analytical stability, steady state, and transient analyses.

Electric and Hybrid Vehicles: Technologies, Modeling and Control: A Mechatronic Approach, First Edition.
Amir Khajepour, Saber Fallah and Avesta Goodarzi.

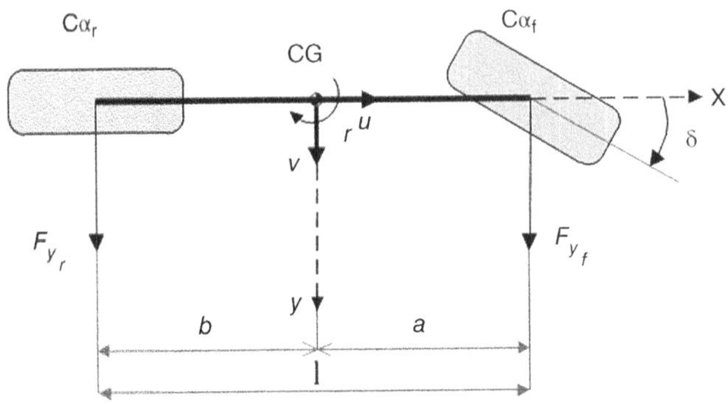

Figure 7.1 Single track linear handling model configuration

7.2.1 Single Track Linear Handling Model

The single track or bicycle model is the simplest linear handling model. In this model, the lateral velocity and yaw rate are the two state variables, and the steering angle is the system input. Many simplifying assumptions are made with this model: (1) the width of the vehicle has no effect; (2) the vehicle motion is planar and roll, pitch and vertical motions are ignored; (3) tire behavior is linear; (4) tier self-aligning moment is ignored; (5) longitudinal dynamics are ignored; (6) longitudinal velocity is constant; (7) tire sideslip angle is small and the lateral acceleration is low; (8) there is no load transfer; and (9) there is no change in tire camber angle. Figure 7.1 illustrates the model configuration.

The bicycle model is formed based on the planar vehicle equations of motion as expressed by Equation 4.13. The lateral and yaw equations of motion are applicable in this case but due to ignoring the longitudinal dynamics, its corresponding equation is not needed. According to the above assumptions and Figure 7.1, the equations of motion for the bicycle model are:

$$\begin{aligned} F_{yr} + F_{yf} &= M(\dot{v} + ur) \\ F_{yf}a - F_{yr}b &= I_z\dot{r} \end{aligned} \tag{7.1}$$

where M and I_z respectively are the mass and moment of inertia of the vehicle about the z axis, and F_{yr} and F_{yf} respectively are the equivalent lateral forces of the front and rear axles. The equivalent lateral force on each axle is the summation of lateral forces of the axle's wheels. Using a linear tire model as expressed by Equation 4.35, they can be represented by:

$$\begin{aligned} F_{yf} &= C_{\alpha f}\alpha_f \qquad & C_{\alpha f} &= 2c_{\alpha f} \\ F_{yr} &= C_{\alpha r}\alpha_r \qquad & C_{\alpha r} &= 2c_{\alpha r} \end{aligned} \tag{7.2}$$

where $c_{\alpha f}$ and $c_{\alpha r}$ respectively are the cornering stiffness of the front and rear tires, and α_f and α_r respectively are the sideslip angles of the front and rear axles, as determined by:

$$\begin{aligned}\alpha_f &= \delta - \tan^{-1}\frac{v_{0f}}{u_{0f}} \approx \delta - \frac{v + ar}{u}\\ \alpha_r &= -\tan^{-1}\frac{v_{0r}}{u_{0r}} \approx -\frac{v - br}{u}\end{aligned} \tag{7.3}$$

By assuming that the value of the sideslip angle is small, the above equations can be linearized as shown in Equation 7.3. Next, by substituting Equation 7.3 into Equation 7.2, and then into Equation 7.1, the final form of single track handling model becomes:

$$\begin{pmatrix}\dot{v}\\ \dot{r}\end{pmatrix} = \begin{pmatrix} -\dfrac{C_{\alpha r} + C_{\alpha f}}{Mu} & \dfrac{(bC_{\alpha r} - aC_{\alpha f})}{Mu} - u \\ \dfrac{(bC_{\alpha r} - aC_{\alpha f})}{I_z u} & -\dfrac{(b^2 C_{\alpha r} + a^2 C_{\alpha f}}{I_z u}\end{pmatrix}\begin{pmatrix}v\\ r\end{pmatrix} + \begin{pmatrix}\dfrac{C_{\alpha f}}{M}\\ \dfrac{aC_{\alpha f}}{I_z}\end{pmatrix}\delta \tag{7.4}$$

Example 7.1

Table 7.1 provides the parameters of a conventional mid-size sedan vehicle. Simulate the vehicle handling responses by using a bicycle model for longitudinal speeds of 70 and 120 km/hr. For both of these maneuvers, set the steering input angle to $\delta = 0.03$ rad. Compare the yaw rate, the lateral velocity, the lateral acceleration, and the tire side slip angle responses for the maneuvers.

Comment on the results and discuss the validity of the bicycle model for the maneuvers considering the bicycle model linearity assumptions. Figure 7.2 shows the tire's lateral force versus sideslip angle for the nominal normal load.

Table 7.1 Vehicle parameters

Parameter	Value
M	1,530 kg
a	1.320 m
b	1.456 m
I_z	4,192 kgm^2
$c_{\alpha f}$	70,000 N/rad
$c_{\alpha r}$	70,000 N/rad

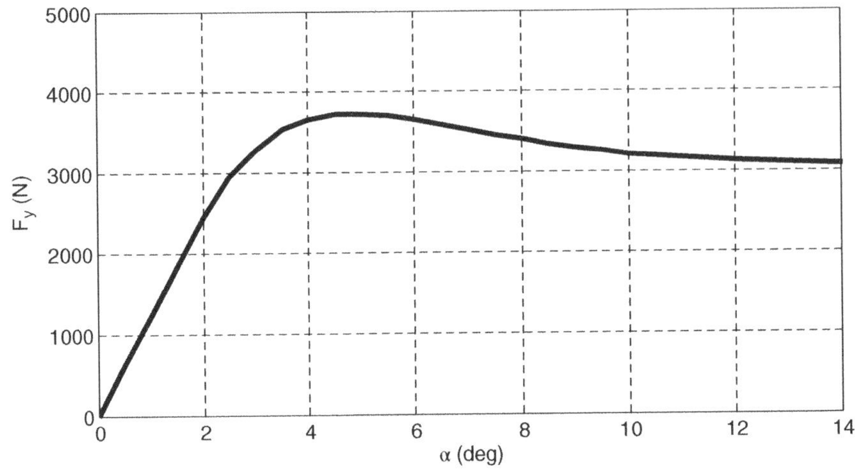

Figure 7.2 Lateral force characteristic curve for the tires in Example 7.1

Solution

The system matrices, Equation 7.4, are calculated as:

For $u = 70$ km/hr:

$$A = \begin{bmatrix} -9.4118 & -18.8044 \\ 0.2336 & -6.6338 \end{bmatrix} \qquad B = \begin{bmatrix} 91.5033 \\ 44.0840 \end{bmatrix}$$

For $u = 120$ km/hr:

$$A = \begin{bmatrix} -5.4902 & -32.9600 \\ 0.1363 & -3.8697 \end{bmatrix} \qquad B = \begin{bmatrix} 91.5033 \\ 44.0840 \end{bmatrix}$$

Figure 7.3 depicts the block diagram of the bicycle model in Equation 7.4.

Figure 7.4 illustrates the simulation results for the yaw rate, the lateral velocity, the lateral acceleration, and the sideslip angles for the front and rear tires. For the longitudinal speed of 70 km/hr, the tire sideslip angles are sufficiently small ($\alpha < 2°$), and also the lateral acceleration is in the linear range ($a_y < 0.4\ g$). Figure 7.5 shows the tire characteristic curve of the bicycle model in Example 7.1. The shaded area represents the linear tire region. The maximum sideslip angle for the first maneuver shown by point A in Figure 7.5 falls in the linear tire region. Therefore, the linear tire model is a valid assumption for the low-speed maneuver and the simulation results are reliable with an acceptable level of accuracy.

For the high-speed maneuver, the maximum sideslip angle for the front wheel is approximately 3.25° shown in Figure 7.5 by point B. However, the actual force from the tire's characteristic curve is indicated by point B$'$, which is considerably different from the assumed force. Such a difference is the underlying reason for inaccuracy of the single-track model when the vehicle undergoes harsh maneuvers.

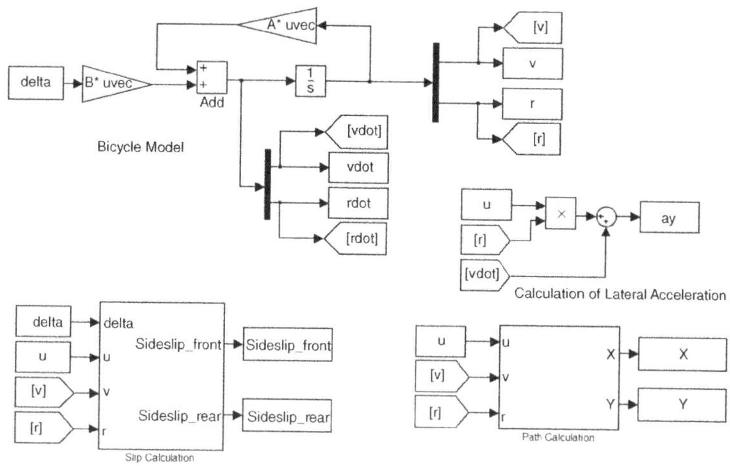

Figure 7.3 Simulink block diagram of the bicycle vehicle model

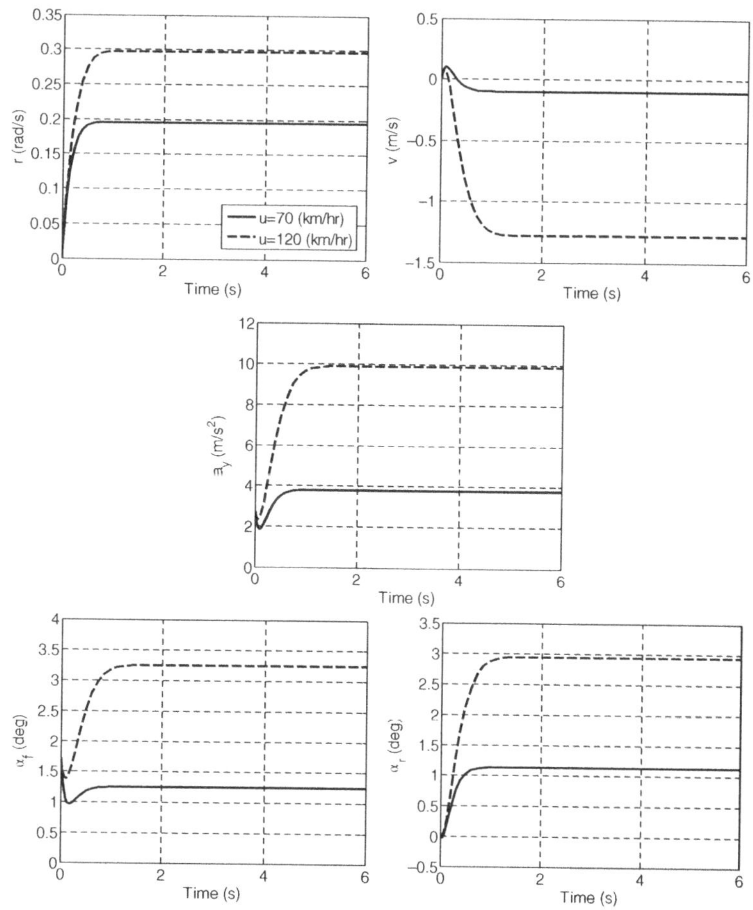

Figure 7.4 Simulation results for the bicycle model

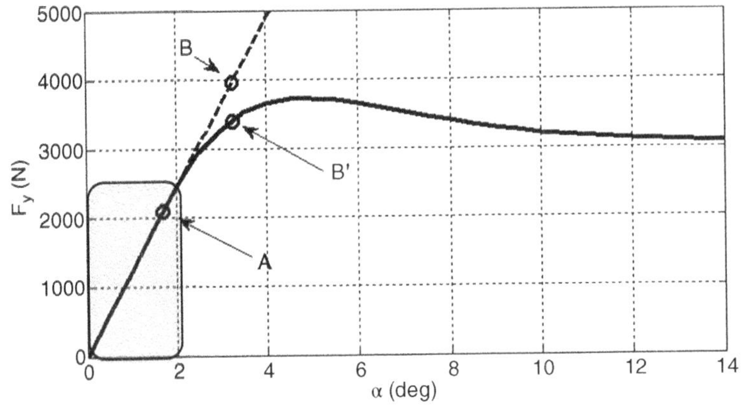

Figure 7.5 Force-slip characteristic curve for the tires in Example 7.1

7.2.2 Analytical Handling Analysis

As can be seen in Example 7.1, the single track handling model can simulate the dynamic behavior of vehicles during low lateral acceleration (up to 0.3–0.4g) manoeuvres [1]. More importantly, the model can be used for analytical studies, such as stability, steady state, and transient analyses.

7.2.2.1 Stability Analysis

For the purpose of stability analysis, the single track handling model presented by Equation 7.4 should be converted to the state space form:

$$\dot{X} = AX + B\delta \tag{7.5}$$

where:

$$X = \begin{pmatrix} v \\ r \end{pmatrix}, \quad A = \begin{pmatrix} -\dfrac{C_{\alpha r} + C_{\alpha f}}{Mu} & \dfrac{(bC_{\alpha r} - aC_{\alpha f})}{Mu} - u \\ \dfrac{(bC_{\alpha r} - aC_{\alpha f})}{I_z u} & -\dfrac{(b^2 C_{\alpha r} + a^2 C_{\alpha f})}{I_z u} \end{pmatrix}, \quad B = \begin{pmatrix} \dfrac{C_{\alpha f}}{M} \\ \dfrac{aC_{\alpha f}}{I_z} \end{pmatrix} \tag{7.6}$$

By applying the Laplace transform to Equation 7.5, we have:

$$sX(s) = AX(s) + B\delta(s) \tag{7.7}$$

Equation 7.7 can be solved for $X(s)$ as:

$$X(s) = (sI - A)^{-1} B\delta(s) \tag{7.8}$$

where I is a 2×2 identity matrix. The system transfer function can be then represented by:

$$G(s) = (sI - A)^{-1}B \tag{7.9}$$

In the above equation, $G(s)$ is a 2×1 vector that relates the system's input δ to each of the system's state variables, v and r. Also the general form of the system's characteristic equation can be represented as:

$$\det(sI - A) = 0 \tag{7.10}$$

By substituting Equation 7.6 into Equation 7.10, the final form of the system's characteristic equation becomes:

$$s^2 + \frac{1}{u}\left(\frac{a^2 C_{\alpha f} + b^2 C_{\alpha r}}{I_z} + \frac{C_{\alpha f} + C_{\alpha r}}{M}\right)s + \frac{1}{MI_z}\left(\frac{l^2 C_{\alpha f} C_{\alpha r}}{u^2} - M\left(aC_{\alpha f} - bC_{\alpha r}\right)\right) = 0 \tag{7.11}$$

The roots of the characteristic equation are known as the system poles. In general, system poles are complex variables having real and imaginary components. The values of the real and imaginary parts have a strong effect on system performance and stability. According to control theory, a dynamic system is stable if all the system poles have negative real components.

Since it is analytically difficult to determine the poles of a system, the Routh–Hurwitz stability criterion is a very popular method used to determine the stability of linear dynamics. Instead of determining the system poles, this method determines the sign of the real part of the poles using the coefficients of the characteristic equation. The sign of the real part of the poles is sufficient to judge whether or not a system is stable. For more information on the Routh–Hurwitz stability criterion, see reference [2].

Applying the Routh–Hurwitz stability criterion to the vehicle handling characteristic equation, presented in Equation 7.11, leads to the following stability condition:

$$\frac{l^2 C_{\alpha f} C_{\alpha r}}{u^2} - M\left(aC_{\alpha f} - bC_{\alpha r}\right) > 0 \tag{7.12}$$

By performing some algebraic operations, the stability condition can be reformulated as:

$$l + k_{us}u^2 \geq 0 \tag{7.13}$$

where k_{us} is known as the under-steer coefficient and defined as:

$$k_{us} = \frac{-M\left(aC_{\alpha f} - bC_{\alpha r}\right)}{lC_{\alpha f} C_{\alpha r}} \tag{7.14}$$

The stability of the vehicle can now be evaluated based on the sign of under-steer coefficient. Three distinct conditions can be considered:

- $k_{us} > 0$: if the under-steer coefficient is positive, the vehicle is known as an under-steer vehicle. In such a situation the vehicle stability condition is always satisfied.
- $k_{us} = 0$: a vehicle with zero under-steer coefficient is a neutral-steer vehicle. According to Equation 7.13, a neutral-steer vehicle is always stable.
- $k_{us} < 0$: a vehicle with negative under-steer coefficient is known as an over-steer vehicle. The stability of an over-steer vehicle is dependent on the vehicle longitudinal speed u. By defining the vehicle critical speed as:

$$u_{cr} = \sqrt{\frac{-l}{K_{us}}} \tag{7.15}$$

A vehicle is stable if $u < u_{cr}$ otherwise, the stability condition is not satisfied and the vehicle becomes unstable.

Example 7.2

Consider the sedan vehicle introduced in Example 7.1. The vehicle is converted to an electric vehicle after replacing the engine with an electric motor and installing a large battery pack in the trunk. Results of the electrification process include increasing both the mass and yaw moment of inertia of the vehicle by 25% and a 20 cm shift of the CG location toward the rear. Table 7.2 summarizes the vehicle parameters for the baseline and the electrified vehicles.

Calculate the under-steer coefficients and critical speeds (if applicable) for both vehicles and comment on the results of the electrification process in terms of stability. Redesign the rear tires of the electrified vehicle to achieve the same under-steer coefficient as the baseline vehicle. Simulate the performance of the vehicles (baseline, electrified and redesigned) for the input step steering angle of 0.03 rad and longitudinal speed of 70 km/hr. Comment on the results in terms of stability and steady state response.

Table 7.2 Parameters of the baseline and electrified vehicles

	Baseline	Electrified
M	1,530 kg	1,912.5 kg
a	1.320 m	1.520 m
b	1.456 m	1.256 m
I_z	4,192 kgm^2	5,240 kgm^2
$c_{\alpha f}$	70,000 N/rad	70,000 N/rad
$c_{\alpha r}$	70,000 N/rad	70,000 N/rad

Solution

For the baseline vehicle, the under-steer coefficient is:

$$k_{us} = \frac{-M\left(aC_{\alpha f} - bC_{\alpha r}\right)}{lC_{\alpha f}C_{\alpha r}} = \frac{-1530(1.320(2\times 70000) - 1.456(2\times 70000))}{2.776(2\times 70000)(2\times 70000)}$$

$$= 5.354\times 10^{-4}\ \frac{\text{rad}}{\text{m/s}^2} = 0.3009\ \frac{\text{deg}}{\text{g}}$$

and, therefore, the vehicle is both stable and under-steer.

For an electrified vehicle, the under-steer coefficient can be calculated as:

$$k'_{us} = \frac{-M'\left(a'C_{\alpha f} - b'C_{\alpha r}\right)}{lC_{\alpha f}C_{\alpha r}} = \frac{-1912.5(1.520(2\times 70000) - 1.256(2\times 70000))}{2.776(2\times 70000)(2\times 70000)}$$

$$= -0.0013\frac{rad}{\frac{m}{s^2}} = -0.7302\frac{\text{deg}}{\text{g}}$$

The negative under-steer coefficient shows that the vehicle is over-steer, indicating that it is less stable than the baseline vehicle. Let us calculate the critical speed:

$$u_{cr} = \sqrt{\frac{-l}{k'_{us}}} = \sqrt{\frac{-2.776}{-0.0013}} = 46.2254\frac{\text{m}}{\text{s}} = 166.41\ \frac{\text{km}}{\text{hr}}$$

In fact, the electrified vehicle is unstable for longitudinal speeds of higher than 166.41 km/hr. This analysis proves the electrification process should be reconsidered.

To overcome such difficulties, one possible solution is to modify the tires and retain the desirable performance of the baseline vehicle even after electrification. In this case, one solution is to keep the front tires the same and modify the rear ones. Using Equation 7.14 to solve for $C_{\alpha r}$ to achieve the same k_{us} as in the baseline vehicle gives:

$$C'_{\alpha f} = \frac{-M'a'C_{\alpha f}}{(lC_{\alpha f}k_{us} - M'b')} = \frac{-1912.5\times 1.520\times(2\times 70000)}{(2.776(2\times 70000)\times 5.354\times 10^{-4} - 1912.5\times 1.256} = 1.885\times 10^{5}\,\frac{\text{N}}{\text{rad}}$$

$$c'_{\alpha f} = \frac{C'_{\alpha f}}{2} = 92748\frac{\text{N}}{\text{rad}}$$

Therefore, a pair of tires with minimum cornering stiffness of 92, 748 N/rad can guarantee a similar performance as the baseline vehicle for the electric one.

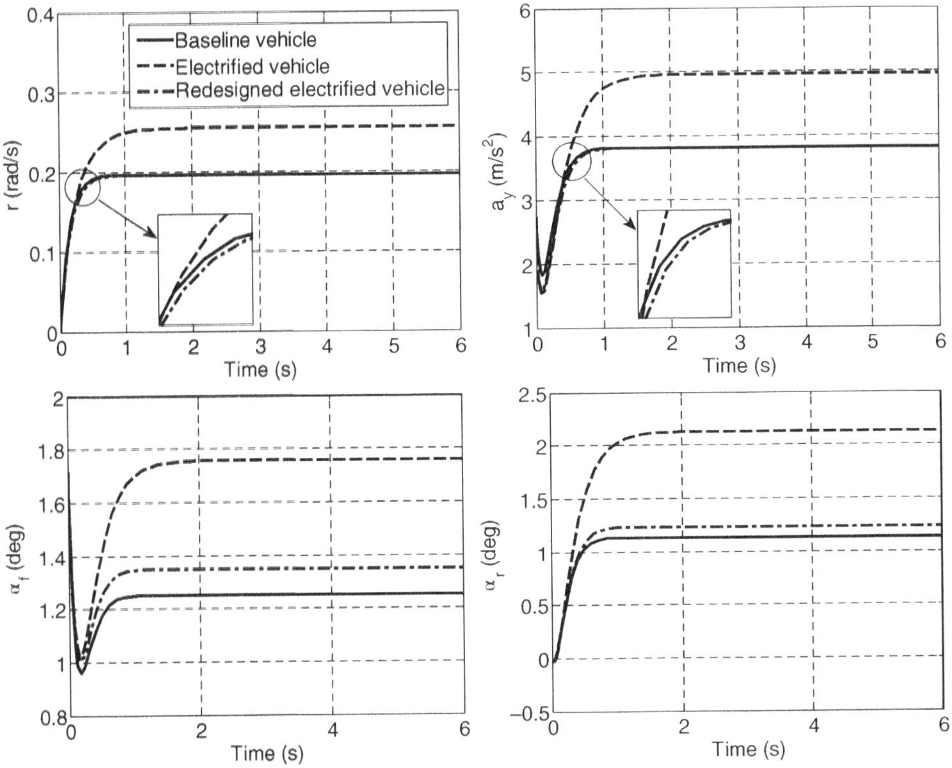

Figure 7.6 Effects of the electrification process on the handling performance

Simulation results compare the performance of the baseline, the electrified, and the redesigned vehicles as illustrated in Figure 7.6. The vehicle velocity is lower than the critical speed, however, the electrified vehicle shows a more sensitive response, with a higher yaw rate, higher lateral acceleration, and higher tire sideslip angles. After redesigning the rear tires, the vehicle retains its under-steer characteristic. However, the performances of the baseline and the redesigned electrified vehicles are not exactly the same, especially considering the sideslip angles in the steady state part of the maneuver.

7.2.2.2 Steady State Analysis

Steady state analysis is used to analytically study the steady state handling behavior of a vehicle. In the steady state, there is no change in the vehicle states and hence their derivatives become zero. Therefore, the steady state equations of the bicycle model using Equation 7.5 become:

$$AX_{ss} + B\delta = 0 \tag{7.16}$$

where the state vector X_{ss} contains the steady state responses of the vehicle and is defined as:

$$X_{ss} = \begin{pmatrix} v_{ss} \\ r_{ss} \end{pmatrix} \tag{7.17}$$

Equation 7.16 is no longer a differential equation and can be simply solved for X_{ss} as:

$$X_{ss} = -A^{-1}B\delta \tag{7.18}$$

After expanding Equation 7.18, the steady state form of the vehicle variables can be found as:

$$r_{ss} = G_r\delta \quad v_{ss} = G_v\delta \tag{7.19}$$

where, G_r and G_v are respectively known as the yaw velocity and lateral velocity gains, and found by:

$$G_r = \frac{u}{l + k_{us}u^2}, \quad G_v = \frac{u\left(b - \frac{Ma}{lC_{\alpha r}}u^2\right)}{l + k_{us}u^2} \tag{7.20}$$

The yaw velocity and the lateral velocity gains represent the sensitivity of the vehicle responses to steering input in a steady state condition. More sensitive and thus more responsive vehicles have larger gains. In general, the yaw velocity and lateral velocity gains are functions of the vehicle's parameters, such as the vehicle mass, inertia, and the tire cornering coefficients, in addition to the vehicle speed. For a vehicle with certain constant vehicle parameters, the gains can vary extensively, based on the vehicle speed.

Figure 7.7 depicts a typical variation of the yaw velocity gain with respect to vehicle speed for under-steer, neutral-steer, and over-steer vehicles. For low speeds, the yaw velocity gains for all types are small and their patterns are similar. By increasing the vehicle speed, the gain, and consequently, the steering sensitivity of vehicles increase, however, this occurs in different patterns. For the neutral steer vehicle, the yaw velocity gain always increases linearly. In the

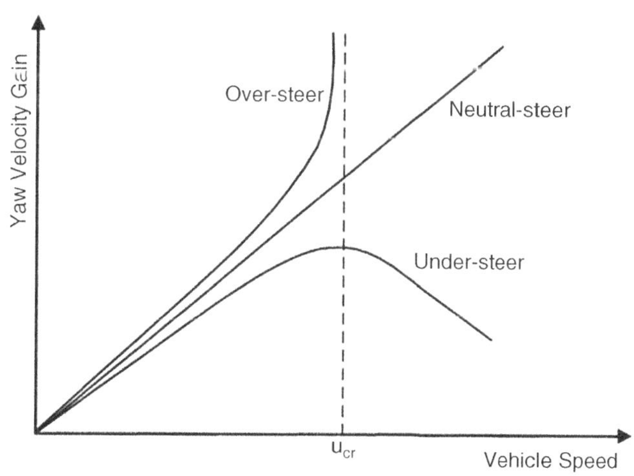

Figure 7.7 Typical variation of yaw velocity gain versus vehicle speed

case of the under-steer vehicle, the gain increases gradually until it reaches its maximum value, at which point the gain slightly decreases. This is the desired vehicle response from the handling and safety perspectives. For the over-steer vehicle, increasing the speed causes the gain to increase sharply, approaching infinity as the speed approaches the critical speed. This means that over-steer vehicles are unstable when near the critical speed.

Example 7.3

Consider the conventional and the electrified vehicles with their parameters, introduced in Table 7.2. As discussed in Example 7.2, the electrification process changes the inertial properties of the vehicle and its CG location. In Example 7.2, we redesigned the rear tires of the electric vehicle to have the same under-steer coefficient as the conventional one. However, the steady state response was still different after the redesign process (see Figure 7.6).

In this example, the objective is to achieve the same steady state response for the electrified vehicle as in the baseline. Calculate the yaw velocity and the lateral velocity gains for the baseline vehicle. Redesign both the front and rear tires using the under-steer coefficient, the yaw velocity gain, and the lateral velocity gain concepts. Simulate the baseline response, for electrified and redesigned vehicles for a step steer maneuver with $\delta = 0.03$ rad and $u = 70$ km/hr. Compare the yaw rate, the lateral velocity, the lateral acceleration, and the sideslip angle responses. Draw the yaw velocity and the lateral velocity gains for a range of vehicle speeds and comment on the results. Are the steady state responses the same for all vehicle speeds? Discuss the results.

Solution

For the baseline vehicle, the under-steer coefficient as calculated as in Example 7.2 is 5.354×10^{-4} rad.s^2/m. For the longitudinal speed of 70 km/hr, the yaw velocity gain and the lateral velocity gain for the baseline vehicle are:

$$G_r = \frac{u}{l + k_{us}u^2} = \frac{70/3.6}{2.776 + 5.354 \times 10^{-4} \times (70/3.6)^2} = 6.5284 \ \frac{1}{\text{s}}$$

$$G_v = \frac{u\left(b - \dfrac{Ma}{lC_{\alpha r}}u^2\right)}{l + k_{us}u^2} = \frac{(70/3.6)\left(1.456 - \dfrac{1530 \times 1.320}{2.776 \times (2 \times 70000)}(70/3.6)^2\right)}{2.776 + 5.354 \times 10^{-4} \times (70/3.6)^2}$$

$$= -3.3214 \ \frac{\text{m}}{\text{rad.s}}$$

In order to have the same steady state behavior after the electrification process, the yaw and lateral velocity gains of the electric vehicle should satisfy:

$$\begin{cases} G_r' = \dfrac{u}{l + k_{us}' u^2} = \dfrac{u}{l + \dfrac{-M'(a'C_{\alpha f}' - b'C_{\alpha r}')}{lC_{\alpha f}' C_{\alpha r}'} u^2} = 6.5284 \dfrac{1}{s} \\ G_v' = \dfrac{u(b' - \dfrac{M'a'}{lC_{\alpha r}'} u^2)}{l + k_{us}' u^2} = \dfrac{u(b' - \dfrac{M'a'}{lC_{\alpha r}'} u^2)}{l + \dfrac{-M'(a'C_{\alpha f}' - b'C_{\alpha r}'}{lC_{\alpha f}' C_{\alpha r}'} u^2} = -3.3214 \dfrac{\text{m}}{\text{rad.s}} \end{cases}$$

Solving this system of equations, the redesigned cornering stiffness values for the front and rear tires are:

$$c_{\alpha f}' = 83154.75 \text{ N/rad}$$

$$c_{\alpha r}' = 112176.43 \text{ N/rad}$$

Simulation results comparing the performance of the baseline, electrified and redesigned vehicles are illustrated in Figure 7.8.

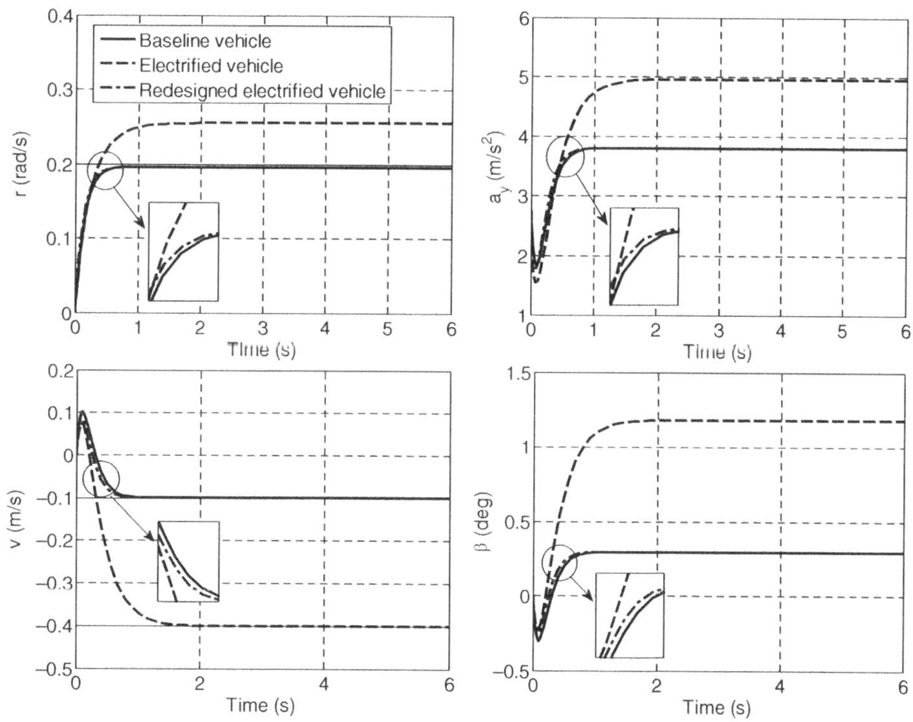

Figure 7.8 Performance comparison for the baseline, electrified and redesigned vehicles

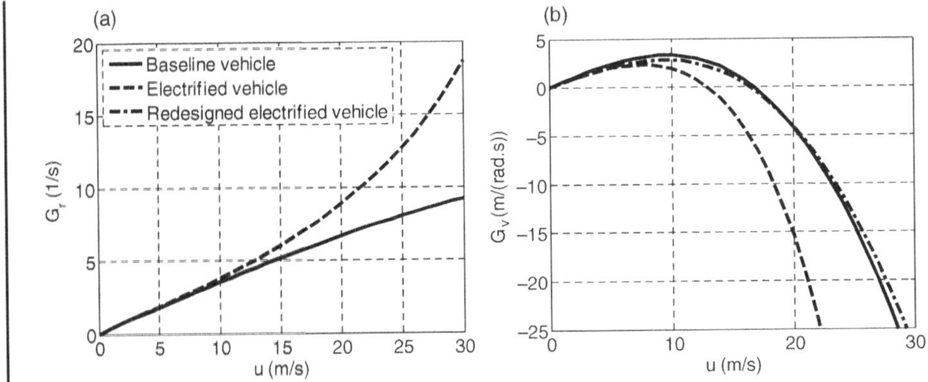

Figure 7.9 Yaw velocity gain (a) and lateral velocity gain (b) for the baseline, electrified and redesigned vehicles

Simulation results verify that the redesigned electric vehicle achieves the same steady state performance as of the baseline vehicle. Note that such performance is only guaranteed for the certain longitudinal speeds considered in the design process. For further clarification, Figure 7.9 plots the yaw and lateral velocity gains for the vehicles versus the longitudinal speed.

From Equation 7.20 and as seen in Figure 7.9, it is clear that the yaw velocity gain for the redesigned and the baseline vehicles should be exactly the same for all longitudinal speeds. However, since the lateral velocity gain is a function of the vehicle speed, the curves only cross each other at the longitudinal speed of zero and 19.44 m/s$\left(70\,\frac{\text{km}}{\text{hr}}\right)$ where the lateral velocity gains were solved to be equal. Using only the tire cornering stiffness values as the design parameters, one cannot have the exact same lateral velocity gains for the vehicles in all longitudinal speeds. However, as seen in Figure 7.9, the results are very close and acceptable.

7.2.2.3 Transient Response Analysis

The quality of the transient response of a vehicle to the steering input affects the vehicle handling performance. To study the vehicle transient response, it is beneficial to first introduce the concepts of the system's damping ratio and natural frequency.

The general form of the characteristic equation of a linear second-order system is:

$$S^2 + 2\xi\omega_n S + \omega_n^2 = 0 \qquad (7.21)$$

where ξ and ω_n are the system's damping ratio and natural frequency, respectively. The value of these parameters significantly affects the system's transient response. Generally, the value of ξ helps classify second-order systems to:

- *Over-damped systems* ($\xi \geq 1$): The system response of the step input is slow and does not have over-shoot.

- *Under-damped system* ($0 < \xi < 1$): The step input response of the system is fast, but it has over-shoot and shows some damped oscillations before converging to its steady state response. The over-shoot and the rate of convergence are influenced by the value of the damping ratio. Generally, a smaller ξ increases the over-shoot, the amplitude of the oscillations and the response's settling time. The system's natural frequency ω_n determines the frequency of the oscillation. A higher value for ω_n decreases the system's response time.
- *Undamped system* ($\xi = 0$): The step input response is undamped with steady state oscillations.

After reviewing the classification of second-order systems, now we are ready to study the transient response of the single track linear handling model. Using the general form of the characteristic equation of a second-order system in Equation 7.21 and the characteristics equation of the single track linear handling model, Equation 7.11, the damping ratio and natural frequency of the vehicle can be found as:

$$\omega_n = \frac{1}{u}\sqrt{\frac{lC_{\alpha f}C_{\alpha r}}{MI_z}}\sqrt{l + k_{us}u^2} \tag{7.22}$$

$$\zeta = \frac{1}{2}\left(\frac{C_{\alpha f} + C_{\alpha r}}{M} + \frac{a^2 C_{\alpha f} + b^2 C_{\alpha r}}{I_z}\right)\frac{1}{\sqrt{l + k_{us}u^2}}\sqrt{\frac{MI_z}{lC_{\alpha f}C_{\alpha r}}} \tag{7.23}$$

As can be seen in Equations 7.22 and 7.23, the damping ratio and natural frequency of the single track handling model are functions of the vehicle's parameters, as well as the vehicle's speed. To achieve the desirable transient response (fast with small over-shoot and high rate of convergence) in a wide range of vehicle speeds, the vehicle's parameters, such as mass, inertia, and tire cornering coefficient should be carefully selected. The following example can help clarify the effects of the vehicle's parameter on the vehicle's transient response.

Example 7.4

Consider the conventional, the electrified, and the redesigned electric vehicles described in the above examples. In order to investigate the effects of the electrification process on the transient response, calculate the natural frequency and damping ratio of the vehicles for the longitudinal speed of 70 km/hr. Are the vehicles over-damped, under-damped, or undamped? Plot the natural frequency and the damping ratio curves as a function of longitudinal speed for each of the vehicles and discuss the results.

Solution

For the conventional vehicle and a longitudinal speed of 70 km/hr, the natural frequency and damping ratio can be calculated as:

$$\omega_n = \frac{1}{u}\sqrt{\frac{lC_{\alpha f}C_{\alpha r}}{MI_z}}\sqrt{l + k_{us}u^2}$$

$$= \frac{1}{\left(\frac{70}{3.6}\right)}\sqrt{\frac{2.776(2\times 70000)(2\times 70000)}{1530\times 4192}}\sqrt{2.776 + 5.354\times 10^{-4}\times (70/3.6)^2} = 8.1748 \text{ rad/sec}$$

$$\zeta = \frac{1}{2}\left(\frac{C_{\alpha f} + C_{\alpha r}}{M} + \frac{a^2C_{\alpha f} + b^2C_{\alpha r}}{I_z}\right)\sqrt{\frac{MI_z}{lC_{\alpha f}C_{\alpha r}}}\frac{1}{\sqrt{l + k_{us}u^2}}$$

$$= \frac{1}{2}\left(\frac{(2\times 70000) + (2\times 70000)}{1530} + \frac{1.320^2(2\times 70000) + 1.456^2(2\times 70000)}{4192}\right)\times$$

$$\sqrt{\frac{1530\times 4192}{2.776(2\times 70000)(2\times 70000)}}\frac{1}{\sqrt{2.776 + 5.354\times 10^{-4}\times (70/3.6)^2}} = 0.9814$$

The calculated damping ratio indicates that the vehicle is under-damped.

Using the same calculations, the natural frequency and damping ratio of the electrified vehicle are:

$$\omega_n = 5.7280 \ rad/sec$$

$$\zeta = 1.1236$$

The damping ratio is more than 1, so the vehicle is over-damped.

For the redesigned vehicle, the transient response characteristics are:

$$\omega_n = 9.0233 \ rad/sec$$

$$\zeta = 0.9836$$

indicating that the redesigned vehicle is under-damped.

Figure 7.10 depicts the damping ratio and natural frequency of the baseline, electrified, and redesigned vehicles for different longitudinal speeds. The results indicate:

- The baseline and redesigned vehicles are over-damped for low velocities and under-damped for high velocities.
- The electrified vehicle is always over-damped.
- The vehicles have lower natural frequencies at higher velocities.

One can verify that for the electrified vehicle, the damping ratio goes to infinity at the speed of 46.2254 m/s. This speed is the critical speed as calculated in Example 7.2. For this speed, the term $l + k_{us}u^2$ in the denominator of the damping ratio equation is zero.

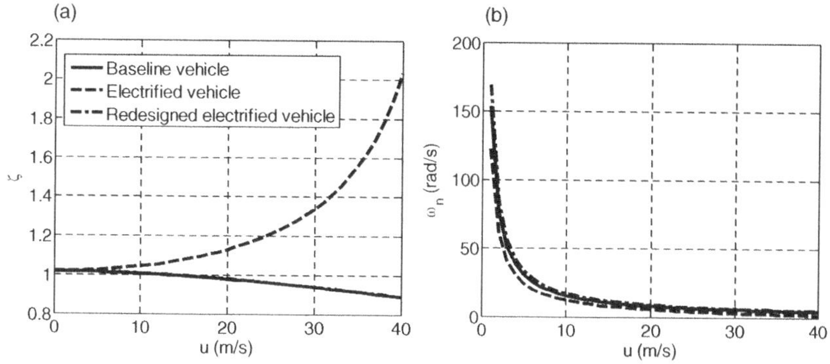

Figure 7.10 Damping ratio (a) and natural frequency (b) for different speeds of the baseline, electrified, and redesigned vehicles

7.2.3 Roll and Pitch Dynamics Models

A vehicle's roll and pitch motions have a direct influence on the vehicle's handling and stability, especially during severe maneuvers. As discussed in Chapter 4, to study the roll and pitch dynamics, one can use decoupled single degree of freedom models or can integrate them with lateral, longitudinal, and yaw dynamics to form comprehensive models. Chapter 4 introduced and discussed both approaches. Examples 7.5 and 7.6 show how simple roll and pitch plane models can work in vehicle dynamics analyses, specifically for the design of electric and hybrid electric vehicles.

Example 7.5

This example focuses on a comparison between the roll dynamics of conventional and electrified vehicles while cornering. This case study is for a conventional mid-size sedan vehicle with parameters listed in Table 7.3.

After the vehicle electrification, the mass, inertia, and center of gravity location change to the parameters listed in Table 7.4.

Assume that the vehicle's lateral acceleration is 0.4g during cornering. Investigate the roll response of the conventional and electrified vehicle and discuss the differences. Redesign the suspension to improve the roll behavior of the electrified vehicle.

Solution

To study the roll motion of the vehicles, we use the roll plane model introduced in Equation 4.14. To use this model, the roll stiffness and damping need to be calculated:

Table 7.3 Vehicle parameters

Parameter	Value
M_s	1,370 kg
I_{xx}	606 kg m^2
h'	0.50 m
T_r	1.550 m
T_f	1.550 m
k_f	15,300 N/m
k_r	18,360 N/m
$k_{ARB,f}$	68,716 N/m
c_f	2,868 Ns/m
c_r	2,868 Ns/m

Table 7.4 Changed parameters after electrification

Parameter	Value
M_s	1712.5 kg
I_{xx}	757.5 kgm^2
h'	0.48 m

$$K_\phi = \frac{1}{2}k_f T_f^2 + \frac{1}{2}k_r T_r^2 + k_{ARB,f} = \frac{1}{2}\left(15300 \times 1.55^2 + 18360 \times 1.55^2\right) + 68716$$
$$= 109150 \text{ N.m/rad}$$

$$C_\phi = \frac{1}{2}\, c_f T_f^2 + \frac{1}{2} c_r T_r^2 = \frac{1}{2}\left(2868 \times 1.55^2 + 2868 \times 1.55^2\right) = 6453.1 \text{ N.m.s/rad}$$

By substituting the above parameters as well as vehicle parameters in Tables 7.3 and 7.4 into roll plan model, Equation 4.14, and solving the resultant differential equation by using either numerical or analytical methods, it is possible to find the roll response of the vehicles.

Figure 7.11 illustrates the roll angle response of the conventional and electrified vehicles. After electrification, due to increased mass and inertia, the steady state roll angle of the vehicle is larger than that of the conventional one. Additionally, in the transient part, the oscillation of the vehicle body has a greater amplitude.

To resolve the differences between conventional and electrified vehicles, we first investigate the steady state response of the vehicles. By substituting all derivatives terms in Equation 4.14 with zero, the general form of roll steady state response becomes:

$$\phi_{ss} = \frac{-M_s a_y h'}{K_\phi - m_s g h'} \tag{7.24}$$

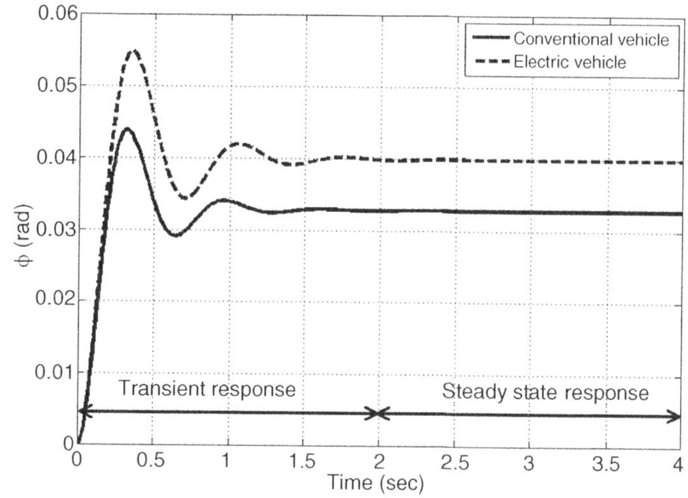

Figure 7.11 Roll angles of the conventional and electrified vehicles

Roll stiffnessK_{ϕ}, in Equation 7.24, depends on the suspension spring's stiffness and the anti-roll bar's torsional stiffness. Since the spring stiffness has a significant effect on the ride quality, we only consider the redesign of the anti-roll bars in this example. Following this approach, let us assume a 30% increase in the front anti-roll bar stiffness. Recalculating K_{ϕ} and solving Equation 4.14, the roll angle of the electrified vehicle can be obtained. Figure 7.12 shows the roll response of the electrified vehicle with the stiffer front anti-roll bar. As seen Figure 7.12, the response is much closer to the conventional car.

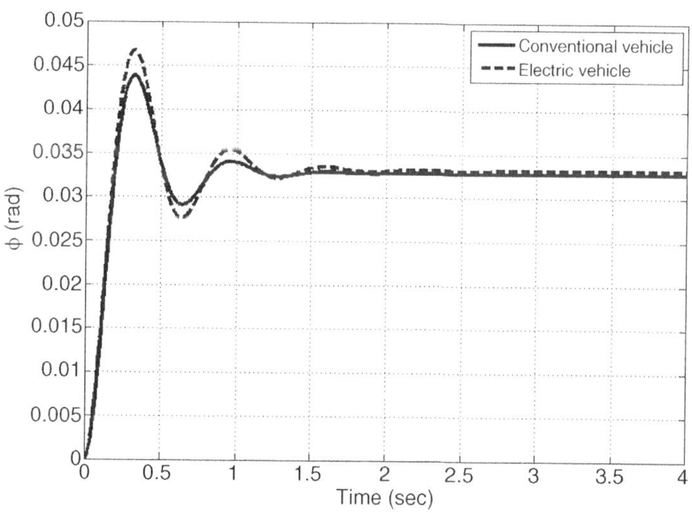

Figure 7.12 Roll angles of the conventional and electrified vehicles after redesigning the anti-roll bar

To calculate the natural frequency and damping coefficient of the vehicles, the roll dynamics equation can be written as:

$$\ddot{\phi}+\frac{C_\phi}{I_{xx}+M_sh'^2}\dot{\phi}+\frac{K_\phi-M_sgh'}{I_{xx}+M_sh'^2}\phi=\frac{M_sa_yh'}{I_{xx}+M_sh'^2} \tag{7.25}$$

According to Equation 7.21, the damping coefficient and natural frequency of vehicle roll dynamics using Equation 7.25 can be found as:

$$\omega_n=\sqrt{\frac{K_\phi-m_sgh'}{I_{xx}+M_sh'^2}} \tag{7.26}$$

$$\xi=\frac{C_\phi}{2\omega_n(I_{xx}+M_sh'^2)} \tag{7.27}$$

As previously discussed, the transient response of the system is under the influence of these two parameters. For the conventional car, the natural frequency and damping ratio are:

$$\omega_{n,conv}=\sqrt{\frac{K_\phi-m_sgh'}{I_{xx}+m_sh'^2}}=\sqrt{\frac{109150-1370\times 9.81\times 0.5}{606+1370\times 0.5^2}}=10.3919\text{ rad/s}$$

$$\xi=\frac{C_\phi}{2\omega_n(I_{xx}+m_sh'^2)}=\frac{6453.1}{2\times 10.3919\times\left(606+1370\times 0.5^2\right)}=0.3273$$

and the corresponding natural frequency and damping ratio for the redesigned electrified vehicle with the new anti-roll bar are 10.27 rad/s and 0.2987. Therefore, the damping ratio of the electrified vehicle is about 10% less than the conventional one. This explains the larger roll angles in the transient response of the electrified vehicle in Figure 7.12. According to Equation 7.27, the only practical way to increase the roll damping ratio ξ is to increase the roll damping C_ϕ and consequently change the suspension system's shock absorber damping. Consequently, if the damping coefficients of the electric vehicle increase by 10%, then the amplitude of oscillations in transient part will be reduced. Figure 7.13 demonstrates the results of these changes. However, the effects of increasing the shock absorber damping on the vehicle ride comfort should be studied.

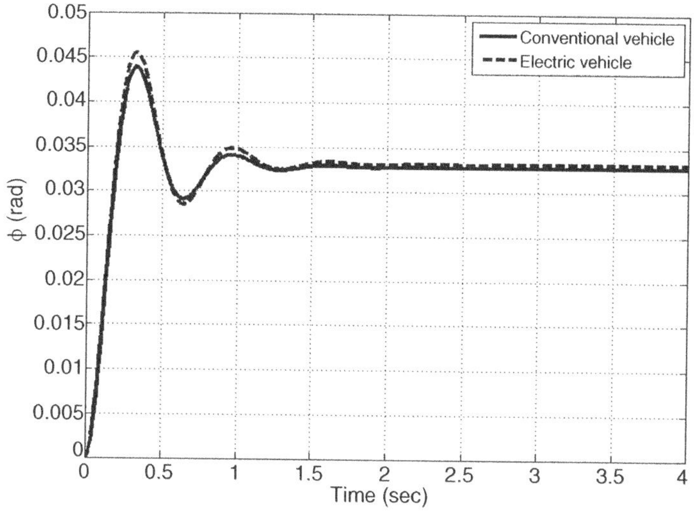

Figure 7.13 Roll angle of the vehicle body after changing the damping coefficient

Example 7.6

This example compares the pitch response of a conventional mid-size sedan vehicle and its electrified version to illustrate the effect of electrification on the pitch dynamics of vehicles. Table 7.5 shows the parameters of the conventional vehicle and Table 7.6 tabulates the electrified vehicle parameters.

Assume the vehicle moves in a straight line with longitudinal acceleration of 0.5 g, calculate the pitch response of the conventional and electrified vehicles and discuss the results.

Table 7.5 Pitch dynamics parameters of the proposed vehicle

Parameter	Value
M_s	1,370 kg
I_{yy}	4,192 kgm^2
h''	0.50 m
L_r	1.660 m
L_f	1.110 m
k_f	15,300 N/m
k_r	1,8360 N/m
c_f	2,868 Ns/m
c_r	2,868 Ns/m

Table 7.6 Changed parameters of the electrified vehicle

Parameter	Value
M_s	1,712.5 kg
I_{yy}	5,240 kgm^2
h''	0.47 m
L_r	1.4662 m
L_f	1.310 m

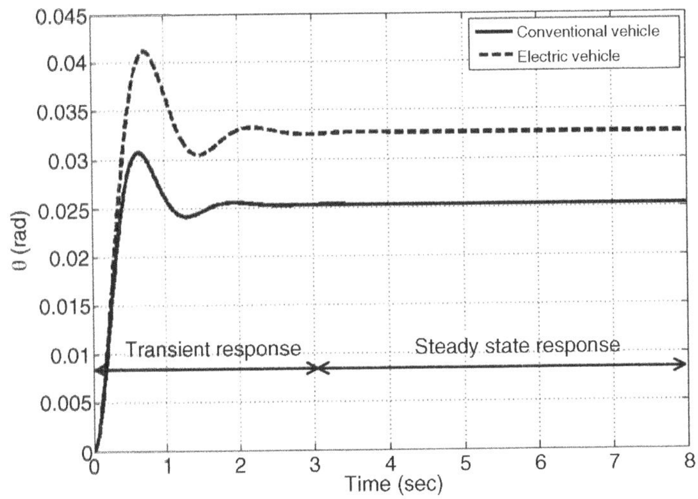

Figure 7.14 Pitch response of the conventional and electrified vehicles

Solution

Consider the pitch plan model expressed by Equation 4.21 and substitute the above-mentioned parameters into it. By solving the resultant differential equation, it is possible to find the pitch response of the conventional and electrified vehicles. Figure 7.14 shows the pitch angle of the vehicle's body during this maneuver for the conventional and electrified vehicles. The pitch angle of the electrified vehicle is larger than that of the conventional vehicle. The pitch angle behavior of the vehicle body needs to be improved by redesigning the suspension system with the same approach discussed in Example 7.5. Again, it must be noted that the effects of suspension redesign on the vehicle ride comfort should be investigated.

7.3 Comprehensive Handling Model of EVs and HEVs

Decoupled handling models such as bicycle, roll and pitch plane models are suitable for linear handling analyses and also for analytical studies, such as investigation of the effects of changes in mass and inertia on the handling and stability of vehicles. The previous section studied the effects of electrification on the vehicle dynamics behavior of electric and hybrid electric

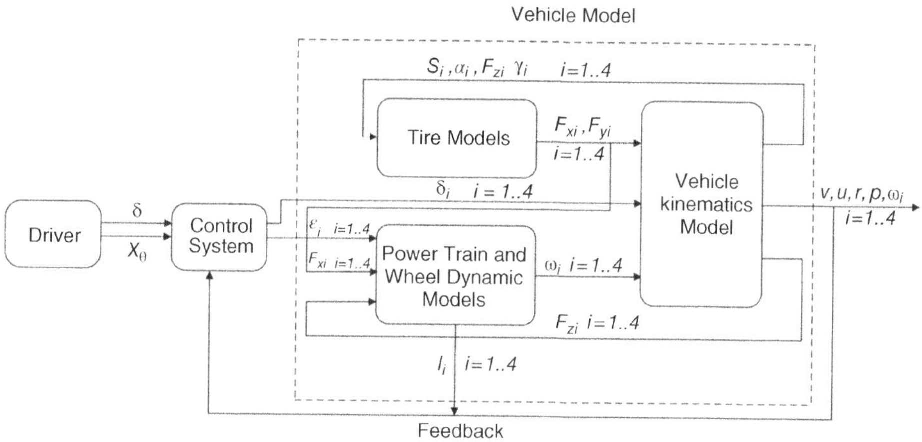

Figure 7.15 EVs and HEVs handling model block diagram [3]

vehicles using such simple models. However, these decoupled models are generally not accurate enough to simulate the dynamic behavior of vehicles in more severe maneuvers.

In this section, we develop a comprehensive handling model for electric and hybrid electric vehicles. This model that is shown in Figure 7.15 is modular and has three main parts: (1) the vehicle kinetics model; (2) the tire models; and (3) the powertrain and the wheel dynamic models [3]. In the following sections, each part of the model and its corresponding variables are first discussed in detail and then the overall model is studied through examples.

7.3.1 Vehicle Kinetics Model

The vehicle is considered as a rigid body in order to derive the equations of motion. As discussed in Chapter 4, four degrees of freedom are considered for the vehicle motion: longitudinal motion (along the x-axis), lateral motion (along the y-axis), yaw motion (rotation about the z-axis), and the roll motion (rotation about the x-axis). Figure 7.16 illustrates the

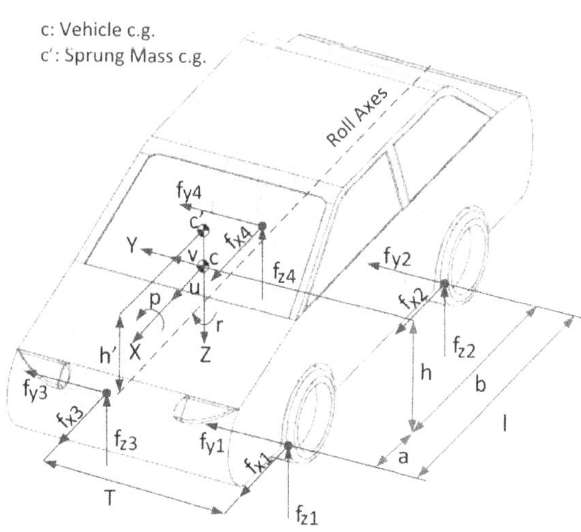

Figure 7.16 The handling model configuration

variables used in these equations. The state variables of the models are: longitudinal velocity u, the lateral velocity v, the yaw rate r, the roll rate p and the wheels rotational speed ω_{wi} which leads to a vehicle model with eight degrees of freedom.

Now, based on the three-dimensional vehicle kinematics model described in Chapter 4, Equation 4.29, the longitudinal, lateral, yaw, and roll equations of motion respectively are found as:

$$M(\dot{u} - vr) = \sum_{i=1}^{4} f_{xi} - \frac{1}{2} C_d A_f \rho_a u^2 - Mg\sin\xi - Mg\mu_R$$

$$M(\dot{v} + ur) + M_s h' \dot{p} = \sum_{i=1}^{4} f_{yi} \qquad (7\text{-}28)$$

$$I_{zz}\dot{r} = a(f_{y1} + f_{y3}) - b(f_{y2} + f_{y4}) + \frac{T}{2}(f_{x1} + f_{x2}) - \frac{T}{2}(f_{x3} + f_{x4})$$

$$I_{xxs}\dot{p} + K_\phi \phi + C_\phi p - M_s g h' \sin\phi - M_s h'(\dot{v} + ur) = 0$$

The terms f_{xi} and f_{yi} are the respective tire forces in the x and y direction as seen in Figure 7.17, that are expressed as a function of the longitudinal and lateral tire forces by:

$$f_{x_i} = F_{x_i}\cos(\delta_{Ti}) - F_{y_i}\sin(\delta_{Ti}) \quad i = 1\ldots4$$

$$f_{y_i} = F_{x_i}\sin(\delta_{Ti}) + F_{y_i}\cos(\delta_{Ti}) \quad i = 1\ldots4 \qquad (7.29)$$

where δ_T is the total steering angle, which is the sum of both driver steering angle δ and the steering angle, due to the roll motion or:

$$\delta_{Ti} = \delta_i + K_{ri}\phi \quad i = 1,4 \qquad (7.30)$$

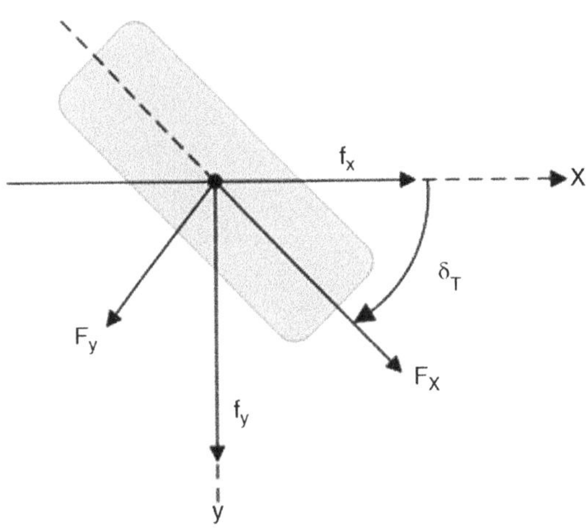

Figure 7.17 The wheel forces

Equation 7.30 is for a four-wheel steering vehicle in which all the wheels are steerable. For front steering cars, only the front wheels are steerable by the driver and the rear wheels' steering angles are due to the roll motion:

$$\delta_{Ti} = \begin{cases} \delta_i + K_{ri}\phi & i = 1, 3 \\ K_{ri}\phi & i = 2, 4 \end{cases} \tag{7.31}$$

The steer-by-roll coefficient of each wheel is determined by the vehicle suspension design.

Normal loads on wheels are used in the modeling process of tires. Using quasi-static lateral and longitudinal load transfer, the normal load expression for each wheel can be written as:

$$\begin{aligned}
F_{z1} &= \frac{W}{2}\left[\frac{b}{l} - \frac{a_x}{g}\left(\frac{h}{l}\right) + K_R\left[\frac{a_{y_s}}{g}\left(\frac{h}{T}\right) - \left(\frac{M_s}{M}\right)\left(\frac{h'}{T}\right)\sin\phi\right]\right] \\
F_{z2} &= \frac{W}{2}\left[\frac{a}{l} + \frac{a_x}{g}\left(\frac{h}{l}\right) + (1 - K_R)\left[\frac{a_{y_s}}{g}\left(\frac{h}{T}\right) - \left(\frac{M_s}{M}\right)\left(\frac{h'}{T}\right)\sin\phi\right]\right] \\
F_{z3} &= \frac{W}{2}\left[\frac{b}{l} - \frac{a_x}{g}\left(\frac{h}{l}\right) - K_R\left[\frac{a_{y_s}}{g}\left(\frac{h}{T}\right) - \left(\frac{M_s}{M}\right)\left(\frac{h'}{T}\right)\sin\phi\right]\right] \\
F_{z4} &= \frac{W}{2}\left[\frac{a}{l} + \frac{a_x}{g}\left(\frac{h}{l}\right) - (1 - K_R)\left[\frac{a_{y_s}}{g}\left(\frac{h}{T}\right) - \left(\frac{M_s}{M}\right)\left(\frac{h'}{T}\right)\sin\phi\right]\right]
\end{aligned} \tag{7-32}$$

where a_{ys} represents the lateral acceleration of the sprung mass, and can be defined by:

$$a_{ys} = \dot{v} + ur + \frac{M_s}{M}h'\dot{p} \tag{7.33}$$

In Equation 7.33, K_R is the front roll stiffness ratio that determines the front/rear distribution of the total lateral load transfer and is calculated by:

$$K_R = \frac{\frac{1}{2}k_s T_{s,f} + k_{ARB,f}}{\frac{1}{2}k_s T_{s,f} + k_{ARB,f} + \frac{1}{2}k_s T_{s,r} + k_{ARB,r}} \tag{7.34}$$

As well, the vehicle governing kinematics equations can be used to find the vehicle path:

$$\begin{aligned} \dot{X} &= u\cos(\psi) - v\sin(\psi) \\ \dot{Y} &= u\sin(\psi) + v\cos(\psi) \end{aligned} \tag{7.35}$$

where ψ is the vehicle yaw angle. The final relationship to consider is the changes in the camber angle of the wheel due to the vehicle roll that is:

$$\gamma_i = K_{\gamma i}\phi + \gamma_{0i} \quad i = 1 \ldots 4 \tag{7.36}$$

where $K_{\gamma i}$ is the camber-by-roll coefficient of each wheel.

7.3.2 The Tire Model

As discussed in Chapter 4, a tire model typically determines the tire longitudinal force F_x and later force, F_y. The linear tire model, used for the single track linear handling model, only considers the lateral force as a linear function of the slip angle and ignores the effects of the longitudinal force. This simplification is only valid for low-g lateral accelerations and the model is not capable of simulating the effect of lateral force saturation in critical conditions [4]. In cases in which both longitudinal and lateral tire forces are applied simultaneously (combined slip condition), we need to incorporate the influence of the longitudinal slip of the tire forces and examine their effects on the tires' side forces.

As discussed in Chapter 4, a more comprehensive model, such as the Magic Formula Pacejka tire model, can present the longitudinal and lateral tire forces as a function of the slip angle α, longitudinal slip s, camber angle γ, and the tire normal force F_z. This nonlinear model is valid over the wide range of the vehicle handling maneuvers, up to the critical condition, and is able to calculate lateral and longitudinal forces of the tire in both the linear and nonlinear regions of the tire working zone. A drawback is that this model is often difficult to use. Chapter 4 presents the formulation of this model.

The normal force and camber angle of the tires are inputs in the tire model, calculated using Equations 7.32 and 7.36. The other input variables, the slip angle of each wheel, is a function of the kinematic variables of the vehicle, written as:

$$\begin{aligned}
\alpha_1 &= \delta_{T1} - \tan^{-1}\left(\frac{v+ar}{u+\frac{1}{2}Tr}\right) \\
\alpha_2 &= \delta_{T2} + \tan^{-1}\left(\frac{br-v}{u+\frac{1}{2}Tr}\right) \\
\alpha_3 &= \delta_{T3} - \tan^{-1}\left(\frac{v+ar}{u-\frac{1}{2}Tr}\right) \\
\alpha_4 &= \delta_{T4} + \tan^{-1}\left(\frac{br-v}{u-\frac{1}{2}Tr}\right)
\end{aligned} \tag{7.37}$$

Moreover, the longitudinal slip can be defined as:

$$s_i = \begin{cases} 1 - \dfrac{u_{ti}}{R\omega_{wi}} & i = 1\ldots4 \quad \text{Accelerating} \\[2ex] -1 + \dfrac{R\omega_{wi}}{u_{ti}} & i = 1\ldots4 \quad \text{Braking} \end{cases} \tag{7.38}$$

where u_{ti} is the *ith* wheel center speed in the direction of the tire heading, written for each wheel as:

$$\begin{aligned} u_{t1} &= \left(u + \frac{1}{2}Tr\right)\cos(\delta_{T1}) + (v + ar)\sin(\delta_{T1}) \\ u_{t2} &= \left(u + \frac{1}{2}Tr\right)\cos(\delta_{T2}) + (v - br)\sin(\delta_{T2}) \\ u_{t3} &= \left(u - \frac{1}{2}Tr\right)\cos(\delta_{T3}) + (v + ar)\sin(\delta_{T3}) \\ u_{t4} &= \left(u - \frac{1}{2}Tr\right)\cos(\delta_{T4}) + (v - br)\sin(\delta_{T4}) \end{aligned} \tag{7.39}$$

7.3.3 Powertrain and Wheel Dynamics Model

Figure 6.14 depicts a wheel with applied driving and braking torques. Based on the labels in Figure 6.14, the general form of the wheel torque equations can be written as:

$$I_{wi}\frac{d\omega_{wi}}{dt} = T_{wi} + T_{bi} - F_{xi}R \quad i = 1\ldots4 \tag{7.40}$$

where T_w and T_b respectively are the driving and braking torques of each wheel, I_w is the wheel rotational inertia, ω_w is the wheel angular velocity, R is the wheel radius, and F_x is the wheel longitudinal force.

The friction brake system generates the braking torque of each wheel T_b while the powertrain system produces the wheels' driving torque T_w. In Equation 7.40, the sign of T_b is always negative due to the nature of the braking torque. The driving torque T_w is usually a positive variable, but can be negative when the powertrain system is in regenerative braking mode and acts similar to a frictional braking torque in order to reduce the vehicle speed.

The formulation of the driving torque of each wheel depends on the powertrain configuration and is different for each distinct type of electric and hybrid electric vehicle. In this section, we develop and discuss the wheels' dynamics of the most common electric and hybrid electric powertrain systems. Chapter 6 already introduced and discussed these powertrain configurations.

7.3.3.1 Two-Wheel Drive Electric Vehicle

As shown in Figure 6.8, the powertrain of a two-wheel drive electric vehicle consists of an electric motor, gearbox, and a differential unit. This is a very common configuration for electric vehicles. Generally, the driving wheels can be either the front or rear wheels; however, the front wheel drive configuration is more common. The electric motor generates the driving torque that is transmitted through the gearbox and differential to the driving wheels. As discussed in

Chapter 6, the driving torque of each wheel for a two-wheel drive electric vehicle can be represented as:

$$T_{wi} = \begin{cases} \frac{1}{2}\left(\eta_g \eta_d \beta_g \beta_d T_m(\omega_m, x_{\theta m}) - (\eta_g \eta_d \beta_g^2 \beta_d^2 I_m + \eta_d \beta_d^2 I_g + I_d)\frac{d\omega_{wi}}{dt}\right) & i = 1,3 \\ 0 & i = 2,4 \end{cases} \tag{7.41}$$

Consequently, the dynamic equations of the wheels can be written as:

$$\begin{cases} I_{eq}\frac{d\omega_{wi}}{dt} = \frac{1}{2}\eta_g \eta_d \beta_g \beta_d T_m(\omega_m, x_{\theta m}) + T_{bi} - F_{xi}R & i = 1,3 \\ I_w \frac{d\omega_{wi}}{dt} = T_{bi} - F_{xi}R & i = 2,4 \end{cases} \tag{7.42}$$

where the wheel equivalent inertia of each wheel is:

$$I_{eq} = I_w + \frac{1}{2}\left(\eta_g \eta_d \beta_g^2 \beta_d^2 I_m + \eta_d \beta_d^2 I_g + I_d\right) \tag{7.43}$$

Moreover, according to the open differential characteristics and simple kinematic rules, the electric motor speed can be found based on the angular velocity of the driving wheels:

$$\omega_m = \frac{1}{2}\beta_g \beta_d(\omega_{w1} + \omega_{w3}) \tag{7.44}$$

7.3.3.2 A Four Motorized Wheel Electric Vehicle

Figure 6.10 illustrates this configuration, in which each wheel is driven by its own electric motor through a fixed reduction gear. In the case of in-wheel-drive vehicles, the reduction gear is eliminated and the wheel is directly driven by an in-wheel-motor. Following the approach discussed in Chapter 6, the driving torque of each wheel can be found as:

$$T_{wi} = \eta_g \beta_g T_{mi}(\omega_{mi}, x_{\theta mi}) - \left(\eta_g \beta_g^2 I_m + I_g\right)\frac{d\omega_{wi}}{dt} \quad i = 1 \ldots 4 \tag{7.45}$$

The wheel dynamic equations can be represented as:

$$I_{eq}\frac{d\omega_{wi}}{dt} = \eta_g \beta_g T_{mi}(\omega_{mi}, x_{\theta mi}) + T_{bi} - F_{xi}R \quad i = 1 \ldots 4 \tag{7.46}$$

where:

$$I_{eq_i} = I_w + I_g + \eta_g \beta_g^2 I_m \tag{7.47}$$

Also the speed of each electric motor is:

$$\omega_{mi} = \beta_g \omega_{wi} \quad i = 1 \ldots 4 \tag{7.48}$$

For an in-wheel-drive configuration without a reduction gear, the wheel driving torques, the wheel dynamic equations, and the motor speed equations, can be simplified to the following forms:

$$\begin{gathered} T_{wi} = T_{mi}(\omega_{mi}, x_{\theta mi}) - I_m \frac{d\omega_{wi}}{dt} \\ \left(I_m + I_g\right) \frac{d\omega_{wi}}{dt} = T_{mi}(\omega_{mi}, x_{\theta mi}) + T_{bi} - F_{xi} R \quad i = 1 \ldots 4 \\ \omega_{mi} = \omega_{wi} \end{gathered} \tag{7.49}$$

7.3.3.3 Four-Wheel Drive Hybrid Electric Vehicle

In this configuration, an internal combustion engine drives the front wheels through a transmission system and electric motors drive each individual rear wheel. Following the same approach as the other configurations, the following equations represent the driving torque and dynamic equations:

$$T_{wi} = \begin{cases} \dfrac{1}{2}\left[\eta_g \eta_d \beta_g \beta_a T_e(\omega_e, x_{\theta e}) - \left(\eta_g \eta_d \beta_g^2 \beta_d^2 I_e + \eta_d \beta_d^2 I_g + I_d\right) \dfrac{d\omega_{wi}}{dt}\right] & i = 1, 3 \\ \eta_g \beta_g T_{mi}(\omega_{mi}, x_{\theta mi}) - \left(\eta_g \beta_g^2 I_m + I_g\right) \dfrac{d\omega_{wi}}{dt} & i = 2, 4 \end{cases} \tag{7.50}$$

$$\begin{cases} I_{eq} \dfrac{d\omega_{wi}}{dt} = \dfrac{1}{2} \eta_g \eta_d \beta_g \beta_d T_e(\omega_e, x_{\theta e}) + T_{bi} - F_{xi} R \quad i = 1, 3 \\ I_{eq} = I_w + \dfrac{1}{2}(\eta_g \eta_d \beta_g^2 \beta_d^2 I_m + \eta_d \beta_d^2 I_g + I_d) \\ \omega_m = \dfrac{1}{2} \beta_g \beta_d (\omega_{w1} + \omega_{w3}) \end{cases} \tag{7.51}$$

$$\begin{cases} I_{eq} \dfrac{d\omega_{wi}}{dt} = \eta_g \beta_g T_{mi}(\omega_{mi}, x_{\theta mi}) + T_{bi} - F_{xi} R \quad i = 2, 4 \\ I_{eq} = I_w + I_g + \eta_g \beta_g^2 I_m \\ \omega_{mi} = \beta_g \omega_{wi} \end{cases} \tag{7.52}$$

7.3.3.4 Torque Coupling Parallel Hybrid Electric Vehicle

Figure 6.11 shows a torque coupling parallel hybrid electric powertrain configuration. In this powertrain configuration, the torque generated by the engine and electric motor come together using a mechanical torque coupling device. As discussed in previous chapters, the torque coupling device can be one of several mechanical devices, but is typically a gearbox with two inputs and one output as illustrated in Figure 6.11. As presented in Chapter 6, the driving torque equations are:

$$T_{wi} = \begin{cases} \frac{1}{2}\left(\eta_{te}\beta_{te}T_e(\omega_e, x_{\theta e}) + \eta_{tm}\beta_{tm}T_m(\omega_m, x_{\theta m}) - I_{eq}^*\frac{d\omega_{wi}}{dt}\right) & i = 1,3 \\ 0 & i = 1,2 \end{cases} \tag{7.53}$$

where:

$$\begin{aligned} \eta_{te} &= \eta_{ge}\eta_{ce}\eta_d, \quad \beta_{te} = \beta_i\beta_{ce}\beta_d \\ \eta_{tm} &= \eta_{gm}\eta_{cm}\eta_d, \quad \beta_{tm} = \beta_g\beta_{ce}\beta_d \end{aligned} \tag{7.54}$$

also the wheel dynamic equations are:

$$\begin{cases} I_{eq}\frac{d\omega_{wi}}{dt} = \frac{1}{2}(\eta_{te}\beta_{te}T_e(\omega_e, x_{\theta e}) + \eta_{tm}\beta_{tm}T_m(\omega_m, x_{\theta m})) + T_{bi} - F_{xi}R & i = 1,3 \\ I_w\frac{d\omega_{wi}}{dt} = T_{bi} - F_{xi}R & i = 1,2 \end{cases} \tag{7.55}$$

where:

$$\begin{aligned} I_{eq}^* = \frac{1}{2}(&\eta_{ge}\eta_{gc}\eta_d\beta_{ge}^2\beta_{ce}^2\beta_d^2I_e + \eta_{gm}\eta_{gm}\eta_d\beta_{gm}^2\beta_{cm}^2\beta_d^2I_m + \eta_{ce}\eta_d\beta_{ce}^2\beta_d^2I_{ge} \\ &+ \eta_{cm}\eta_d\beta_{cm}^2\beta_d^2I_{gm} + \eta_d\beta_d^2I_c + I_d) \end{aligned} \tag{7.56}$$

$$I_{eq} = I_w + I_{eq}^* \tag{7.57}$$

Now that all the corresponding equations of the three main components of the vehicle model shown in Figure 7.15 have been developed and discussed, the variables and how the components are connected to each other are clear. Using the above equations and the model shown in Figure 7.15, the response of a vehicle can be simulated for any given set of vehicle parameters and driver inputs. In the following, the model is used to simulate vehicle dynamics in various case studies.

7.3.4 Simulation Study

The vehicle model discussed above can be used to study the handling dynamics of electric and hybrid electric vehicles. As an example, consider a four motorized wheel electric vehicle. For such vehicles, the traction torque of each wheel can be adjusted, individually. Therefore, the total driver's torque command (determined by the accelerator pedal position) can be distributed to the wheels in different ways.

A simulation analysis is presented here to investigate the effects of traction torque distribution on the handling performance of a four motorized wheel electric vehicle. This concept is important for developing torque vectoring control strategies which will be discussed in Chapter 9. Such controllers monitor the performance of the vehicle and intelligently distribute the traction torques on the wheels to stabilize the vehicle and improve performance.

7.3.4.1 Case Study 1: Driving on a Dry Road

For a mid-sized, four motorized electric sedan, consider the following traction torque distribution scenarios:

- Equal traction torques applied to all wheels: 25% of the total demand torque applied to each wheel.
- Traction torque applied to the right wheels: 50% of the total demand torque applied to the front-right wheel, and 50% of the total demand torque applied to the rear-right wheel.
- Traction torque applied to the left wheels: 50% of the total demand torque applied to the front-left wheel, and 50% of the total demand torque applied to the rear-left wheel.

For each of the torque distribution strategies, the handling performance is simulated for a right-hand step-steer maneuver with an initial speed of 90 km/hr on a dry road ($\mu = 1$). Figure 7.18 illustrates the simulation results for: (a) yaw rate, (b) side-slip angle, (c) lateral acceleration, (d) vehicle path, (e) total longitudinal force, and (f) external yaw moment. The external yaw moment is the result of the traction torque distribution. The simulations are based on using Equations 7.28 to 7.40 together with Equations 7.45 to 7.49 for the four motorized wheel electric powertrain.

As can be seen from the graphs in Figure 7.18, the total longitudinal force is approximately similar for all maneuvers. This is to be expected, noticing that the total driver demand torque is the same for the maneuvers. However, the yaw rate, the lateral acceleration, the side-slip angle and the vehicle path are different from one case to another. Applying the torque to the right wheels resists the steering command of the driver and consequently the vehicle shows more under-steer behavior compared to the equal torque distribution scenario. Therefore, the vehicle shows lower yaw rate, slip angle, and lateral acceleration responses. The negative external yaw moment originating from the traction force difference is the underlying reason for this behavior.

On the other hand, when the traction torque is applied to the left wheels, the vehicle shows more over-steer behavior. In this case, the positive external yaw moment empowers the steering command and increases the yaw rate, lateral acceleration and side-slip angle, as can be seen in Figure 7.18.

7.3.4.2 Case Study 2: Driving on a Wet Road

As mentioned earlier, the effects of the traction torque distribution on the handling performance are helpful in developing effective vehicle dynamics control and driver assistance systems. Such systems can effectively distribute the torques to avoid instability or help achieve a more desirable vehicle performance regardless of the road conditions.

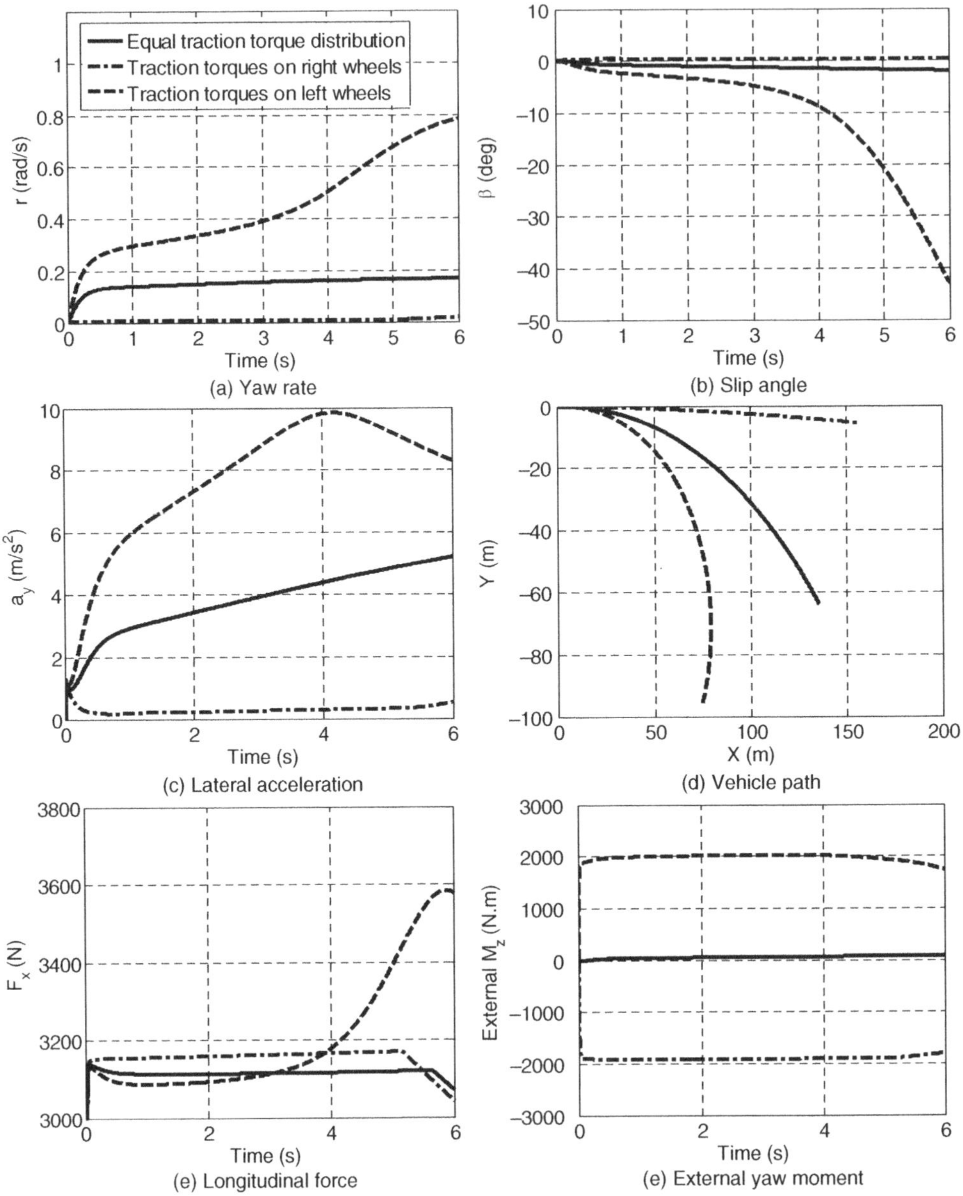

Figure 7.18 Handling performance for different traction torque distribution strategies

To show this potential capability, consider the four motorized wheel electric vehicle with equal torque distribution strategy and the following simulation scenarios:

- Step steer maneuver on a dry road ($\mu = 1$)
- Step steer maneuver on a wet road ($\mu = 0.5$)

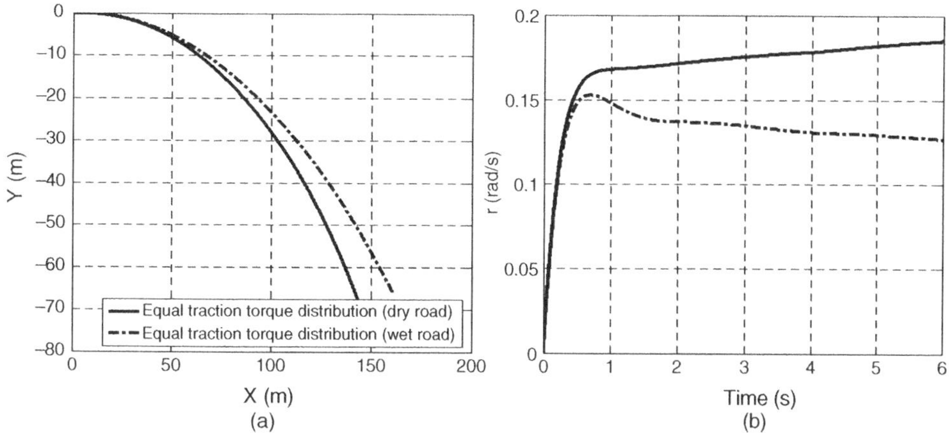

Figure 7.19 Comparison of the vehicle path (a) and yaw rate (b) on the dry and wet roads

The vehicle path and yaw rate for these two maneuvers are shown in Figure 7.19. The graphs show that on the wet road, the yaw rate decreases and the vehicle is less sensitive to the steering input. Therefore, the vehicle shows more under-steer behavior compared to the dry road conditions.

The torque distribution strategy can be used to improve the performance on the wet road and to achieve a behavior similar to that of the dry road condition. Assume that 32.5% of the total demand traction torque is applied to each of the left wheels, and 17.5% of the total demand torque is applied to each of the right wheels. Simulation results for this case on the wet road are shown in Figure 7.20 and compared with the equal traction torque distribution strategy for both dry and wet road conditions. The results show that on the wet road, the vehicle with the adjusted traction torques can follow a path close to that of the dry road condition. The yaw rate has increased compared to the equal torque distribution on the wet road, which shows that the vehicle can better deliver the requested yaw rate demanded by the driver.

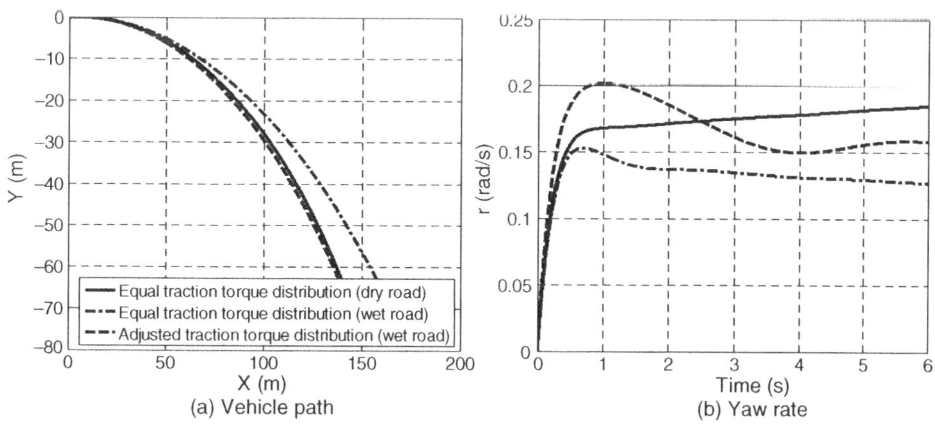

Figure 7.20 Effects of the torque distribution strategy on improving the vehicle performance

Problems

1. The parameters of an electric vehicle are summarized in Table 7.7.
   (a) Formulate the single track handling model for this vehicle in the state-space format.
   (b) Simulate the response of the vehicle for a constant step steering input of 0.04 rad and a constant longitudinal velocity of 50 km/hr. Comment on the validity of the single track handling model for the vehicle.
2. Consider the electric vehicle with the parameters described in Question 1.
   (a) Calculate the under-steer coefficient for the vehicle and show that the vehicle is over-steer.
   (b) At what longitudinal speed does the vehicle become unstable?
   (c) Assume that during the design process, the battery pack can be installed in several configurations such that one can keep the mass and yaw moment of inertia the same, and change the CG position. Calculate the maximum allowable distance between the front axle and the CG for which the vehicle is stable for all longitudinal velocities.
3. During the design process of an electric vehicle, the effects of the battery pack location on the handling performance are being investigated. The parameters of the vehicle when the battery pack is installed in the front box, central box and the rear box are listed in Table 7.8. Assume that the tire cornering stiffness for each tire is 30 000 (N/rad).
   (a) Using the lateral and yaw velocity gains for the steady state conditions, explain which configuration makes the vehicle more responsive to the steering input at the longitudinal speed of 50 km/hr.
   (b) For each of the configurations, plot the yaw velocity gain for a range of vehicle speeds and discuss stability of the vehicle for different speeds.
   (c) Calculate the under-steer coefficient for the three configurations. Discuss the stability of the vehicles and compare the results with (b).

Table 7.7 Vehicle parameters

Parameter	Value
M	1,100 kg
a	1.4 m
b	1.1 m
I_z	2,000 kgm2
$c_{\alpha f}$	24,000 N/rad
$c_{\alpha r}$	26,000 N/rad

Table 7.8 Vehicle inertial parameters for different battery locations

Parameter	Value		
	Front box	Central box	Rear box
M	1,200 kg	1,200 kg	1,200 kg
a	1 m	1.15 m	1.3 m
b	1.3 m	1.15 m	1 m
I_z	2,400 kgm^2	2,300 kgm^2	2,450 kgm^2

4. For the electric vehicle with different battery location options summarized in Table 7.8, analyze the transient response characteristics.
   (a) For each configuration, calculate the natural frequency and the damping ratio at the longitudinal speed of 50 km/hr. Determine whether each configuration is under-damped or over-damped.
   (b) Plot the natural frequency and the damping ratio as a function of time for each configuration. Discuss the transient response characteristics for different range of speeds.

References

1. Smith, D.E., and Starkey, J.M. (1995) Effect of model complexity on the performance of automated vehicle steering controllers: model development, validation and comparison. *Journal of Vehicle System Dynamics*, **24**, 163–181.
2. Ogata, K. (2008) *Modern Control Engineering*, 5th edn, Pearson Education, Harlow.
3. Esmailzadeh, E., Vossoughi, G.R., and Goodarzi, A. (2001) Dynamic modeling and analysis of a four motorized wheels electric vehicle. *International Journal of Vehicle System Dynamics*, **35** (3), 163–194.
4. Pacejka, H.B. (2002) *Tire and Vehicle Dynamics*, Elsevier Butterworth-Heinemann, Oxford.

8

Energy/Power Allocation and Management

8.1 Introduction

As the electrical and electronic content in vehicles increases, the on-board electric power requirement for non-propulsion loads will increase from about 1 kW to 5 kW, and vehicle propulsion loads will exceed 100 kW. This increase in power demand is primarily a result of electric and hybrid drivetrains, however, other automotive technologies for internal combustion engine vehicles contribute to this trend, such as variable engine valve and active suspension [1]. The huge amount of required power complicates the issue of appropriate power management and distribution. In conventional ICE vehicles, an alternator and voltage regulator manage electric power. In early electric and hybrid vehicles, the on-board power management strategy controlled the state of charge of the battery. However, in modern EVs and HEVs, the purpose of a strategic power and energy management system is to control and coordinate the power generation, energy storage, and power flow within subsystems to achieve maximum overall system efficiency. In addition, it must prioritize real-time power requests from loads and allocate the available power resources from generation and storage devices to maximize the vehicle's efficiency and performance. The use of energy storage devices in vehicles has altered the paradigm of power management to include energy management as well. Managing power and energy improves vehicle performance and system reliability, and makes it possible to reduce vehicle weight, size, and cost.

Power and energy loads in an EV/HEV/PHEV require a management strategy to allocate power and energy resources. This strategy should maximize the vehicle's energy efficiency, provide a high level of vehicle performance, and minimize emission levels. The ultimate aim is to achieve energy and power efficiency without compromising the important aspects of the vehicle's performance, such as driving range, acceleration, comfort, convenience, and emission levels.

Electric and Hybrid Vehicles: Technologies, Modeling and Control: A Mechatronic Approach, First Edition.
Amir Khajepour, Saber Fallah and Avesta Goodarzi.

This chapter explains the concept of power/energy management in EVs and HEVs. It also describes available rule-based and optimization-based control strategies used in EV and HEV applications.

8.2 Power/Energy Management Controllers

Typically, the terms power management and energy management are incorrectly applied interchangeably. However, power and energy management refer to different concepts. The fundamental difference between power management and energy management of a vehicle is that energy management refers to the accumulation (integration) of power over a given time period. That is, energy management deals with energy consumption and recuperation over a trip. Alternatively, power management deals with instantaneous power distribution and power flow control between electric and mechanical powertrain components to satisfy the power demands.

Many control strategies have tried to achieve the optimal power/energy management solution for EVs and HEVs. Although the ultimate objective of all available power management optimization techniques is to achieve higher fuel economy and lower emissions while maintaining the requirements for individual mobility (such as range, acceleration and speed), they may be classified in various ways, such as: controller performance, real-time implementation, and overall objectives. For example, one can classify these strategies based on their dependency on knowledge of future demands and situational awareness. Some non-causal control methods suggest strategies based on mission profiles. These strategies provide benchmarks, but are not readily implementable in practice as the complete mission profile is unknown before departure. Furthermore, total vehicle power demands are sensitive to variations in path, altitude, gradient, and velocity profile. Strategies based on heuristics are much more practical to implement. These strategies use Boolean rules or fuzzy logic and offer a qualitative rather than quantitative description of the system. Beginning with a set of static rules and an expert knowledge of the system's threshold settings, these strategies can be tuned in an empirical recursive manner or using offline optimization. The classification shown in the flowchart of Figure 8.1 organizes energy/power management techniques in terms of control strategy [2,3].

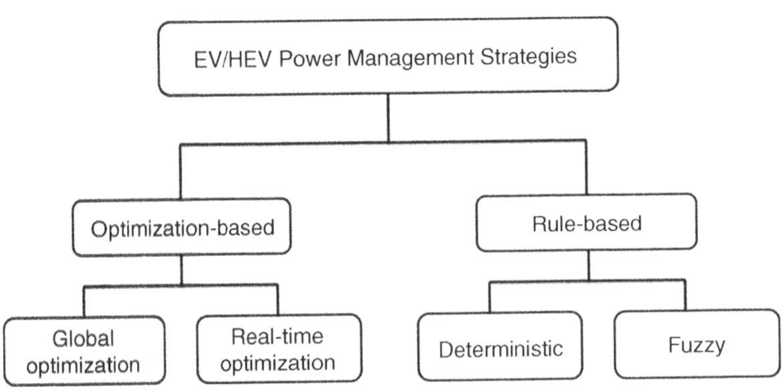

Figure 8.1 Classification of power management optimization techniques

8.3 Rule-Based Control Strategies

HEVs and PHEVs primarily use rule-based control strategies. These controllers function by switching between predefined heuristic operation rules and thresholds, which are set to allow the combustion engine and the electric motor to operate within their most efficient regions and to maximize the recuperation of vehicle, kinetic, and potential energies. Rule-based controllers are static controllers in which the heuristic rules are determined based on vehicle operation modes and variables where the transition decisions are based on instantaneous inputs. The variables include, but are not limited to, vehicle velocity, battery SoC, and the power generation of the engine and motor. The rules are independent of any prior knowledge of the trip and drive cycle. The rule-based control strategies are simple, robust to vehicle parameter uncertainties, and computationally efficient, thereby making them appropriate for real-time implementation. However, since the rules and thresholds have to be defined in advance, they may not provide an optimal solution for different driving cycles, as their performance can be sensitive to change of vehicle operation conditions. There are two types of rule-based controllers: deterministic rule-based methods and fuzzy logic rule-based methods.

8.3.1 Deterministic Rule-Based Control Strategies

By design, deterministic rule-based methods operate on a set of rules defined by an analysis of power flow in a hybrid drivetrain, efficiency/fuel maps of ICE, and human experiences. Typically, the algorithms are implemented through look-up tables. Examples of deterministic rule-based strategies are: thermostat controller, maximum SoC strategy, and state-machine based.

8.3.1.1 Thermostat Controller

The thermostat controller strategy is the most primitive method of a rule-based controller. Other names for the thermostat strategy include: thermostatic SoC, the "bang-bang" controller, and the on/off controller. The objective of the controller is to maintain the battery's SoC within its pre-set upper and lower limits by turning the engine off and on while considering the torque demands. The thermostat control strategy is mostly suited for series HEVs and PHEVs, although it can be applied on a parallel configuration as well. As mentioned in Section 2.4.3, in a series hybrid powertrain, the total traction force demanded by the driver is provided through the electric motor, while the IC engine keeps the battery's SoC within its predefined boundaries.

In this strategy, when the SoC approaches the upper limit, the engine turns itself off and the battery provides the power demand. When the SoC approaches the lower limit, the engine turns on and the battery starts charging at a predetermined power level, set by the controller that runs the engine at its most efficient level. The "on" operating point of an IC engine can be determined according to the controller objectives and powertrain specifications such as: emissions, fuel efficiency, and the electric motor/generator characteristics.

In a thermostat control of series HEVs/PHEVs, the priority is to set the IC engine operating mode at its most efficient point. Those series HEVs and PHEVs, which commute in prescheduled routes, for example, such as city buses with specific driving conditions, driving for a long time (with a low load) on a highway at constant speed, are the most practical application of the thermostat controller. The reason is that in such driving cycles, the high peak transient power demands, such as rapid acceleration is limited, thereby making the average power demand low enough that an electric motor can provide the required traction force at all times. If the objective of a secondary power source is to improve the vehicle performance (e.g., acceleration, gradeability, etc.), this control strategy might not be suitable as there is no guarantee that the battery has enough SoC to satisfy the demand requirements at all times or it has enough charge to run the vehicle entirely in electric mode.

Example 8.1

Investigate the effectiveness of the thermostat control strategy on a series HEV in the FTP-75 and HWEFT driving cycles (Figures 5.45 and 5.25).

For this example and those to follow in this chapter, the subsequent information and assumptions are used:

- *Vehicle*: The general specifications of the vehicle are shown in Table 8.1. Using the information in this table along with vehicle drive cycle, the power demand can be calculated through Equation 6.1 that is the vehicle longitudinal dynamics discussed in Chapter 6.
- *ICE*: The ICE performance characteristics are shown in Figure 8.2. This figure shows the maximum torque and power of the engine with respect to engine speed. The engine

Table 8.1 General specifications of the hybrid vehicle

Parameters	Unit	Value
Curb weight	kg	1336
Frontal area	m^2	2.26
Drag coefficient	–	0.32
Wheel radius	m	0.31

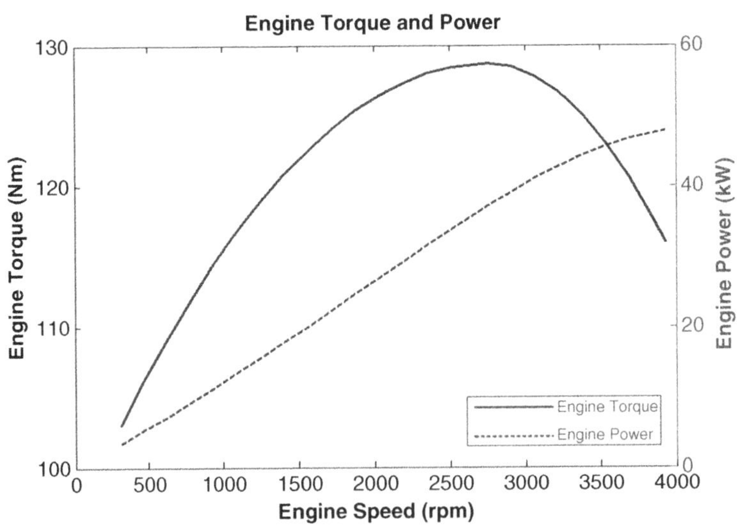

Figure 8.2 Engine torque (solid line) and engine power (dashed line) vs. engine speed

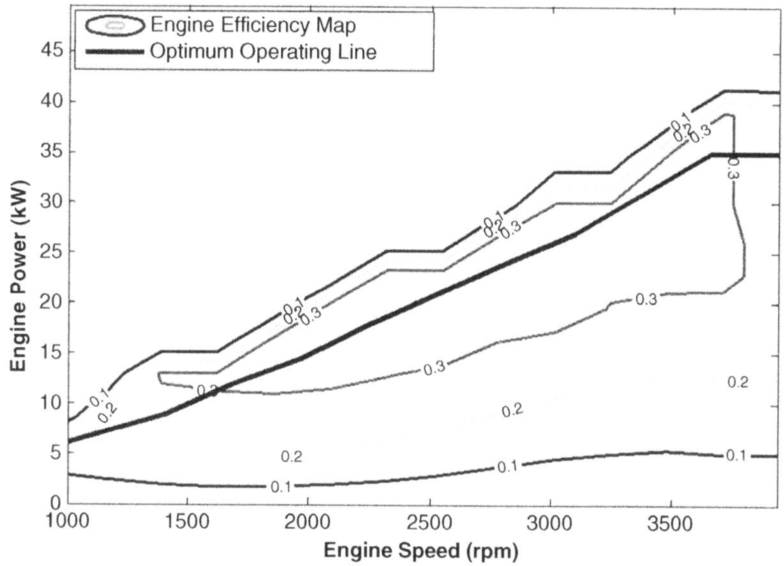

Figure 8.3 Engine efficiency as a function of engine speed and engine torque

efficiency map as a function of engine speed and engine power is shown in Figure 8.3. This figure also illustrates optimum operating line of the engine.

- *Electric motor/generator*: In the examples, permanent-magnet AC motors are used. Performance characteristics of the electric motor, including maximum propulsion and regeneration torque with respect to the motor speed, as well as motor efficiency map as a function of motor torque and motor speed are shown in Figure 8.4. It is assumed that the characteristics of generator are similar to those of the electric motor.

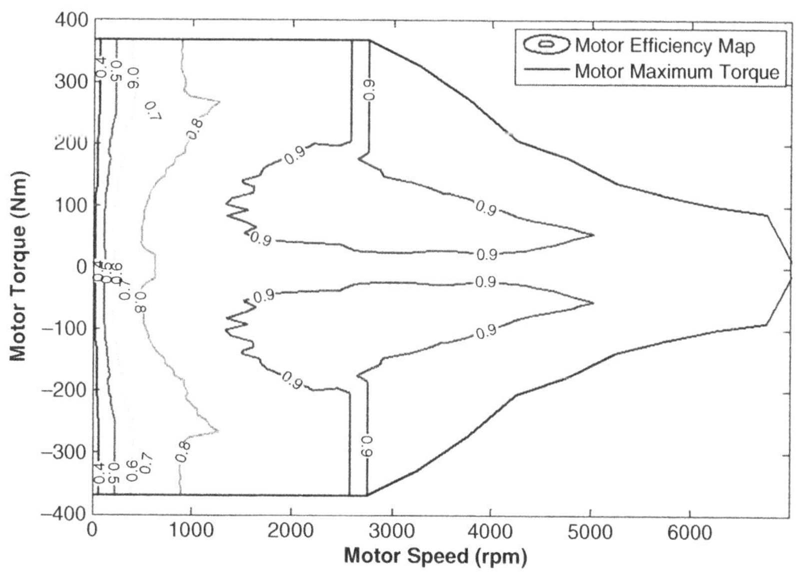

Figure 8.4 Motor efficiency as a function motor speed and motor torque

Table 8.2 Specifications of the battery

Parameters	Unit	Value
Number of battery pack (parallel connection)	–	3
Number of cells in the battery pack (Series connection)	–	100
Nominal voltage of each cell	V	3.3
Capacity of each cell	Ah	14.4

Table 8.3 Gear ratios of the transmission, final drive and engine/generation gear

Gears	Ratio
1st gear	5
2nd gear	3.5
3rd gear	1
Engine/generator gear	1.3

- *Battery*: Lithium-ion battery packs are used in the HEV. Specifications of the battery are summarized in Table 8.2. Battery current and SoC are calculated by equations described in Section 5.3.
- *Conversion*: To compare total energy consumption of HEVs with that of a conventional vehicle, a conversion factor between electrical and fossil fuel is required. In this book, the conversion factor suggested by the United States Environmental Protection Agency (EPA) is used. The EPA considers 33.7 kWh of electricity equivalent to one gallon of gasoline.
- *Transmission*: The gearbox of the HEV consists of three gear sets. In addition, the series HEV is equipped with an engine/generator gear. The ratios of these gears are shown in Table 8.3.

Consider the following rules for a series HEV in which the improvement of fuel efficiency is targeted:

1. When the SoC approaches the lower limit, the engine turns on, and the generator starts charging batteries. In this case, the engine always operates in the optimal operating region.
2. When the SoC approaches the upper limit, the engine turns itself off, and the batteries provide the demand power for the electric motor.

The engine operating map is depicted in Figure 8.5. For this case study, assume that the engine always operates at its optimal operating point during both the FTP-75 and HWFET driving cycles.

Figures 8.6 and 8.7 show the SoC variation and engine on/off cycles for the FTP-75 and HWFET driving cycles, respectively. Figures 8.8 and 8.9 represent electric motor operating points in the driving cycles. Note that in a series HEV, there is no control on electric motor performance since the electric motor supplies the power. However,

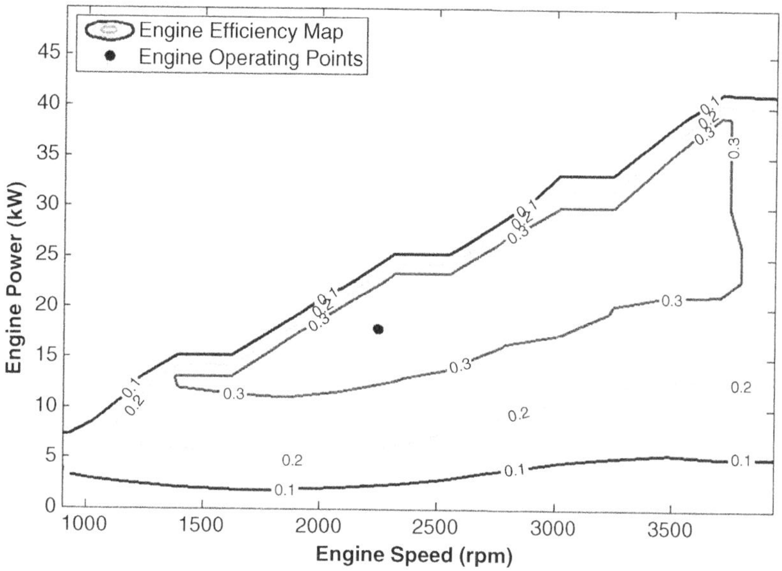

Figure 8.5 Engine operating points in the FTP-75 and HWFET driving cycles

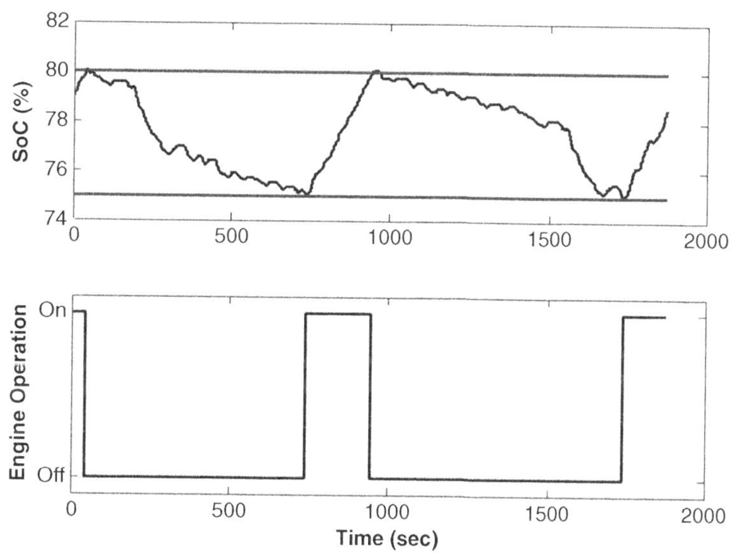

Figure 8.6 Battery SoC in the FTP-75 driving cycle

Figure 8.9 shows that the electric motor mostly operates at its efficient region. The reason is that HWFET represents a highway driving cycle in which the variation of the power demand is less than that of the FTP-75. In addition, an electric motor is more efficient during high speed-low torque driving conditions. Tables 8.4 and 8.5 compare the fuel

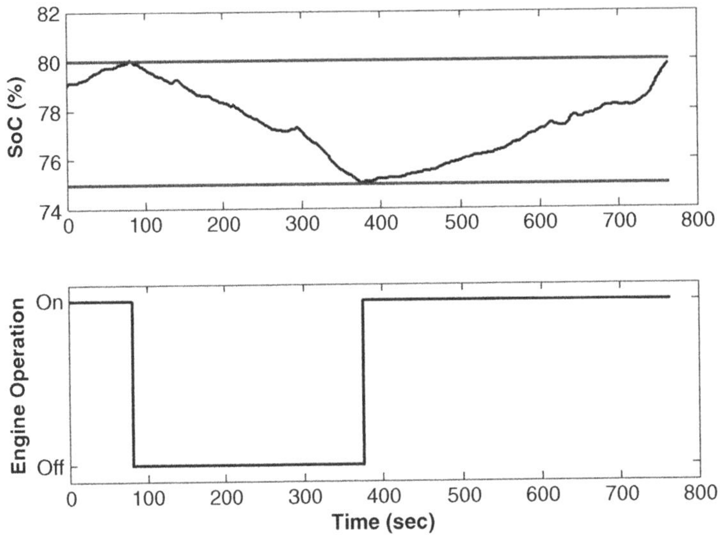

Figure 8.7 Battery SoC in the HWFET driving cycle

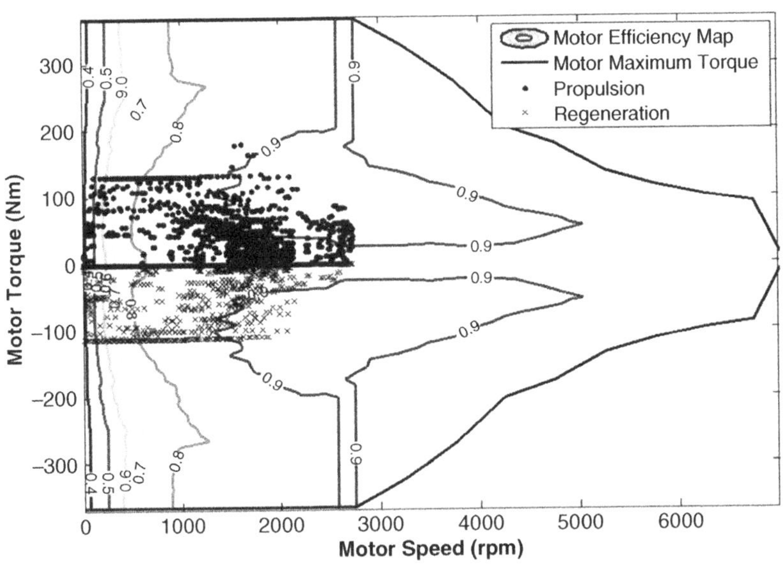

Figure 8.8 Motor operating points in the FTP-75 driving cycle

efficiency of the series HEV with a conventional ICE vehicle. Since the initial and final SoC are not the same, the total fuel consumptions in the tables are calculated by converting the change in the battery SoC to equivalent fuel using the EPA energy conversion factor discussed above. According to the data, in both driving cycles, the series

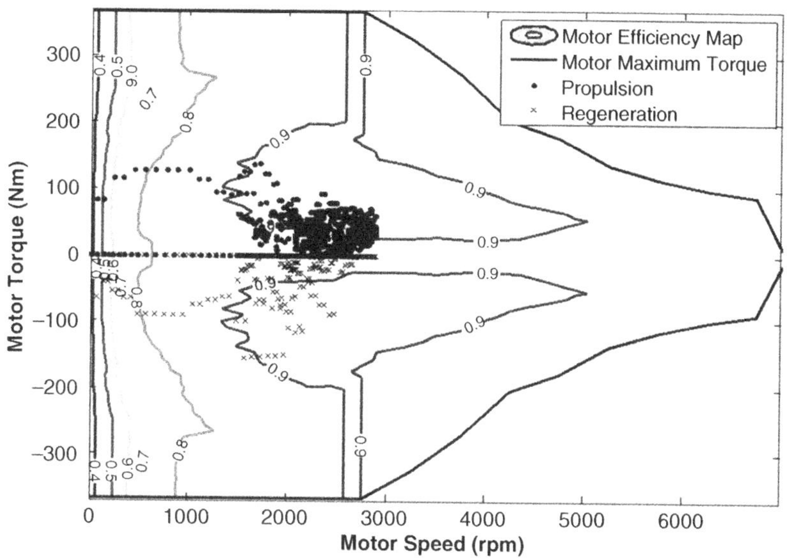

Figure 8.9 Motor operating points in the HWFET driving cycle

Table 8.4 Comparison of the fuel consumption between a series hybrid vehicle with thermostat control strategy and a conventional vehicle in the FTP-75 driving cycle

Fuel consumption	Series hybrid vehicle (thermostat control strategy)(L/100 km)	Conventional vehicle (L/100 km)
Engine fuel consumption	3.32	9.11
Equivalent fuel consumption of the battery	0.11	–
Total fuel consumption	3.43	9.11

Table 8.5 Comparison of the fuel consumption between a series hybrid vehicle with thermostat control strategy and a conventional vehicle in the HWFET driving cycle

Fuel consumption	Series hybrid vehicle (thermostat control strategy)(L/100 km)	Conventional vehicle (L/100 km)
Engine fuel consumption	4.37	6.38
Equivalent fuel consumption of the battery	−0.24	–
Total fuel consumption	4.13	6.38

HEV is more efficient than the conventional ICE vehicle. As seen in the tables, fuel consumption of the HEV in FTP-75 is less than that of the HWFET cycle. This is because the FTP-75 is a city driving cycle with more stop and go intervals, resulting in recapturing the vehicle kinetic energy through regenerative braking.

Example 8.2

Investigate the effectiveness of the thermostat control strategy on a parallel HEV in the FTP-75 and HWEFT driving cycles. In this example, we use the same information given in the series HEV in Example 8.1, however, we assume another speed reduction with a gear ratio of 2.4 right after the transmission.

Consider the following rules for the power management of the parallel HEV.

During traction:

1. When the SoC approaches the lower limit, the engine turns on, and two scenarios may happen:
   i. When the power demand is greater than the optimum engine power, the engine operates at its optimum operating line, and the electric motor provides the remaining demanded power.
   ii. When power demand is less than the optimum engine power and SoC is below its top line, the engine operates at its optimum line to provide demanded power. The motor uses the remaining power generated by the engine to generate electric energy for charging the batteries.
2. When the SoC approaches the upper limit, the engine turns off, and the electric motor provides all demanded power.

During braking:

1. When demanded braking power is less than the maximum power that the motor can regenerate, all braking power is used to charge the batteries.
2. When demanded braking power is greater than the maximum power that the motor can regenerate, the friction brake produces the remaining amount.

The FTP-75 Driving Cycle

Figure 8.10 illustrates the simulation results for SoC variation and the engine on-off cycle for the parallel full hybrid vehicle with thermostat control strategy in the FTP-75 driving cycle. Figures 8.11 and 8.12 show the operating points of the IC engine and electric motor. Note that unlike the series configuration, it is not possible to set the IC engine to work on its optimum efficiency point in a parallel configuration. However, as reflected in Figure 8.12, when the IC engine is "on," its operating points are located on or near to optimal efficiency line of its map. The electric motor operating modes are not close to its efficiency region due to the frequent stop and go conditions in an urban driving cycle.

Table 8.6 depicts the fuel consumption of the engine, and equivalent fuel consumption of the battery in an FTP-75 driving cycle for both hybrid and conventional vehicles.

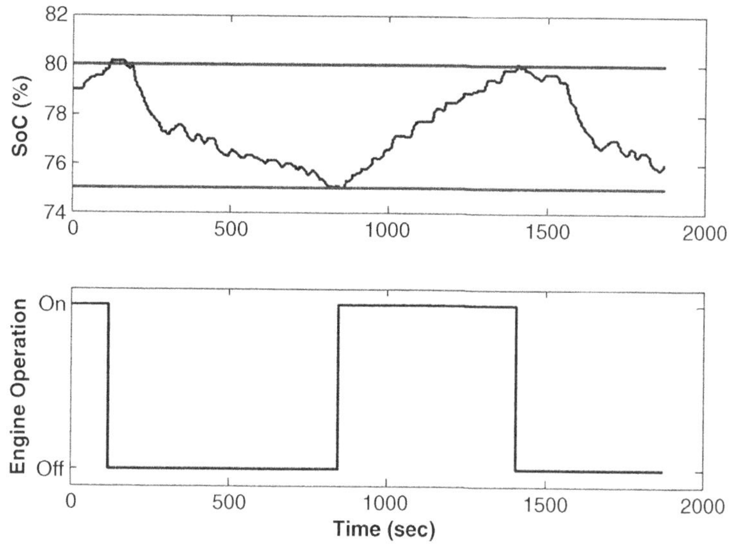

Figure 8.10 Battery SoC in the FTP-75 driving cycle

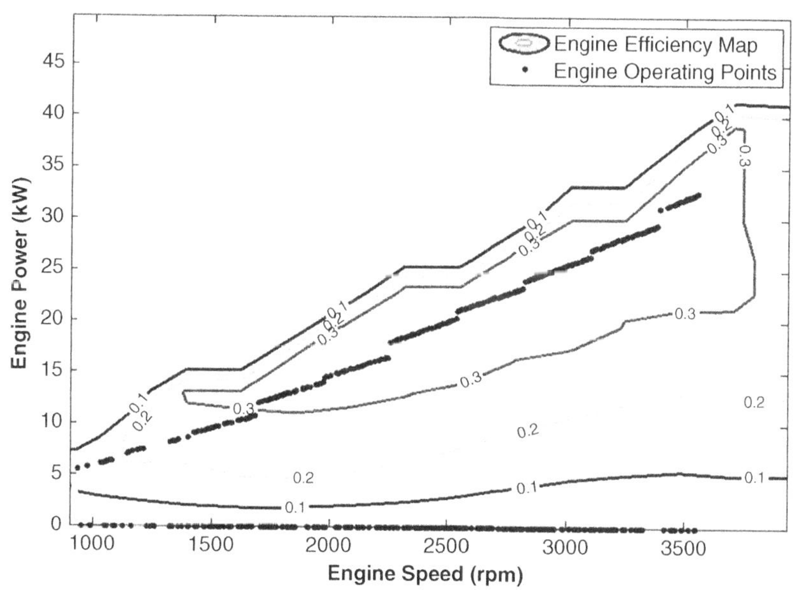

Figure 8.11 Engine operating points in the FTP-75 driving cycle

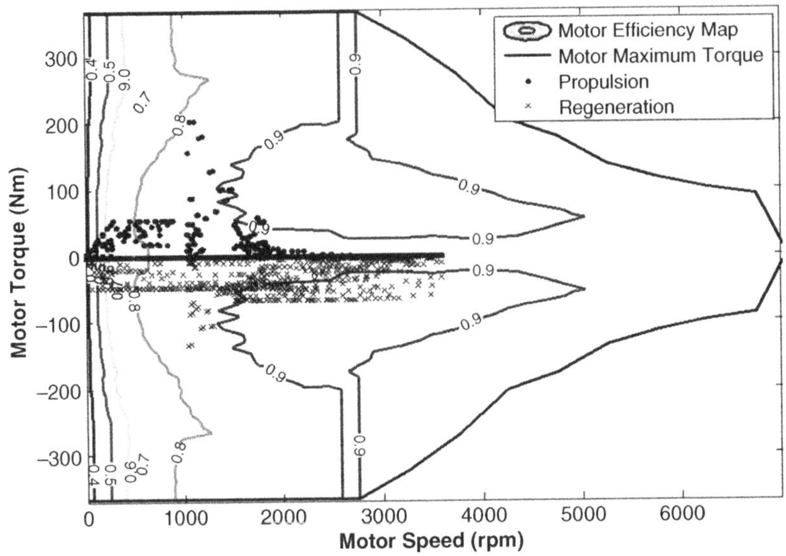

Figure 8.12 Motor operating points in the FTP-75 driving cycle

Table 8.6 Comparison of the fuel consumption between a parallel hybrid vehicle with thermostat control strategy and a conventional vehicle in the FTP-75 driving cycle

Fuel consumption	Parallel hybrid vehicle (thermostat control strategy) (L/100 km)	Conventional vehicle (L/100 km)
Engine fuel consumption	3.99	9.11
Equivalent fuel consumption of the battery	0.90	–
Total fuel consumption	4.89	9.11

The HWFET Driving Cycle

Figures 8.13 to 8.15 show the simulation results for the parallel full HEV with thermostat control strategy in the HWFET driving cycle. As the simulation results show, in this case, the engine is always "on," since there is not frequent high power demand in the HWEFT driving cycle. Note that though the engine is always "on," the electric motor and regenerative braking system help the IC engine to operate close to its efficiency points. Table 8.7 shows the fuel consumption of the engine and equivalent battery fuel consumption in the HWFET driving cycle for both hybrid and conventional vehicles.

Maximum SoC Control Strategy

The Maximum SoC control strategy uses the engine as a primary source of torque, while the electric motor supplies additional power when required. Due to the charge-sustaining operation, the power source SoC remains within its upper and lower limits. This control

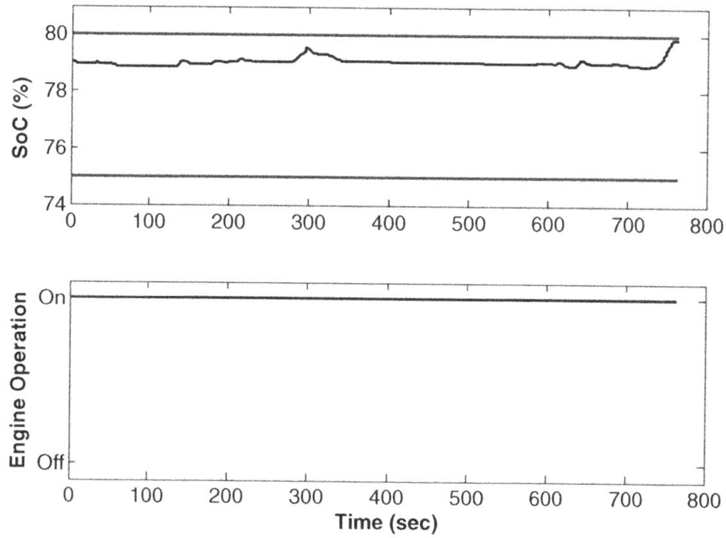

Figure 8.13 Battery SoC in the HWFET driving cycle

strategy is considered efficient for vehicles with frequent stop-go driving patterns. Keeping the SoC of the power source in its maximum operation boundaries will guarantee consistent high vehicle performance (speed, acceleration, gradeability, etc.).

The Toyota Prius and the Honda Insight are examples of mass-produced HEVs that use power management systems based on the Maximum SoC control strategy. It is possible to

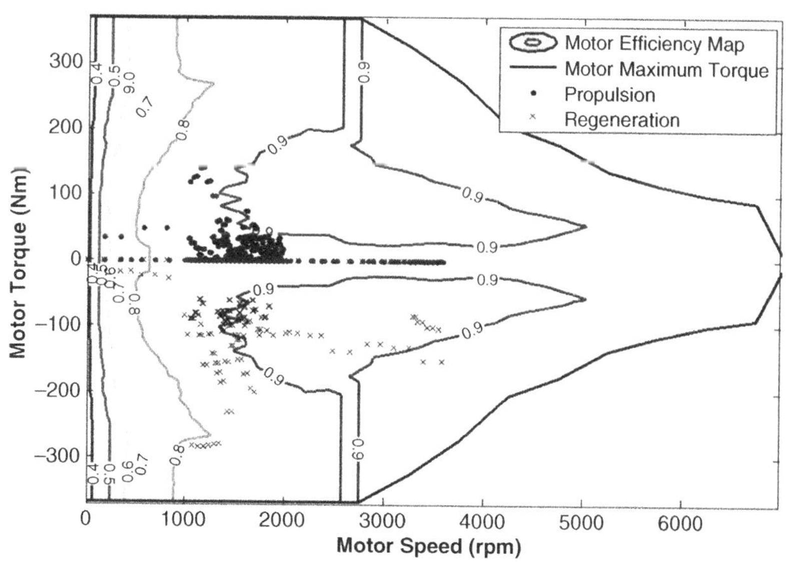

Figure 8.14 Motor operating points in the HWFET driving cycle

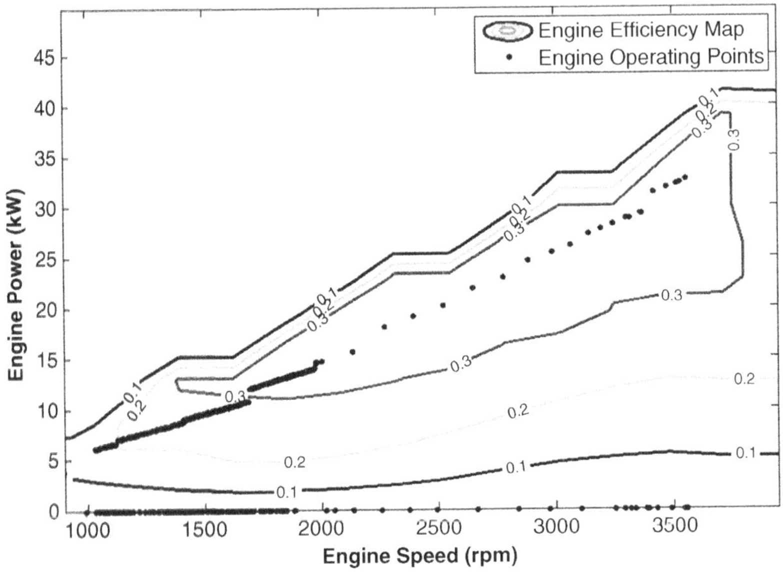

Figure 8.15 Engine operating points in the HWFET driving cycle

Table 8.7 Comparison of the fuel consumption between a parallel hybrid vehicle with thermostat control strategy and a conventional vehicle in the HWFET driving cycle

Fuel consumption	Parallel hybrid vehicle (thermostat control strategy) (L/100 km)	Conventional vehicle (L/100 km)
Engine fuel consumption	4.92	6.38
Equivalent fuel consumption of the battery	−0.30	–
Total fuel consumption	4.62	6.38

extend this strategy to the series hybrid powertrains by defining rules based on the operational modes of a series drivetrain. Even though this approach is popular and widely implemented, it lacks the optimization of the entire powertrain efficiency, and ignores direct emission minimization.

Example 8.3

Investigate the effectiveness of the Maximum SoC control strategy on the parallel HEV described in Example 8.2 in the FTP-75 and HWEFT driving cycles.

Consider the following rules for a parallel HEV in which improving vehicle performance is the objective.

During traction:

1. In low vehicle speeds in which the engine needs to run at lower than its idle speed, an electric motor is the only source of generating demanded power.
2. When the power demand is greater than the optimum engine power, the engine operates at its optimum operation line, and the electric motor provides the remaining demanded power.
3. When the power demand is less than the optimum engine power and the SoC is below its top line, the engine operates at its optimum line to provide the demanded power. The motor uses the remaining power generated by the engine to generate electrical energy to charge the batteries.
4. The engine is the only source of power generation when the power demand is less than the optimum engine power and the SoC is over its top line. In this case, the power produced by the engine is equal to the power demanded, so the engine cannot operate at its optimum operating line.

During braking:

1. When the demanded braking power is less than the maximum power that the motor can regenerate, all of the braking power charges the batteries.
2. When the demanded braking power is greater than the maximum power that the motor can regenerate, the friction brake produces the remaining power.

The FTP-75 Driving Cycle

Figure 8.16 shows the battery's SoC variation. According to this result, the controller first charged the battery through the IC engine up to the pre-defined maximum SoC level of 80%. If the SoC is higher than the threshold, the engine will switch to traction-only mode. However, the battery will continue charging while the engine is in traction-only mode. This is due to the regenerative braking system keeping the SoC above the pre defined threshold, through which the battery is charging. Figures 8.17 and 8.18 illustrate the simulation results for the engine and electric motor operating points on their efficiency maps. As the plots represent, neither the electric motor nor the IC engine is working in the efficiency regions. This occurs when the power generated by the engine in its optimal efficiency operating points is greater than the load power while the battery's SoC is maximum, the Maximum SoC controller pushes the engine to operate further than its optimal operating point, thereby reducing the efficiency of the overall system. Table 8.8 shows the fuel consumption of the engine in an FTP-75 driving cycle for both hybrid and conventional vehicles. As the data show, the Maximum SoC controller still improves the fuel efficiency of the vehicle due to the regenerative braking system and motor assistance in engine operation during peak power demands.

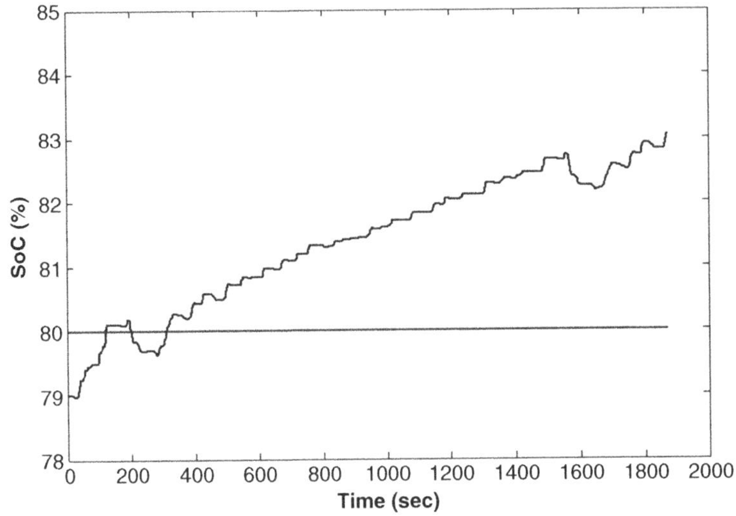

Figure 8.16 Battery SoC in the FTP-75 driving cycle

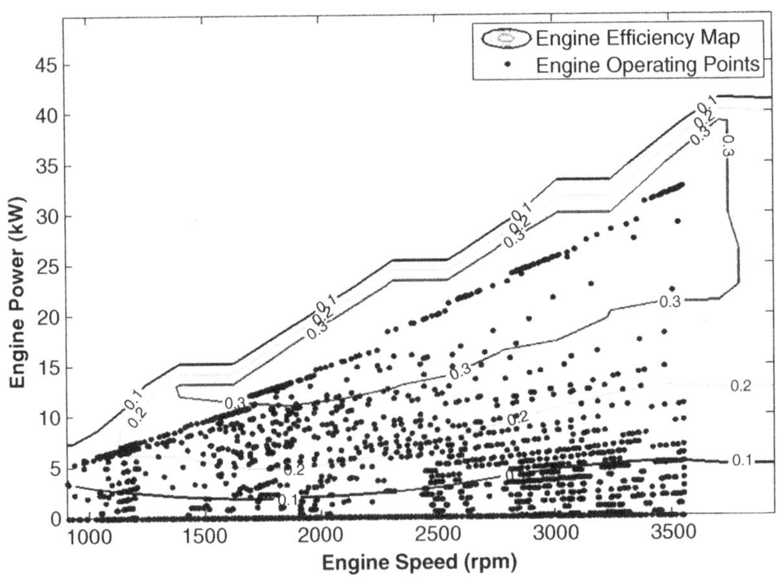

Figure 8.17 Engine operating points in the FTP-75 driving cycle

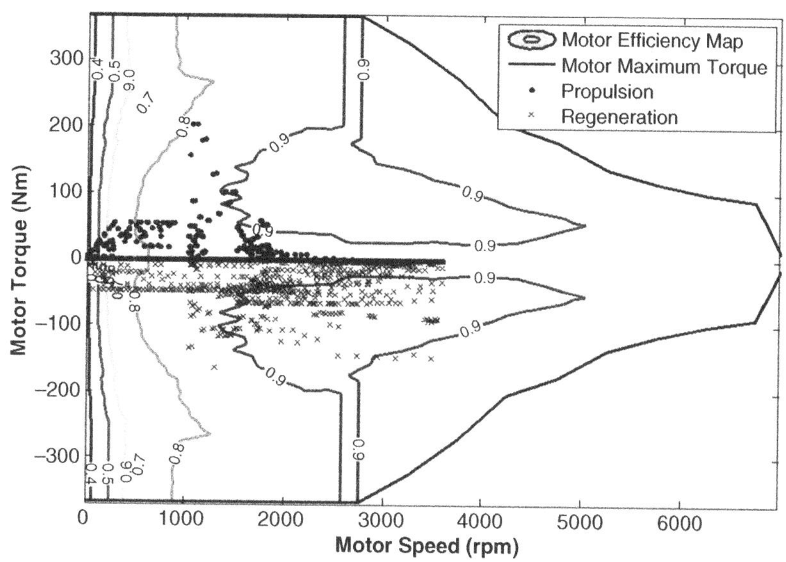

Figure 8.18 Motor operating points in the FTP-75 driving cycle

Table 8.8 Comparison of the fuel consumption between a parallel hybrid vehicle with Max. SoC control strategy and a conventional vehicle in the FTP-75 driving cycle

Fuel consumption	Parallel hybrid vehicle (Max. SoC control strategy) (L/100 km)	Conventional vehicle (L/100 km)
Engine fuel consumption	7.03	9.11
Equivalent fuel consumption of the battery	−1.24	–
Total fuel consumption	5.79	9.11

The HWFET Driving Cycle

Figures 8.19 to 8.21 depict the simulation results for the parallel vehicle with the Maximum SoC control strategy in the HWFET driving cycle. According to Figure 8.19, a Maximum SoC controller keeps the variation of the battery's SoC close to the maximum value. In this situation, since the battery is not fully charged and the average power load is relatively high, the IC engine's operating points are close to its efficient operating points. Similarly, the electric motor works predominantly in its efficient operation regions. Table 8.9 shows the fuel consumption of the engine, and the equivalent fuel consumption of the battery in an HWFET driving cycle for both hybrid and conventional vehicles.

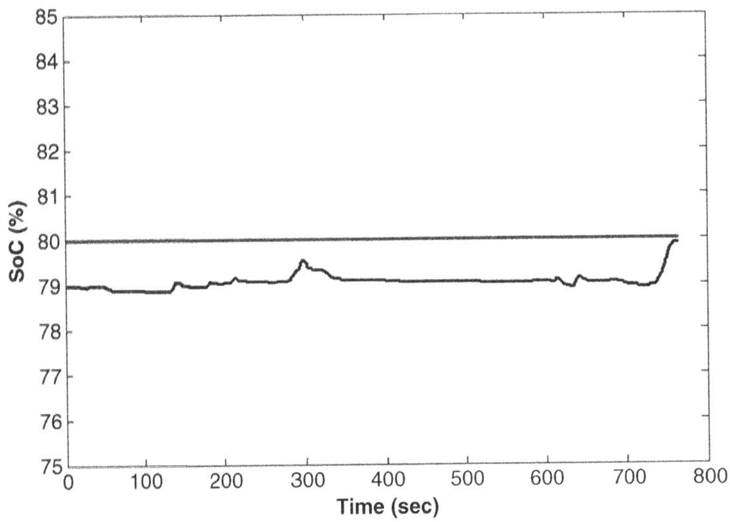

Figure 8.19 Battery SoC in the HWFET driving cycle

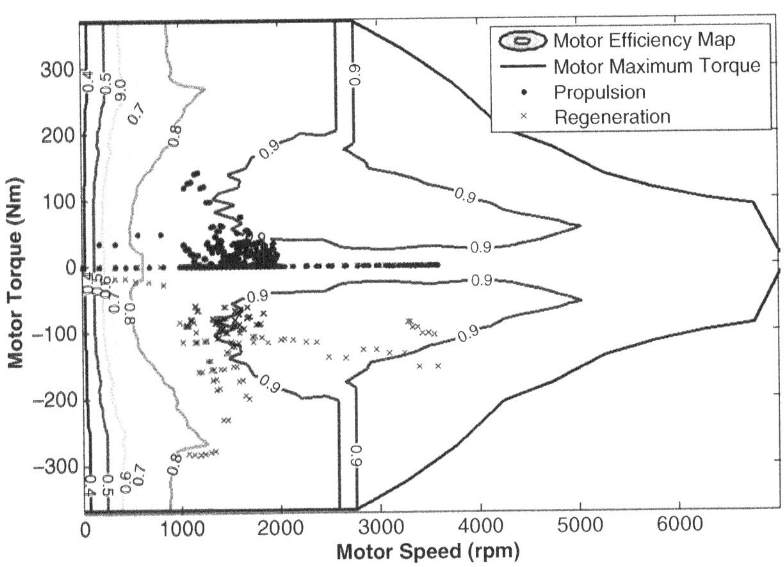

Figure 8.20 Motor operating points in the HWFET driving cycle

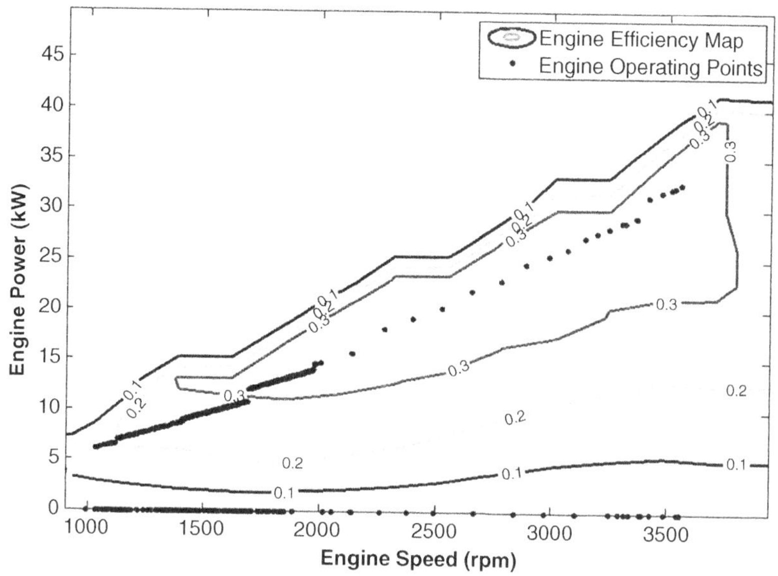

Figure 8.21 Engine operating points in the HWFET driving cycle

Table 8.9 Comparison of the fuel consumption between a parallel hybrid vehicle with Max. SoC strategy and a conventional vehicle in the HWFET driving cycle

Fuel consumption	Parallel hybrid vehicle (Max. SoC control strategy) (L/100 km)	Conventional vehicle (L/100 km)
Engine fuel consumption	4.92	6.38
Equivalent fuel consumption of the battery	−0.30	–
Total fuel consumption	4.62	6.38

Example 8.4

Investigate the effectiveness of the Maximum SoC control strategy on a series HEV in the FTP-75 and HWEFT driving cycles. The vehicle components and parameters are the same as those in Example 8.1.

During traction:

1. When the power demand is greater than the optimum engine power and the SoC is over its top line, the engine operates at its optimum region, and the batteries provide the remaining power.

2. When the power demand is greater than the optimum engine power and the SoC is below its top line, the engine operates at maximum load to meet the power demand, and the engine uses the remaining generated power to charge the batteries.
3. When the power demand is less than the optimum engine power and the SoC is over its top line, the engine is the only source for generating demanded power. In this mode, the engine cannot operate within its optimum region, and while the engine speed is constant, the engine torque changes according to demanded power.
4. When the power demand is less than the optimum engine power and the SoC is below its top line, the engine operates at its optimum region to provide demanded power, and the remaining generated power by the engine charges the battery.

During braking:

1. When the demanded braking power is less than the maximum power that the motor can regenerate, all the braking power charges the battery.
2. When the demanded braking power is greater than the maximum power that the motor can regenerate, the friction brake produces the remaining amount.

Figures 8.22 and 8.23 illustrate the battery's SoC variations in the FTP-75 and HWFET driving cycles, respectively. After the battery reaches its threshold, the engine switches to traction-only mode. However, the regenerative braking system further charges the battery. Compared to the FTP-75, there is less variation in the battery's SoC in the HWFET driving cycle. This is a result of the lack of frequent stop and go conditions in highway driving cycles. Figures 8.24 to 8.26 represent IC engine and electric motor operating points in both drive cycles. Note that in both cases, the battery is easily charged to its full level. This causes the engine/generator to operate with power output smaller than its optimum, thereby reducing the efficiency of the drivetrain.

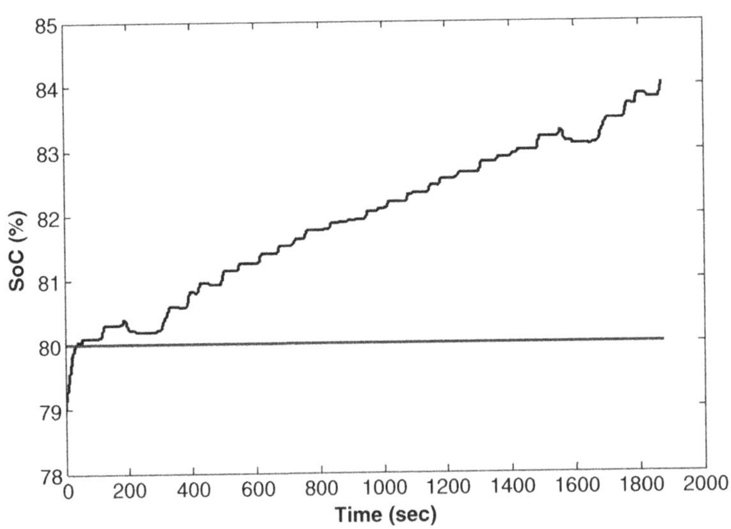

Figure 8.22 Battery SoC in the FTP-75 driving cycle

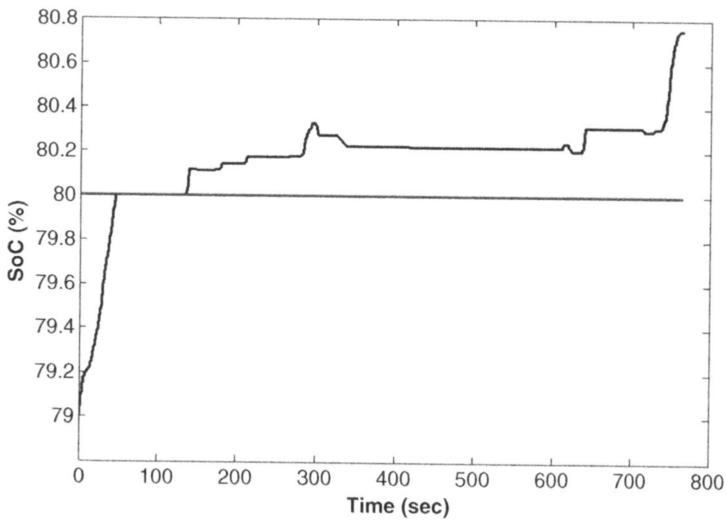

Figure 8.23 Battery SoC in the HWFET driving cycle

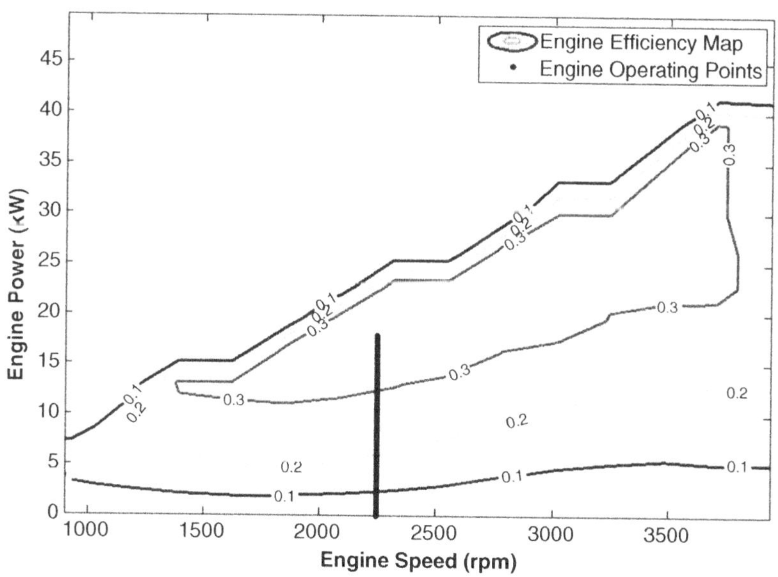

Figure 8.24 Engine operating points in the FTP-75 and HWFET driving cycles

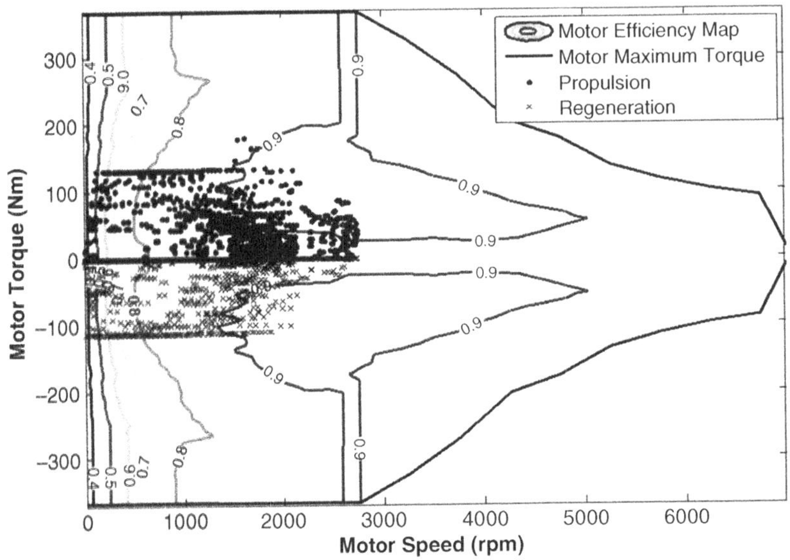

Figure 8.25 Motor operating points in the FTP-75 driving cycle

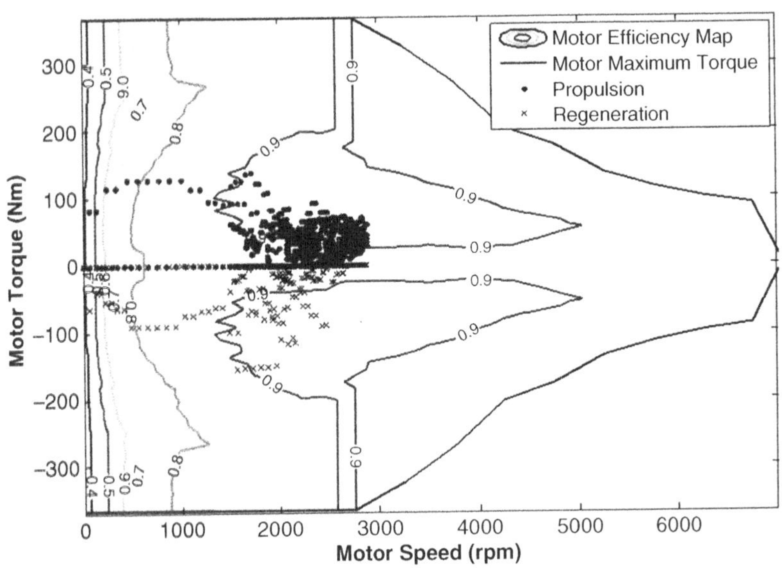

Figure 8.26 Motor operating points in the HWFET driving cycle

Table 8.10 Comparison of the fuel consumption between a series hybrid vehicle with Max. SoC control strategy and a conventional vehicle in the FTP-75 driving cycle

Fuel consumption	Series hybrid vehicle (Max. SoC control strategy) (L/100 km)	Conventional vehicle (L/100 km)
Engine fuel consumption	6.41	9.11
Equivalent fuel consumption of the battery	−1.80	–
Total fuel consumption	4.61	9.11

Table 8.11 Comparison of the fuel consumption between a series hybrid vehicle with Max. SoC strategy and a conventional vehicle in the HWFET driving cycle

Fuel consumption	Series hybrid vehicle (Max. SoC control strategy) (L/100 km)	Conventional vehicle (L/100 km)
Engine fuel consumption	4.86	6.38
Equivalent fuel consumption of the battery	−0.55	–
Total fuel consumption	4.31	6.38

Tables 8.10 and 8.11 represent the fuel consumption of the engine, and the equivalent fuel consumption of the battery in the FTP-75 and HWFET driving cycles for both hybrid and conventional vehicles, respectively. Compared to HWFET, there is a considerable improvement in fuel efficiency due to the higher performance of the regenerative braking system.

The following summarizes the results of Examples 8.1 to 8.4:

- Series and parallel hybrid vehicles with both Maximum SoC and a thermostat have lower fuel consumption than conventional vehicles, especially during urban driving cycles due to recovery from expending braking energy.
- With the thermostat control strategy, the engine always works on optimum operating points; whereas with the Maximum SoC control strategy, the engine has to operate on less efficient points when power demand is less than optimum engine power and the SoC is over its top line. As a result, the thermostat control strategy always has lower fuel consumption than the Maximum SoC.
- When the main goal of the powertrain system design is performance (rather than fuel consumption), the Maximum SoC strategy is appropriate because it keeps the SoC at the maximum level and batteries can consistently provide the demanded power.
- Based on simulation results, series hybrid vehicles with a thermostat control strategy have the lowest fuel consumption among all rule-based control strategies, as engine speed and engine torque are set to constant values that consume the least amount of fuel.

- In the rule-based control strategies, motor operating points cannot be controlled, so it cannot be guaranteed that it works in the most efficient region at all times.
- The performance of both controllers can be improved by applying appropriate defining rules and tuning the control system for certain specific driving cycles.

8.3.1.2 The State-Machine Based Strategy

The state-machine based strategy is composed of a set of states, a set of transitions, and actions. In this strategy, first, all possible vehicle operation modes are identified and defined as controller states. Next, the transitions between these states are determined by considering a change in driver demand, a change in vehicle operating condition, or a system or a subsystem fault, in conjunction with the performance and drivability objectives. In the final stage, transitions are analyzed for exclusivity in order to guarantee single-valued decisions within the state machine. Dynamic control or optimization algorithms may be used to generate output commands to each powertrain subsystem according to desired power demand and vehicle states. This strategy is applicable to any hybrid powertrain configuration. Although it has a simple and robust structure for practical implementation, its use cannot guarantee optimization of the performance objectives, such as fuel economy or emissions.

8.3.2 Fuzzy-Rule-Based Control Strategies

Fuzzy logic controllers are another form of rule-based controllers that handle the multi-domain, nonlinear, and time-varying structure of the power split problem in all forms of hybrid vehicles. These controllers can provide a real-time and suboptimal power split using their decision-making property, rather than using deterministic rules. The main advantages of fuzzy logic controllers are their robustness against inaccurate measurements and component uncertainties, and that they are easily adaptable to component variations. However, compared to deterministic rule-based controllers, fuzzy logic controllers consume more memory and controller time, resulting in the integration of faster microprocessors with larger memory. Additionally, they still use predetermined rules based on specific drive cycles, thereby making them sensitive to changes in driving conditions. Fuzzy logic controllers can also be combined with a state machine algorithm to make discrete decisions (such as turning the engine on or off), or to calculate the power split ratio in the hybrid mode of the state machine algorithm. Variations of fuzzy logic controllers developed to improve its performance include: the adaptive fuzzy logic controller, the predictive fuzzy logic controller, and the neuro-fuzzy controller at the cost of increasing the computational burden.

8.3.3 Rule-Based Control Strategies for PHEVs

It is not possible to directly apply the rule-based control strategies explained in Sections 8.3.1 and 8.3.2 to a PHEV. In a PHEV, the primary traction source is the electric motor, and it is possible to drive the PHEV in electric mode until the battery is almost depleted, then switch to a charge-sustaining mode in which the PHEV operates similar to an HEV. Another option is to blend the use of the IC engine and electric motor throughout the trip in a way that nearly depletes the battery at the charging station [4]. Thus, two operational modes are possible for a PHEV: charge-depletion and charge-sustaining modes.

- *Charge-depleting mode*: The PHEV can operate in this mode when the battery's state of the charge is high. It allows the battery's charge to deplete to its minimum value with either pure–electric operation or a blended operation of the electric motor and IC engine. Due to the lower cost of the electric grid energy compared to fossil fuel energy, it is cheaper to drive the PHEV in its pure-electric mode for short trips between two charging stations. In charge-depletion mode, the engine only assists the electric motor when the requested power is more than what the battery can deliver. The regenerative braking system solely charges the battery.
- *Charge-sustaining mode*: This mode is activated when the battery SoC is approaching its minimum value. In this mode, the PHEV operates by blending the IC engine and the electric motor. The battery's state of the charge allows for small deviations from a set value, similar in operation to a conventional HEV. Thus, all the rule-based power/energy managements are transferable to PHEVs in this mode.

8.4 Optimization-Based Control Strategies

One way to approach optimizing a power/energy management system is to use a mathematical model where a set of decision variables represents the operation of each power system or component. Each measure of system performance can be the expressed as a function of the decision variables and the overall performance of the system can be expressed as a weighted combination of the functions representing the measures of system performance. This is called the objective function.

In optimal controllers, the design procedure starts by defining a proper multi-objective cost function with which an optimal control action will be formulated with or without prior knowledge of drive cycles and with consideration of physical constraints and limitations. The ultimate objective of the controllers is to minimize the vehicle emission and fuel consumption, while considering the torque demand, and the component and energy limitations. The solution to the optimization problem yields the "optimal" distribution of power demand between the engine, motor, generator, and brakes. Optimal control systems provide an optimal solution for the power management of electric and hybrid vehicles. However, their formulation is based on dynamic and static mathematical models of the powertrain system components, resulting in sensitivity to system parameter variation, system uncertainties, imprecise measurement, and noises.

In optimization-based controllers, constraint equations set restrictions on the possible values of decision variables. By minimizing (or maximizing, if relevant) the objective function, it is possible to determine values of the decision variables, for a specific set of constraints. Several key factors impose constraints upon the system: dynamic resource allocation, physical constraints of vehicle components, uninterruptible power availability, power quality, system stability, and fault diagnosis and prognosis [1]:

- *Dynamic resource allocation*: There is a significant difference between peak and average power demands for electric loads in a vehicle. As a result, it is impractical to design a vehicle's power system to meet the peak power demand for all loads. The vehicle's power and energy management system must match the power sources with loads, dynamically prioritize and schedule load requests, and allocate power and energy based on need and priority.

- *Physical constraints of vehicle components*: Physical properties impose constraints on power and energy management optimization algorithms. Batteries are sensitive to temperature, have a limited lifespan, and a limited capacity for power and energy. The power output and efficiency of motors and generators are functions of machine speed, torque, and temperature.
- *Uninterruptible power availability*: Certain electric loads are essential for vehicle operation or for safety reasons. Such loads include: steer-by-wire, brake-by-wire, safety restraint subsystems, and some military systems. They must run at all times and must receive uninterruptible power. In order to ensure uninterrupted power for these loads, it is a good practice to have one or more back-up batteries in case of a primary power source interruption. To minimize weight and cost, the back-up power source usually has a limited energy and power capacity. In order to maximize the back-up battery's operational life, charging and discharging cycles should be kept to a minimum, also the battery should be disconnected when not in use to maintain its state of charge.
- *Power quality*: Steady power quality of power buses is necessary to guarantee the safety and proper operation of electric loads and power control units. There are many noisy and transient automotive environments as a result of switching large currents through inductive loads. Disconnecting a battery while it is charging can cause load dumps, which result in voltage spikes. This spike would appear on the power bus and could potentially damage all the electronic devices connected to the bus. Low temperatures or sudden overload can cause the bus voltage to drop below the desirable range.
- *System stability*: DC power systems with multiple switch-mode power converters are sensitive to parameter and load variation. A concern regarding automotive power systems is whether the system is able to move from one steady-state condition to another after a change in power demand, loss of power sources, or other disturbances, such as a short circuit or open circuit. In automotive power systems, system stability depends on several factors including: switching-off power electronic converters, nonlinearities of magnetic components, self-protection operation of power electronic circuits, and temperature variations. Managing loads according to the operating conditions of power sources and distribution systems helps ensure that the system operates around its nominal power source.
- *Fault diagnosis and prognosis*: Due to the complexity of electrical/electronic architectures, having a way of identifying faults and their causes through the automotive power management and distribution system is essential. This allows for the detection and mitigation of conditions such as short circuits and open circuits. By incorporating this into the power management and distribution system, it is possible to monitor and isolate fault conditions and reconfigure the system to minimize the impact of a fault condition. In a more sophisticated system, historical data can predict the state of components and subsystems to provide early warnings for future component failure.

The optimization-based controllers can be classified into global and real-time power management control systems according to their solutions. Time-invariant global optimum solutions can be obtained by performing optimization with the complete knowledge of past and future power demands and driving conditions. The on-line implementation of such strategies is not directly possible due to the associated computational burden. Additionally, as in practice, there are changes to the drive cycles, and as a result, global optimal systems are not able to adapt to the real-time changes, thereby limiting their efficiency in varying driving conditions. In contrast, the time-dependent cost functions of real-time optimal control systems rely on powertrain system variables at the current time and optimize the power distribution between the system components instantaneously. Although real-time algorithms are implementable on-board, they do not provide the global optimal solution.

In general, it is important to develop an optimization-based controller for power/energy management of EVs/HEVs/PHEVs that is:

- computationally efficient
- predictive in nature
- adaptive to changes in the plant operating point and external disturbances
- systematically in tune with less parameter sensitivity.

8.4.1 Optimization Problem Formulation

8.4.1.1 HEVs and PHEVs

The main objective of energy/power management of HEVs and PHEVs is to minimize the fuel consumption over a trip. It is important to minimize the total fuel consumed during a driving mission, rather than to minimize the fuel mass-flow rate at all times of the driving mission [1]. In order to design a power/energy control strategy for an HEV, the following considerations need to be taken into account: optimal engine operation; engine speed, fluctuation, and on/off cycling; battery capacity and voltage; power distribution; and geographical policy. Optimal engine operation describes the engine operating point, line, and region, where the engine has the maximum fuel economy for different power demands. Engine speed, fluctuation, and on/off cycling reduce fuel efficiency and increase vehicle emissions. Maintaining battery capacity at a level that is able to both provide sufficient power for acceleration, and accept regenerative power while braking or going downhill is necessary. Maintaining the battery voltage is required to prevent damage to the battery. The distribution of power demand between the battery and the engine is important to ensure that there will always be enough power available to meet demand. Certain cities or areas have low emission or zero emission zones so geographical policies must be taken into account when deciding whether to operate an HEV from battery power only.

The simplest cost function defined for HEV power/energy management optimization problem to minimize the fuel mass over a vehicle trip as in [5] is:

$$J = \int_0^t \dot{m}_f(\tau, u(\tau))d\tau \tag{8.1}$$

where $u(t)$ is the vector of control input, and $\dot{m}_f$ is the fuel mass flow rate. In HEVs, this includes engine and motor powers (torques). Pollutant emissions can also be included in the cost function, J, as:

$$J = \int_0^t \left(\dot{m}_f(\tau, u(\tau)) + \sum \lambda_i \dot{m}_{x_i}(\tau, u(\tau))\right)d\tau \tag{8.2}$$

where $\dot{m}_x$ is the mass flow rate of interested emission, the subscript x denotes the type of emission, and λ_i is the weighting factor, this value depends on the penalty given to the associated emission. However, as mentioned in Section 1.3, with the advent of new emission control technologies and sophisticated modifications on IC engine designs, it is usually possible to reduce the pollutant emissions to negligible levels. Driving and comfort issues, such as noise and vibration, can also be included in cost function by defining an appropriate weighting factor.

- *Battery SoC constraint*: In HEVs, sustaining the energy storage system is imperative as the battery can only recharge through the on-board regenerative braking system and engine. As a result, it only permits small deviations from the nominal value of SoC [5]. Thus, in HEVs energy/power management controllers must ensure small SoC variations over the vehicle trip. Two types of battery sustenance constraints are soft and hard constraints. The battery sustenance constraint can be applied either as soft constraint or hard constraint. A soft constraint can be applied by penalizing deviations from the initial value of the energy stored at the end of a trip while a hard constraint can be applied by requiring that the energy stored at the end of trip equal the value at the start of trip.

A penalty function $\phi(SoC(t))$ can be added to the cost function in Equation 8.2 in order to formulate a charge-sustaining cost function of the form

$$J = \phi(SoC(t)) + \int_0^t \left(\dot{m}_f(\tau, u(\tau)) + \sum \lambda_i \dot{m}_{x_i}(\tau, u(\tau))\right) d\tau \tag{8.3}$$

Assuming a full charge battery at the start of trip, according to Equation 7.53, the hard constraint can be defined as:

$$SoC(t) = SoC(0) \Rightarrow \phi(SoC(t)) = \int_0^t i(\tau) d\tau = 0 \tag{8.4}$$

A soft constraint can be formulated by a function of difference between the final and initial state of charges. One formulation is a quadratic representation of this difference as:

$$\phi(SoC(t)) = \lambda(SoC(0) - SoC(t))^2 \tag{8.5}$$

Here λ is a positive weighting factor. Quadratic SoC constraint penalizes deviations from the nominal SoC, regardless of the sign of SoC. The penalty function in Equation 8.5 can be added to the cost function in Equation 8.2 to formulate the optimization problem. Another formulation may represent the SoC constraint as a linear function of SoC differences as:

$$\phi(SoC(t)) = \lambda(SoC(0) - SoC(t)) \tag{8.6}$$

In the linear formulation, the penalty function penalizes battery use at the cost of allowing the energy stored in the battery to be used to save fuel in the future. The penalty function in

Equation 8.6 can be reformed as:

$$\phi(SoC(t)) = \lambda \int_{0}^{t} i(\tau)d\tau \tag{8.7}$$

Adding the above penalty function to the cost function in Equation 8.2 produces:

$$J = \int_{0}^{t} \left(\dot{m}_f(t, u(\tau)) + \sum \lambda_i \dot{m}_{x_i}(\tau, u(\tau)) + \lambda i(\tau)\right) d\tau \tag{8.8}$$

It is possible to choose λ arbitrarily, however, its recommended value in the regulatory standard SAE J1711 is 38 kWh per gallon of gasoline. Another option is to represent the SoC penalty function as a piecewise-linear function:

$$\phi(SoC(t)) = \begin{cases} \lambda_{dis}(SoC(0) - SoC(t)), & SoC(t) < SoC(0) \\ \lambda_{chg}(SoC(0) - SoC(t)), & SoC(t) < SoC(0) \end{cases} \tag{8.9}$$

It is possible to impose other constraints, which primarily concern physical operation limits, on the state and control variables. The mechanical constraints on both the speed and torque of the engine and electric motor are as follows.

Constraints on speeds:

$$0 < \omega_m(t) < \omega_{m,max} \tag{8.10}$$

$$\omega_{e,min} < \omega_e(t) < \omega_{e,max} \tag{8.11}$$

Constraints on torques:

$$T_{m,min} < T_m(t) < T_{m,max} \tag{8.12}$$

$$T_{e,min} < T_e(t) < T_{e,max} \tag{8.13}$$

In addition, if the powertrain (for example, a parallel powertrain), includes a reductor to ensure that both the motor and the engine achieve their maximum speed simultaneously, or if the powertrain includes a gear set, the imposed constraint is:

$$k(t) \in K(t) \tag{8.14}$$

Here $k(t)$ is gear number and $K(t)$ is a set of admissible gear numbers at a given sample time.

8.4.1.2 EVs

Since the battery is the only energy source of EVs, the objective of EV energy/power management is to minimize the energy consumption of the vehicle and its auxiliary electronic systems during the trip, in order to extend the vehicle driving range as much as possible. In EV applications, kWh is unit of energy, and kWh/km is energy per unit distance is a measure of vehicle energy consumed. To measure the energy performance of EVs, the energy efficiency, η_e, is defined as the ratio of output energy, E_{out} to input energy, E_{in}:

$$\eta_e = \frac{E_{out}}{E_{in}} \tag{8.15}$$

Similarly, power efficiency is:

$$\eta_P = \frac{P_{out}}{P_{in}} \tag{8.16}$$

Although applied interchangeably for vehicle applications, the terms energy efficiency and power efficiency are significantly different. Note that depending on the speed and torque conditions of electric motors integrated in an EV, the power efficiency of each EV is different. In an EV, the associated efficiency map of its electric motor, such as in Figure 7.6, helps estimate the power efficiency at each operating point (e.g., rated power at rated torque and rated speed). Energy efficiency is the summation of powers over a given time period [6]. In other words, power efficiency deals with vehicle performance characteristics (e.g., acceleration, deceleration, gradeability), while energy efficiency refers to the vehicle travel range.

Energy consumption of an electric vehicle is equal to the integration of the power output at the battery terminals over the vehicle travel interval. The battery output power is equal to the power delivered to the driving wheels plus the power consumed by the auxiliary electronic systems (e.g., A/C, radio, windshield, window power, etc.). To calculate battery output power:

$$P_{b,out} = \frac{V}{\eta_t \eta_m} \sum f_x + \sum p_{aux} \tag{8.17}$$

Here η_t and η_m are the transmission and motor drive efficiencies. $\sum f_x$ is the summation of longitudinal forces given in Equation 6.1 while $\sum p_{aux}$ is the summation of power consumption by the auxiliary electronic systems.

However, it is possible to compensate for some of this energy consumption by regenerating wasted kinetic energy during braking through regenerative braking in which the motor is operating as a generator and restoring energy into the batteries. The regenerated braking power can be expressed as:

$$P_{b,regen} = \frac{\beta V}{\eta_t \eta_m} \sum f_x \tag{8.18}$$

Called the regenerative braking factor, $0 < \beta < 1$ is the percentage of the total braking power (energy) that can be regenerated. This factor is a function of the applied braking strength and the design and control of the braking system [7]. Thus, the total energy consumption from the batteries is:

$$E_{consumption,EV} = \int P_{b,out} dt - \int P_{b,regen} dt \tag{8.19}$$

When the value of energy consumption approaches the total energy in the batteries measured at their terminals, it indicates that the battery is almost completely flat and needs to be charged.

As previously mentioned, the objective of an energy management system of an EV is to extend the vehicle driving range using available energy from the batteries. Accordingly, the cost function, J, of the energy management optimization problem for EV applications is to minimize vehicle energy consumption over a trip of duration t. Thus, the cost function is:

$$J = \int_0^t P_{b,out} d\tau - \int_0^t P_{b,regen} d\tau \tag{8.20}$$

subject to the limitations of the drivetrain components, such as those mentioned for HEVs. Note that there is no sustenance SoC constraint in EV applications, since the battery is the only energy source of the system. Thus, an EV needs to use the battery energy until depletion. Also, note that if the EV benefits from a hybrid storage system, such as hybrid battery-ultracapacitor, the cost function in Equation 8.20 needs to be modified by adding the energy capacity of an ultra-capacitor. However, it complicates the optimization problem.

8.4.2 Global Energy/Power Management Optimization

In general, there are solutions to power management optimization problems that are locally optimal, but not globally optimal. The task of a global power management optimization is to find the best set of parameters that optimize a power management cost function as defined using historical data over a given driving cycle. Such controllers are typically optimized offline and the solution is implemented on electric or hybrid vehicles. Despite providing a global optimal solution, requisite knowledge of the entire driving pattern (e.g., battery SoC, driving conditions, driver response, and route prediction) and the high computational burden prevent the real-time implementation of such controllers. However, these issues are declining with the rapid development of technologies such as Intelligent Transportation Systems (ITS), Geographical Information Systems (GIS), Global Positioning Systems (GPS), and Internet maps, as well as available wireless technologies. The on-board GPS reports the vehicle location in real time, while roadside sensors capture historic traffic information in real time. Traffic flow information may be obtained in real time with historical traffic flow data and advanced traffic flow modeling techniques. Combining this information will significantly reduce the uncertainty of trip prediction. If the trip becomes significantly predictable, global optimization techniques will then be implementable for real-time purposes.

In addition, using modern technologies, global power/energy management systems can be a path planner for the electric and hybrid vehicles using the traffic information and other data obtained through sensors. Note that the shortest path found using a global power/energy managment system may not be necessarily the path with minimum energy consumption, since the traffic flow, downhill, uphill, path surface conditions, and the like can affect the energy consumption of a vehicle significantly.

As another advantage, the results of global control strategies can be a design reference for evaluating the quality of other control strategies, and as a basis to define rules for online implementation. EVs and PHEVs are the principal potential applications of global power management systems where the battery depletion is important. Additionally, in public transportation vehicles along fixed routes, with advance knowledge of travel path and conditions, such controllers may be applicable. Linear programming and dynamic programming are common methods for a global power management.

8.4.3 Real-Time Energy/Power Management Optimization

Instead of solving the whole control problem globally, real-time power management systems find the local optimal solution for the proportion of power between power sources at each time step. The main advantage of real-time optimization controllers over the afore-mentioned controllers refers to their ability to optimize and to adapt themselves in real time to varying driving conditions. The cost function of real-time controllers depends on the system variables at the current time, such as present power source and SoC. The system variables are defined based on historic data, such as past drive cycles and past traffic reports. In such strategies, the control actions typically are calculated by minimizing the time-dependent local cost function at each instant and by using a mathematical model of the powertrain and its components. Real-time optimal controllers generally have a simple structure in order to be implementable on-board microprocessors with limited computation and memory resources. However, in order to complete online optimization in real time, such strategies typically require more computational power than rule-based methods. Additionally, they do not guarantee a behavior that is sufficiently close to the global optimal solution. Model predictive control (MPC) and energy consumption minimization system (ECMS) are examples of the real-time control strategies.

As shown in Figure 8.27, with the expansion of geographical information technologies, and high-resolution maps of terrains, it is possible to obtain global optimized performance of a power/energy controller while adapting the control actions in real time by properly combining

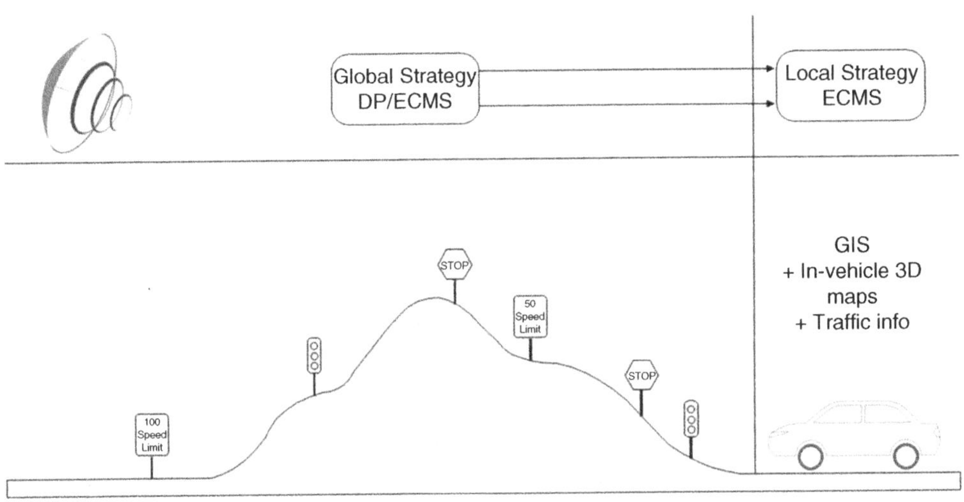

Figure 8.27 Combined global and local power/energy management controller

the global and real-time optimization strategies. For example, it is possible to use dynamic programming and model predictive controller simultaneously to achieve a reliable and global optimized performance.

8.4.4 Optimization Techniques

8.4.4.1 Linear and Dynamic Programming

Linear programming is a mathematical method for the optimization of linear cost functions subject to linear constraints and linear mathematical models of the powertrain system and components. However, the power management of electric vehicles and hybrid vehicles is a nonlinear optimization problem. The problem can be approximated as a linear problem using linearization methods such as piece-wise affine linearization. The need for linearization reduces the application of linear programming to complicated powertrain architectures and complex power management problems.

As a numerical algorithm based on Bellman's optimality principle, dynamic programming (DP) is the most popular technique for global power management optimization problems [8]. This optimization approach transforms a complex, constrained problem into a sequence of simpler sub-problems. By discretization of the time, control inputs and state variables of the system, DP evaluates a given cost function for every possible state and control input combination at each time step. The global optimal solution is a control signal that guides the system to the final time interval with minimum cost expended.

In power management applications, using mathematical dynamic models of the powertrain system and its components, DP can provide an optimal control law minimizing fuel consumption, or combined fuel consumption and emissions for a given driving cycle. However, this approach is not an option for real-time implementation due to computation time and requisite prior knowledge about the entire driving cycle. Despite this, DP is a valuable analytical tool that provides several important advantages for electric and hybrid electric power management applications, including:

- DP can determine the global optimal solution and is a valuable benchmark while developing online power management methods.
- In the design process of electric and hybrid vehicles, DP is a valuable asset to help with the optimal component selection and sizing.
- Evaluation of the effects of the problem requirements and constraints can be done using DP.

Formulate the general form of the DP approach as follows:

Find the optimal control sequence:

$$\pi(x_0) = \{u_0, u_1, \ldots, u_{N-1}\} \tag{8.21}$$

that minimizes the cost function:

$$J(x_0) = g_N(x_N) + \sum_{k=0}^{N-1} g_k(x_k, u_k, w_k) \tag{8.22}$$

such that:

$$x_{k+1} = f(x_k,\ u_k,\ w_k) \tag{8.23}$$

$$x(0) = x_0 \tag{8.24}$$

$$x_k \in X_k \subset \Re^{\mathrm{n}} \tag{8.25}$$

$$x_k \in [x_{min}, x_{max}] \tag{8.26}$$

$$u_k \in U_k \subset \Re^{\mathrm{m}} \tag{8.27}$$

$$u_k \in [u_{min}, u_{max}] \tag{8.28}$$

where k represents sample-time, N is the final time step, u_k is the control input, x_k is the state vector, w_k is the vector of time-varying model parameters, f represents (linear or nonlinear) dynamics of the system, m is the dimension of input spaces, and n is the dimension of the state spaces. Figure 8.28 shows a schematic of the DP approach. The algorithm starts from the final time and proceeds backwards in time ($k = N,\ N-1, \ldots,\ 0$) in the discrete state-space. In the backward process, DP evaluates the optimal cost-to-go and the optimal control sequence that drives the system from x_k to x_N. Possible state values at each time step $\left(x_k^1,\ x_k^2, \ldots,\ x_k^{n_x}\right)$ are conceptualized as nodes in a grid, with n_x nodes that cover $x_k \in [x_{min}, x_{max}]$.

The sub-problem for the $(N-1)th$ step is to find the best control input u_{N-1} that drives the system from x_{N-1} to x_N and results in the minimum cost function:

$$J^*_{N-1}(x_{N-1}) = g_N(x_N) + g_{N-1}(x_{N-1},\ u_{N-1},\ w_{N-1}) \tag{8.29}$$

For every step k ($0 < k < N-1$), the sub-problem is to find u_k that drives the system from x_k to x_{k+1} with the minimum cost of:

$$J^*_k(x_k) = g_k(x_k,\ u_k,\ w_k) + J^*_{k+1}(x_{k+1}) \tag{8.30}$$

$J^*_k(x_k)$ is called optimal cost-to-go function.

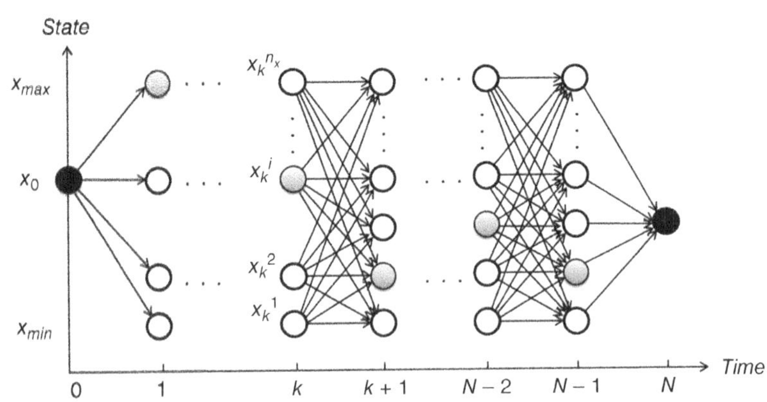

Figure 8.28 DP evaluation of the global optimal solution

The optimal path-to-go from the initial state x_0 to the final state x_N (gray nodes in Figure 8.28) and the corresponding control law ($\pi(x_0) = \{u_0, u_1, \ldots, u_{N-1}\}$) that enables this optimal trajectory is the solution for the global optimization problem.

Such a control law guarantees optimality of the control solution considering that the algorithm assesses every possible trajectory. Note that during the recursive evaluation of the optimal trajectory, the solution discards every path that violates the problem constraints.

DP Algorithm for power management applications: Recent research has considered the DP approach both for optimal design of EVs and HEVs, and as a benchmark for assessing the performance of other power management methods [7–11]. Depending on the vehicle type, choice of the states, inputs, cost function and the problem constraints, it is possible to formulate the power management problem and find the optimal control solution using the DP approach [12].

One possible formulation of the optimization problem to be solved by the DP algorithm for HEVs is summarized as follows:

Find the engine power (P_e) that minimizes:

$$J = \sum_{k=0}^{N} \Delta m_f(P_e(k),\ \omega_e(k)) \tag{8.31}$$

Subject to:

$$SoC(k+1) = f(P_{demand}(k), P_e(k),\ SoC(k)) \tag{8.32}$$

$$P_{demand}(k) = P_e(k) + P_b(k) \tag{8.33}$$

$$SoC(0) = SoC_0 \tag{8.34}$$

$$SoC_{min} \le SoC(k) \le SoC_{max} \tag{8.35}$$

$$0 \le P_e(k) \le P_{e,max} \tag{8.36}$$

$$P_{b,min} \le P_b(k) \le P_{b,max} \tag{8.37}$$

$$\omega_{e,min} \le \omega_e(k) \le \omega_{e,max} \tag{8.38}$$

where the demand power ($P_{demand}(k)$) is a given function, P_b is the battery power, m_f is the fuel mass and ω_e is the engine speed. The following example clarifies application of the DP approach for fuel consumption minimization in a PHEV.

Example 8.5

Design a DP energy management control system for a plug-in series HEV shown in Figure 8.29, considering the following assumptions and constraints:

$$SoC_0 = 60$$

$$SoC_{min} = 20$$

$$SoC_{max} = 90$$

$$\omega_{gen} = \omega_e$$

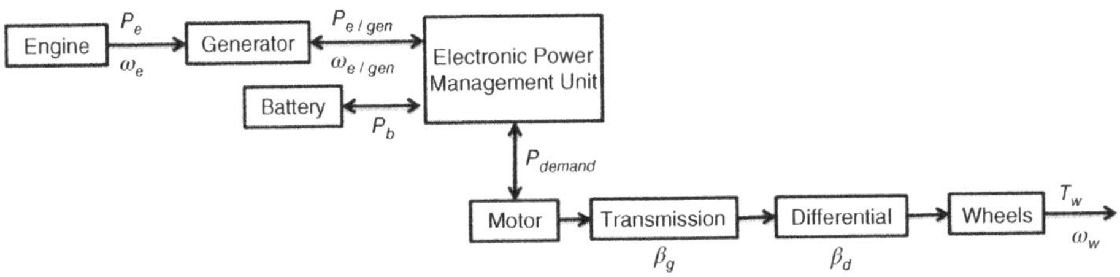

Figure 8.29 A plug-in series hybrid electric vehicle

where ω_{gen} is the generator speed. Assume that the PHEV is working on charge depletion mode. Evaluate the fuel consumption for the HWFET driving cycle, and compare that with the fuel consumption of a conventional vehicle. Consider the energy regeneration for charging the batteries in the algorithm.

Solution

This example discusses the global optimization of the power management control strategy for a plug-in series hybrid electric vehicle. In such HEVs, both the battery and the engine/generator unit provide the demand power:

$$P_{demand}(k) = P_b(k) + P_{e/gen}(k)$$

where $P_{e/gen}$ is the output power available from the engine/generator unit. When braking, the demand power is negative and the regenerative braking charges the batteries. When accelerating, assume that the engine/generator unit is the main source to provide the demand power. Such an assumption is valid considering that the DP algorithm evaluates the cost function for all possible combinations of contributions from the battery and engine/generator sources to provide the demand power.

Therefore, at each time interval:

$$P_b(k) = P_{demand}(k) - P_{e/gen}(k)$$

The output power from the generator depends on the efficiency maps of the engine and the generator for every specific operating point:

$$P_{e/gen}(k) = P_e(k)\eta_e(\omega_e, P_e)\eta_{gen}(\omega_{gen}, P_{gen})$$

where η_e and η_{gen} are the engine and generator efficiency, respectively. Assuming the same speeds for the generator and the engine:

$$\omega_{gen} = \omega_e = \omega_{e/gen}$$

The cost function can be defined as:

$$J = \sum_{k=0}^{t_f} \Delta m_f(P_e(k),\ \omega_e(k))$$

At each time interval, the algorithm receives the demand power and evaluates the cost function for all possible magnitudes of $P_{e/g}$ and $\omega_{e/g}$. Let us discuss the optimization process considering the demand power conditions:

Positive demand power (Traction mode): Depending on the chosen operating point, two scenarios are possible for positive demand power:

$$P_{demand} \leq P_{e/gen}$$

In this case, the engine/generator unit provides the demand power. The rest of the generated power charges the batteries.

$$P_{demand} > P_{e/gen}$$

All the power generated goes to the electric motor. Batteries provide the remaining required power.

Negative demand power: For a negative demand power, regenerative power charges the batteries. The engine works in idling mode to minimize the fuel consumption. Discrete grids represent the time in the optimization domain, SoC as the state variable, and $P_{e/gen}$ and $\omega_{e/gen}$ as the inputs are considered as:

- Time grid: one second intervals between $t = 0$ and $t = t_f$
- State of charge grid: 10 grid points between SoC_{min} and SoC_{max}
- Power grid: 10 grid points between 0and $P_{e,max}$
- Speed grid: 10 grid points between $\omega_{e,min}$ and $\omega_{e,max}$.

Note that increasing the number of grid points results in a more accurate solution, but increases the computation time. In the optimization process, the DP algorithm discards routes from the solution set that violate the problem constraints. A generic Matlab DP function is available in [13] and is used in this example.

The described DP power management strategy minimizes the fuel consumption for the HWFET driving cycle. Figure 8.30 illustrates the demand power for the HWFET driving cycle.

Figures 8.31 and 8.32 illustrate the optimal results from the DP for the engine/generator power ($P_{e/gen}$) and battery power. The summation of the battery and engine/generator power is equal to the demand power as required in the problem. During the time frames when the DP algorithm has selected the battery as the main power source (e.g., during the first 300 seconds of the driving cycle), the engine power is minimum in order to minimize the fuel consumption. Whenever the demand power is negative (i.e., during braking), the battery is charged (i.e., battery power is negative) as can be seen in Figure 8.32.

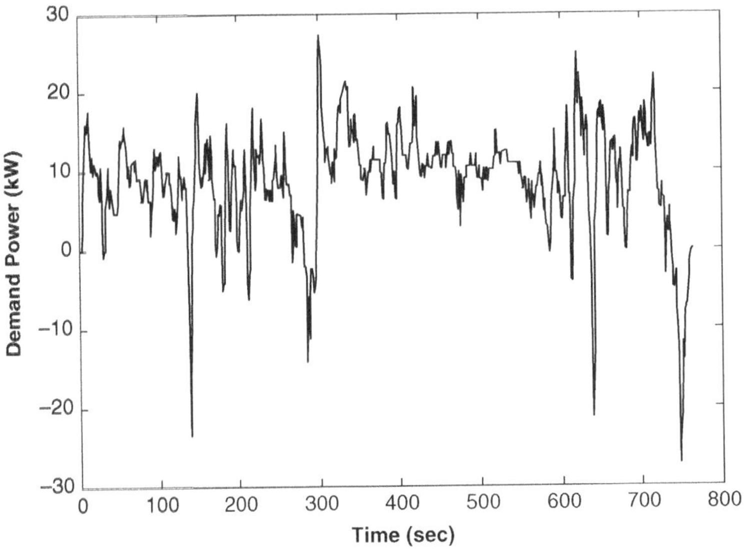

Figure 8.30 Demand power in the HWFET driving cycle

The optimal state (*SoC*) trajectory in Figure 8.33 shows how the state evolves through time in the optimal solution.

Figure 8.34 illustrates the engine operating points in the optimal solution. According to the results, all the requested power from the motor is put in the region with the highest

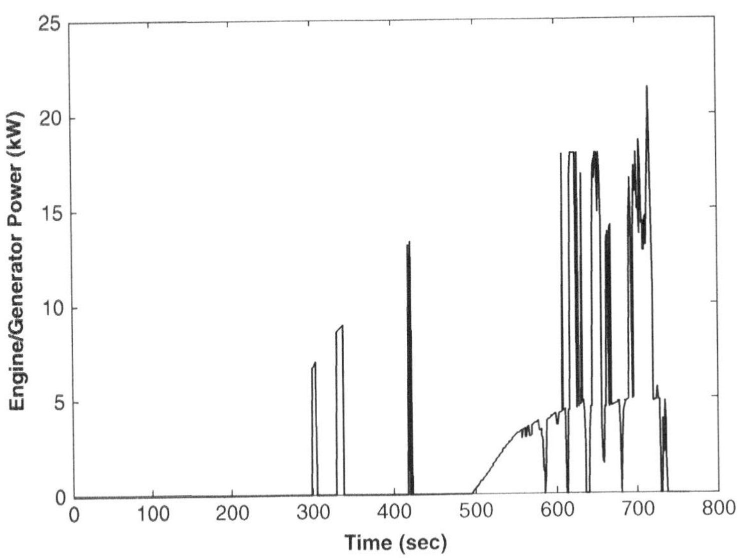

Figure 8.31 Optimal engine/generator power trajectory for the HWFET driving cycle

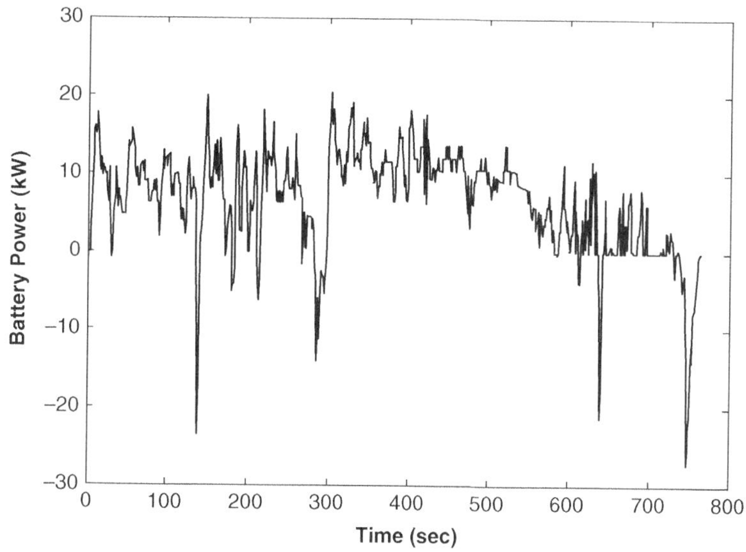

Figure 8.32 Optimal battery power trajectory for the HWFET driving cycle

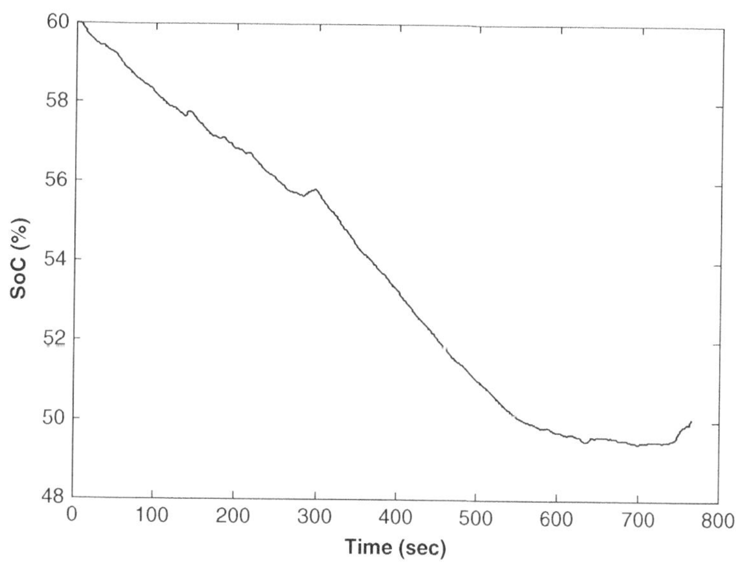

Figure 8.33 Optimal SoC trajectory for the HWFET driving cycle

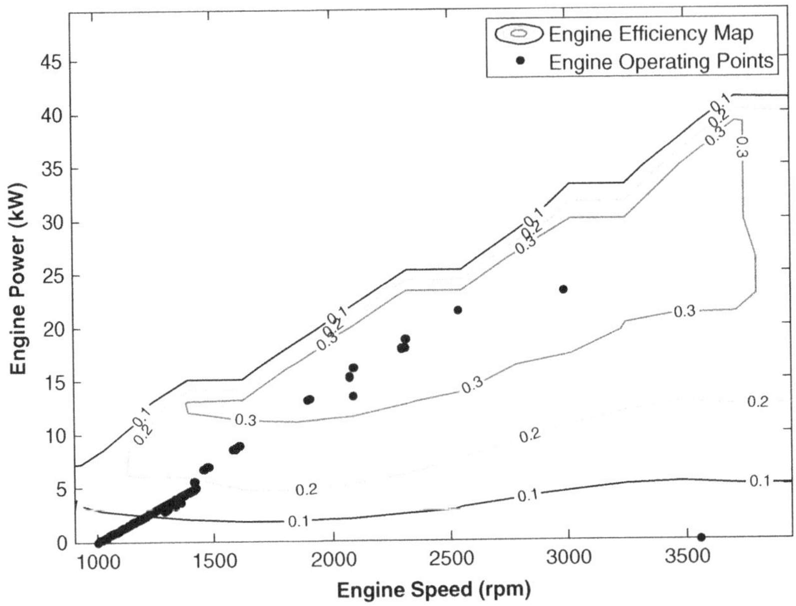

Figure 8.34 Optimal engine operating points for the HWFET driving cycle

Table 8.12 Comparison of the fuel consumption between a PHEV Dynamic Programming strategy and a conventional vehicle in the HWFET driving cycle

Fuel consumption	Plug-in series hybrid vehicle (DP control strategy) (L/100 km)	Conventional vehicle (L/100 km)
Engine fuel consumption	1.58	6.38
Equivalent fuel consumption of the battery	2.52	–
Total fuel consumption	4.10	6.38

efficiency. In other words, for all requested power from the engine/generator unit, the algorithm has adjusted the engine speed so that the operating point is put in the region with the highest possible efficiency. Such an outcome verifies the minimized fuel consumption for the assigned value of the $P_{e/gen}$ at every time step.

In order to assess the performance of the DP approach, assess the fuel consumption for the PHEV with this power management strategy compared to that of the conventional vehicle in Table 8.12. The results show that the HWFET driving cycle achieved a significant reduction in the required fuel.

8.4.4.2 Model Predictive Control (MPC)

Model predictive control solves the power management optimization problem online at each time interval in a future time frame, based on the predicted states and inputs, while respecting the limitations and time-varying constraints of powertrain components such as the engine, motor, generator, and battery. In each time interval, new optimizations are completed using updated predictions and new measurement data. In such controllers, instead of using information about the future drive cycle, a mathematical model is used to estimate the torque demand and the resulting velocity over a future prediction horizon. The main advantage of MPC algorithms is their ability to handle constraints directly in the design procedure.

8.4.4.3 Energy Consumption Minimization System (ECMS)

Originally proposed by Paganelli *et al.* [14], the concept of ECMS is based on findings that, in a hybrid vehicle, running the engine replenishes the energy consumed from the battery. As such, discharging the battery at any time is equivalent to some fuel consumption in the future and vice versa. Therefore, this strategy calculates the proper control signal by minimizing an instantaneous cost function where the combination of energy used in the battery and the ICE fuel consumption through an equivalent factor represents the total energy consumed. For hybrid vehicles, ECMS minimizes the instantaneous fuel consumption while keeping the power source SoC around a constant reference point.

Figure 8.35 represents the energy/power flows through the main components of a hybrid powertrain. It shows the consequences of the SoC constraint over the fuel consumption. Figure 8.35(a) depicts the energy/power flows corresponding to a battery discharge at the present (current sampling time), and the power flows for future use, which keep the SoC at its required level of charging. In this mode, both motor and engine provide the required power/energy for vehicle traction. To maintain the battery SoC, the same amount of power/energy will be consumed in the future for the battery recharge by converting the power/energy provided by the IC engine to electricity by the electric motor in its generator mode. At each instant, for each

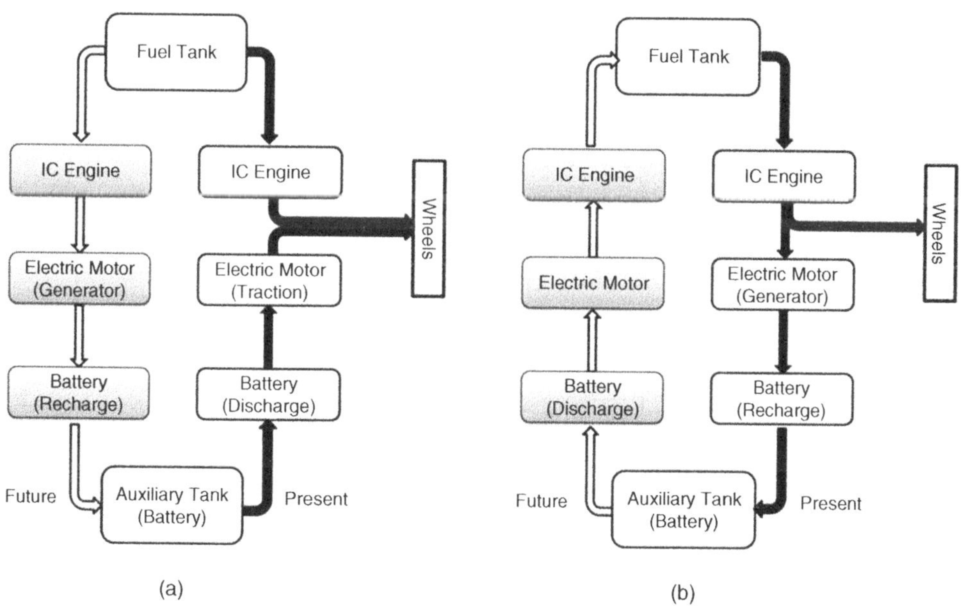

Figure 8.35 Energy path for ECMS during the (a) battery discharge and (b) recharge

possible set-point of the electric motor, the extra fuel consumption used to recharge the battery is defined as the equivalent fuel consumption. Note that this recharge will happen in the future. However, since the all set-points of the components are unknown at the current time, mean efficiencies are used in ECMS.

Figure 8.35(b) indicates the energy/power flow corresponding to the battery recharge mode at the present (current sampling time) and the power flows necessary in the future to keep the SoC at its required charging level. In this mode, the engine provides the required power/energy for vehicle traction and the battery recharging. To maintain the battery at its pre-defined level of SoC, discharge the battery in the future. This discharge should happen when the electric motor provides additional mechanical power/energy to the system. Assume that the mechanical power generated by electric motor is equivalent to a real fuel consumption reduction. In other words, the IC engine will need to generate less mechanical power.

ECMS Algorithm: As mentioned above, ECMS is an instantaneous optimization method, which depends solely on the system variables at each instant. In ECMS, the minimization of the integral cost in Equation 8.8 occurs at each instant, resulting in the following cost function:

$$J_t = \dot{m}_e(T_e(t)) + \dot{m}_{b,eq}(T_e(t)) \tag{8.39}$$

where $\dot{m}_e(t)$ is the fuel mass flow rate at each time, and $\dot{m}_{b,eq}(t)$ is the instantaneous equivalent fuel rate consumption by discharging/charging the battery given by:

$$\dot{m}_{b,eq}(t) = \left(s_{dis}\,\alpha \frac{1}{\eta_m \eta_b} + s_{chg}(1-\alpha)\eta_m \eta_b \right) \frac{P_m}{H_{LHV}} \tag{8.40}$$

where η_m is the efficiency of electrical motor, s_{dis} and s_{chg} is the equivalence factor for the discharge and charge cases, respectively. H_{LHV} is the low heating value of fuel, and α is:

$$\alpha = \frac{1 + \text{sign}\,(T_m)}{2} \tag{8.41}$$

The optimization problem in Equation 8.39 can be re-formulated as follows:

$$\left[T_e^{opt}(t), T_m^{opt}(t)\right] = \arg\min_{[T_e(t),\, T_m(t)]} J_t \qquad if \quad T_w \geq 0 \tag{8.42}$$

$$T_e^{opt}(t) = 0,\; T_m^{opt}(t) = T_w \qquad if \quad T_w < 0 \tag{8.43}$$

subject to:

$$0 \leq T_m(t) \leq T_{m,max}(t) \tag{8.44}$$

$$0 \leq T_e(t) \leq T_{e,max}(t) \tag{8.45}$$

$$SoC_{min} < SoC(t) < SoC_{max} \tag{8.46}$$

The relation between engine and electric motor torques is:

$$T_w(t) = \beta_g \beta_d \left(T_e(t) + \gamma_{m,e}\, T_m(t)\right)\, \eta_g \tag{8.47}$$

where β_g is the reduction ratio of the gearbox, β_d is the reduction ratio of the differential, β_g is the efficiency of gearbox, and $\gamma_{m,e}$ is:

$$\gamma_{m,e} = \frac{\omega_m}{\omega_e} \tag{8.48}$$

ω_m is electric motor speed and ω_e is the engine speed. These speeds are limited to:

$$0 \leq \omega_e \leq \omega_{e,max} \tag{8.49}$$

$$0 \leq \omega_m \leq \rho \omega_{e,max} \tag{8.50}$$

The wheel speed, ω_w, is:

$$\omega_w = \frac{\omega_e}{\beta_g \beta_d} = \frac{\omega_m}{\gamma_{m,e}\, \beta_g \beta_d} \tag{8.51}$$

The proper selection of the equivalent factors, s_{dis} and s_{chg} plays an important role in overall performance of the controller. Performance is dependent on driving conditions, it might be suitable for certain driving cycles but lead to poor performance for other driving cycles. On-line estimation of an equivalent factor, based on available vehicle and driving information, may improve the performance of the controller in a wider range of driving conditions. Low computational burden and near-optimal characteristics of ECMS make it appropriate for real-time power management control of HEVs and PHEVs.

Example 8.6

Design an ECMS energy management control for a double-shaft parallel HEV shown in Figure 8.36 according to the assumptions below.

$SoC_{min} = 60\%$

$SoC_{max} = 80\%$

$\gamma_{me} = 1$ (Ratio between motor speed and engine speed)

$H_{LHV} = 44.001\ (MJ/kg)$

$\eta_b = \eta_g = 1$

Evaluate the fuel consumption for the FTP-75 and HWFET driving cycles, and compare that with the fuel consumption of conventional vehicle. In addition, investigate the effect of ECMS gains, s_{dis} and s_{chg}, in both driving cycles. In ECMS design, consider the possibility of energy regeneration.

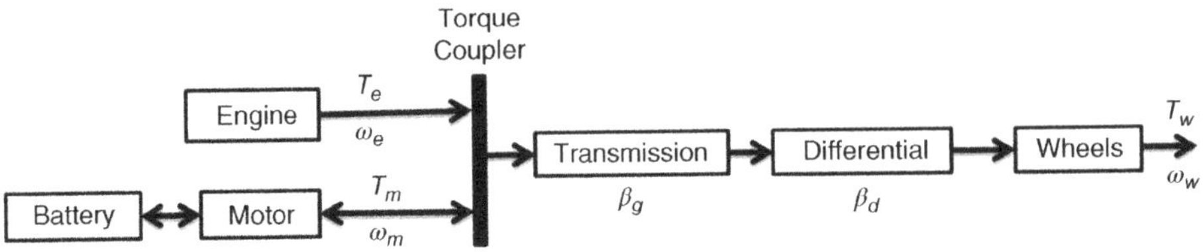

Figure 8.36 Double-shaft parallel powertrain

Solution

According to Equation 8.41, the outputs of the ECMS are motor and engine torques at each instant. The first step is to determine the possible engine and motor torques that could be optimal torques. Figure 8.37 shows the possible engine and motor torques. In this figure, by using engine speed and motor speed, the maximum torques for both motor and engine are calculated. For these calculations, look-up tables of maximum torque versus angular velocity of both engine and motor are used. However, these possible torques should satisfy the constraints in Equations 8.49 and 8.50. Due to these constraints, some possible torques may be removed and are not used for the rest of the ECMS algorithm.

After extraction of the possible optimal torques, feed them to the ECMS algorithm to find the optimal engine and motor torques which satisfy Equation 8.42.

The calculations of the optimal torques depend on the wheel torque, which can be either positive (traction) or negative (braking). Both possible conditions are discussed here:

- *Positive wheel torque (traction)*: in this condition, calculation of optimal torques depends on the SoC of the battery:
  - If $SoC_{min} \leq SoC \leq SoC_{max}$, use the ECMS algorithm to calculate the optimal torques for both the motor and the engine.
  - If $SoC < SoC_{min}$, the battery needs to be charged. In this case, the electric motor should not be used, as $T_m^{opt} = 0$. In addition, a part of engine torque known as the regenerative torque, T_{regen} functions to charge the battery. Therefore, in this condition, the engine torque is used for both traction and to charge the battery.

    We calculate the engine optimal torque, T_e^{opt}, by using the graph demonstrating optimal engine torque (in terms of fuel consumption) versus engine speed. Figure 8.38 shows that the demanded torque can be less or more than $T_{e,f}^{opt}$ which represents the optimal engine torque in terms of fuel consumption.

    If $T_{e,f}^{opt} > \dfrac{T_w}{\beta_g \beta_d \, \eta_g}$, then:

$$T_e^{opt} = \frac{T_w}{\beta_g \beta_d \eta_g},$$

$$T_{regen} = T_{e,f}^{opt} - T_w,$$

When T_{regen} is used to charge the battery.

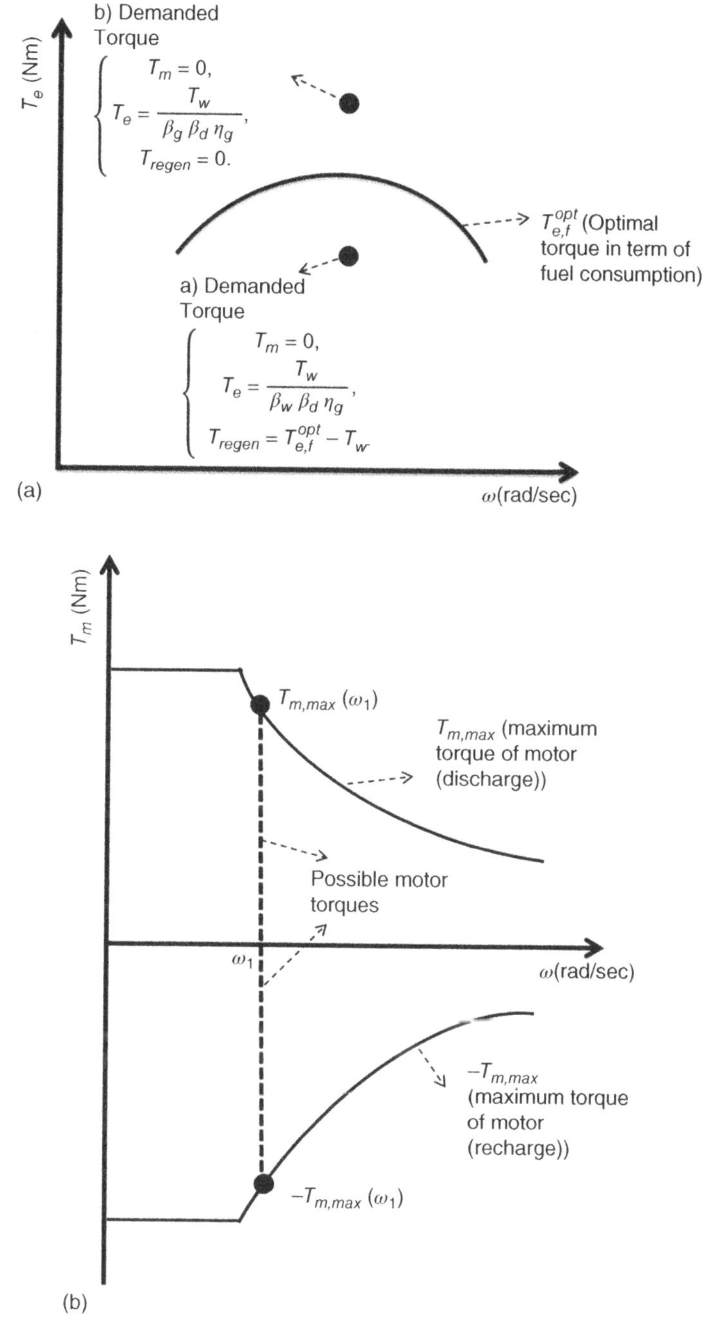

Figure 8.37 Possible torques for (a) engine and (b) motor

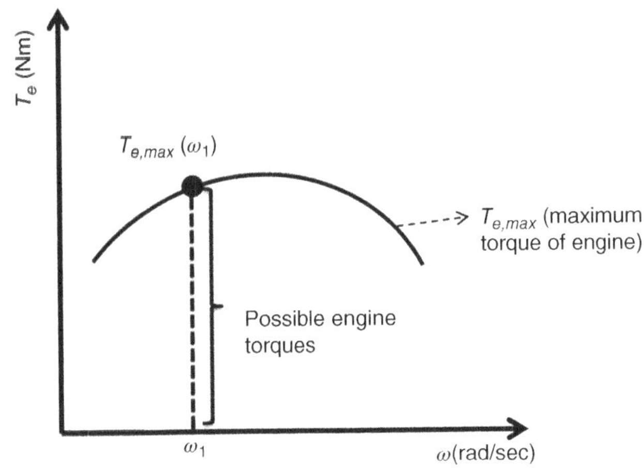

Figure 8.38 Optimal torque of an engine in terms of fuel consumption

If $T_{e,f}^{opt} \leq \dfrac{T_w}{\beta_g \beta_d \eta_g}$ then:

$$T_e^{opt} = \frac{T_w}{\beta_g \beta_d \eta_g},$$
$$T_{regen} = 0.$$

- If $SoC > SoC_{max}$, then only the electric motor needs to discharge the battery, it means that $T_e = 0$ and:

$$T_m^{opt} = \frac{T_w}{\beta_g \beta_d \eta_g}$$
$$T_{regen} = 0.$$

- *Negative wheel torque (braking)*: in this condition, optimal torques of motor and engine are:

$$T_m^{opt} = 0,$$
$$T_e^{opt} = 0,$$
$$T_{regen} = \frac{T_w}{\beta_g \beta_d \eta_g}.$$

The final optimal torque of the electric motor is the summation of T_m^{opt} and T_{regen}. In addition, whenever $T_e^{opt} = 0$, the engine speed is at idle speed, which is 900 (rpm). Whenever the wheel speed is less than the idle speed, assume that only the motor is working.

An ECMS with the above conditions functions to see its performance for two standard driving cycles including the FTP-75 and the HWFET.

Table 8.13 Two different settings of the ECMS control strategy for the FTP-75 driving cycle

	s_{dis}	s_{chg}
Setting 1	3.5	4.0
Setting 2	2.5	4.5

FTP-75

The performance of an ECMS control strategy with two settings, as reflected in Table 8.13, is discussed below.

Figure 8.39 shows the engine operating points for the above ECMS settings. This figure shows that the engine operating points for both settings are more distributed in the same region of the engine efficiency between 0.30 and 0.35, which is a suitable approximation of the maximum region of engine efficiency.

Figure 8.40 depicts the electric motor operating points. In the ECMS control strategy with setting 1, the electric motor has more operating points with negative torques, as a result, this setting recharges the battery more than the other setting. However, in the ECMS control strategy with setting 2, the electric motor discharges more. In addition, the motor operates in its efficiency regions more often when ECMS is in setting 2. Figure 8.41 shows the comparison between the battery SoCs of electric motor for these two settings.

Table 8.14 compares the fuel consumption of a parallel hybrid vehicle and a conventional vehicle when applying an ECMS control strategy in both settings. The total fuel consumption with Setting 1 is 3.81 (L/100 km), which is less than the total fuel consumption with Setting 2. In setting 1, the IC engine contributes more to the power demand than the electric motor. Since the power demand was less than the power provided by the engine at its optimal efficient operating point, the rest of the power charges the battery. In addition, the regenerative braking system recovers more braking energy in setting 1. In contrast, setting 2 uses the electric motor as the dominant source of providing the power demand, resulting in less battery charging and higher electric energy consumption. These reasons justify the lower SoC and higher fuel consumption with setting 2 compared with setting 1. Note that both settings decrease the total fuel consumption in comparison with the conventional vehicle.

HWFET: Similar to the case study for the FTP-75 driving cycle, Table 8.15 displays two selected settings for the ECMS control strategy to investigate the performance of this strategy in an HWFET driving cycle.

Simulation results shown in Figures 8.42 to 8.44 indicate that with engine setting 1 as the dominant power source, the regenerative braking system recuperates more braking energy than in setting 2. Since the average power demand is less than the torque generated at the optimum efficient operating point of an IC engine, the battery uses some of the generated power by the IC engine to recharge. However, as HWFET represents a highway driving cycle that does not include a frequent stop-and-go driving pattern, the recuperated braking energy is not significant. Thus, the battery SoC drops more here than in the FTP-75 driving cycle. In addition, in setting 2, the electric motor generates the majority of power required for vehicle propulsion; more electric energy is used thereby depleting the battery more quickly. Additionally, as reflected in Table 8.16, when using setting 1, the

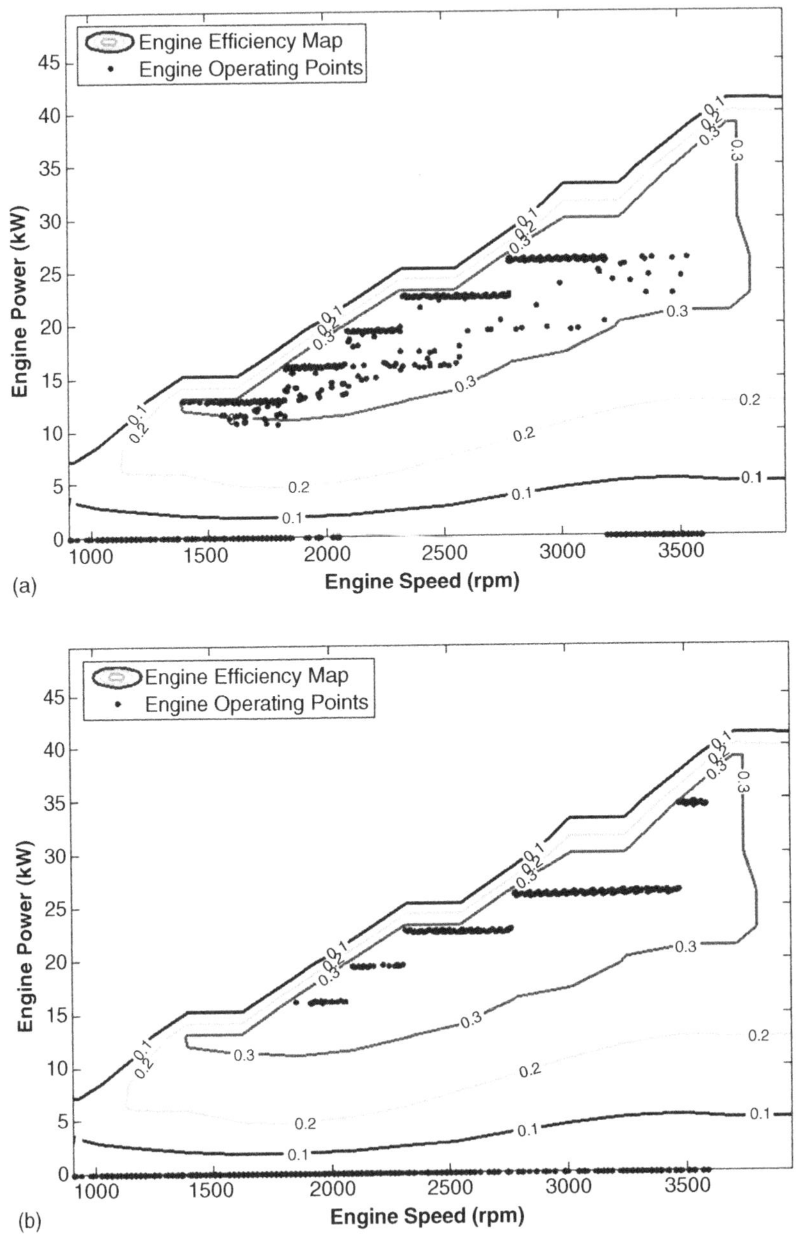

Figure 8.39 Engine operating points in the FTP-75 driving cycle with ECMS for (a) setting 1, (b) setting 2

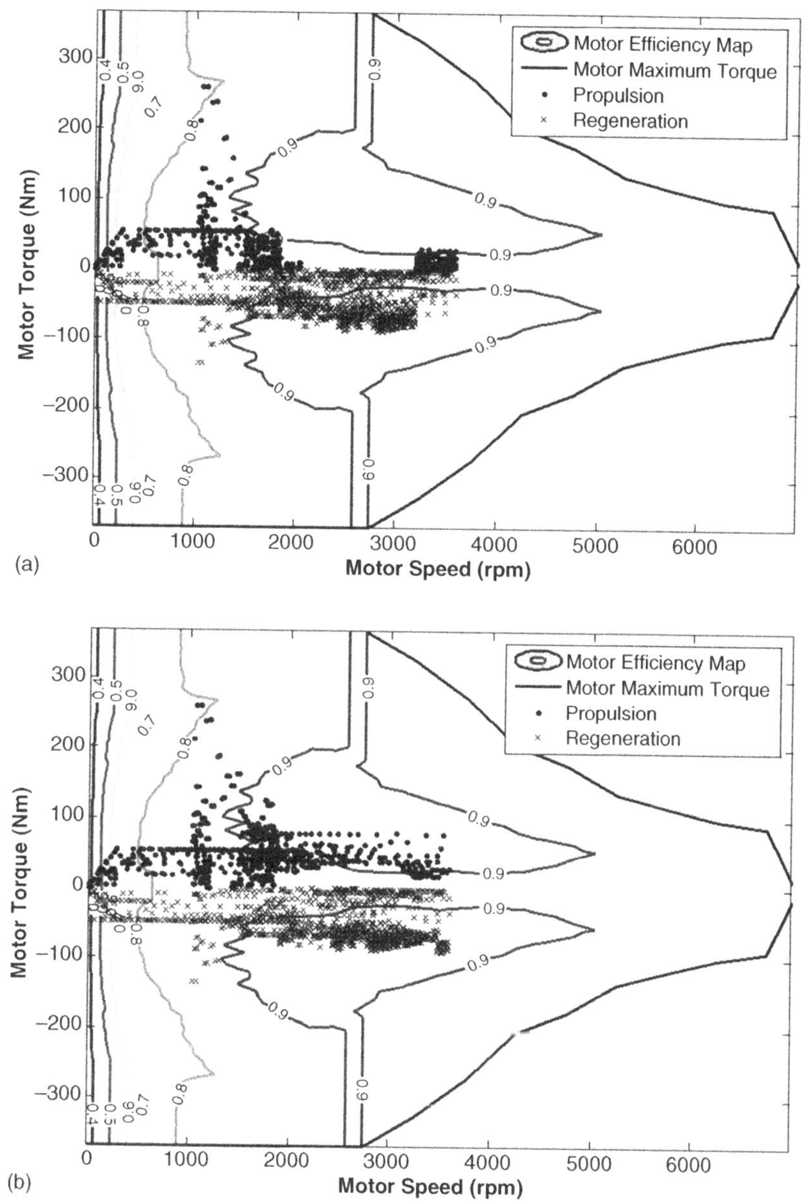

Figure 8.40 Motor operating points in the FTP-75 driving cycle with ECMS for (a) setting 1, (b) setting 2

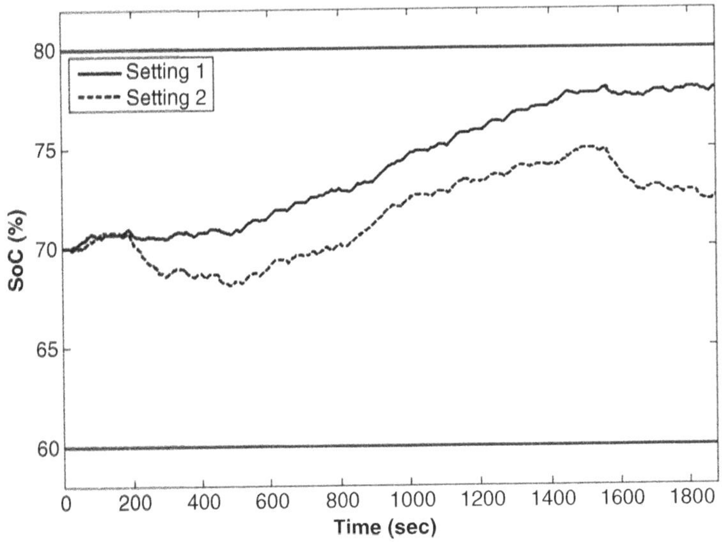

Figure 8.41 Comparison between the battery SoCs of (a) setting 1, (b) setting 2

Table 8.14 Comparison between the fuel consumption of a parallel hybrid vehicle with two different ECMS control strategy settings and a conventional vehicle in the FTP-75 driving cycle

Fuel consumption	Parallel hybrid vehicle (ECMS – Setting 1) (L/100 km)	Parallel hybrid Vehicle (ECMS – Setting 2) (L/100 km)	Conventional vehicle (L/100 km)
Engine fuel consumption	6.20	5.56	9.11
Equivalent fuel consumption of the battery	–2.39	–0.74	–
Total fuel consumption	3.81	4.82	9.11

engine's fuel consumption is 0.56 (L/100 km), and engine fuel consumption when using setting 2 is 2.38 (L/100 km), confirming the calculation in the discussion above. It needs to be mentioned that with setting 2, the motor makes more of a contribution than the engine. This means that the final level of SoC is lower than the initial one, and hence should be converted to the equivalent fuel consumption as discussed in Example 8.1. The equivalent

Table 8.15 Two different settings of the ECMS control strategy for the HWFET driving cycle

	s_{dis}	s_{chg}
Setting 1	2.6	6.0
Setting 2	2.5	5.0

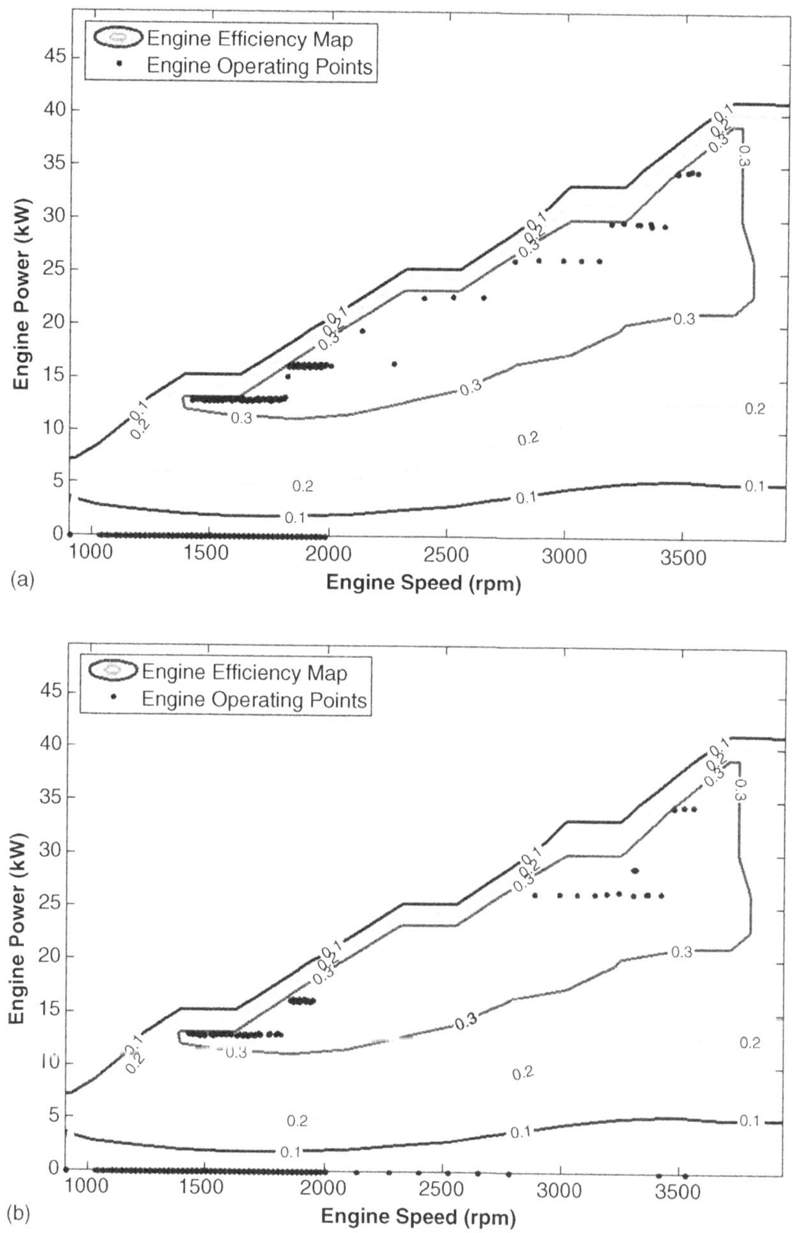

Figure 8.42 Engine operating points in the HWFET driving cycle with ECMS for (a) setting 1, (b) setting 2

fuel in this case is 2.94 (L/100 km). The total fuel consumption of a hybrid vehicle is 3.43 (L/100 km) and 3.50 (L/100 km) when using setting 1 and setting 2 in the ECMS control strategy, respectively. Both settings are lower than the conventional vehicle's fuel consumption.

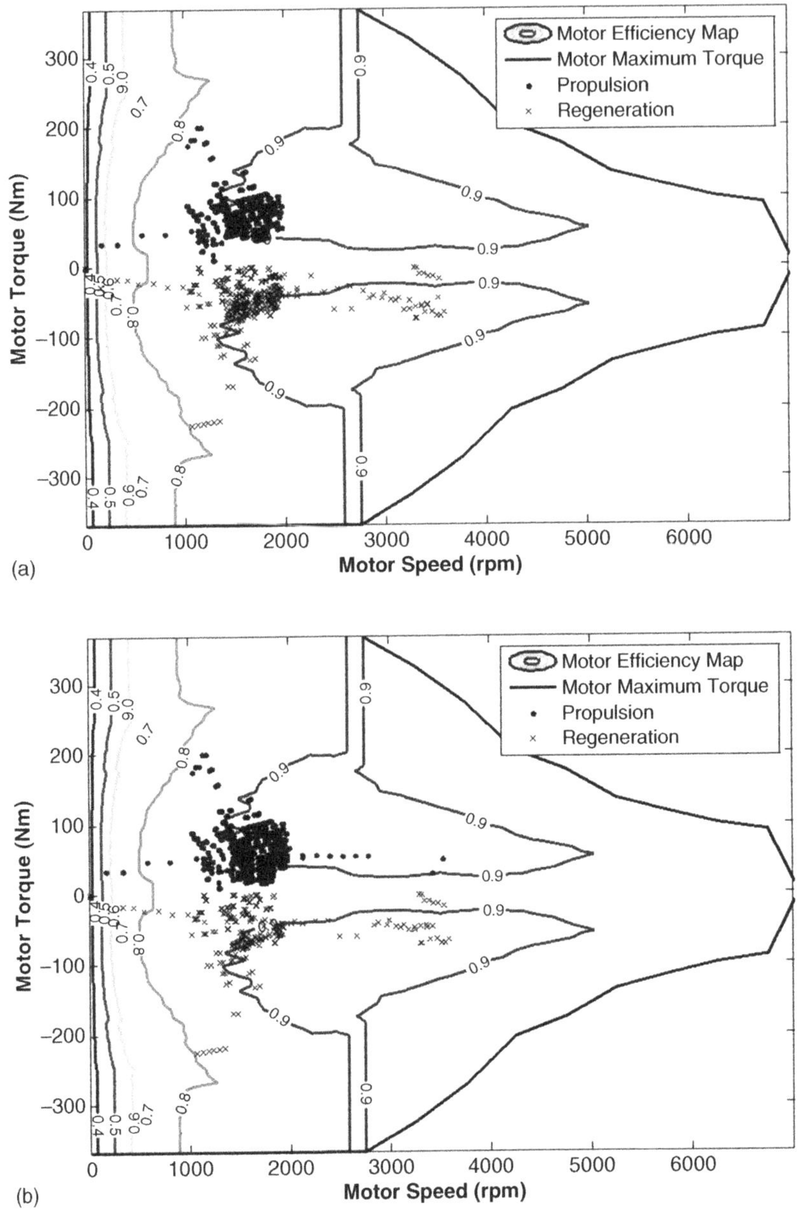

Figure 8.43 Motor operating points in the HWFET driving cycle with ECMS for (a) setting 1, (b) setting 2

The example shows that the ECMS control strategy with different settings has different performances for a specific driving cycle, so for each driving cycle, ECMS must be tuned properly. To enhance performance of ECMS controller in different driving cycles, an adaptive approach [15] can be applied to tune the ECMS gains on-line, subject to driving pattern specifications.

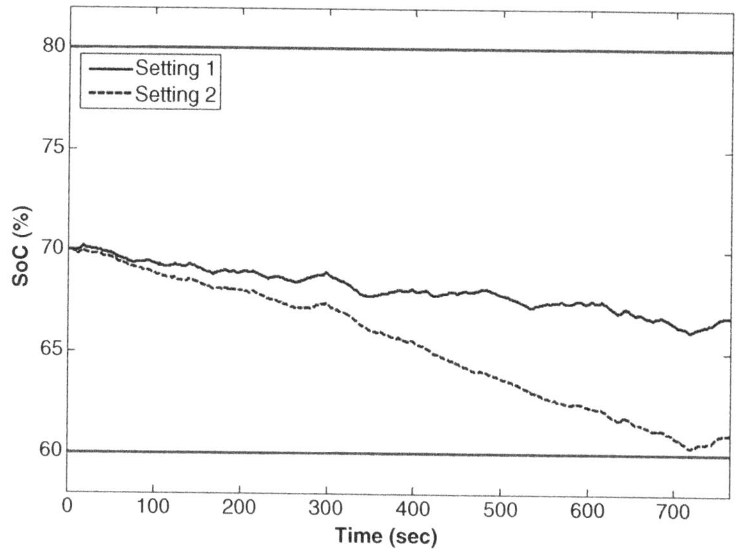

Figure 8.44 Comparison between the battery SoCs of (a) setting 1, (b) setting 2

Table 8.16 Comparison between the fuel consumption of a parallel hybrid vehicle with two different ECMS control strategy settings and a conventional vehicle in an HWFET driving cycle

Fuel consumption	Parallel hybrid vehicle (ECMS – Setting 1) (L/100 km)	Parallel hybrid vehicle (ECMS – Setting 2) (L/100 km)	Conventional vehicle (L/100 km)
Engine fuel consumption	2.38	0.56	6.38
Equivalent fuel consumption of the battery	1.05	2.94	–
Total fuel consumption	3.43	3.50	6.38

References

1. Emadi, A. (2005) Optimal power management and distribution in automotive systems, in *Handbook of Automotive Power Electronics and Motor Drives*, CRC Press, Boca Raton, FL.
2. Salmasi, F.R. (2007) Control strategies for hybrid electric vehicles: evolution, classification, comparison, and future trends. *IEEE Transactions on Vehicular Technology*, **56**(5).
3. Wirasingha, S.G., and Emadi, A. (2011) Classification of control strategies for plug-in hybrid electric vehicles. *IEEE Transactions on Vehicular Technology*, **60**(1).
4. Zhang, C., and Vahidi, A. (2012) Route preview in energy management of plug-in hybrid vehicles. *IEEE Transactions on Control Systems Technology*, **20**(2), 546–553.
5. Sciarretta, A., and Guzzella, L. (2007) Control of hybrid electric vehicles. *Control Systems, IEEE*, **27**(2), 60–70.
6. Chan, C.C., and Chau, K.T. (2001) *Modern Electric Vehicle Technology*, Oxford University Press, Oxford.
7. Ehsani, M., Gao, Y., and Emadi, A. (2010) *Modern Electric, Hybrid Electric, and Fuel Cell Vehicles: Fundamentals, Theory, and Design*, 2nd edn, CRC Press, Boca Raton, FL.

8. Bellman, R. (1957) *Dynamic Programming*, Princeton University Press, Princeton, NJ.
9. Zhang, X., and Mi, C. (2011) *Vehicle Power Management: Modeling, Control and Optimization*, Springer, New York.
10. Pérez, L.V., Bossio, G.R., Moitre, D., and García, G.O. (2006) Optimization of power management in a hybrid electric vehicle using dynamic programming. *Mathematics and Computers in Simulation*, **73**(1), 244–254.
11. Moura, S.J., Fathy, H.K., Callaway, D.S., and Stein, J.L. (2011) A stochastic optimal control approach for power management in plug-in hybrid electric vehicles. *IEEE Transactions on Control Systems Technology*, **19**(3), 545–555.
12. Scordia, J., Desbois-Renaudin, M., Trigui, R., Jeanneret, B., Badin, F., and Plasse, C. (2005) Global optimisation of energy management laws in hybrid vehicles using dynamic programming. *International Journal of Vehicle Design*, **39**(4), 349–367.
13. Sundstrom, O., and Guzzella, L. (2009) A generic dynamic programming Matlab function, *in 18th IEEE Conference on Control Applications, (CCA) & Intelligent Control (ISIC)*, Saint Petersburg, Russia, July, vol. 8, pp. 1625–1630.
14. Paganelli, G., Delprat, S., Guerra, T., Rimaux, J., and Santin, J. (2002) *Equivalent Consumption Minimization Strategy for Parallel Hybrid Powertrains*. Proceedings of IEEE Vehicular Technology Conference, vol. 4, pp. 2076–2081.
15. Musardo, C., Rizzoni, G., and Staccia, B. (2005) *A-ECMS: An Adaptive Algorithm for Hybrid Electric Vehicle Energy Management*. Proceedings of 44th IEEE Conference on Decision and Control, Seville, Spain, 12–15 December.

9

Control of Electric and Hybrid Electric Vehicle Dynamics

9.1 Introduction

Vehicles are normally controlled by driver commands that are applied to the vehicle through the use of the steering wheel, as well as through the brake and accelerator pedals. Vehicles consistently respond to these driver commands during normal driving conditions; however, under harsher conditions such as slippery roads or during severe maneuvers, vehicles may not adequately respond to these driver commands. In such situations, controlling the vehicle becomes a difficult task for the driver, and may even result in the driver completely losing control of the vehicle. In other words, during harsh driving conditions, vehicle dynamics is mostly nonlinear, and control of the vehicle through driver input is not easy. In such situations, the vehicle should be stabilized and controlled through methods other than traditional driver inputs.

An active control system can be used to control vehicle dynamic behavior during harsh driving conditions. Technically, such systems are categorized as vehicle dynamic control (VDC) systems. The early attempts to develop VDC systems date back to early 1980s. Since then, both auto-makers and academics have conducted extensive research in this field, resulting in the implementation of commercial systems such as the anti-lock braking system (ABS), the traction control system (TCS), and the electronic stability program (ESP) in mass-produced cars.

In parallel with the idea of developing modern electric and hybrid electric vehicles, the practice of implementing more precise and effective vehicle dynamic control systems has been considered. Based on the innovative powertrain architecture of modern electric and hybrid electric vehicles, the performance of current VDC systems can be considerably improved; this makes electric and hybrid electric vehicles safer and more controllable than their conventional counterparts.

Electric and Hybrid Vehicles: Technologies, Modeling and Control: A Mechatronic Approach, First Edition.
Amir Khajepour, Saber Fallah and Avesta Goodarzi.

The design of VDC systems for electric and hybrid vehicles is the main focus of this chapter. However, to make the subject easier to understand, we will first discuss the fundamentals and working principles of VDC systems.

9.2 Fundamentals of Vehicle Dynamic Control (VDC) Systems

Before the mid-1970s, vehicles were controlled solely based on driver inputs. The introduction of modern anti-lock braking systems (ABS) in the early 1980s opened the door to vehicle dynamic control systems. Since then, numerous systems have been studied and developed. Regardless of the differences between performance, functionality, and complexity, all VDC systems work on similar principles. These principles will be examined in closer detail in the section below.

9.2.1 Driver, Vehicle, and Environment

When a car is driven, three main elements are at work: the driver, the vehicle, and the environment. Together, they form a dynamic system that can be shown in Figure 9.1. The driver provides inputs to the vehicle in terms of steering, braking, and accelerating commands, then receives visual and sensual feedback from the vehicle response and compares them with environmental inputs such as the road path, ongoing traffic, and road signs. Based on the difference between the desired and actual vehicle response, the driver may correct the commands. Conversely, the vehicle is also under the influence of environmental effects such as road conditions, as well as wind and weather conditions.

As long as all of the system elements work well, system performance will be satisfactory. However, if one of them is not working normally, the system performance will decrease or even becomes completely unstable. Generally, an inexpert driver, a defective car, or a slippery road all have similar consequences and can be the cause of partial loss of control or even a fatal accident.

The afore-mentioned idea can be represented mathematically using the single track handling model. According to the discussions outlined in Chapter 7, the single track handling model has been developed for normal road and driving conditions; now we will improve it for slippery road conditions and severe maneuvers. According to Figure 9.2, tire behavior is no longer linear in such conditions, and its working point is shifted to the nonlinear and eventually saturation zone.

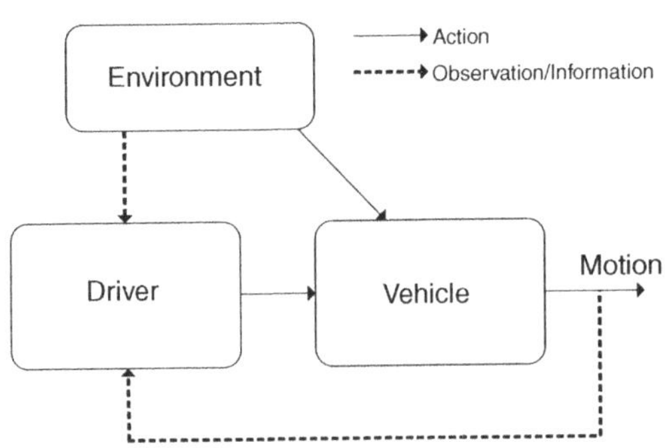

Figure 9.1 Block diagram of driver/vehicle/environment system

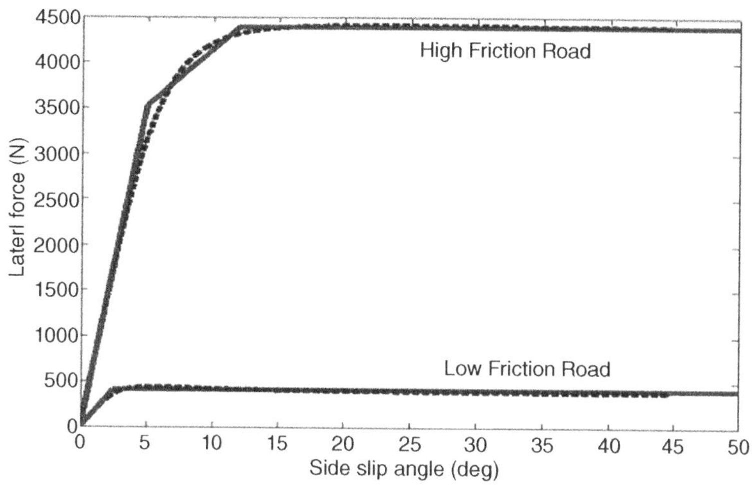

Figure 9.2 The general form of a piece-wise linear tire model

To achieve suitable accuracy and to benefit from the simplicity of linear theories, a piece-wise linear approach can be used for tire modeling. In order to cover the entire tire working region, the lateral force versus slip angle curve can be estimated by some lines. The general form of each linearized piece is written as:

$$F_y = c_\alpha \alpha + f_0 \tag{9.1}$$

By using the above tire model and following the same approach discussed in Chapter 7, the generalized single track handling mode can be found as:

$$\begin{pmatrix} \dot{v} \\ \dot{r} \end{pmatrix} = \begin{pmatrix} -\dfrac{C_{\alpha r} + C_{\alpha f}}{Mu} & \dfrac{(bC_{\alpha r} - aC_{\alpha f})}{Mu} - u \\ \dfrac{(bC_{\alpha r} - aC_{\alpha f})}{I_z u} & -\dfrac{(b^2 C_{\alpha r} + a^2 C_{\alpha f})}{I_z u} \end{pmatrix} \begin{pmatrix} v \\ r \end{pmatrix} + \begin{pmatrix} \dfrac{C_{\alpha f}}{M} \\ \dfrac{aC_{\alpha f}}{I_z} \end{pmatrix} \delta + \begin{pmatrix} \dfrac{f_{0f} + f_{0r}}{M} \\ \dfrac{af_{0f} - bf_{0r}}{I_z} \end{pmatrix} \tag{9.2}$$

where $C_{\alpha f}$, $C_{\alpha r}$ respectively are cornering stiffness of front and rear axles and f_{0f},f_{0r} are the constant parameters. By considering Equation 9.2, one can see that the last term has been added to the standard form of single track handling model. This generalized form of the single track model is capable of describing the vehicle dynamics behavior either in linear or nonlinear tire working zones.

Now assume, due to reasons such as adverse road/driving conditions, or a lack of the driver's skills, the tires' working point reaches the saturated zone. Consequently, the cornering stiffness of the tire gradually approaches zero ($C_\alpha \to 0$), and Equation 9.1 is converted to:

$$F_y = f_0 \tag{9.3}$$

According to Equation 9.3, the vehicle handling model, Equation 9.2, in the saturation zone can be rewritten as:

$$\begin{pmatrix} \dot{v} \\ \dot{r} \end{pmatrix} = \begin{pmatrix} 0 & -u \\ 0 & 0 \end{pmatrix} \begin{pmatrix} v \\ r \end{pmatrix} + \begin{pmatrix} \dfrac{f_{0f} + f_{0r}}{M} \\ \dfrac{af_{0f} - bf_{0r}}{I_z} \end{pmatrix} \tag{9.4}$$

It is interesting to note that in the above equation, the term that includes the steering input has disappeared. This means that in the saturation zone, the vehicle is no longer controllable by the driver.

Example 9.1

Consider that a single track handling model for a vehicle with the pricewise linear tire model is shown in Figure 9.3. The vehicle specifications are provided in Table 9.1. Simulate a sinusoidal steering input maneuver with a speed of 100 km/h; plot the tires' working points, vehicle path, yaw rate, sideslip angle, and lateral acceleration of the vehicle in the following cases:

(a) $\delta = 0.04\sin(t)$rad
(b) $\delta = 0.06\sin(t)$rad

Are the working points of the tire in the linear or saturation zones? Discuss how it affects the vehicle's behavior.

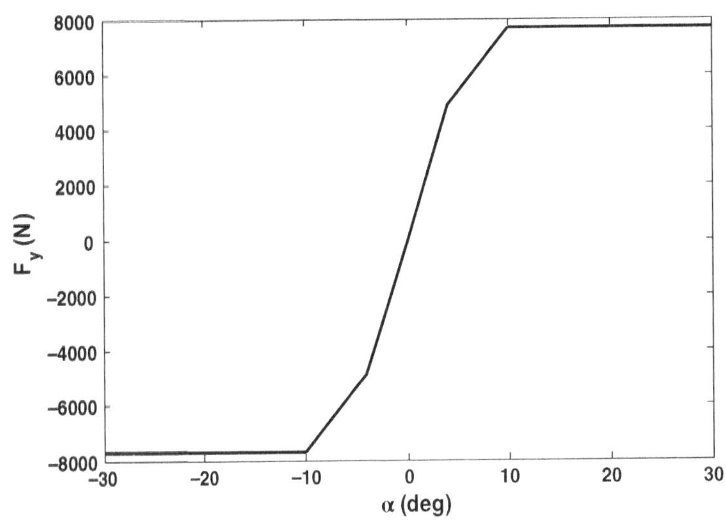

Figure 9.3 Lateral force vs. slip angle

Table 9.1 Vehicle specifications

Parameter	Value
M	1530 (kg)
a	1.320 (m)
b	1.456 (m)
I_z	4192 (kgm^2)
$c_{\alpha f}$	70000 (N/rad)
$c_{\alpha r}$	70000 (N/rad)

Solution

By substituting the vehicle parameters, tire model and the steering input in Equation 9.2, a system of linear differential equations for each tire working zone is formed. Then to find the vehicle's response and the tires' working point, one can solve them either numerically or analytically.

As illustrated in Figure 9.4, the working point of the front tire in case (a) is in the linear zone. However, in case (b), it goes into the saturation zone. As a result, in the first case, the vehicle can be controlled through steering input and has normal behavior (Figure 9.5). In the second case, however, tires work in the saturation zone, and steering input is no longer effective; the vehicle drives off the path (Figure 9.6). According to Figures 9.5 and 9.6, the vehicle has an unstable behavior and all the parameters show that the vehicle is out of control.

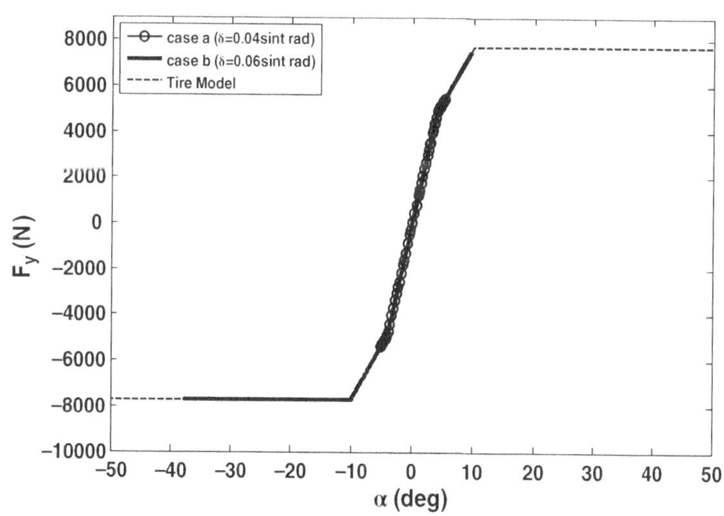

Figure 9.4 Front tire's working points in cases (a) and (b)

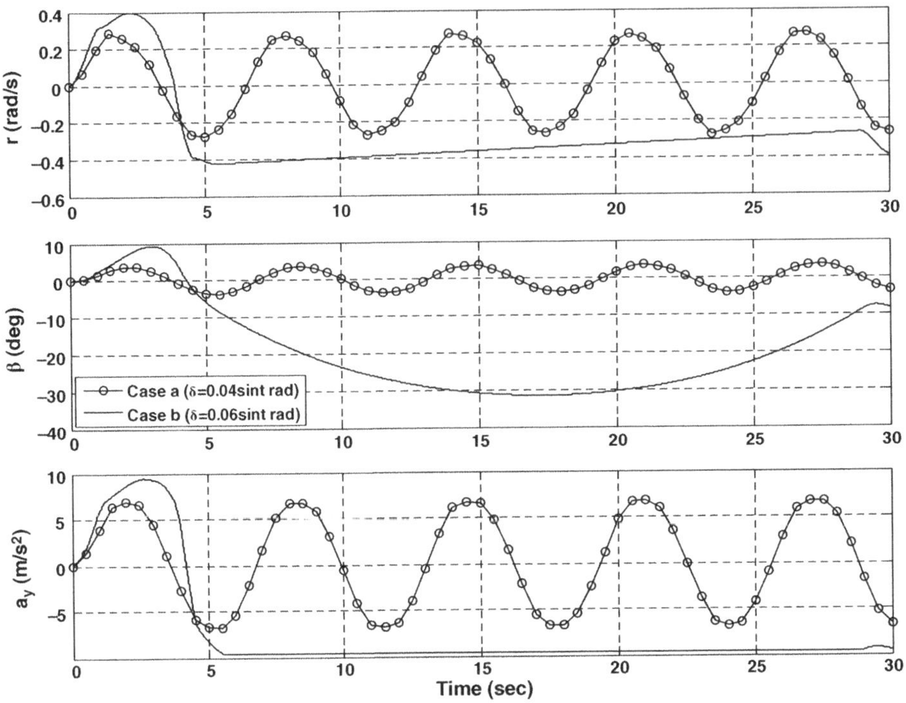

Figure 9.5 Yaw rate (r), vehicle sideslip angle (β), and lateral acceleration (α_y) for cases (a) and (b)

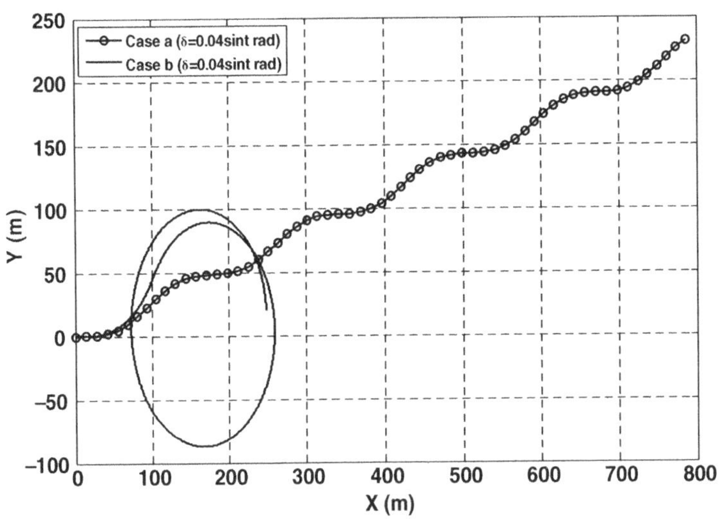

Figure 9.6 Vehicle path for cases (a) and (b)

The analysis above shows that driving conditions and environmental effects can make the vehicle uncontrollable. In other words, due to the inherent characteristics of vehicle subsystems such as tires, the desired vehicle response is not achievable in some conditions, even by skilled drivers. In such conditions, the vehicle response is no longer under the influence of the driver steering input, and could become worse if controlled by a panicking or unskilled driver.

9.2.2 *Working Principle of VDC systems*

Generally, VDC systems can improve the vehicle dynamic response by utilizing sensors, actuators, and an electronic control unit, especially in harsh environmental and driving conditions. A VDC system is added to the driver/environment/vehicle system as the fourth element, as illustrated in Figure 9.7.

Generally, a VDC system receives information such as the vehicle velocity, acceleration in the three directions, forces, moments, etc., plus the driver inputs such as steering angle, brake pedal force, and accelerator pedal position. Moreover, advanced systems take measurements from surrounding environments such as the road path, road irregularities, and the position of surrounding vehicles. Based on the input data, and by using a control algorithm, the VDC system assists the driver by applying additional control signals to the vehicle's systems (e.g., steering, braking, and powertrain). By applying these control actions, VDC systems try to minimize the difference between the existing dynamic behavior of the vehicle and its desired behavior.

The desired vehicle behavior can be defined in terms of four different factors, as illustrated in Figure 9.8. Initially, the vehicle should consistently follow the driver's commands without requiring him/her to exert much mental and physical effort. The steering, braking, and accelerating should feel smooth, quick, and predictable, and the quality of the response should be less sensitive to the vehicle speed. Second, the vehicle should remain stable and controllable in different driving conditions. Third, the vehicle's dynamic behavior should be less sensitive to environmental effects such as road conditions. Finally, the vehicle with a VDC system should correct any driver error, especially in panicking or emergency situations.

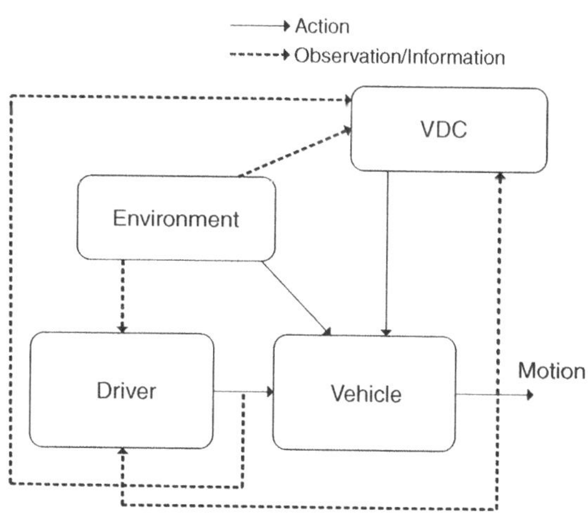

Figure 9.7 Vehicle dynamics control system as the fourth element of a driver/vehicle/environment system

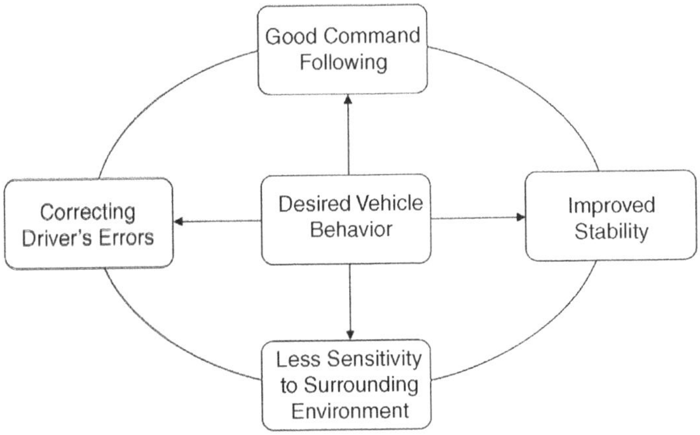

Figure 9.8 Four elements of the desired vehicle behavior

The afore-mentioned goals cannot all be achieved easily. VDC systems that are used in modern vehicles can only achieve some of these goals. However, continual research, development, and production of modern VDC systems will help progress towards realizing all aspects of these desired vehicle behaviors.

Currently, different types of VDC systems, as discussed in the next section, are used in modern cars. More advanced systems are in the research and development stages.

9.2.3 VDC Systems Classification

Generally, VDC systems can be classified as follows into different systems.

9.2.3.1 Longitudinal Control Systems

The anti-lock brake system (ABS) and the traction control system (TCS) are the earliest VDC systems that came onto the market. As mentioned in Chapter 3, both ABS and TCS try to optimize the braking and acceleration behavior of the vehicle. As such, they can be categorized in the class of longitudinal control systems. While control of the wheel's longitudinal slip is the working principle of both of these systems, they differ slightly in some respects. For example, while the ABS controls the slip during braking and prevents the wheel from locking, the TCS controls the slip during acceleration and prevents the wheel from spinning excessively.

As illustrated in Figure 3.27, during hard braking and severe accelerating, the tire's longitudinal force is maximized by controlling the wheel longitudinal slip in its optimum zone, where the tire capacity for lateral force generation is still acceptable. Consequently, the vehicle's stopping distance is minimized and the vehicle remains stable and steerable until it comes to a complete stop. On the other hand, during accelerating, the vehicle's tractive force is maximized and the vehicle can effectively accelerate or climb from steeper slopes. All the above improved performances are achievable regardless of the road conditions.

From the control engineering point of view, both ABS and TCS are relatively simple feedback control systems. Both systems sense the rotational speed of a wheel and estimate the longitudinal slip. They then control the slip through adjusting the delivered braking or tractive torque to the wheel. The configuration of the anti-lock brake and traction control systems was discussed in Chapter 3.

Cruise control (CC) and active cruise control (ACC) are also categorized in this class. When these systems are activated by the driver, the system controls the accelerator pedal in order to keep the vehicle velocity constant. As such, driving becomes easier and more comfortable for the driver, allowing him/her to concentrate more on the road traffic, thus improving driving safety. ACC is a more advanced system which, in addition to controlling vehicle velocity, also detects the speed of the vehicle ahead of it using a radar system. It then tries to automatically keep a safe distance between the current vehicle and the vehicle ahead.

9.2.3.2 Vertical Control Systems

As mentioned in Chapter 3, active and semi-active suspension systems can control the vertical dynamics of vehicles. Specifically, they can control the vibration of sprung and unsprung masses. By controlling the sprung mass (body) vibration, the ride comfort of the passengers is improved. Likewise, control of the unsprung mass (wheel) vibration can improve the vehicle's road-holding performance, consequently enhancing its handling and performance. In addition to the vertical motion, active, and to some extent semi-active, suspension systems can also be beneficial in controlling the body roll and pitch angular motions. Application of vertical control systems in mass-produced cars is still limited due to the cost of semi-active and especially active suspension systems.

9.2.3.3 Roll Control Systems

As discussed in Chapters 4 and 6, a vehicle's angular motion (particularly roll motion) strongly affects the vehicle's handling and safety including the influences on load transfer and wheel alignment. Therefore, active control of roll motion can be useful in improving vehicle handling and safety. As seen in the previous section, active suspension systems and to some extent, semi-active suspension systems can control the roll and pitch motions. However, the added costs, especially for active suspensions, prevent their use in mass-produced vehicles.

Active roll control (ARC) systems are dedicated and cheaper systems for controlling vehicle roll angle. As illustrated in Figure 9.9, the key element of this system is a controllable anti-roll bar. A classic anti-roll bar is a torsional spring, but an active anti-roll bar includes a hydraulic

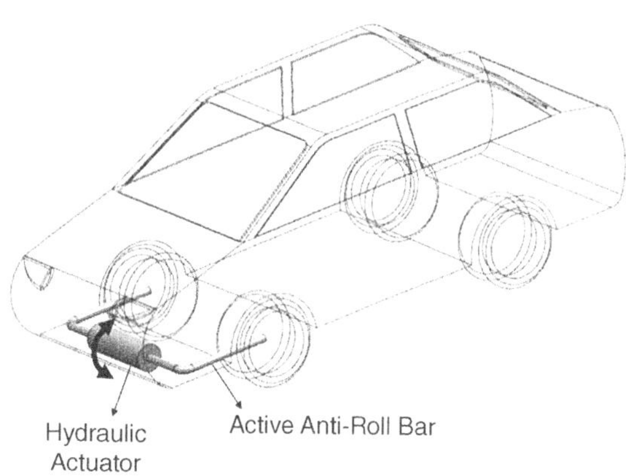

Figure 9.9 Active roll control (ARC) system with a controllable anti-roll bar

actuator in the middle of the bar that can change the effective stiffness of the anti-roll bar depending on the vehicle situation. The system includes sensors and an electronic control unit. During sharp turns, the actuator increases the effective stiffness of the bar in order to decrease the body roll angle by applying a roll moment to the vehicle body. Conversely, while driving on rough roads, the system decreases the anti-roll bar's effective stiffness to ease the independent movement of the wheels, thus enhancing ride comfort.

9.2.3.4 Yaw Control Systems

The VDC systems mentioned above can control traction, braking, ride, or roll dynamics. However, they can only indirectly affect the handling and stability of the vehicle. The handling and stability can be effectively managed if yaw and lateral motion are directly controlled using active yaw moment control systems. To correct the yaw and lateral motions, the total acting yaw moment on the vehicle should be automatically controlled by a feedback control system through control of longitudinal or lateral forces of the tires.

Essentially, the yaw moment is produced by lateral and/or longitudinal tire forces. According to Figure 7.16, the total vehicle yaw moment is:

$$M_z = a\left(F_{y1} + F_{y3}\right) - b\left(F_{y2} + F_{y4}\right) + \frac{T}{2}(F_{x1} + F_{x2}) - \frac{T}{2}(F_{x3} + F_{x4}) \tag{9.5}$$

The first two terms in Equation 9.5 represent the produced yaw moment by tire lateral forces, while the last two terms are the yaw moment that is produced by the unequal longitudinal forces of the left and right side tires. Based on the formulation of the vehicle yaw moment, yaw control systems can be categorized into two main groups, as follows:

Direct Yaw Moment (DYC) Control systems: In vehicles controlled directly by the driver, the yaw moment is governed by the lateral forces of the tires, as in such vehicles, the longitudinal forces of the tires on the left/right are normally equal to each other. The driver controls the steering angle and hence the side slip angle of the steered wheels is changed. Consequently, the lateral forces of the tires and eventually the vehicle yaw moment are changed.

The vehicle yaw moment can be alternatively controlled by unequal transversal distribution of longitudinal force that is called "direct yaw moment control." Generally, in a vehicle that is equipped with a DYC system, the vehicle yaw moment and consequently the vehicle's motion can be controlled in two different ways: by the driver through the lateral forces and by the DYC system through unequal distribution of left and right longitudinal forces.

For a better understanding of how a DYC system can be beneficial in the control of vehicle dynamics, let us reformulate the generalized form of the single track handling model in Equation 9.2 as follows:

$$\begin{pmatrix} \dot{v} \\ \dot{r} \end{pmatrix} = \begin{pmatrix} -\dfrac{C_{\alpha r} + C_{\alpha f}}{Mu} & \dfrac{(bC_{\alpha r} - aC_{\alpha f})}{Mu} - u \\ \dfrac{(bC_{\alpha r} - aC_{\alpha f})}{I_z u} & -\dfrac{(b^2 C_{\alpha r} + a^2 C_{\alpha f})}{I_z u} \end{pmatrix} \begin{pmatrix} v \\ r \end{pmatrix} + \begin{pmatrix} \dfrac{C_{\alpha f}}{M} \\ \dfrac{aC_{\alpha f}}{I_z} \end{pmatrix} \delta + \begin{pmatrix} \dfrac{f_{0f} + f_{0r}}{M} \\ \dfrac{af_{0f} - bf_{0r}}{I_z} \end{pmatrix} + \begin{bmatrix} 0 \\ 1 \end{bmatrix} M_z^e \tag{9.6}$$

where M_z^e is DYC yaw moment that is the two last terms of Equation 9.5:

$$M_z^e = \frac{T}{2}(F_{x1} + F_{x2}) - \frac{T}{2}(F_{x3} + F_{x4}) \tag{9.7}$$

As mentioned above, DYC systems work based on unequal distribution of left/right longitudinal forces. In such situations, the first and second terms of Equation 9.7 are no longer equal and an external yaw moment is applied to the vehicle by the DYC system.

Equation 9.6 again shows that the vehicle can be controlled by two inputs: steering angle (δ) and DYC yaw moment (M_z^e). Under normal driving conditions when the system is still linear, the vehicle can be effectively controlled by the steering angle input and the DYC input is not needed. However, as shown by Equation 9.4, during harsher conditions such as when tires are saturated, the steering input no longer affects vehicle dynamics, but the DYC input is still available:

$$\begin{pmatrix} \dot{v} \\ \dot{r} \end{pmatrix} = \begin{pmatrix} 0 & -u \\ 0 & 0 \end{pmatrix}\begin{pmatrix} v \\ r \end{pmatrix} + \begin{pmatrix} \dfrac{f_{0f} + f_{0r}}{M} \\ \dfrac{af_{0f} - bf_{0r}}{I_z} \end{pmatrix} + \begin{bmatrix} 0 \\ 1 \end{bmatrix} M_z^e \tag{9.8}$$

It should be noted that a tire with a saturated lateral force can still generate a longitudinal force (Figure 4.21). Therefore, by knowing that DYC yaw moment is generated by longitudinal forces, even under harsh driving conditions, it remains workable and can control the vehicle effectively.

Example 9.2

Suppose a proportional controller is used to calculate the external yaw moment, M_Z^e, in a DYC system based on the vehicle's yaw rate, lateral velocity, and the steering angle as follows:

$$M_Z^e = k_v v + k_r r + k_\delta \delta$$

where k_v, k_r, and k_δ are the gains of control law. Reformulate the generalized form of single track handling model using the above-mentioned control law. How does the model change when tires are in the saturation zone?

In Example 9.1, it was shown that in a maneuver with $\delta = 0.06\sin(t)$ and $u = 100$ km/hr, the vehicle became unstable. Investigate the behavior of vehicle in Example 9.1 during this maneuver if it is equipped with a DYC system with $k_v = 100$, $k_r = -10\,000$, and $k_\delta = -35\,000$.

Solution

By substituting the given control law in Equation 9.6, the single track vehicle handling model with a DYC is found:

$$\begin{pmatrix} \dot{v} \\ \dot{r} \end{pmatrix} = \begin{pmatrix} -\dfrac{C_{\alpha r} + C_{\alpha f}}{Mu} & \dfrac{(bC_{\alpha r} - aC_{\alpha f})}{Mu} - u \\ \dfrac{(bC_{\alpha r} - aC_{\alpha f})}{I_z u} & -\dfrac{(b^2 C_{\alpha r} + a^2 C_{\alpha f})}{I_z u} \end{pmatrix} \begin{pmatrix} v \\ r \end{pmatrix} + \begin{pmatrix} \dfrac{C_{\alpha f}}{M} \\ \dfrac{aC_{\alpha f}}{I_z} \end{pmatrix} \delta$$

$$+ \begin{pmatrix} \dfrac{f_{0f} + f_{0r}}{M} \\ \dfrac{af_{0f} - bf_{0r}}{I_z} \end{pmatrix} + \begin{bmatrix} 0 \\ 1 \end{bmatrix} (k_v v + k_r r + k_\delta \delta)$$

$$\begin{pmatrix} \dot{v} \\ \dot{r} \end{pmatrix} = \begin{pmatrix} -\dfrac{C_{\alpha r} + C_{\alpha f}}{Mu} & \dfrac{(bC_{\alpha r} - aC_{\alpha f})}{Mu} - u \\ \dfrac{(bC_{\alpha r} - aC_{\alpha f})}{I_z u} + k_v & -\dfrac{(b^2 C_{\alpha r} + a^2 C_{\alpha f})}{I_z u} + k_r \end{pmatrix} \begin{pmatrix} v \\ r \end{pmatrix}$$

$$+ \begin{pmatrix} \dfrac{C_{\alpha f}}{M} \\ \dfrac{aC_{\alpha f}}{I_z} + k_\delta \end{pmatrix} \delta + \begin{pmatrix} \dfrac{f_{0f} + f_{0r}}{M} \\ \dfrac{af_{0f} - bf_{0r}}{I_z} \end{pmatrix}$$

When the tires are in the saturation zone, the preceding model can be written as:

$$\begin{pmatrix} \dot{v} \\ \dot{r} \end{pmatrix} = \begin{pmatrix} 0 & -u \\ k_v & k_r \end{pmatrix} \begin{pmatrix} v \\ r \end{pmatrix} + \begin{pmatrix} 0 \\ k_\delta \end{pmatrix} \delta + \begin{pmatrix} \dfrac{f_{0f} + f_{0r}}{M} \\ \dfrac{af_{0f} - bf_{0r}}{I_z} \end{pmatrix}$$

In this equation, the steering angle is no longer missing. As such, despite the fact that the tires are in the saturation zone, the steering angle still influences the vehicle dynamics.

To study the stability of the system with a DYC, consider the system's characteristics equation as discussed in Chapter 7:

$$S^2 - k_r S + k_v u = 0$$

As discussed in Chapter 7, based on the Routh–Hurwitz stability criterion, the system is stable if $k_r < 0$ and $k_v > 0$.

By substituting the vehicle parameters, steering input and the above-mentioned DYC control law in Equation 9.6 one can easily simulate the vehicle responses. According to the simulation results, with the given gains for the DYC system, the working points of the tires are in the linear zone as illustrated in Figure 9.10 (only those of the front tires are illustrated; the working points of the rear tires are almost the same as the front ones). Therefore, a vehicle can be controlled through the steering input and can have normal dynamic behavior (Figures 9.11 and 9.12). According to Figure 9.12, the DYC system generates maximum yaw moment around 2400 Nm to be able to prevent the vehicle from being unstable.

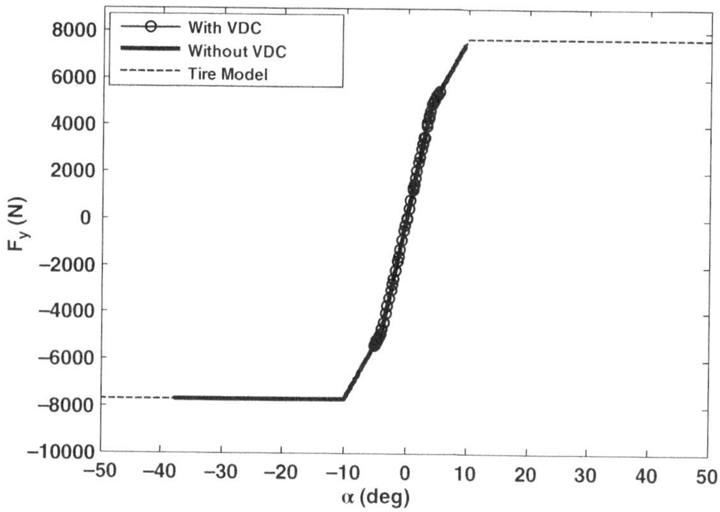

Figure 9.10 Working points of front tires for uncontrolled and controlled vehicles

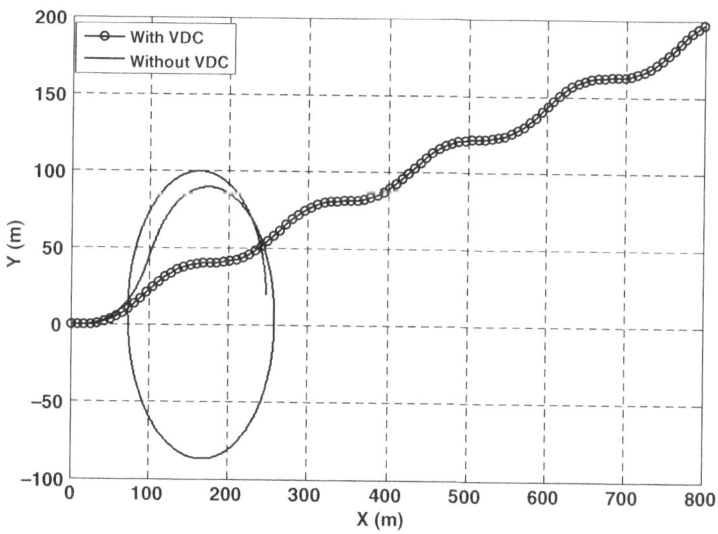

Figure 9.11 Vehicle paths for uncontrolled and controlled vehicles

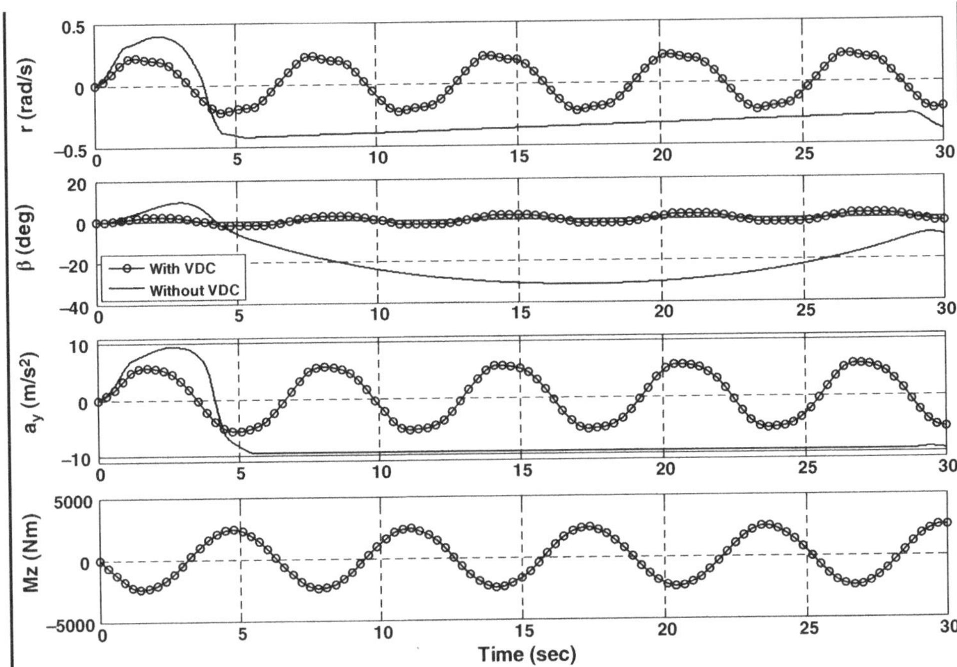

Figure 9.12 Yaw rate (r), vehicle sideslip angle (β), lateral acceleration (a_y) and external yaw moment for uncontrolled and controlled vehicles

Direct yaw moment control systems can be divided into two groups: differential traction and differential braking systems:

- *Differential braking systems:* Unequal distribution of vehicle braking forces between the left and right wheels of an axle is the most practical approach to generating the external yaw moment (Figure 9.13(a)). Differential braking systems can be achieved by modifying the common ABS systems and the addition of a DYC controller.

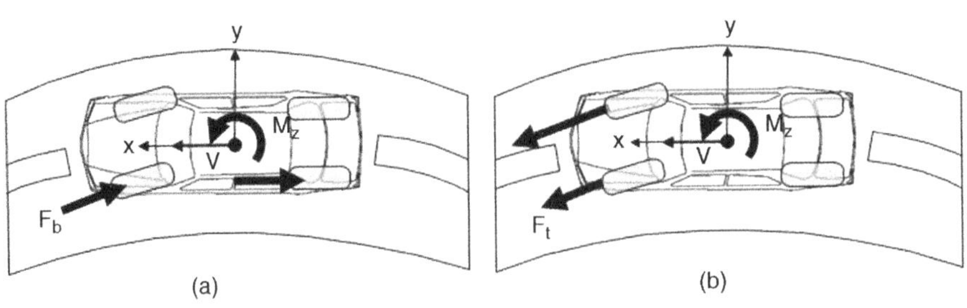

Figure 9.13 Different types of DYC systems; (a) differential braking, (b) differential traction systems

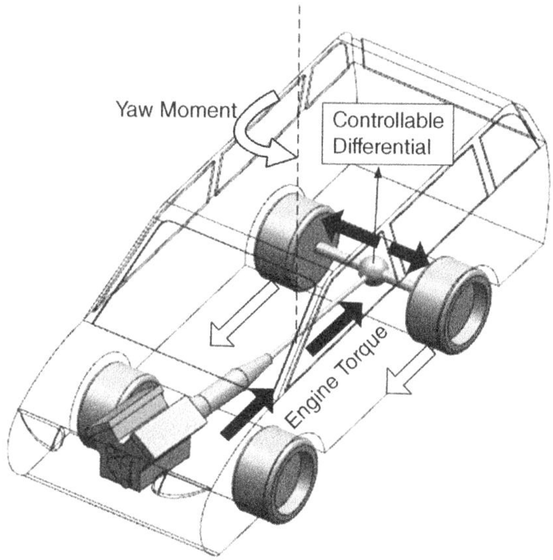

Figure 9.14 Differential traction system for an internal combustion engine vehicle using controllable differential

Differential braking DYC systems are available under different commercial names such as Electronic Stability Control (ESC), Electronic Stability Program (ESP), Dynamic Stability Control (DSC), and Vehicle Stability Control (VSC). These systems are widely implemented in most modern cars as an effective active safety system.

- *Differential traction systems:* Alternatively, DYC systems can work based on controlling the vehicle traction forces rather than the braking forces, as illustrated in Figure 9.13(b). In this technique, different traction forces are applied to the wheels of the drive axle to generate a stabilizing yaw moment. For internal combustion engine vehicles (ICEV), differential traction can be realized by using a controllable torque biasing device that can control the amount of torque delivered to each axle wheel. It can be a controllable differential that can vary the left and right driving torque based on an electronic control signal. As shown in Figure 9.14, the torque generated by the engine passes through the gearbox, and then a controllable differential distributes the torque unequally between the wheels of the drive axle. The differential traction in the motorized wheels electric or hybrid electric vehicles can be utilized more easily. This is because the driving torque of each wheel in these vehicles can be controlled individually without the use of an additional device.

Unlike differential braking, differential traction does not have negative effects on the longitudinal responsc of the vehicle, as it does not slow down the vehicle's longitudinal motion (the vehicle speed is inevitably reduced during differential braking). However, the maximum achievable yaw moment by differential traction especially in ICEVs is usually lower than that of differential braking. This is because the produced yaw moment is dependent on the current amount of tractive force, which in many cases such as cruising, is only a small portion of the engine torque capacity. Consequently, the maximum achievable yaw moment will be limited in these cases.

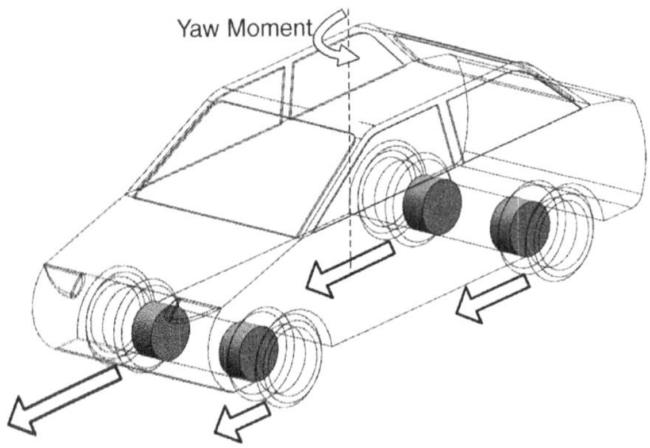

Figure 9.15 Torque vectoring in a motorized four-wheel electric drive

A more advanced differential traction system is known as the torque vectoring system. In these systems, not only are the tractive forces of the left and right wheels controllable, but the tractive force distribution between the front and rear axles can be controlled as well. By combining the left/right and front/rear controllability, the tractive force of each wheel can be controlled individually. In the case of ICEVs, the complete implementation of a torque vectoring system requires a complicated transmission system. The system includes two controllable differentials for the front and rear axles, and a central torque biasing device controlled by an electronic control unit. On the other hand, a motorized four-wheel electric drive can easily realize the torque vectoring process (Figure 9.15).

Generally, differential braking and differential traction can be combined to generate higher amounts of stabilizing yaw moment. However, such a goal can only be achieved by using advanced drive configurations, such as motorized wheels in electric drive systems.

Example 9.3

Consider an electric vehicle equipped with four electric in-wheel motors. The torque of the electric motors can be calculated using the following characteristic equation:

$$T_m = \begin{cases} 260\, x_{\theta m}\ \text{Nm if} & \omega \leq 105\ \text{rad/s} \\ \dfrac{27300}{\omega} x_{\theta m}\ \text{Nm if} & 105 \leq \omega \leq 210\ \text{rad/s} \end{cases}$$

where $-1 \leq x_{\theta m} \leq 1$ is the control signal, which is a function of the accelerator pedal position, and can also be independently controlled by the DYC system. Suppose the absolute torque of the electric motors during braking is also calculated by the above equation.

As mentioned above, under normal driving conditions, all electric motors generate the same amount of torque. However, during harsh driving conditions, the DYC system can change the torque of electric motors to create a yaw moment to improve the vehicle handling.

Table 9.2 Vehicle specifications

Parameter	Value
M	1530 (kg)
a	1.320 (m)
b	1.456 (m)
μ	0.98
T	1.62 (m)
R	0.3 (m)

Consider a maneuver at a speed of 130 km/h, where the total requested vehicle traction force is 1300 N. Considering small steering angles, calculate the maximum possible generated yaw moment by the DYC system for the following cases:

- friction brake
- traction of electric motors
- traction and braking of electric motors.

To solve the problem, use the information provided in Table 9.2.

Solution

First, we need to calculate the speed and torque of each electric motor at the given instant:

$$\omega = \frac{u}{R} = \frac{(130/3.6)}{0.3} = \frac{120.4\ rad}{s} = 1149.5 \text{ rpm}$$

$$T_m = \frac{\left(\sum_{i=1}^{4} F_{xi}\right)R}{4} = \frac{1300 \times 0.3}{4} = 97.5 \text{ Nm}$$

(a) As was explained in Chapter 4, the maximum force that can be applied on each tire by a friction brake is μF_z. In order to generate the maximum yaw moment, braking is applied on both wheels of one side (Figure 9.16):

$$F_{x3} = F_{x4} = 0$$

$$M_z = \frac{T}{2}(F_{x1} + F_{x2}) = \frac{T}{2}(\mu F_{z1} + \mu F_{z2}) = \frac{1.62}{2}(0.98 \times (765 \times 9.81)) = 5957.2 \text{ Nm}$$

(b) As Figure 9.17 indicates, to maximize the generated yaw moment, all required traction must be provided by the electric motors on one side. Next, each electric motor should create 650 N traction as half of the total traction (equal to 195 Nm driving torque). Then, the generated yaw moment by the DYC system is:

$$M_z = \frac{T}{2}(F_{x3} + F_{x4}) = \frac{1.62}{2}(650 + 650) = 1053 \text{ Nm}$$

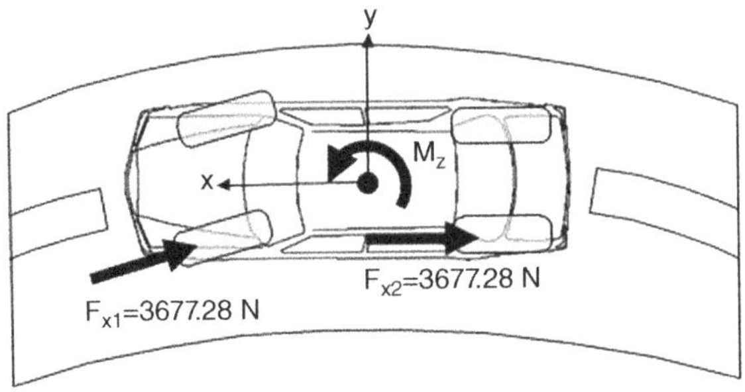

Figure 9.16 Yaw moment generated by differential friction braking

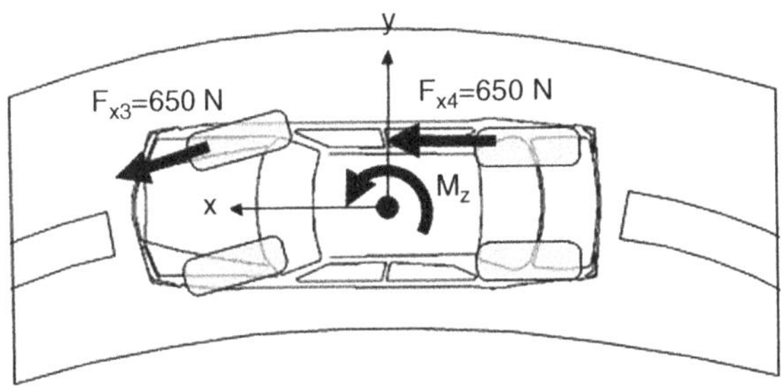

Figure 9.17 Yaw moment generated by differential traction

(c) At a speed of 120.4 rad/s, the maximum torque of the electric motor is:

$$T_m = \frac{27300}{120.4} = 226.7\ \text{Nm}$$

If the maximum motor torque capacity is used, the vehicle traction force is equal to:

$$F_x = 2 \times \left(\frac{T_m}{R}\right) = 2 \times \left(\frac{226.7}{0.3}\right) = 1511.3\ \text{N}$$

which is greater than the given required traction force. As a result, the electric motors on the other side should produce braking torque in order to keep the vehicle traction force unchanged. As such, each electric motor should generate 105.65 N braking force (equal to 31.7 Nm negative torque) as seen in Figure 9.18.

$$M_z = -\frac{T}{2}(F_{x1} + F_{x2}) + \frac{T}{2}(F_{x3} + F_{x4}) = \frac{1.62}{2}(1511.3) - \frac{1.62}{2}(105.65 + 105.65)$$
$$= 1395.3\ \text{Nms}$$

Figure 9.19 shows the operating points of the electric motors in cases (b) and (c).

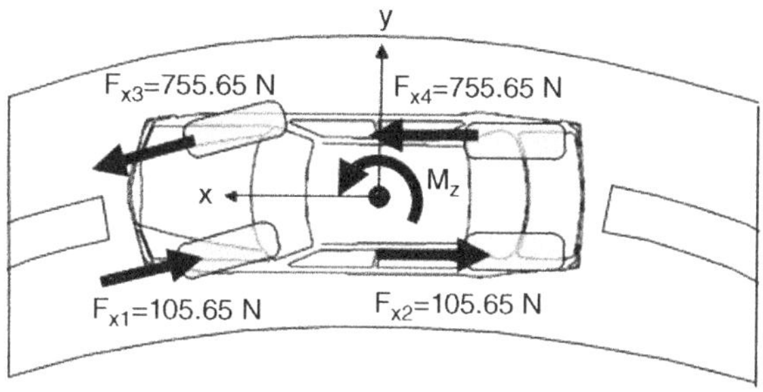

Figure 9.18 Generating yaw moment by electric motors actuation

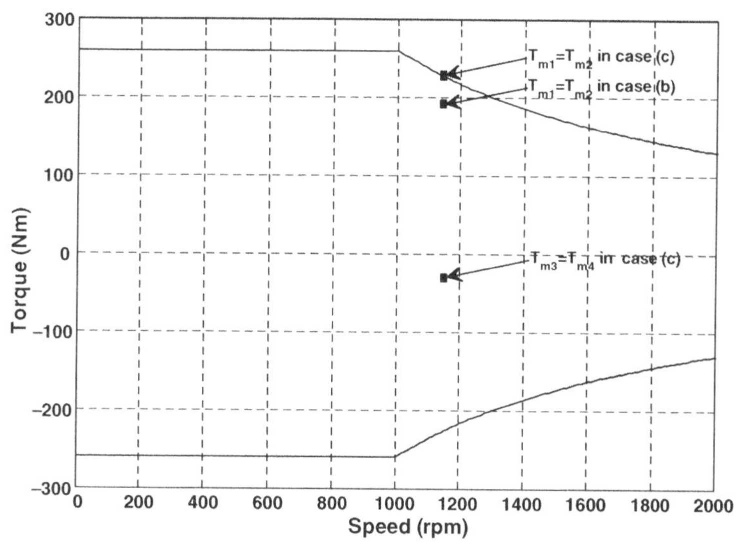

Figure 9.19 Operating points of electric motors in cases (b) and (c)

Indirect yaw moment (IDYC) control systems: According to Equation (9.5), the vehicle yaw moment can also be controlled through the tires' lateral forces. This is the natural way for the driver to control the vehicle's motion. In fact, by applying a steering angle to the front wheels through the steering system, the tires' side slip angles are varied, and consequently, the lateral forces are changed. Changing the lateral forces of the front tires excites the vehicle's lateral dynamics, changes the vehicle yaw moment, and affects the vehicle yaw motion.

This method of yaw moment control can be employed by VDC systems. A category of VDC systems, known as the active steering control (ASC) system, works based on controlling the steering angle. ASC systems can be categorized as active front steering (AFS), and active four-wheel steering (A4WS) systems. In AFS systems, the driver's steering input is automatically adjusted by adding or subtracting a correction steering angle. In other words, the driver's steering angle is modified by an additional steering input through the AFS system. In the case of A4WS systems, both the front and rear wheel steering angles are controlled by the VDC system.

While there are some mechanical and electro-mechanical types of AFS and A4WS systems, steer-by-wire (SBW) systems can easily and precisely realize AFS and A4WS control. However, SBW systems are still under development and have not yet been implemented in mass-produced vehicles. More information about AFS, 4WS, and SBW systems can be found in Chapter 3.

Regardless of the technological aspects of ASC systems, it is necessary to discuss how active steering control can affect the vehicle dynamics. For this purpose, we can once again use the generalized form of the single track handling model that is represented by Equation 9.2. It can be reformulated for a vehicle that is equipped with an active steering control system. In such a vehicle, the steering angle of the front and rear axles can be expressed as:

$$\delta = \begin{cases} \text{Front} & \delta + \Delta\delta \\ \text{Rear} & \delta_r \end{cases} \tag{9.9}$$

where δ is the driver's steering angle, and $\Delta\delta$, δ_r are the steering angle corrections for the front and rear wheels that are applied by the ASC system. By having the steering angles for both axles, the model formulation is changed to:

$$\begin{pmatrix} \dot{v} \\ \dot{r} \end{pmatrix} = \begin{pmatrix} -\dfrac{C_{\alpha r} + C_{\alpha f}}{Mu} & \dfrac{(bC_{\alpha r} - aC_{\alpha f})}{Mu} - u \\ \dfrac{(bC_{\alpha r} - aC_{\alpha f})}{I_z u} & -\dfrac{(b^2 C_{\alpha r} + a^2 C_{\alpha f})}{I_z u} \end{pmatrix} \begin{pmatrix} v \\ r \end{pmatrix} + \begin{pmatrix} \dfrac{C_{\alpha f}}{M} \\ \dfrac{aC_{\alpha f}}{I_z} \end{pmatrix} \delta + \begin{pmatrix} \dfrac{f_{0f} + f_{0r}}{M} \\ \dfrac{af_{0f} - bf_{0r}}{I_z} \end{pmatrix}$$
$$+ \begin{pmatrix} \dfrac{C_{\alpha f}}{M} \\ \dfrac{aC_{\alpha f}}{I_z} \end{pmatrix} \Delta\delta + \begin{pmatrix} \dfrac{C_{\alpha r}}{M} \\ -\dfrac{bC_{\alpha r}}{I_z} \end{pmatrix} \delta_r \tag{9.10}$$

As seen above, the system is affected by three inputs: δ, $\Delta\delta$, δ_r. During normal driving conditions, as long as the driver can apply the appropriate steering angle, the ASC inputs are zero. In harsh maneuvers or slippery roads where the expected vehicle response differs from its actual response, the ASC system will inject correcting steering angles to stabilize the vehicle motion. At any given time, the expected vehicle response is estimated from the driver's steering angle and vehicle velocity, and the actual vehicle response is obtained from the vehicle's sensors, including the yaw rate and accelerometer sensors.

In addition, ASC systems can change the handling feel of a vehicle. Vehicle sensitivity to driver input can be increased or decreased as a result of ASC inputs. For instance, an ASC system can magnify the driver's steering input to have a more responsive car. Conversely, an ASC system can reduce the steering input of the driver to have a less-responsive driving experience. The following example is useful for a better understanding of this subject.

Example 9.4

Suppose in an ASC system, the steering angle corrections for the front and rear wheels are obtained based on a proportional controller as:

$$\begin{aligned}\Delta\delta &= k_{v1}v + k_{r1}r + k_{\delta 1}\delta \\ \delta_r &= k_{v2}v + k_{r2}r + k_{\delta 2}\delta\end{aligned}$$

where k_{v1}, k_{r1}, $k_{\delta 1}$, k_{v2}, k_{r2}, and $k_{\delta 2}$ are the control gains.

(a) Using the above control laws, develop the single track handling model for a vehicle equipped with an ASC system. Discuss the system performance in the linear and saturation zones of tires.

(b) Using the vehicle specifications stated in Table 9.1, plot the yaw velocity gain of the vehicle with and without the ASC system, and discuss how it can affect vehicle handling. Use two subsequent sets of the gains for the ASC system:

I) $k_{v1} = -0.003, k_{r1} = -0.007, k_{\delta 1} = -0.1, k_{v2} = -0.007, k_{r2} = -0.001, k_{\delta 2} = 0.1$

II) $k_{v1} = -0.003, k_{r1} = -0.007, k_{\delta 1} = -0.1, k_{v2} = -0.002, k_{r2} = -0.001, k_{\delta 2} = -0.2$

Solution

(a) By substituting the control laws in Equation 9.10, the single track handling model of a vehicle with an ASC system becomes:

$$\begin{pmatrix}\dot{v}\\ \dot{r}\end{pmatrix} = \begin{pmatrix} -\dfrac{C_{\alpha r}+C_{\alpha f}}{Mu} & \dfrac{(bC_{\alpha r}-aC_{\alpha f})}{Mu}-u \\ \dfrac{(bC_{\alpha r}-aC_{\alpha f})}{I_z u} & -\dfrac{(b^2C_{\alpha r}+a^2C_{\alpha f})}{I_z u}\end{pmatrix}\begin{pmatrix}v\\ r\end{pmatrix} + \begin{pmatrix}\dfrac{C_{\alpha f}}{M}\\ \dfrac{aC_{\alpha f}}{I_z}\end{pmatrix}\delta + \begin{pmatrix}\dfrac{f_{0f}+f_{0r}}{M}\\ \dfrac{af_{0f}-bf_{0r}}{I_z}\end{pmatrix}$$

$$+\begin{pmatrix}\dfrac{C_{\acute{a}f}}{M}\\ \dfrac{aC_{\acute{a}f}}{I_z}\end{pmatrix}(k_{v1}v+k_{r1}r+k_{\delta 1}\delta)+\begin{pmatrix}\dfrac{C_{\acute{a}r}}{M}\\ -\dfrac{bC_{\acute{a}r}}{I_z}\end{pmatrix}(k_{v2}v+k_{r2}r+k_{\delta 2}\delta)$$

$$\begin{pmatrix}\dot{v}\\ \dot{r}\end{pmatrix} = \begin{pmatrix} -\dfrac{C_{\alpha r}+C_{\alpha f}}{Mu}+\dfrac{C_{\acute{a}f}}{M}+k_{v1}+\dfrac{C_{\acute{a}r}}{M}k_{v2} & \dfrac{(bC_{\alpha r}-aC_{\alpha f})}{Mu}-u+\dfrac{C_{\acute{a}f}}{M}k_{r1}+\dfrac{C_{\acute{a}r}}{M}k_{r2} \\ \dfrac{(bC_{\alpha r}-aC_{\alpha f})}{I_z u}+\dfrac{aC_{\acute{a}f}}{I_z}k_{v1}-\dfrac{bC_{\acute{a}r}}{I_z}k_{v2} & -\dfrac{(b^2C_{\alpha r}+a^2C_{\alpha f})}{I_z u}+\dfrac{aC_{\acute{a}f}}{I_z}k_{r1}-\dfrac{bC_{\acute{a}r}}{I_z}k_{r2}\end{pmatrix}\begin{pmatrix}v\\ r\end{pmatrix}$$

$$+\begin{pmatrix}\dfrac{C_{\alpha f}}{M}+\dfrac{C_{\acute{a}f}}{M}k_{\delta 1}+\dfrac{C_{\acute{a}r}}{M}k_{\delta 2}\\ \dfrac{aC_{\alpha f}}{I_z}+\dfrac{aC_{\acute{a}f}}{I_z}k_{\delta 1}-\dfrac{bC_{\acute{a}r}}{I_z}k_{\delta 2}\end{pmatrix}\delta+\begin{pmatrix}\dfrac{f_{0f}+f_{0r}}{M}\\ \dfrac{af_{0f}-bf_{0r}}{I_z}\end{pmatrix}$$

When the tires are in the linear zone, the preceding model can be rewritten as:

$$\begin{pmatrix} \dot{v} \\ \dot{r} \end{pmatrix} = \begin{pmatrix} -\dfrac{C_{\alpha r}+C_{\alpha f}}{Mu}+\dfrac{C_{\text{áf}}}{M}k_{v1}+\dfrac{C_{\text{ár}}}{M}k_{v2} & \dfrac{(bC_{\alpha r}-aC_{\alpha f})}{Mu}-u+\dfrac{C_{\text{áf}}}{M}k_{r1}+\dfrac{C_{\text{ár}}}{M}k_{r2} \\ \dfrac{(bC_{\alpha r}-aC_{\alpha f})}{I_z u}+\dfrac{aC_{\text{áf}}}{I_z}k_{v1}-\dfrac{bC_{\text{ár}}}{I_z}k_{v2} & -\dfrac{(b^2C_{\alpha r}+a^2C_{\alpha f})}{I_z u}+\dfrac{aC_{\text{áf}}}{I_z}k_{r1}-\dfrac{bC_{\text{ár}}}{I_z}k_{r2} \end{pmatrix} \begin{pmatrix} v \\ r \end{pmatrix}$$

$$+\begin{pmatrix} \dfrac{C_{\alpha f}}{M}+\dfrac{C_{\text{áf}}}{M}k_{\delta 1}+\dfrac{C_{\text{ár}}}{M}k_{\delta 2} \\ \dfrac{aC_{\alpha f}}{I_z}+\dfrac{aC_{\text{áf}}}{I_z}k_{\delta 1}-\dfrac{bC_{\text{áf}}}{I_z}k_{\delta 2} \end{pmatrix}\delta$$

As can be seen, in the linear zone, the structure of the vehicle model is not fundamentally changed compared to a vehicle without an ASC system. The control gains, however, modify the vehicle's handling behavior based on their tuning and magnitudes.

When the tires are in the saturation zone, the model changes to:

$$\begin{pmatrix} \dot{v} \\ \dot{r} \end{pmatrix} = \begin{pmatrix} 0 & -u \\ 0 & 0 \end{pmatrix} \begin{pmatrix} v \\ r \end{pmatrix} + \begin{pmatrix} \dfrac{f_{0f}+f_{0r}}{M} \\ \dfrac{af_{0f}-bf_{0r}}{I_z} \end{pmatrix}$$

In this case, as seen in the above equation, the driver and correcting steering angle inputs have disappeared and the vehicle is no longer controllable. In other words, the ASC systems are not able to affect the vehicle's dynamics when all the tires are saturated.

(b) As explained in Chapter 7, yaw velocity gain can be obtained by Equation 7.20. To show how the ASC systems can affect the vehicle's dynamics in the linear handling region, an analysis of this factor is helpful. Figure 9.20 depicts the yaw velocity gain of the vehicle without control and compares it to the vehicles with an ASC system for set (I) and (II) of gains. Figure 9.20 clearly shows that the vehicle handling performance changes when the vehicle is equipped with an ASC system. The changing of the handling response depends on the tuning of ASC gains.

For the gain set (I), the vehicle with the ASC system has lower yaw velocity gain. This shows that the vehicle becomes less sensitive to the driver's steering input. On the other hand, for the gain set (II), the vehicle is more sensitive to the steering input, indicating a larger yaw response for a given steering angle. This is a preferred response in sport cars while lower yaw velocity gain is more appropriate in family cars.

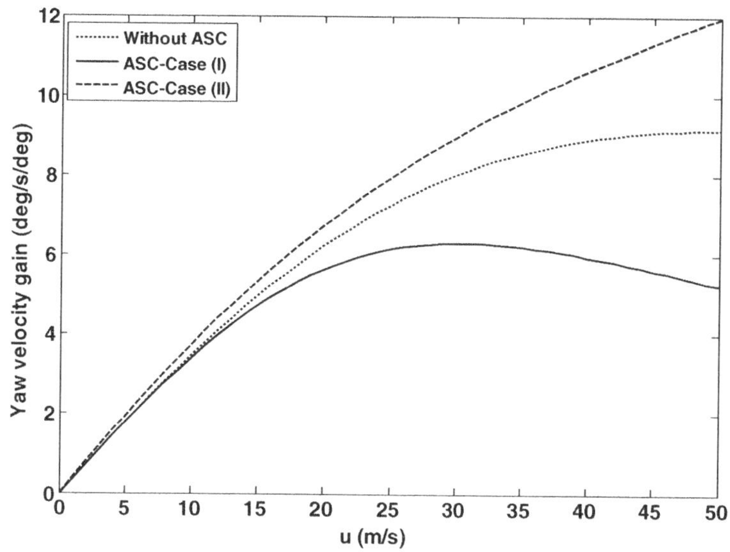

Figure 9.20 Yaw velocity gain versus vehicle speed for uncontrolled vehicles, and two types of ASC vehicles

Based on the above example, one can see that ASC systems can provide variable handling characteristics; moreover, they can prevent tires from saturation to improve their stability. It should, however, be noted that despite the ability of the ASC systems to improve the vehicle's handling characteristics in the linear tire zone, it is less helpful when the tires are saturated.

9.2.3.5 Integrated Systems

Each of the above-mentioned VDC systems has its own advantages and disadvantages; therefore, no individual system can provide perfect performance characteristics under every single road and driving condition. For instance, while differential braking systems can effectively stabilize vehicles during severe maneuvers, their efficiency deteriorates on roads with different friction of coefficient on each side of the vehicle. This is because on the low friction side, the maximum available braking force and consequently the generated external yaw moment are considerably lower than those of the high friction side. Likewise, while ASC systems can provide adjustable and pleasant handling behavior under normal driving conditions, they are less effective in hazardous emergency situations. Additionally, from a practical point of view, implementing different individual control systems with different architectures and configurations is problematic and their outputs might conflict each other.

To resolve such issues, the idea of using an integrated VDC system instead of multiple individual systems has been introduced in the last decade. Several conceptual and actual systems have been reported under "integrated or global chassis control," "integrated vehicle dynamic control," or "vehicle dynamic management" [1]. All these systems try to coordinate vehicle dynamic functions via integrated control of the active chassis systems, including braking, steering, and suspension systems. It is important to emphasize that different control systems, even without direct coupling, may be coupled via the vehicle causing unwanted effects. As a result, it is strongly believed that an integrated control strategy can provide enhanced performance compared to individual control systems.

9.3 VDC Implementation on Electric and Hybrid Vehicles

Electric and hybrid electric vehicles with a central powertrain can have almost the same VDC systems (differential braking, active steering, and active suspension systems) as those of conventional vehicles. However, there are other powertrain configurations for EVs and HEVs that could enhance VDC's abilities. For example, consider a four-wheel drive hybrid configuration, where the front and rear axles are driven by an ICE and an electric motor, respectively. In such a vehicle, in addition to the afore-mentioned systems, vehicle dynamics can be controlled by managing the front/rear distribution of the tractive force. This can be easily implemented in this vehicle because each axle is driven independently. Additionally, we can equip the rear axle with two independent electric motors to make it capable of differential traction.

Also, a completely motorized wheel configuration with four independently controlled wheel motors is the other possible alternative. As mentioned in the previous chapters, such a design provides a packaging advantage because the single large motor is replaced by a number of smaller motors. More importantly, using this configuration, the traction of each wheel can be controlled independently, and the VDC system becomes capable of complete torque vectoring. This capability can dramatically improve the vehicle handling and stability compared to conventional vehicles.

This section discusses the design of a dedicated VDC system to demonstrate the improved ability of modern EVs and HEVs in controlling vehicle dynamics.

Suppose that the four wheels of an EV vehicle are independently steerable and are driven by independent motors. While it is the most sophisticated configuration for vehicle dynamics control, the general form of the proposed control strategy can easily be adapted to simpler vehicle types. This includes a four independent-wheel drive system without active steering control, or the afore-mentioned hybrid four-wheel drive configuration.

9.3.1 Structure of the Control System

As mentioned earlier, the main objective of VDC systems is to achieve the desired vehicle behavior. Therefore, the essential task of the proposed control system is to minimize the difference between the desired and the existing motion of the vehicle. To achieve this goal, it is necessary to design the control system based on an appropriate architecture.

Different types of architectures for VDC systems have been proposed and implemented in recent years, but the large majority of them use the multi-layer structure [2–8] or a hierarchical control system that is used in industrial and vehicular control systems. A multi-layer control system is generally made up of a set of controllers that are arranged in a hierarchical tree. In this system, each layer operates independently based on the input signals from the upper layer, and sends its output signals to the lower layer. The layers also can receive information from sensors, or from other layers.

In fact, multi-layer control systems have been inspired by man-made social, military, or administrative systems. Such systems with complex behavior are often organized as a hierarchy

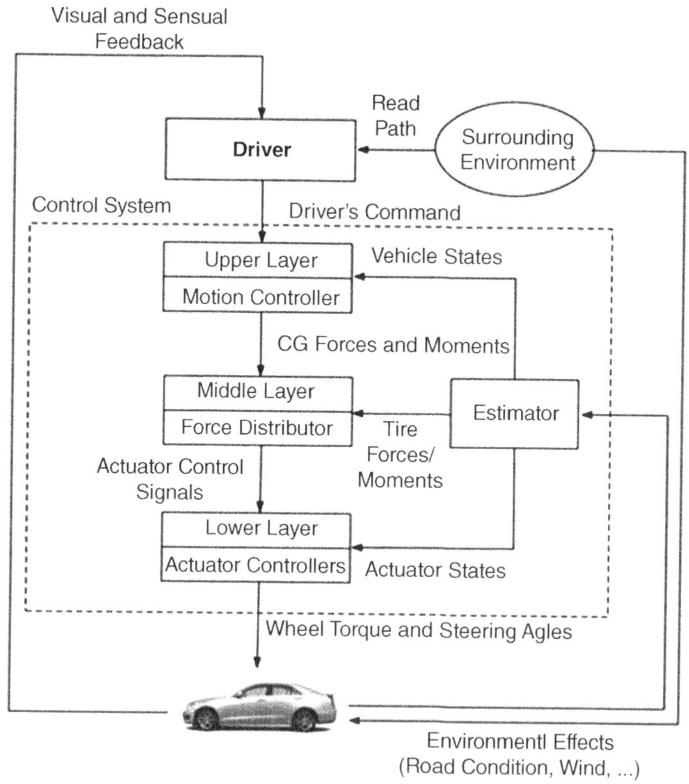

Figure 9.21 A typical multi-layer vehicle dynamic control system's architecture

to facilitate the decision-making process by dividing it between different layers. In this approach, a large complex control process is divided into smaller and simpler processes (layers) to improve system reliability, and simplify the control system design.

As illustrated in Figure 9.21, a typical multi-layer VDC system consists of three layers that are discussed in detail below.

9.3.1.1 Upper Layer: Motion Controller

The motion controller is at the top level of the control system. For a better understanding of the motion controller's functions, let us suppose the vehicle is a planar rigid body that moves in a XY plan. The vehicle's existing motion is specified by its longitudinal, lateral, and yaw velocities. These are respectively symbolized by u, v, and r. The desired motion of the vehicle is denoted by the desired longitudinal, lateral, and yaw velocities u_d, v_d, and, r_d. Assuming these velocities are known (they will be defined later), the role of a VDC system is to minimize the difference between the existing and desired vehicle velocities.

Now consider $\overline{F}_x, \overline{F}_y$, and $\overline{M}_z$ are the forces and moment that need to be applied to the vehicle's center of gravity in the x, y, and about the z axis to generate the desired motion. For simplicity, these are called CG forces and moment. CG forces and moment represent the summation of forces and moments that are produced by the tires and applied to the vehicle.

The main contribution of the motion controller is to calculate the desired value of CG forces and moment in such a manner that the difference between the existing and desired vehicles' velocities is minimized. For this purpose, the motion controller obtains the CG forces based on the driver's inputs, including the steering angle (δ), the brake pedal force (F_b), and the accelerator pedal position (x_θ). It also uses other sensory data or estimated information of the vehicle and its surrounding for the CG forces calculation. The calculated desired CG forces and moment are then sent to the mid-layer for distribution among the tires.

9.3.1.2 Mid-layer, the Force Distributor

This layer distributes the dictated desired CG forces and moment by the motion controller between the tires. It determines the desired longitudinal and lateral forces of each wheel (F^d_{xi}, F^d_{yi}) in such a manner that the difference between the resultant CG forces and moment, and their corresponding desired values become minimal. If the vehicle does not have an active steering control system, the mid-layer controller only calculates the desired longitudinal force of each wheel.

For this purpose, the distributer needs to know the existing longitudinal, lateral, and normal forces of each wheel. However, their direct measurements are not possible; therefore, the estimated value of these forces should be received through the estimator unit. The calculated desired forces of each wheel are sent to the lower layer.

9.3.1.3 Lower-Layer, Actuator Controllers

The lower layer of the control system consists of the independent actuator controllers, which control different actuators of the vehicle. These could include the wheel motors, the brake system, and the steering system controllers. Each one of these local controllers receives its corresponding desired force(s) and then attempts to actualize it.

9.3.1.4 Estimator

In many cases, direct measurement of a variable or parameter is expensive or even impossible. In vehicles, lateral and longitudinal velocities and tire forces are among those variables of which their direct measurements are impractical. As a result, an estimator unit is needed in the system's architecture to provide an estimation of such variables. The estimation process relies on other vehicle sensory data and estimation algorithms.

9.3.2 Control System Design

As mentioned above, the proposed VDC system includes three main parts. In this section, each part is designed and discussed in detail.

9.3.2.1 Motion Controller Design

The role of the motion controller is to calculate the desired CG forces and moment based on the driver's input and vehicle states. In the following, a method to calculate the desired forces of the CG based on the driver's inputs is presented.

- *Longitudinal CG force:* According to Equation 4.13, the governing equation of the longitudinal dynamics of a vehicle can be written as:

$$\frac{du}{dt} = \frac{1}{M}\left(\overline{F}_x - F_L\right) + rv \tag{9.11}$$

where, $\overline{F}_x$ is the longitudinal CG force and F_L is the cumulative vehicle resistance forces, including the aerodynamic drag, and the road gradient resistance as defined by Equation 6.2. The term rv represents the cross-coupling effect between the lateral and the longitudinal dynamics.

The driver operates the accelerator and brake pedals to maintain his/her targeted longitudinal velocity (u_d). The motion controller calculates the desired longitudinal CG force ($\overline{F}_{xd}$) based on the accelerator pedal position (x_θ) and the brake pedal force (f_b). It can be formulated as:

$$\overline{F}_{xd} = \overline{F}_a(x_\theta, u) - \overline{F}_b(f_b) \tag{9.12}$$

Generally, $\overline{F}_a(x_\theta, u)$ and $\overline{F}_b(f_b)$ are defined based on the pedal's positions and vehicle speed and also the desired accelerating and braking characteristics of the vehicle. Considering marketability issues, the drivers' expectation of the vehicle response must not be violated. Therefore, the longitudinal dynamic behavior of the proposed electric vehicle must be similar to that of conventional vehicles. To satisfy this, the following form can be considered for $\overline{F}_a$:

$$\overline{F}_a = k_a(u)x_\theta \tag{9.13}$$

where $k_a(u)$ can be characterized by the curve shown in Figure 9.22. The curve represents a constant traction at lower speeds and a constant power characteristic at higher speeds. As discussed in Chapter 6, this characteristic is a typical traction-speed curve for vehicles.

$\overline{F}_b$ can be formulated as:

$$\overline{F}_b = \begin{cases} k_{b1}f_b & f_b \le f_b^* \\ k_{b2}f_b + f_{b0} & f_b > f_b^* \end{cases} \tag{9.14}$$

This formula is illustrated by Figure 9.23. It shows a typical progressive braking characteristic curve that is regularly used by conventional vehicles. The curve shows that for more comfortable driving in city traffic, the brake is less sensitive at the beginning of its stroke. Then, its sensitivity increases sharply to provide more safety during emergency braking.

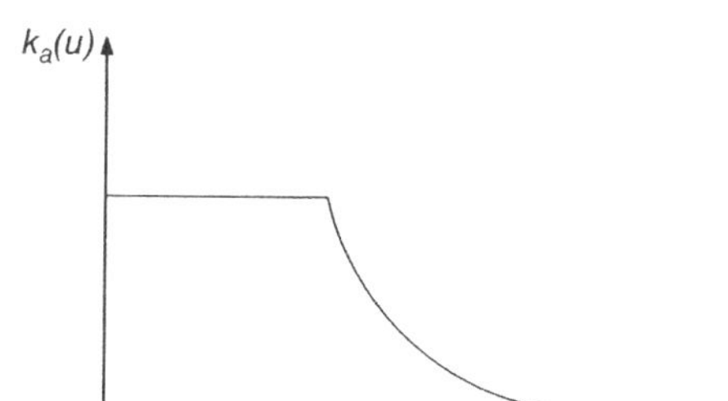

Figure 9.22 Characteristic curve for $k_a(u)$

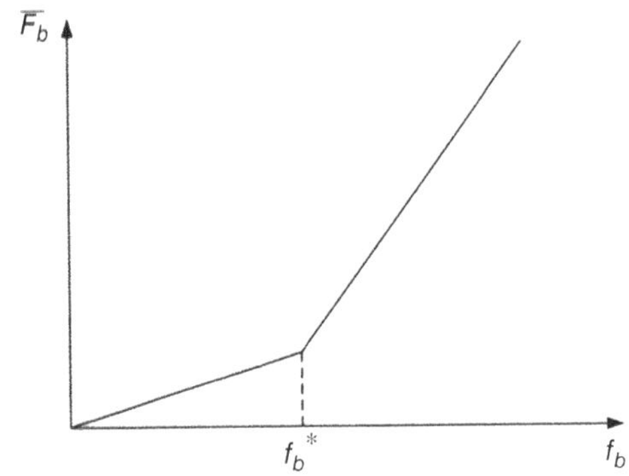

Figure 9.23 Characteristic curve for $\overline{F}_b$

The driver operation can be simply modeled by a proportional and integrator law:

$$\begin{aligned} x_\theta &= k_p^a e_u + k_i^a \int e_u dt \\ \overline{F}_b &= k_p^b e_u + k_i^b \int e_u dt \end{aligned} \tag{9.15}$$

Where k_p^a, k_i^a, k_p^b and k_i^b are the drivers' gain, x_θ and $\overline{F}_b$ are the accelerator pedal position and the brake pedal force, and the term e_u is the longitudinal velocity tracking error, defined as:

$$e_u = u_d - u \tag{9.16}$$

- *Lateral CG force and yaw CG moment:* Unlike longitudinal velocity, drivers do not have any idea about the desired lateral and yaw velocities. The driver simply applies the steering input and expects perfect handling behavior in terms of command following and stability. Therefore, desired lateral and yaw velocities must first be determined to calculate the desired lateral CG force and yaw moment. For this purpose, we shall make a reference vehicle model and calculate the vehicle's desired lateral and yaw velocities by using this model.

The following points should be considered in order to develop an appropriate vehicle reference model:

- The best vehicle handling responses are achieved during driving on dry and high friction roads. The lateral force capacity of tires is dramatically reduced on slippery low-friction surfaces, and the vehicle handling behavior is seriously deteriorated. Therefore, the road condition should be assumed to be perfect to define the desired lateral and yaw velocities. Consequently, the vehicle's handling behavior is not affected by the road condition in defining the desired lateral and yaw velocities by the proposed VDC system.
- The dynamic behavior of the vehicle is linear during normal driving conditions as seen in Chapter 7. However, the vehicle's behavior becomes nonlinear during severe maneuvers. This means that there is no longer a linear and consistent relationship between the driver steering input and the vehicle's response. The nonlinear behavior of the vehicle is initiated from different sources especially from the tire dynamics. The desired lateral and yaw velocities are defined based on the linear vehicle dynamics.

- Transient handling behavior may result in undesirable effects such as overshoot, delay, or lag in the vehicle's response time. To define the desired lateral and yaw velocities, the steady state part of the reference model responses are considered.
- The vehicle's handling behavior can be simulated by using a bicycle model. Although such a simple model cannot accurately represent the actual vehicle behavior in various situations, it is ideal for defining the desired behavior of the vehicle. This is because we are looking for the desired or ideal behavior of a vehicle regardless of the maneuver severity or road condition.

Considering the discussions above, the desired lateral and yaw velocities are the steady state values of the lateral and yaw velocity responses of a bicycle model with a linear tire model. The model's parameters, including mass, inertia, and CG position, are set to the vehicle's nominal values. The tires' cornering stiffness is set to the corresponding values of dry roads. Accordingly, based on Equations 7.19 and 7.20, the desired lateral and yaw velocities can be formulated as:

$$
\begin{aligned}
r_d &= \frac{\delta u}{l + k_{us}^* u^2} \\
v_d &= \frac{\delta u \left(b - \dfrac{Ma}{lC_{\alpha f}^*} u^2 \right)}{l + k_{us}^* u^2}
\end{aligned}
\tag{9.17}
$$

where k_{us}^* is the under-steer coefficient of the reference model that is calculated by Equation 7.14.

The analogous quantity of the lateral velocity is the vehicle sideslip angle. Vehicle stability is primarily determined based on the magnitude of the vehicle sideslip angle. Generally, a smaller sideslip angle means increased stability. Therefore, to achieve more stability and simplify the motion controller design process, the desired value of the lateral velocity can be modified and simply set to zero:

$$v_d = 0 \tag{9.18}$$

After defining the desired yaw and lateral velocities, the controller design can be started. For this purpose, first, the lateral and yaw velocities errors are defined as the difference between the desired and actual values as:

$$
\begin{aligned}
e_v &= v_d - v \\
e_r &= r_d - r
\end{aligned}
\tag{9.19}
$$

The main goal of the motion controller is to minimize these errors. There are different types of control theories that can be used for this purpose, but a proportional/integrator/derivative (PID) control is a simple and effective way to minimize the errors. By applying a PID controller to each of the errors in Equation (9.19), the lateral CG forces and yaw CG moment are defined as:

$$
\begin{aligned}
\overline{F}_{yd} &= k_{P1}^v e_v + k_{P1}^v \frac{de_v}{dt} + k_{I1}^v \int e_v dt + k_{P1}^r e_r + k_{D1}^r \frac{de_r}{dt} + k_{I1}^r \int e_r dt \\
\overline{M}_{zd} &= k_{P2}^v e_v + k_{D2}^v \frac{de_v}{dt} + k_{I2}^v \int e_v dt + k_{P2}^r e_r + k_{D2}^r \frac{de_r}{dt} + k_{I2}^r \int e_r dt
\end{aligned}
\tag{9.20}
$$

where $k^v_{P1}, k^v_{D1}, k^v_{I1}, k^r_{P1}, k^r_{D1}, k^r_{I1}, k^v_{P2}, k^v_{D2}, k^v_{I2}, k^r_{P2}, k^r_{D2}$ and k^r_{I2} are the control gains. To achieve the best performance, the control gain can be a function of the vehicle speed [2].

Since a proportional control action usually provides satisfactory results, for simplicity, we use the proportional controllers in Equation (9.20) resulting in:

$$\begin{aligned} \overline{F}_{yd} &= k^v_{p1} e_v + k^r_{p1} e_r \\ \overline{M}_{zd} &= k^v_{p2} e_v + k^r_{p2} e_r \end{aligned} \tag{9.21}$$

By substituting Equations (9.17), (9.18) and (9.19) into Equation (9.21) we have:

$$\begin{aligned} \overline{F}_{yd} &= -k^v_{P1} v - k^r_{P1} r + k^\delta_{P1} \delta \\ \overline{M}_{zd} &= -k^v_{P2} v - k^r_{P2} r + k^\delta_{P2} \delta \end{aligned} \tag{9.22}$$

where

$$k^\delta_{P1} = \frac{k^r_{P1} u}{l + k^*_{us} u^2} \quad k^\delta_{P2} = \frac{k^r_{P2} u}{l + k^*_{us} u^2} \tag{9.23}$$

According to Equation (9.22), the CG force and moment are found by the vehicle state variable feedback and the driver's steering angle feed-forward. This type of the control law has been used in previous examples, and its positive performance has been demonstrated.

9.3.2.2 The Force Distributor

As illustrated by Figure 9.24, the actual forces and moment at the vehicle's center of gravity are a direct result of the longitudinal and lateral forces of the tires. The force distributor receives the desired value of CG forces and moment from the upper layer, and calculates the desired longitudinal and lateral forces of each tire. For this purpose, the following equations should be satisfied:

$$\begin{aligned} & F^d_{x1} + F^d_{x2} + F^d_{x3} + F^d_{x4} - \overline{F}_{xd} = 0 \\ & F^d_{y1} + F^d_{y2} + F^d_{y3} + F^d_{y4} - \overline{F}_{yd} = 0 \\ & \frac{T}{2}\left(F^d_{x1} + F^d_{x2} - F^d_{x3} - F^d_{x4}\right) + a\left(F^d_{y1} + F^d_{y3}\right) - b\left(F^d_{y2} + F^d_{y4}\right) - \overline{M}_{zd} = 0 \end{aligned} \tag{9.24}$$

By ignoring steering angle effects in the above equations, the longitudinal and lateral forces of the tires are considered to be in the x and y directions, respectively. This can be a valid assumption because during most moderate to high speeds maneuvers, the wheels' steering angle is very small.

According to Equation (9.24), one can see that the numbers of unknown variables are more than the number of equations. This results in an infinite number of solutions for this system of algebraic equations. Therefore, the problem can be reformulated as an optimization problem. In other words, in addition to satisfying the necessary conditions that are listed in Equation (9.24), other constraints can be added to find a solution for improved vehicle performance. For this purpose, let us define the tire workload as:

$$m_i = \frac{F^2_{xi} + F^2_{yi}}{F^2_{zi}}, \quad i = 1 \ldots 4 \tag{9.25}$$

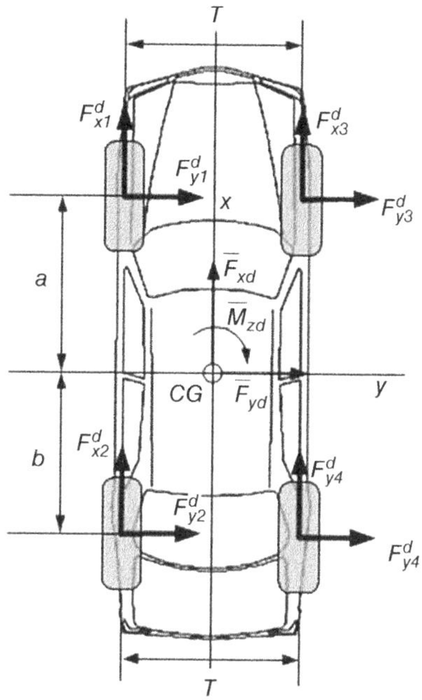

Figure 9.24 The tires' forces and the resultant forces and moment at the vehicle's CG

According to the friction circle theory discussed in Chapter 4, the smaller tire workload means that the tire's working point is far from its saturation limit. Thus, minimizing the tire workload can improve the tire reserve capacity, and consequently, the handling and safety of the vehicle.

To formulate Equation (9.25) as an optimization problem, the performance index or cost function should be defined. In defining the performance index, the main objectives are minimizing the CG forces and moment error, and minimizing the tire workload usage. Hence, the following performance index can be defined for our proposed optimization problem [9]·

$$J=\sum_{i=1}^{4}C_i\left(\frac{F^{d^2}_{xi}+F^{d^2}_{yi}}{F^2_{zi}}\right)+C_5\left(\sum_{i=1}^{4}F^d_{xi}-\overline{F}_{xd}\right)^2+C_6\left(\sum_{i=1}^{4}F^d_{yi}-\overline{F}_{yd}\right)^2$$
$$+C_7\left(\frac{T}{2}(F^d_{x1}+F^d_{x2}-F^d_{x3}-F^d_{x4})+a(F^d_{y1}+F^d_{y3})-b(F^d_{y2}+F^d_{y4})-\overline{M}_{zd}\right)^2 \qquad (9.26)$$

In the above equation, F^d_{xi} and F^d_{yi} are the optimization variables and F_{zi} is the tire's vertical force that is supposed to be estimated by the estimator unit. Furthermore, C_i's are weighting factors – their value determines the relative importance of different terms.

In the case of vehicles that are not equipped with an active steering control system, the steering angles cannot be controlled independently. Thus, lateral forces are no longer controllable. However, the lateral forces are still needed in the optimization process. They are the known parameters that are estimated by the estimator. In this case, the performance index can be rewritten as:

$$J = \sum_{i=1}^{4} C_i\left(\frac{F_{xi}^{d^2} + F_{yi}^2}{F_{zi}^2}\right) + C_5\left(\sum_{i=1}^{4} F_{xi}^d - \overline{F}_{xd}\right)^2 + C_7\left[\frac{T}{2}(F_{x1}^d + F_{x2}^d - F_{x3}^d - F_{x4}^d)\right.$$
$$\left. + a(F_{y1} + F_{y3}) - b(F_{y2} + F_{y4}) - \overline{M}_{zd}\right]^2 \tag{9.27}$$

Following the standard approach of optimization problems, the necessary conditions to minimize the performance index are:

$$\frac{\partial J}{\partial F_{xi}^d} = 0, \quad \frac{\partial J}{\partial F_{yi}^d} = 0 \quad i = 1 \ldots 4 \tag{9.28}$$

The above equations are formed in a system of algebraic equations that can be solved by numerical techniques.

9.3.2.3 The Actuators Controllers

The calculated desired values of longitudinal and lateral forces of each tire are received by the actuator controllers to be actualized through the motor, brake, and steering systems.

- *Wheel torque control:* To control the longitudinal force, consider the wheel dynamics Equation (7.40), with the desired longitudinal force for wheel i as:

$$I_{w_i}\frac{d\omega_{wi}}{dt} = T_{d_i} + T_{b_i} - F_{xi}^d R \quad i = 1, \ldots, 4 \tag{9.29}$$

where T_d and T_b are the motor and friction brake torques of each wheel. By considering T_{wi} as the total torque of each wheel, it can be found as:

$$T_{wi} = T_{d_i} + T_{b_i} = I_{w_i}\frac{d\omega_{wi}}{dt} + F_{xi}^d R \quad i = 1, \ldots, 4 \tag{9.30}$$

In the above equation, the angular acceleration, normal force, inertia, and radius of each wheel are supposed to be known, measurable, or estimated parameters. As such, T_{wi} can be calculated by this open-loop control law. However, in practice, unmodeled dynamics, changing system parameters, and inaccurate measurement or estimation of parameters/variables can reduce the effectiveness and performance of the above control law. This problem can be resolved by using the estimated value of the tire longitudinal force $\tilde{F}_{xi}$ and defining the force error as:

$$e_{xi} = F_{xi}^d - \tilde{F}_{xi} \tag{9.31}$$

Now by adding a control action, a closed-loop control system can be formed. Here, for simplicity, a PI (proportional-integrator) control action is used:

$$T_{wi} = I_{w_i}\frac{d\omega_{wi}}{dt} + F_{xi}^d R + k_P^x e_{xi} + k_I^x \int e_{xi} dt \quad i = 1, \ldots, 4 \tag{9.32}$$

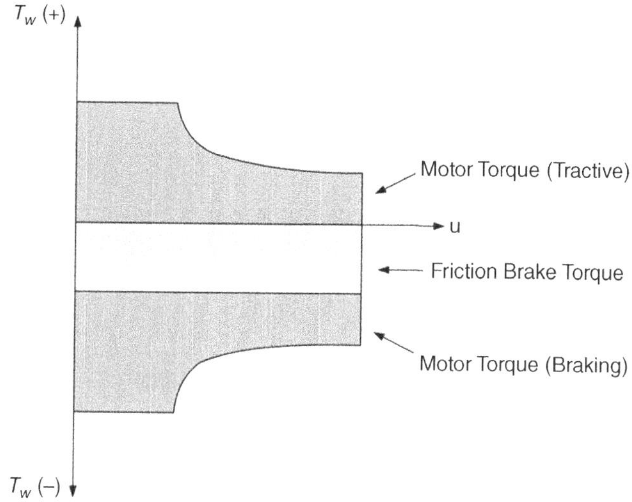

Figure 9.25 Tractive and braking torque capacity

Friction brakes and electric motors are two sources of negative wheel torque; thus, the calculated total torque should be distributed between them. Figure 9.25 shows the logic of this process. As displayed, for tractive torques ($T_{wi} > 0$), the electric motor is the only source. However, in the case of braking ($T_{wi} < 0$), the negative torque can be generated by the motor and/or the friction brake. Therefore, different strategies can be implemented for braking torque management. Nevertheless, if the electric motors can regenerate vehicle kinetic energy, then activating them has more priority. However, as can be seen in Figure 9.26, the torque capacity of motors is limited. Thus, if the requested torque is higher than the motor capacity, the remaining torque should be generated by the friction brake.

Eventually, the requested motor torque is implemented through the motor driver unit and the friction brake torque is controlled by the brake line hydraulic pressure modulator.

- *Steering angle control:* The lateral force of the tires can be adjusted through the control of each wheel's steering angle. For this purpose, the tire model is necessary. Consider the

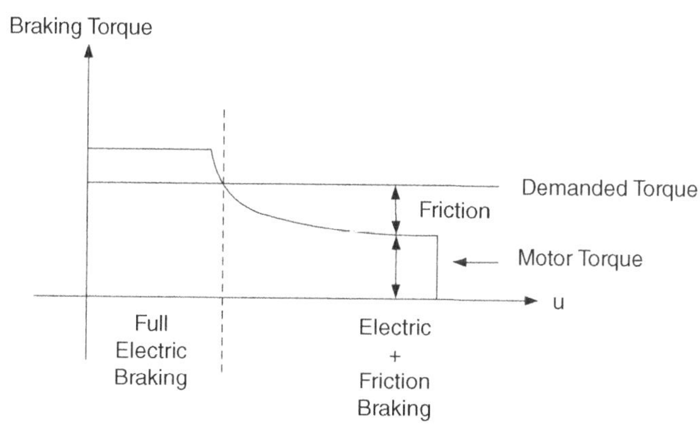

Figure 9.26 Contribution of the electric motor and friction brake in generating braking torque

general form of the tire lateral force model (Equation 4.42) and substitute the desired value of lateral force, the estimated value of normal force and longitudinal slip, and also the camber angle. This results in:

$$F_{yi}^{d} = F_y\left(\alpha_i, \tilde{s}_i, \tilde{F}_{zi}, \gamma_{0i}\right) \tag{9.33}$$

Equation (9.33) forms an algebraic equation that can be solved for side slip angle α_i. Suppose that the solution for the side slip angle is α_i^*, the steering angle of each wheel can be the found by using Equation (7.37) as:

$$\begin{aligned} \delta_1 &= \alpha_1^* + \arctan\left(\frac{v+ar}{u+\frac{1}{2}Tr}\right) \quad & \delta_2 &= \alpha_2^* - \arctan\left(\frac{br-v}{u+\frac{1}{2}Tr}\right) \\ \delta_3 &= \alpha_3^* + \arctan\left(\frac{v+ar}{u-\frac{1}{2}Tr}\right) \quad & \delta_4 &= \alpha_4^* - \arctan\left(\frac{br-v}{u-\frac{1}{2}Tr}\right) \end{aligned} \tag{9.34}$$

Based on the same reasons previously discussed, the values calculated above for steering angles can be improved by adding a feedback control system.

We can define the lateral force error as:

$$e_{yi} = F_{yi}^{d} - \tilde{F}_{yi} \tag{9.35}$$

where $\tilde{F}_{yi}$ is the estimated value for each tire's lateral force. Next, we can add a PI control action to Equations (9.34) [4]:

$$\begin{aligned} \delta_1 &= \alpha_1^* + \arctan\left(\frac{v+ar}{u+\frac{1}{2}Tr}\right) + k_P^y e_{y1} + k_I^y \int e_{y1} dt \\ \delta_2 &= \alpha_2^* - \arctan\left(\frac{br-v}{u+\frac{1}{2}Tr}\right) + k_P^y e_{y2} + k_I^y \int e_{y2} dt \\ \delta_3 &= \alpha_3^* + \arctan\left(\frac{v+ar}{u-\frac{1}{2}Tr}\right) + k_P^y e_{y3} + k_I^y \int e_{y3} dt \\ \delta_4 &= \alpha_4^* - \arctan\left(\frac{br-v}{u-\frac{1}{2}Tr}\right) + k_P^y e_{y4} + k_I^y \int e_{y4} dt \end{aligned} \tag{9.36}$$

Table 9.3 The vehicle parameters

Parameter	Value	Parameter	Value
m	1,300 kg	I_w	2.1 kgm^2
m_s	1170 kg	h'	0.2 m
a	1.2 m	K_ϕ	45000 N/m
b	1.3 m	C_ϕ	2600 Ns/m
I_z	2,500 kgm^2	T	1.4 m
I_x	750 kgm^2	R	0.3 m

In Equation (9.36), it is assumed that each wheel's steering angle can be controlled independently. In conventional cars, however, only the steering angle of front wheels that are dependent is controlled by one actuator. In the following, the above analysis is used in an example for further clarification.

9.3.3 Simulation Study

A simulation study is presented here to study the effectiveness of the described VDC system in different driving conditions. The VDC system is implemented on an electric vehicle with four motorized wheels. The motors' torque is formulated by:

$$T_m = \begin{cases} 260\, x_{\theta m} \text{ Nm if } & \omega \leq 105 \text{ rad/s} \\ \dfrac{27300}{\omega} x_{\theta m} \text{ Nm if } & 105 \leq \omega \leq 210 \text{ rad/s} \end{cases} \tag{9.37}$$

It is assumed that the vehicle is not equipped with an active steering system. Therefore, traction/brake torque distribution is the only way to stabilize and control the vehicle. The vehicle model described in Chapter 7 is used here to simulate the vehicle behavior and evaluate the control performance. The main vehicle parameters are summarized in Table 9.3.

The Magic Formula Pacejka tire model is used in the simulation to generate the longitudinal and lateral tire forces. Tire characteristic curves in longitudinal and lateral directions for the nominal vertical load are shown in Figure 9.27.

Parameters used in the motion controller and tire force distribution algorithms are set as shown in Table 9.4. Desired yaw rate of the vehicle at each longitudinal velocity is calculated by Equation (9.17), and the desired lateral velocity is assumed to be zero.

The objective function described by Equation (9.27) is used in the simulation with an optimization procedure. Vertical longitudinal and lateral tire forces are assumed to be available. In real-world applications, these forces need to be estimated.

Two different case studies are presented here to show the VDC system performance in improving the handling and stabilization of the vehicle. In each case study, the performance of the vehicle equipped with the VDC system is compared to the uncontrolled vehicle. For the uncontrolled vehicle, the total vehicle tractive force is equally distributed to all wheels.

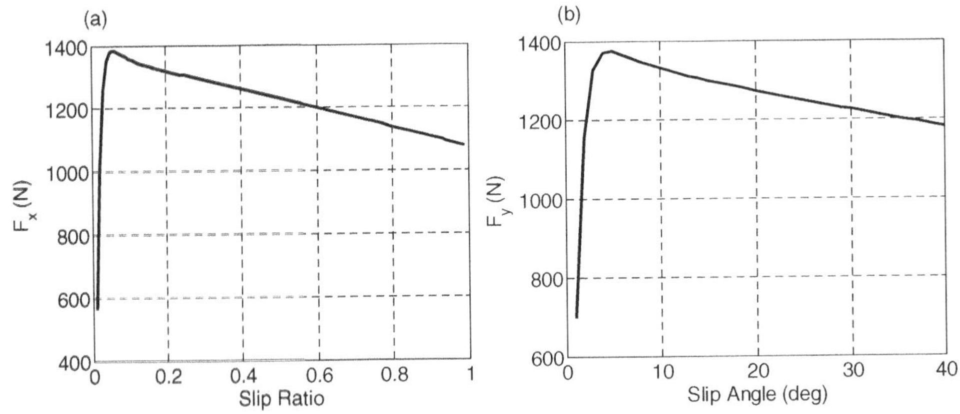

Figure 9.27 Tire longitudinal (a) and lateral (b) forces characteristic curves for the nominal vertical load

9.3.3.1 Case Study 1: Constant Steer Maneuver on a Slippery Road

For the first case study, it is assumed that the vehicle is performing a constant steer maneuver ($\delta = 0.03$ rad) on a slippery road ($\mu = 0.4$). Initial velocity is set to 90 km/hr and accelerator pedal position is 0.2 (constant) throughout the maneuver. The vehicle response is simulated by solving the vehicle differential equations numerically with the inputs from the VDC system.

First, let us look at the vehicle response for the uncontrolled and the controlled vehicles. Simulation results for the vehicle path and yaw rate are shown in Figure 9.28. Thanks to the control system and the torque distribution strategy, the controlled vehicle can desirably follow the desired yaw rate and the desired path. The desired path is calculated using Equation (7.35) together with the desired yaw rate, the desired lateral velocity and the simulated longitudinal velocity at every time instance. The uncontrolled vehicle demonstrates an oscillatory yaw rate and a more over-steer behavior which indicate less stability compared to the controlled vehicle.

Better performance of the controlled vehicle can also be verified using the simulation results for other important handling variables as shown in Figure 9.29. While the longitudinal velocities for both cars are comparable, the controlled vehicle shows less sideslip angle, less lateral velocity and less lateral acceleration, which all indicate better stability and handling performance compared to the uncontrolled counterpart. Such an improvement in the stability and handling performance is the result of the effective torque distribution dictated by the control system.

Table 9.4 VDC system parameters

Parameter	Value	Parameter	Value
C_1	10^6	C_6	50000
C_2	10^6	k^v_{P2}	2.1 kgm^2
C_3	10^6	k^r_{P2}	0.2 m
C_4	10^6	k^*_{us}	3 deg/g
C_5	10		

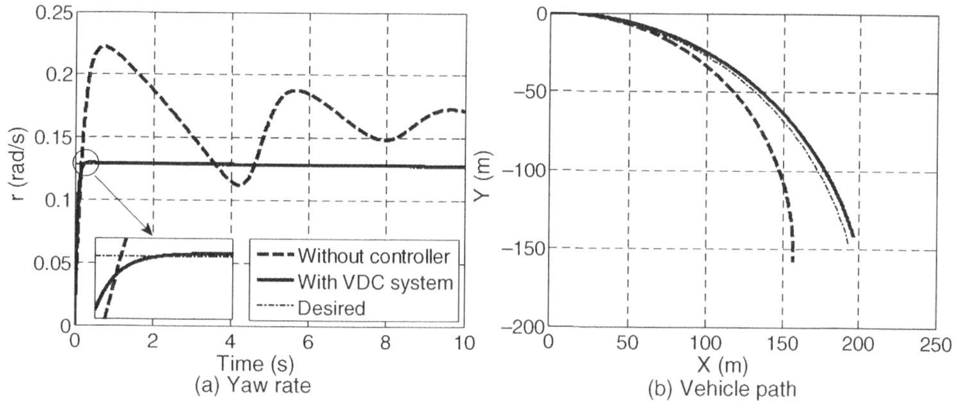

Figure 9.28 Simulation results for the step-steer maneuver

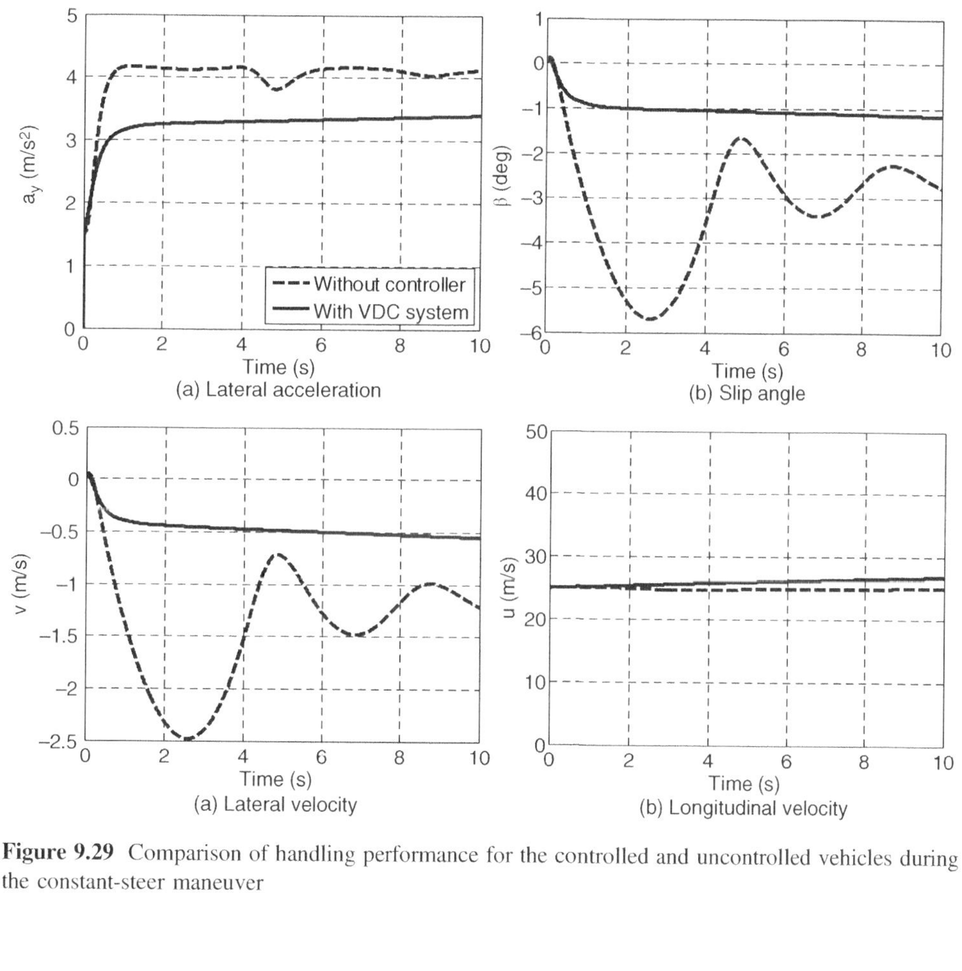

Figure 9.29 Comparison of handling performance for the controlled and uncontrolled vehicles during the constant-steer maneuver

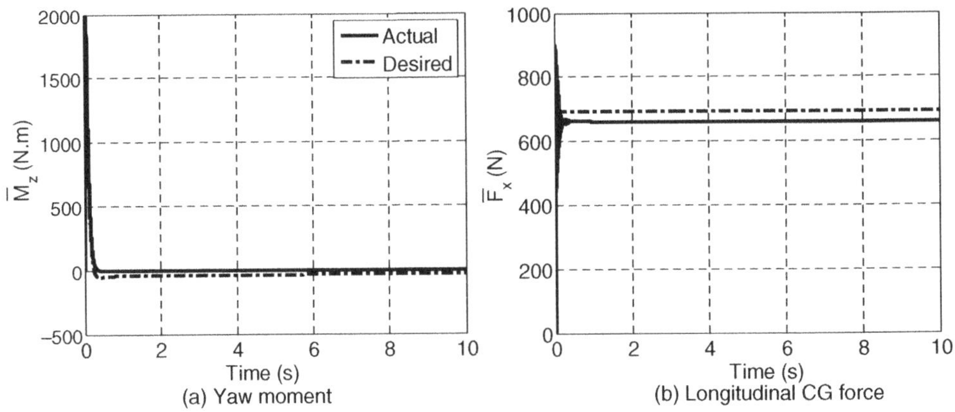

Figure 9.30 Control performance for the constant-steer maneuver

To investigate the control performance, the desired and actual of the total yaw moment and total longitudinal force are shown in Figure 9.30. The desired yaw moment is large at the beginning of the maneuver due to large difference between the desired and the actual yaw rate. As the vehicle's yaw rate gets closer to the desired reference, the yaw rate error decreases and the desired yaw moment approaches zero. This all happens while the total longitudinal force remains close to the driver's desired command and hence the longitudinal performance is also satisfactory throughout the maneuver.

The responsibility of the control system in this application is to generate an external yaw moment which can stabilize the vehicle and improve the handling performance. Initially, as shown in Figure 9.31, the control system generates a positive external yaw moment to help reach the desired yaw rate. For the steady state part of the maneuver, a negative external yaw moment is generated to make the vehicle more under-steer to improve the stability and handling performance.

The described external yaw moment is realized via traction/brake torque distribution to the vehicle wheels. The wheel torques shown in Figure 9.32 demonstrate how the external yaw

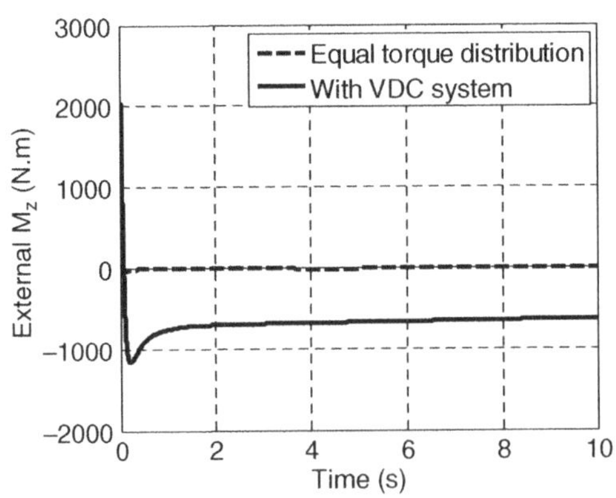

Figure 9.31 External yaw moment for a constant-steer maneuver

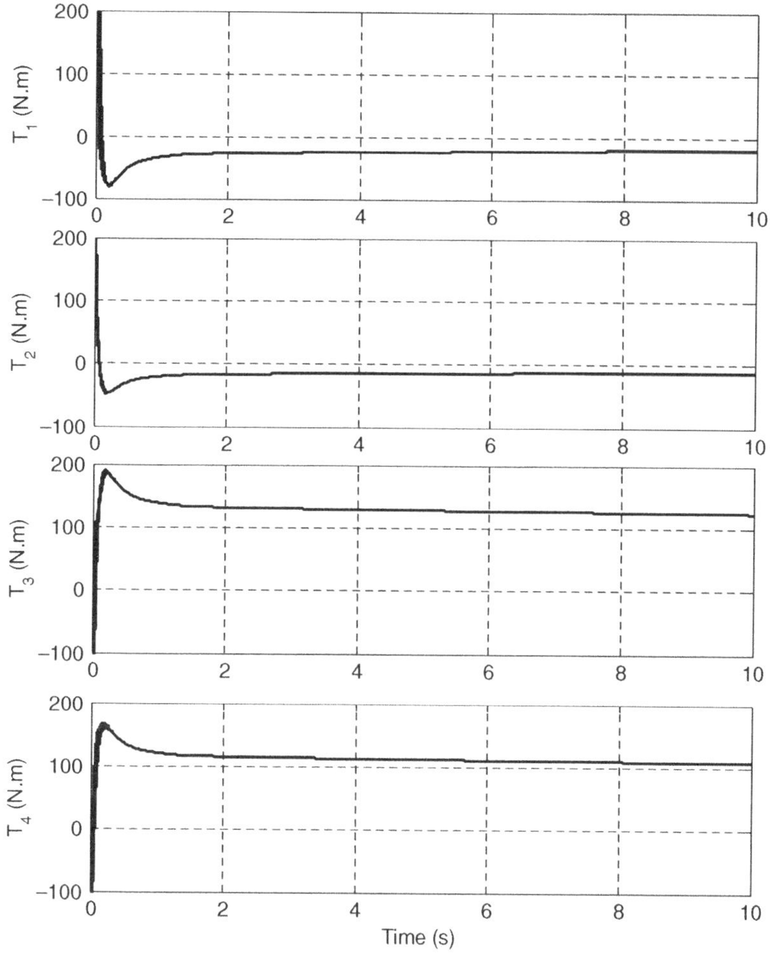

Figure 9.32 Applied wheel torques by the VDC system for a constant-steer maneuver

moment is generated. Initially, a positive torque (traction) is applied to each of the left wheels and a negative torque (braking) is applied to each of the right wheels to generate the required positive external yaw moment. Afterwards and during the steady state part of the maneuver, a negative torque is applied to the left wheels and a positive torque is applied to the right wheels to generate a negative external yaw moment which is required to make the vehicle more understeer to improve the handling performance.

9.3.3.2 Case Study 2: Sine Steer Maneuver on a Dry Road

For the second case study, let us investigate the control performance in a more demanding transient maneuver on a dry road ($\mu = 0.9$). The driver is assumed to command a sine wave steering input (Figure 9.33) and a 0.1 accelerator pedal position. Initial velocity of the vehicle is set to 120 km/hr to replicate harsh driving conditions.

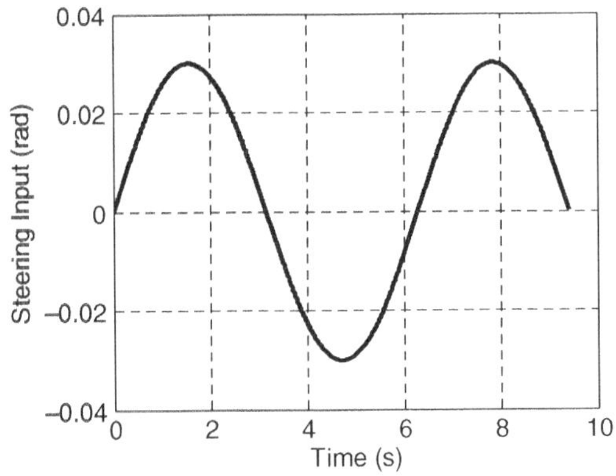

Figure 9.33 Steering input for the second case study

Similar to what discussed in the previous case study, performance of the vehicle in this maneuver is simulated for the uncontrolled and controlled vehicles. Simulation results for yaw rate and vehicle path are shown in Figure 9.34. The results show that the uncontrolled vehicle is not capable of performing the maneuver, while the controlled vehicle can track the desired trajectory.

Also, the slip angle, yaw rate and lateral velocity plots in Figure 9.35 demonstrate that the uncontrolled vehicle is unstable in this maneuver. Again, the vehicle equipped with the VDC system demonstrates a desirable handling performance which is the result of effective torque distribution on the wheels by the control system.

The desired and actual yaw moment and longitudinal CG force are shown in Figure 9.36. The desired longitudinal force is requested by the driver's pedal position command and is an input to the control system. The desired yaw rate is calculated using Equation (9.17) and is also actualized by the VDC system for the maneuver.

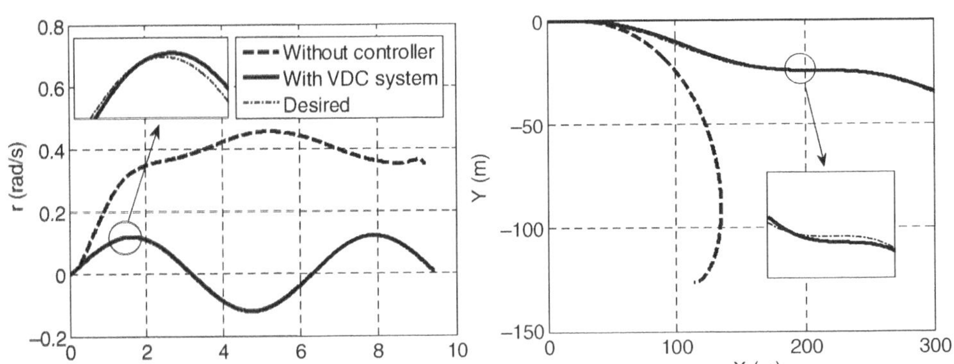

Figure 9.34 Simulation results for the sine-steer maneuver

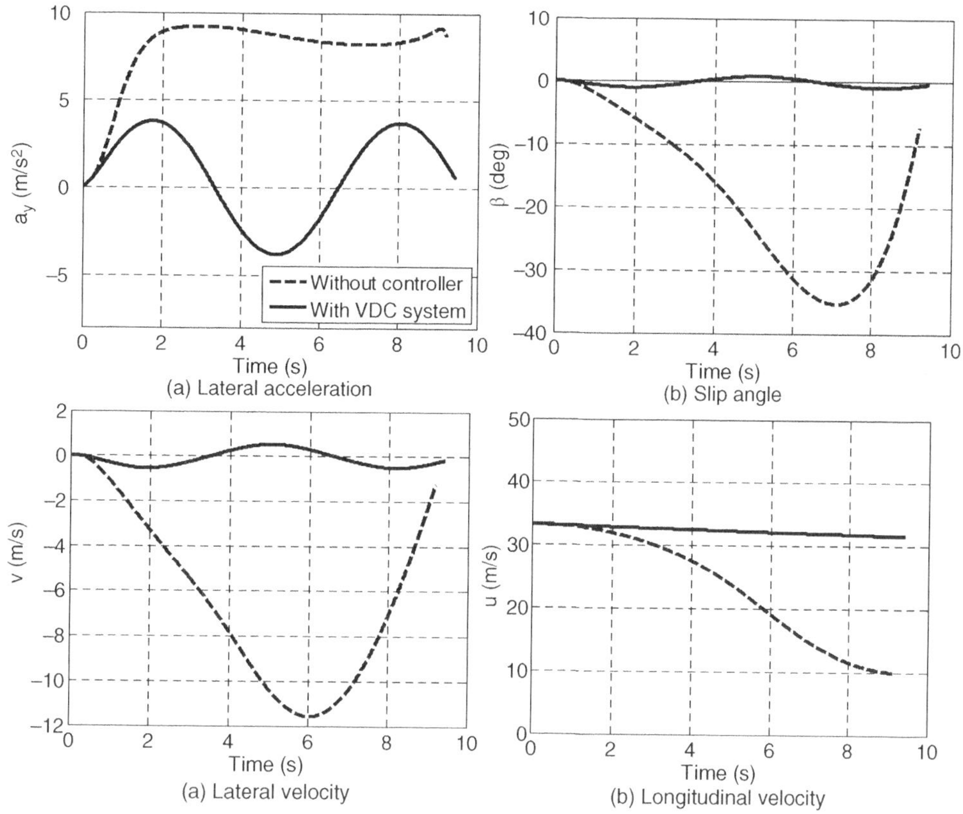

Figure 9.35 Comparison of handling performance for the controlled and uncontrolled vehicles during a sine-steering maneuver

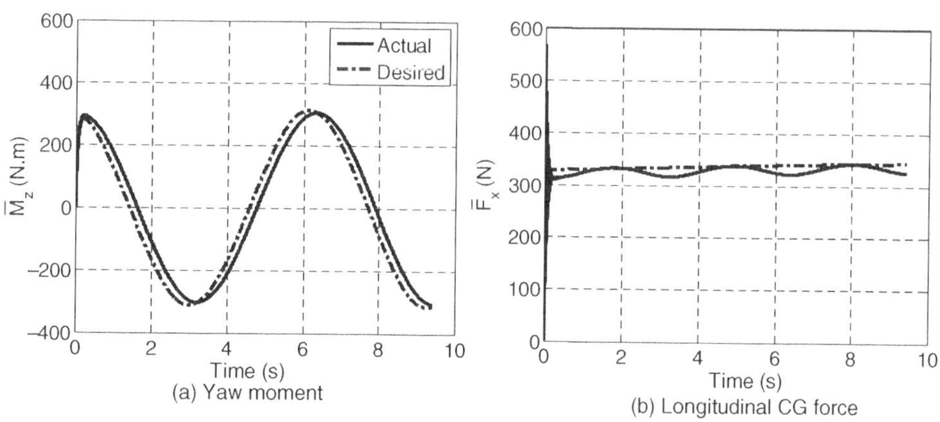

Figure 9.36 Control performance for the sine-steer maneuver

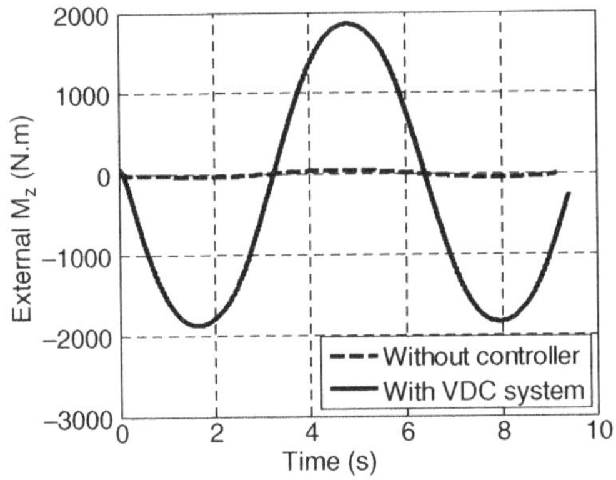

Figure 9.37 External yaw moment for the sine-steer maneuver

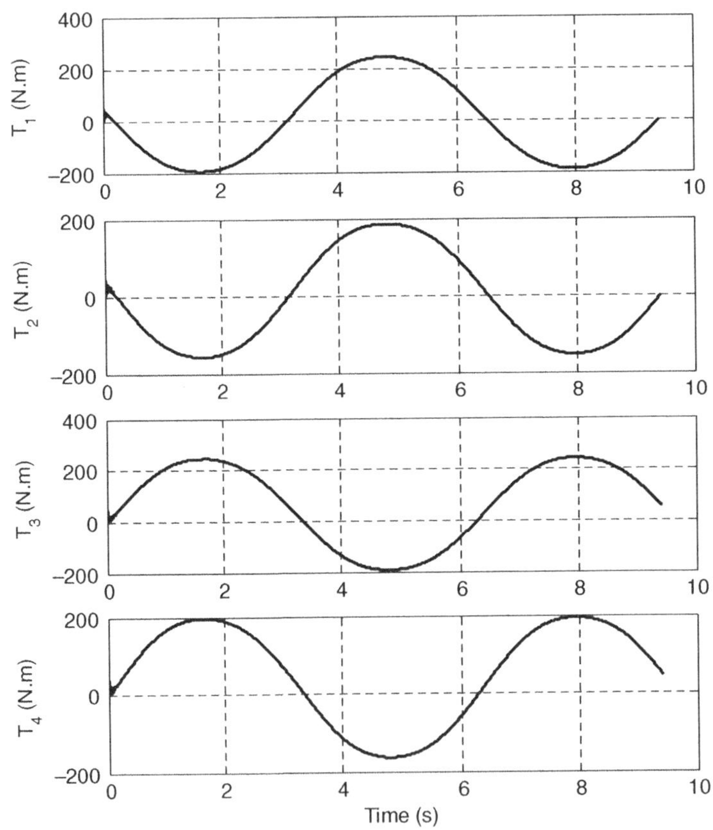

Figure 9.38 Applied wheel torques by the VDC system for the sine-steer maneuver

The external yaw moment produced by the control system is shown in Figure 9.37. One can note that the external yaw moment is generated in a way to make the vehicle more under-steer. In other words, a negative external yaw moment is applied when the steering input is positive, and vice versa.

To achieve such an external yaw moment, traction/brake torques are applied to the wheels as shown in Figure 9.38. During the first 3 seconds of the maneuver, brake torques (negative torques) are applied to the right wheels and tractive torques (positive torques) are applied to the left wheels. This produces the negative external yaw moment required in this period. Similarly, whenever the required external yaw moment is positive, tractive torques are applied to the left wheels and brake torques are applied to the right wheels. This torque distribution strategy can significantly help to make the vehicle safer for various driving conditions as shown by these two case studies.

Problems

1. Explain the driver/vehicle/environment dynamic system and describe how a vehicle dynamic controller can improve the system responses as the fourth element.
2. Explain the working principles of vehicle dynamics control systems, what is the desired vehicle motion and how it is defined?
3. Draw a diagram to classify the different vehicle dynamics control systems.
4. What are the two main sources of yaw moment control in a vehicle? Explain the differences between the indirect and direct yaw moment control and list the advantages and disadvantages of each method.
5. Describe the differences between differential braking, differential traction and torque vectoring.
6. Explain the special capabilities of modern electric and hybrid electric vehicles such as motorized wheel electric vehicles for more effective vehicle dynamics control.
7. Describe the structure of a multi-layer vehicle dynamics control system for EVs and HEVs and explain the function of each part.

References

1. A. Goodarzi, and M. Alirezaei, "Integrated Fuzzy/Optimal Vehicle Dynamic Control," *International Journal of Automotive Technology*, **10**(5) (2009), 567–575.
2. E. Esmailzadeh, A. Goodarzi, and G.R. Vossoughi, "Optimal Yaw Moment Control Law for Improved Vehicle Handling," *International Journal of Mechatronics*, **13**(7) (2003), 659–675.
3. T. Acarman, "Nonlinear Optimal Integrated Vehicle Control Using Individual Braking Torque and Steering Angle with On-Line Control Allocation by Using State-Dependent Riccati Equation Technique," *Vehicle System Dynamics*, **47**(2) (2009), 155–177.
4. A. Goodarzi, and E. Esmailzadeh, "Design of a VDC System for All-Wheel Independent Drive Vehicles," *IEEE/ASME Transaction on Mechatronics*, **12**(6) (2007), 632–639.
5. A.T. Van Zanten, "Evolution of Electronic Control Systems for Improving the Vehicle Dynamics," *Proceedings of the AVEC '02*, 2002.

6. P. Falcone, F. Borrelli, J. Asgari, H. E. Tseng, and D. Hrovat, "Predictive Active Steering Control for Autonomous Vehicle Systems," *IEEE Transactions of Control Systems Technology*, January 2007.
7. M. Abe and O. Mokhiamar, "An Integration of Vehicle Motion Controls for Full Drive-By-Wire Vehicle," *Proceedings of IMechE*, Vol. **221** Part K: J. Multi-body Dynamics, JMBD72 – ImechE, 2007.
8. T.H. Hwang, K. Park, S.-J. Heo, S. H. Lee, and J. C. Lee "Design of Integrated Chassis Control Logics for AFS and ESP," *International Journal of Automotive Technology*, **9**(1) (2008), 17–27.
9. A. Goodarzi, and V. Daneshmand,"A Novel Algorithm for Optimum Distribution of Tire Forces in an Integrated Chassis Control System," paper presented at 21st International Symposium on Dynamics of Vehicles on Roads and Tracks (IAVSD09), Stockholm, Sweden, August 17–21, 2009.

Index

Printed and bound by CPI Group (UK) Ltd, Croydon, CR0 4YY
17/09/2026
14964270-0005